PRINCIPLES OF
COMPARATIVE ANATOMY
OF INVERTEBRATES

VOLUME 2 ORGANOLOGY

W. N. Beklemishev

PRINCIPLES OF COMPARATIVE ANATOMY OF INVERTEBRATES

VOLUME 2   ORGANOLOGY

Translated by Dr J. M. MacLennan
Edited by Dr Z. Kabata

THE UNIVERSITY OF CHICAGO PRESS

Standard Book Number: 226–04175–1
Library of Congress Catalog Card Number: 70–97749

THE UNIVERSITY OF CHICAGO PRESS CHICAGO 60637
OLIVER AND BOYD LTD, EDINBURGH

This is a translation of the 3d (1964)
edition of ОСНОВЫ СРАВНИТЕЛЬНОЙ
АНАТОМИИ БЕСПОЗВОНОЧНЫХ Vol. II
by В. Н. Беклемишев published by
NAUKA, Moscow

English translation © by Oliver and Boyd 1969
Printed in the United States of America

CONTENTS

Chapter I   STRUCTURAL AND ORGANOLOGICAL DIFFERENTIATION OF THE LOWER METAZOA

1. GENERAL REMARKS

The most characteristic feature of the organisation of Metazoa, in contrast to most colonies of Protozoa and to lower (filamentous) algae, is the multi-layered arrangement of their cells (or energids), in which the outer cells form a continuous layer separating the animal's body from the external environment. This creates an internal environment of the multicellular animal, in which all the cells of its body are wholly or partially embedded. The appearance in Metazoa of an internal environment of the organism, partly insulated from the external environment in which it lives, is an important progressive phenomenon in the history of life. It represents a substantial step forward in the body integration of Metazoa, in the unification of all their constituent cells into one organism; to realise that, one need only compare a sponge with a branched colony of Choanoflagellata or of *Vorticella*. Moreover, the multilayered arrangement of the body cells in Metazoa became a powerful stimulus to differentiation of these cells, to their separation into a number of specialised cell types, and also to the formation of various symplasts. At first, however, e.g. in coelenterates and sponges, the separation of the internal from the external environment of the organism is still very slight; the unity of the internal environment is not organised, interaction by the cells of the organism is relatively small, and consequently the cells are still very independent. The increasing insulation of the organism from the direct action of its external environment, the constantly-developing unification of its internal environment and stabilisation of its physical and chemical properties, the division of labour among the cells and their morphological differentiation, the combination of cells into more complex and efficient units (organs and apparatuses), the increasing unification and integration of the whole organism—all these form a series of very important steps in the progressive evolution of Metazoa.

The function of separation of the external from the internal environment is performed by the animal's integument; the function of maintaining the stability of internal environmental conditions is performed by the digestive apparatus and the respiratory and excretory organs; integration of the organism, its unification into a single whole, is attained by the work of the circulatory, endocrinological, and nervous systems. The most characteristic feature of animals is active procurement of food, and consequently the work of the digestive apparatus in all animals is made possible by the existence of a locomotor apparatus, which assumes also a number of other

1

functions; the supporting apparatus fills the purely subsidiary role of reinforcing the mechanical stability of the body. Finally, sexual reproduction is accomplished by means of the genital apparatus.

Such allocation of functions to morphologically-individualised apparatuses, however, is achieved to a greater or lesser extent only in the higher ranks of the animal kingdom: we find it in the higher members of various evolutionary stems. In the commencing stages of evolution some of the above functions (respiratory, excretory, circulatory, endocrinological) may have no specialised apparatus morphologically designed to fulfil them, or may be fulfilled by an assemblage of discrete organs, not morphologically combined into a single apparatus (e.g. respiration and excretion in polychaetes, see Chaps. II and VII). The development of separate organs to perform particular functions, their combination into apparatuses, and the further development of these apparatuses constitute the content of *organology*. All the chief functions and apparatuses have to a considerable extent evolved independently in most of the main stems of the animal kingdom, and similarly-named organs in members of different phyla are often not homologous. So one has very often to use analogies as connecting links, and to supplement comparative-anatomical analysis with the data of comparative physiology and comparative ecology.

The terms 'apparatus' and 'system of organs' are often taken to be synonymous, such expressions as 'digestive system' or even 'genital system' being used. That terminology is incorrect, or at least unsuitable. Since the beginning of the 19th century the term 'system of organs' has denoted a combination of organs of identical or similar function or structure: for instance, the dental system of mammals or the skeletal muscle system of insects, or the excretory system of earthworms, which consists of separate nephridia. A system of organs is a purely tectological concept (see Vol. 1, Chap. I). On the other hand, an apparatus is a collection of organs, similar or dissimilar, which together take part in performing a single common function and form a unified whole constructed on a specific plan. From the point of view of comparative anatomy an apparatus is an architectonical concept. For instance, the digestive apparatus of *Ascaris* consists of the pharynx, which swallows food; the mid-gut, joined to the posterior end of the pharynx, which digests and absorbs food; and finally the hind-gut, which carries away undigested residues and is joined to the posterior end of the mid-gut; all these are united in fulfilment of a common function, nutrition, and are arranged in a definite order and connection. In the simplest case an apparatus may consist of a single organ, e.g. the digestive apparatus of *Hydra;* but siliceous sponges, whose digestion is handled by separate phagocytes, have no digestive apparatus: they have only a collection of digestive cells.

As we have seen in Vol. 1, Chap. II, the Metazoan body is built up of cells or energids and their secretions (intercellular substances, cuticle, etc.). Cells of a single type and performing similar functions may be combined directly into a single apparatus providing for fulfilment of that function throughout the body. An example is the muscular apparatus of small turbellarians, which consists of a large number of separate muscle and epithelial-muscle cells distributed among numerous cells of other kinds but arranged among them on a definite plan, which ensures the contractility of the animal's body. In other cases cells performing an identical function may be grouped into tissues from which elementary organs are formed: the latter compose complex organs, which in turn combine to form apparatuses. The higher Bilateria are constructed on this principle.

Study of the types of cells that compose multicellular organisms is the subject of a special branch of cytology; study of tissue structures is the subject of histology; on that account we shall touch on only a few questions pertaining to these, concerning the most important structural features of the bodies of Metazoa. Study of organs and apparatuses is the subject of organology. We shall dwell in somewhat greater detail on organological questions in subsequent chapters after having, at the end of the present chapter, discussed the first appearance of structural and organological differentiation in the lower Metazoa.

2. PRINCIPAL WAYS IN WHICH METAZOA BECOME STRUCTURALLY COMPLEX: PARENCHYMATOUS AND TISSUE STRUCTURE, AMORPHOUS AND POLARISED STRUCTURE

When once we have accepted the cell as the basic structural unit in the bodies of Metazoa, the most primitive structure is a homogeneous union of an indefinitely large number of identical cells, located without any definite repetitive order in relation to one another, not fused into syncytia and not separated by intercellular matter. Such a structure might be called *primary parenchyma* (W. N. Beklemishev, 1925). The word 'parenchyma' is borrowed from plant anatomy, where it denotes a structure formed by the union of identical cells arranged without any regularity in relation to one another.

We must state at once that no adult multicellular animal possesses primary parenchymatous structure, although we find it in embryos. The best example is the embryo of the hydroid *Tubularia mesembryanthemum* (A. Tikhomirov, 1887) or of *Clava squamata* (K. Harm, 1902) at the morula stage, when it is a compact sphere composed of identical cells arranged without any definite structural regularity and almost without any intercellular matter (Fig. 1, *A*).

A basic structural feature of adult Metazoa is polymorphism of their cellular elements, the presence in their bodies of many kinds of cells, or, as it is sometimes expressed, many cell systems. By a cell system we must understand an accumulation of similar cells in a multicellular organism, regardless of how these cells are arranged, i.e. whether they are mixed with elements of other systems or are grouped together. In this sense we speak, for instance, of a system of reticulo-endothelial cells.

It is because of the polymorphism of their cells that adult Metazoa cannot possess primary parenchymatous structure. The most primitive structure conceivable in adult Metazoa is *compound parenchyma*. By compound parenchyma we understand the structure of an animal constructed of heterogeneous cells, the cells of different systems being equally mixed together, not grouped in homogeneous complexes, and without structural regularity in their arrangement. Compound parenchymatous structure in its pure form also does not occur in any adult Metazoa, but there are two groups of animals whose structure is close to it: these are sponges and acoelous turbellarians.

In the bodies of sponges there are three main categories or systems of cells: *archaeocytes*, *dermacytes*, and *choanocytes* (E. Minchin, 1900). Choanocytes are collared flagellate cells with a protoplasmic body containing a nucleus and a blepharoplast, with a flagellum arising from the

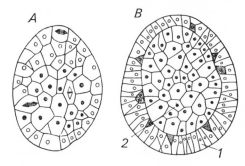

Fig. 1. *An example of primary parenchymatous structure in Metazoa. The morula of the hydroid* Clava squamata *and its further development by means of parenchymatous delamination.*

A—morula. B—parenchymula (after Harm, from Davydov). 1—ectoderm or kinetoblast; 2—endoderm or phagocytoblast.

latter; the flagellum is surrounded by a collar at its point of origin. In their assemblage of cellular organellae and in the method of division of their nucleus choanocytes are precisely similar to the vegetative stages of Choanoflagellata (Protomonadina). In calcareous sponges the choanocytes may be converted into amoeboid cells—phagocytes; they also give rise to primary sexual cells and consequently form a single class of cells together with archaeocytes. In siliceous sponges these apparently form two separate classes. There are several varieties of archaeocytes and dermacytes, but for the most part these varieties are not strictly defined, and in certain conditions one may pass into the other. Archaeocytes have particularly wide potentialities, and on regeneration they may differentiate, giving rise to all types of dermacytes. In the bodies of siliceous sponges archaeocytes are normally represented by digestive cells (phagocytes), primary sexual cells, and granular cells. The dermacytes vary considerably and constitute the greater part of the body; they are of the following main categories: (*i*) pinacocytes, covering the external surface of the body; in calcareous sponges they do not form a separate epithelium but only the bounding layer of the parenchyma, not separated from the rest of its mass; (*ii*) porocytes—tubular cells carrying water from the afferent canals into the flagellate chambers; (*iii*) myocytes—fusiform cells, possessing some contractility, although they are not true muscle cells; (*iv*) scleroblasts, secreting skeletal elements; (v) glandular cells; and so on. Dermacytes are therefore represented by various types of cells, but true muscle cells and true nerve cells, as possessed by other Metazoa, are not found in sponges.

The sponge body consists of a large number of different kinds of cells distributed within fairly abundant intermediate matter. In the parenchyma formed by these cells there are well-defined flagellate chambers lined with choanocytes. Their presence distinguishes sponge structure from typical heterogeneous parenchyma. Because of their presence, not all the body cells are mixed and located without order. Some of them have become differentiated into separate combinations of regularly-arranged elements. Many species, however, have resting stages, when the flagellate chambers disappear: their cells either scatter through the parenchyma or degenerate, to be later regenerated from the archaeocytes. In the resting state a sponge consists of almost true heterogeneous parenchyma, differing from the latter mainly in its possession of abundant intermediate matter. Besides, various categories of cells in the sponge body are mixed, but not quite uniformly. Myocytes accumulate around the orifices of the irrigation system; spongoblasts (i.e. scleroblasts of keratose sponges) form epithelial masses around the fibres that they construct, but these masses are purely temporary. They all become disorganised at the edges, and the different areas of the parenchyma are distinguished only by the degree of their saturation with different types of cells. Sponges therefore possess only incipient tissue-formation and not true tissues. At the same time architectonical regularity and architectonical integration are well expressed in the structural plans of the irrigation system and of the skeleton of sponges.

Comparing acoelous turbellarians with sponges, we first find great differences between them in the composition of their cell elements. We find in turbellarians all the main types of cells possessed by higher Metazoa: integumental, sustentacular, muscular, neural, glandular, digestive cells, etc.[1] In contrast, sponges actually occupy a very isolated position in the animal kingdom, being sharply distinguished from other Metazoa in the direction of specialisation of their cells.

We shall now discuss the structure of acoelous turbellarians, using as an example the most primitive members of that group (Fig. 2). Externally the body is covered with a layer of integumental cells. These cells are covered with cilia, and in many Acoela myofibrillae pass through them, so that the integumental cells of these Acoela are ciliated epithelial-muscle cells. In young *Oligochoerus erythrophthalmus* the apical ends of these cells

---

[1] The creation of a natural system of the cell elements of Metazoa and the study of their evolution during the history of the animal kingdom are problems of a special branch of cytology; solution of these problems will be of very great assistance in the development of animal morphology and physiology. The works of zoologists and histologists contain much material for solving these problems; but all that material is still disconnected, and comprehensive attempts to synthesise it—at least since the time of E. Mechnikov, E. Haeckel, and O. and R. Hertwig—have not been undertaken. Not only that, but only a few biologists clearly understand the task of studying the evolution of types of cell elements.

combine into a smooth surface, whereas their basal ends are frayed and do not form a distinct boundary with the rest of the parenchyma (Fig. 10, *A*). Like the pinacocytes of calcareous sponges they constitute merely a boundary layer of the parenchyma (W. N. Beklemishev, 1937). Deeper down, beneath the covering-cell layer, lies the rest of the parenchyma. Its most primitive form has been described by L. von Graff (1891) in *Proporus venenosus* and *Convoluta sordida* (Fig. 2), and by A. V. Ivanov (1952c) in *Oxyposthia praedator*. The main mass of parenchyma is there formed of supporting and digestive elements. The latter, as in sponges, are represented by migratory phagocytes (Fig. 2, *4*) scattered among the cells of other systems. That parenchyma is traversed by muscle cells in various directions. The bodies of glandular cells, whose ducts open to the exterior, lie partly within the boundary layer of covering cells and partly within the mass of parenchyma. Neural cells also are partly represented by sensory cells lying in the boundary layer (epidermis), while others form a subcutaneous plexus, which in places may form strands passing through the parenchyma. There are also sexual elements · the male elements in all Acoela are scattered in small groups; in young specimens the female elements also are sometimes scattered, but with increasing age they form compact masses in most Acoela; but in *Xenoturbella* (E. Westblad, 1949) and *Oxyposthia* (Acoela) (A. Ivanov, 1952c) the female sexual cells are scattered diffusely even when mature.

The principles of heterogeneous parenchyma are here violated by the existence of external epithelium, which, however, is sometimes ill-defined. Moreover, other types of cells also are not evenly distributed through the animal's body: digestive elements predominate in the central, axial, part of the body, and neural elements at its front end; male sex cells occur mainly near the dorsal side and female near the ventral side. Therefore even in the most primitive Acoela there already appears some tendency to segregation, i.e. to individualisation and to grouping together of cells of a single system. As in sponges, however, here we may speak only of incipient segregation, incipient tissue-formation.

When we say that the body cells of primitive Acoela are mixed irregularly in heterogeneous parenchyma, that relates only to the absence of repetitive structural regularity in their arrangement. Architectonical regularity of arrangement does exist. The muscle fibres run in all directions and are mixed with elements of other types, and so, strictly speaking, are irregularly arranged. But at the same time all the muscle fibres in the body are arranged on a single general plan, forming a single contractile apparatus consisting of cutaneous muscles, dorso-ventral muscles, front-end-retractor muscles, etc. The same architectonical regularities of arrangement are displayed by the neural elements combined into the neural apparatus, and to a lesser degree by other systems of elements.

Syncytial structures are widespread in Acoela. All the digestive elements are usually coalesced into a digestive syncytium (endocytium, in E. Westblad's terminology). Often, but far from always, the external epithelium and the peripheral parenchyma (which in such cases Westblad calls epicytium and exocytium) are syncytial. In an extreme case the

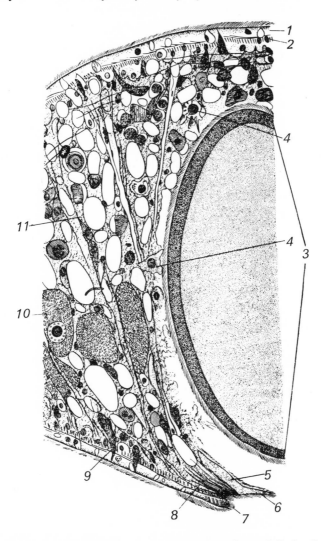

Fig. 2. *Primitive type of histological structure in the acoelous turbellarian* Convoluta sordida, *part of a cross-section in the oral region.*

1—epithelial layer of epidermis containing some of the nuclei of the latter, the other nuclei being embedded in the parenchyma; 2—longitudinal muscles of the integument; 3—ingested food; 4—amoeboid cells (phagocytes); 5—pharynx; 6—diaphragm of edge of mouth; 7—outer edge of mouth; 8—mouth-dilator muscles; 9—cutaneous glands; 10—oogonia; 11—dorso-ventral muscle fibres (after von Graff).

whole body, except for the genital, glandular, and neural cells, assumes a syncytial structure. O. Steinböck (1954) believes that all Acoela have a fully-syncytial structure and that it is primitive for them (see Vol. 1,

Chap. II). In both respects, however, he is in error. Syncytial structure is not dominant in all Acoela, and it is not fully syncytial in any of them; and the derivation of Acoela directly from polyenergid Protozoa encounters unsurmountable obstacles, as we have seen in Vol. 1, Chap. II. Generally speaking, one should not attach too much morphological importance to the syncytial structures so widespread in Acoela. Such structures arise and disappear too easily in all Metazoa, and in the most varied stages of ontogeny.

Such is the structure of the most primitive Acoela. In contrast, the external covering layer of most turbellarians is very distinctly differentiated from the rest of the parenchyma, taking the form of cutaneous epithelium. The basal ends of the covering cells and of some glandular cells are levelled off, and a boundary layer is secreted between them and the remaining body mass. This cutaneous epithelium forms a separate tissue, consisting of several different kinds of cells: covering, sensory, glandular. The first two of these classes of cells do not exist elsewhere in the body. They are distinct from all other cells and form a tissue crystallised out, as it were, from the rest of the parenchyma.

By a *tissue* we understand a finite part of an organism, constructed of cells and cell products of one or several different types and distinguished by their morphological composition from adjacent parts of the body. The concept of 'tissue' does not imply any definite form: each tissue is one of the kinds of structure found in any organism. Every tissue mass of definite form is no longer simply tissue but is a definite structural unit, an elementary organ, constructed of that tissue.[1] Tissues may be divided into 'simple' and 'compound'. Simple tissues, such as the intestinal epithelium of *Ascaris*, consist of components of a single kind; compound tissues, such as the cutaneous epithelium of turbellarians, consist of elements of several different kinds.

The word 'tissue' is often used in another sense, that of a system of cells. For instance, we speak of the connective tissue of vertebrates, signifying thereby elements of the corresponding system regardless of whether they form part of compound muscular, glandular, or other tissues or form regions of simple tissue, the connective tissue in the strict sense of that word. The absence of a clear distinction between the concept of 'tissue' and 'system of cells' (class of cells) introduces much vagueness and inconsistency into many works and textbooks on histology.

With the appearance of the first definite tissue, the structure of acoelous turbellarians begins to pass from heterogeneous-parenchymatous into *tissue structure*. In some Acoela the digestive elements also are separated from the others and form a single digestive syncytium and a true tissue

---

[1] A. N. Studitskii (1947) gives the following definition of an organ: '... part of an organism adapted to fulfilment of a definite function by means of the organisation of its component elements (cells and tissues) in the form of a special structure.' In other words, an organ is characterised morphologically by a definite structural plan, and a tissue only by a definite structure.

compound, since the syncytium is permeated by muscle fibres. In this way the progress of structure within the order Acoela consists in successive differentiation of separate tissues, one after another, from the original compound parenchyma, i.e. in transition from parenchymatous to tissue structure.

Tissue structure is found in all higher Metazoa; consequently many morphologists overlook the very existence of parenchymatous structure and therefore do not notice the essential differences existing between the structure of lower Metazoa—sponges, coelenterates, and flatworms—and that of higher Metazoa.

Thus one type of modification of the general structural plan of an animal consists in the grouping of cells of different types in separate tissues, i.e. the animal passes from parenchymatous to tissue structure. Such grouping of uniformly-specialised cells into large complexes greatly increases the efficiency of their work.

A second type of modification of structure consists in the development of orderliness or regularity in the arrangement of the cells themselves. To understand the nature of that regularity we must apply to tissue cells the concepts of symmetry that we apply to the whole organism. From the point of view of symmetry we may divide the cells of Metazoa into the following classes: (*i*) anaxonic cells, in which we cannot distinguish any elements of symmetry because of the irregularity or inconstancy of their shapes; these include, for instance, various amoebocytes; (*ii*) polyaxonic cells, e.g. the stellate cells of connective tissue; (*iii*) stauraxonic homopolar cells, e.g. smooth-muscle cells; these cells have a single main axis, both poles of which are identical; (*iv*) stauraxonic heteropolar cells, a typical example of which is any epithelial cell, e.g. one of ciliated epithelium. Stauraxonic heteropolar cells may be of two types: multiradiate, e.g. most epithelial cells, and biradiate, e.g. the epithelial-muscle cells of *Hydra* or the muscle cells of *Ascaris*. Through cells of the latter type we can draw only two planes of symmetry, one along the muscle fibre and the other cutting it transversely into two equal parts. Nerve cells also represent different forms of the monaxonic-heteropolar type.

The symmetry of tissue cells is not usually maintained very strictly, but it is generally not difficult to ascertain to which of the above types the organisation of a cell belongs.

Comparing the tissue cells of Metazoa with the monoenergids of Protozoa, we find almost the same types of symmetry in each case; the only substantial difference is the absence of spiral symmetry in tissue cells, which is due to the absence of free-moving flagellate forms among them; all free-moving tissue cells of Metazoa are amoeboid, and flagellate and ciliate cells always form part of epithelium. The only free-moving flagellate cells of Metazoa are spermatozoa, and among these spiral symmetry appears, for instance, in the sperms of *Haploposthia viridis* (Turbellaria Acoela) (Fig. 3) and a few others.

The existence of different types of symmetry in cells means that the structure of any tissue or parenchyma may be either amorphous or polarised (W. N. Beklemishev, 1925). A tissue composed of anaxonic or polyaxonic elements in which a main axis cannot be distinguished is necessarily amorphous, or unpolarised; but a tissue composed of monaxonic elements is equally amorphous if the axes of the latter are not regularly arranged but lie in various directions without any structural order.

Fig. 3. *Sperm of* Haploposthia viridis (*Turbellaria Acoela*): *an example of spiral symmetry among the sperms of Metazoa* (after E. Westblad).

In contrast to such amorphous structure we may speak of polarised structure. The simplest type of polarised structure is represented by muscle bundles (Fig. 4, *A*). In a bundle of smooth muscle cells fusiform, monaxonic-homopolar cells are all arranged with their main axes parallel to one another. Within such a tissue various directions can be distinguished. One direction, parallel to the main axes of the cells, is longitudinal with relation to that tissue, and the others are transverse or oblique.

A second type of polarisation is epithelial polarisation. Take any columnar epithelium, e.g. the usual type of epithelium in the mid-gut of higher Metazoa (Fig. 4, *B*). At their apical ends all the cells possess a special cuticle traversed by pores, which is not found at their basal ends. The nucleus, and at its apical side the centrosome, lie along the main axis of the cell.

In epithelium such heteropolar cells are located not only with their main axes parallel to one another but with their similarly-named poles pointing in the same direction. The cells are arranged in the same way in multi-layered epithelium in invertebrates (Fig. 4, *C*). Epithelial polarisation is therefore a double regularity in the arrangement of cells: regularity in the directions of the main cell axes and regularity in those of their poles.

A third type of polarisation is epithelial-fascicular polarisation. It occurs with monaxonic-heteropolar cells of biradiate type, e.g. in epithelial-muscle tissue (Fig. 4, *D*). For instance, in the ectoderm of *Hydra* each of the epithelial-muscle cells consists of an epithelial part and a muscular outgrowth rising perpendicular to its main axis. In this tissue the main

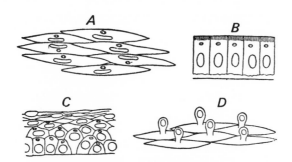

Fig. 4. *Types of polarised structure.*
A—fascicular polarisation (smooth muscles). *B*—epithelial polarisation (single-layer columnar epithelium). *C*—the same, multi-layered epithe-lium. *D*—epithelial-muscle polarisation (from Beklemishev, modified).

axes of the cells are parallel to one another and the same poles point in the same direction, and at the same time the planes of biradiate symmetry of all the cells are parallel to one another. That is because the muscular outgrowths of the cells have a fascicularly-polarised arrangement; the axes of the outgrowths, being secondary axes of the epithelial-muscle cells, lie parallel to one another.

We find this type of polarisation in the musculature of ascarids and in some other tissues, as well as in the ectoderm of *Hydra* and in other myoepithelia.

The physiological significance of polarisation of tissues is clear from the above examples. Polarisation, i.e. a regular, symmetrical, parallel arrangement of cells, is the morphological expression of unity of their functions: they are arranged in regular formation because they work together. That is clearly seen in a muscle bundle: the contraction of the muscle as a whole is built up of contractions of separate fibres, and the latter must be parallel to one another to produce that effect. It is clear also with regard to a passive function of epithelium. If epithelium covers a certain surface and the basal ends of the cells must be attached to a boundary membrane, while the apical ends must form a continuous surface separating the body from the external environment or lining a cavity, it is clear that all the homologous ends of the cells must point in the same direction.

From the structural-morphological point of view it is easily seen that amorphous structure is the most primitive, and in comparison with it every polarised structure is derivative.

3. PRIMARY BODY LAYERS OF THE LOWER METAZOA

We now move on to discussion of the morphological significance of the first tissues of the lower Metazoa and of the organs formed by them, and shall first dwell on those primary organs whose homologues are found in all higher forms and which therefore form part of the basic structural plan of all Metazoa.

We have seen that in sponges the only morphologically-differentiated tissue is flagellate epithelium; from the moment of its formation all the remaining parenchyma becomes, strictly speaking, a second tissue, but that tissue is so diversified in composition that it retains an extremely close resemblance to parenchyma.

To what extent does the flagellate epithelium of adult sponges correspond to any of the chief layers composing the larval body?

The commencing phases of sponge development, as we have seen in Vol. I, Chap. III, 2, are of different forms: blastula, parenchymula, amphiblastula; the blastula is transformed into a parenchymula as it develops. In all sponge larvae, however, kinetoblast and phagocytoblast (E. Mechnikov's terms) are characteristic formations. Kinetoblast cells have flagella, but no collars; in the structure of their kinetia they resemble Protomonadina. They form the external epithelially arranged layer of the larva (in a parenchymula) or its front part (in a typical amphiblastula). Phagocytoblast cells are located inside the larva or at its rear end, and in the former case form unpolarised tissue. The sponge parenchymula is essentially similar in its differentiation to the hydroid parenchymula (see below), and the amphiblastula may be relatively easily derived from it. Therefore we may regard the primary layers of the larvae as homologous in the two groups.

According to the classic data of Minchin (1900), in metamorphosis the phagocytoblast of the larva gives rise to all the dermacytes and archaeocytes, and the kinetoblast to the choanocytes. According to O. Duboscq and O. Tuzet (1933, 1935, 1937), however, in calcareous sponges the phago-cytoblast gives rise only to various classes of dermacytes, and the kinetoblast to choanocytes and archaeocytes. But in any case the metamorphosis of sponges differs so much from that of hydroids that the structural plan of the adult sponge has very little in common with that of hydroids and other Metazoa.

In coelenterates, especially hydroids, we also find structural differentia-tion into compound tissues that form two principal organs of the animal and two body layers—integument and gut. The methods of differentiation of these primary layers in coelenterates are very diverse, but Mechnikov succeeded in reducing all of the variations into a single consecutive series.

As A. A. Zakhvatkin (1949) has shown by analysis of published embryological data, the primary type of ontogeny of multicellular animals still bears clearly-marked protozoan features. The cleavage of the ovum in many metagenetic medusae and sponges resembles in detail the same process in Volvocidae; it takes place by means of the longitudinal division of the zygote typical of Flagellata, and later also of the blastomeres, and it results in the formation of a blastula consisting of flagellate cells and again strikingly like adult Volvocidae. The chief difference is that the cells composing the body of *Volvox* are of phytomonad type, while the blastula cells are similar in structure to the migratory stages of Protomonadina. The flagellate blastula of the lower Metazoa is also a migratory stage in their life cycle, and A. A. Zakhvatkin calls it a *synzoospore*, i.e. a formation arising through non-separation of a family of migratory cells—zoospores. In the hydroids whose development includes a flagellate blastula, its formation concludes embryonic development (embryogeny) and all of its subsequent transformations represent post-embryonic development.

The structure of a blastula is very primitive: it consists of cells of a single type, but these cells are combined into epithelium and consequently assume structural regularity of arrangement, or polarisation.

A primitive type of later development of the blastula is *multipolar immigration:* this type is exemplified in the narcomedusa *Solmundella* (synonym *Aeginopsis*) (E. Metschnikoff, 1886). The formation of primary layers begins there with some of the blastula cells losing their connection with the rest of the epithelium and migrating inward, filling the blastocoel (Fig. 5, *A*). When this process ends, the remaining external epithelium is changed from blastoderm to kinetoblast, and the newly-formed internal layer is phagocytoblast. The resultant product is a two-layered larva— a parenchymula, already possessing tissue structure. Of its two tissues, the kinetoblast retains the polarised structure of the former blastoderm, and the phagocytoblast is amorphous.

In many hydroids the parenchymula is later transformed into a planula, the deep layers of phagocytoblast break down and one cell-layer remains, adjoining the kinetoblast and giving rise to the polarised digestive epithelium of the adult hydroid. The kinetoblast of the planula is often differentiated, and besides flagellate cells glandular, stinging, sensory, and nerve cells develop in it; from being a simple tissue it becomes a compound tissue. In Leptolida the further metamorphosis of the larva consists in its attachment to the substrate by its animal pole, while the mouth breaks through at the vegetative pole and tentacles are formed by evagination of the body walls.

A second type of formation of primary layers is *unipolar immigration,* typified by the hydromedusa *Aequorea*. Here also the initial stage is a blastula, and the primary layers are formed by immigration (and proliferation). The immigration, however, is not multipolar but unipolar, i.e. it involves not the whole surface of the blastula but only its posterior, vegetative pole (Fig. 5, *B*). As a result the blastocoel is filled with dense, amorphous phagocytoblast and a parenchymula is formed; its further development takes place as in the preceding type.

A third type of formation of primary layers (the hydroid *Clava* type) occurs in the hydroids whose embryos develop within the gonophore. As a result of cleavage, instead of a blastula a compact embryo (morula)

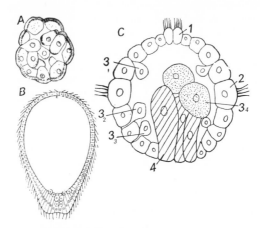

Fig. 5. *Comparison of methods of formation of embryonic layers in typical Proto-stomia and hydroids.*
A—multipolar immigration in the hydromedusa *Aeginopsis* (after Mechnikov). *B*—beginning of process of unipolar immigration in the medusa *Aequorea* (after Claus). *C*—diagram of formation of embryonic layers with spiral cleavage (original). 1—apical plate; 2—prototroch; 3_1–3_4—immigrating cells of the first to fourth quartets, becoming part of the peripheral phagocytoblast; 4—immigrating endoderm cells at the vegetative pole.

is formed directly, representing the perfect type of primary-parenchymatous structure. The primary body layers are also formed by a simplified method—in a typical case, by *parenchymatous delamination* (Fig. 1); the cells of the external layer of the morula assume a prismatic shape and heteropolar structure and therefore become distinct from the other cells, building smooth epithelium, while the initial cell mass retains polygonal shapes and amorphous arrangement for some time. The kinetoblast and phagocytoblast are formed in the same way in the *Tubularia* embryo (A. Tikhomirov, 1887). With this type of development free-living existence usually begins with the planula stage. Development within the gonophore therefore leads to embryonisation of a number of stages and lengthening of the embryogeny period at the expense of the first stages of post-embryonic development (A. Zakhvatkin, 1949).

A fourth type of formation of the primary layers, *invagination*, never occurs in hydroids, but is found in several Scyphomedusae and Actiniaria. Here also a blastula is formed by the conclusion of cleavage, but later invagination of some of the blastoderm takes place at the vegetative pole instead of immigration of separate cells. As a result, by-passing the parenchymula stage, a fully-epithelised two-layered embryo, a gastrula, is produced directly from the blastula. The kinetoblast is formed from the blastoderm of the animal hemisphere and the phagocytoblast from that of the vegetative hemisphere. At the same time the gut cavity and the mouth are formed. As compared with hydroid development, the whole process is extremely simple and rational. Epithelial structure is retained at all stages of development.

This simplicity is not a sign of primitiveness, but the product of extreme rationalisation. The orderly, simple, direct method of development by invagination is not the first but the final stage in the evolution of methods of development of the principal layers in coelenterates. We must remark, however, that this theory, originated by Mechnikov and Tikhomirov, is far from being universally accepted in spite of its correctness. Even today many still adhere to the views of E. Haeckel (1874, etc.), according to which invagination was the primitive method of formation of the primary layers. All the facts stated above show the invalidity of that point of view.

In this connection we must say a few words on the suggested beginning of the evolution of Metazoa. All former theories were based on the premise that the free-swimming larvae of sponges and coelenterates recapitulate a free-swimming adult ancestral stage, or in other words that the primary type of ontogeny of the lower Metazoa was that in which all stages of the life cycle led a free-swimming mode of life. From that point of view sponges and sessile coelenterates arose as regressive branches of the main stem of free-swimming lower Metazoa from which all Bilateria sprang. A. A. Zakhvatkin (1949), however, comparing the development of the lower Metazoa with the life cycles of Protozoa, has demonstrated that it is more probable that the primary free-swimming larvae of sponges and hydroids are purely distributive stages and that therefore they cannot recapitulate any adult, feeding stage of their ancestors. If that is so, any assurance that primary Metazoa led a free-swimming mode of life as adults vanishes. With no less probability we may surmise that they were sessile animals from the very beginning, like the overwhelming majority of modern primitive Metazoa.

The relationship of Bilateria to coelenterates that swam by means of ciliary movement and were provided with an aboral sense organ is beyond doubt; among modern coelenterates ctenophores are nearest to that structural plan, and we find traces of it in Narcomedusae and Siphonophora Disconantae (see Vol. 1, Chap. III). From the point of view of the dominant theories all these forms would have to be regarded as relics of a main stem of primary free-swimming coelenterates. If, however, following A. A. Zakhvatkin, we assume that the adult stage of the primary Metazoa was represented by a sessile animal, we can imagine that transfer of such adult features as an open mouth and grasping tentacles to earlier stages of the life cycle, while ciliary locomotion was retained, could very soon produce adult coelenterates swimming by means of flagella or cilia. The planula with its histological differentiation, its complex kinetoblast composition, and its epithelised phagocytoblast already constitutes a first step in that direction, and the actinulae of Tubulariidae (Athecata) and the related larvae of Narcomedusae and Disconantae, perhaps, a second step. The primary sessile coelenterates could thus pass to a planktonic mode of life without losing their ciliary method of locomotion, and give rise to the main progressive stems of the animal kingdom.

At present it is hardly possible to consider that any of the hypotheses about the original mode of life of primary Metazoa is at all conclusive. On such questions any expression of opinion can be no more than a guess. As our knowledge of the structure and development of Protozoa and lower Metazoa improves, however, our picture of their interrelationships will no doubt become more complete and precise, and many hypotheses based on inadequate material or incorrect reasoning, such as E. Haeckel's gastraea theory or the Hadži-Steinböck theories of the origin of turbellarians from infusorians, must be gradually discarded.

In any case, the possession of two primary layers—kinetoblast and phagocytoblast, already separated in the larva—is characteristic of coelenterates.. They constitute the primary organs of coelenterates and are formed by the primary tissues—cutaneous and intestinal—of these animals.

The Metazoan group third in order of primitiveness is ciliate worms (Turbellaria). Among them formation of the primary layers appears most clearly in Polyclada (it apparently takes place similarly in Acoela, but there it has not been adequately studied).

As a result of quadriradiate, and, on the whole, fairly primitive, spiral cleavage in Polyclada a single-layered embryo, similar to a blastula, is formed. Later certain cells migrate within that embryo, namely, all the macromeres, all the cells of the fourth quartet, some of the offspring of cells 2a–2d, and apparently a certain number of cells of the third quartet. All these cells together give rise to the phagocytoblast. The latter thus arises here by means of multipolar immigration, with a tendency to become unipolar (Fig. 5, C). This tendency is expressed in the fact that the number of cells immigrating at the vegetative pole greatly surpasses the number immigrating at any other part of the body.

Essentially, therefore, the method of formation of the primary layers in Polyclada corresponds to one method of their formation in hydroids. Other animals with typical spiral cleavage resemble Polyclada in this respect. Both of them differ from hydroids in the constant and relatively-small number of immigrant cells. Polyclada have only about 16, while in other animals with spiral cleavage the number of these cells is still smaller. Moreover, in hydroids the fate of all the immigrant cells is the same, whereas in animals with spiral cleavage each of the cells has its own sharply-defined prospective significance. In particular, in Polyclada the four macromeres perish, cells 4a–4d give rise to the whole of the central phago-cytoblast (the intestinal epithelium) and some of the peripheral phago-cytoblast, and all the other cells form peripheral phagocytoblast.

The method of formation of phagocytoblast by invagination arose independently in many stems of Protostomia, as it also arose among several higher Cnidaria, but in both cases it was a secondary simplification process. In Protostomia invagination is neither a widespread nor a primitive method of forming the embryonic layers. Calling every process of formation of embryonic layers gastrulation, and every two-layered embryo a gastrula, is a relic of Haeckel's gastraea theory.

The division of mesoderm (peripheral phagocytoblast) into ectomeso-derm and endomesoderm (mesenchyma and mesoblast) is artificial and certainly does not have the great significance attributed to it in embryology. In other words, the theory of the dual nature of mesoderm, at least with regard to Protostomia, should be rejected.

As we have seen, in adult coelenterates all or almost all the phagocyto-blast is polarised and gives rise almost exclusively to the epithelium of the gut. Unpolarised peripheral phagocytoblast is only slightly developed in coelenterates. On the other hand, in most platodes the peripheral part of the phagocytoblast remains unepithelised and forms a more-or-less-differentiated layer of amorphous tissue between the gut and the integu-ment. In this way further separation of the primary phagocytoblast into two layers takes place—into central and peripheral phagocytoblast, which are constructed of two different tissues—gut epithelium and amorphous mesenchyma.

In Acoela all the phagocytoblast remains amorphous, as in sponges; in some of them, as we have seen, it forms a single layer, but one unusually diversified in its cell composition. These are forms that remain two-layered in the adult state.

Thus the three most primitive groups of Metazoa (sponges, coelenterates, and turbellarians) are characterised by development in them of the same primary layers, kinetoblast and phagocytoblast, and by a tendency to further differentiation of both of them, especially the latter. That tendency, incipient in sponges, is well expressed in coelenterates and reaches full development in turbellarians and all higher forms.

In primitive forms kinetoblast is characterised by the presence of flagella or, later, of cilia, forming its primary locomotor apparatus (whence the name 'kinetoblast'). It is also characterised by its location: it always forms the external layer of the body (whence its second name 'ectoderm'), the only exception being provided by adult sponges (see Chap. I). With the polar type of formation of primary layers kinetoblast is formed from the animal half of the embryo.

Phagocytoblast is disintinguished by great development of the digestive function: the phagocytes or digestive syncytia of Acoela and the in-testinal epithelium and phagocytes of other Metazoa are formed from it; but a number of other types of cells (sustentacular, muscle, etc.) are also formed from it. In addition, phagocytoblast is characterised by its location on the inner side of the kinetoblast; almost the only exception is again found in adult sponges. With the polar type of formation of primary layers phagocytoblast is formed from the vegetative half of the embryo.

According to Duboscq and Tuzet, in calcareous sponges not only choanocytes but also archaeocytes arise from the kinetoblast. The archaeocytes are intermingled with dermacytes, and therefore in adult sponges the derivatives of phagocytoblast and kinetoblast are not separated from each other territorially and do not form the primary tissues and primary layers of the animals. The same applies to the choanocytes on one hand and to all the rest of the 'parenchyma' of sponges on the other hand. The difference between sponges and other Metazoa is therefore wider than is usually believed.

It is true that kinetoblast and phagocytoblast constitute the principal body layers of adult coelenterates and turbellarians and form their primary organs: integument and intestine in coelenterates, and integument and parenchyma in acoelous turbellarians. Differentiation and separation of the two layers, however, especially the phagocytoblast, has already begun in coelenterates and has advanced further in turbellarians. The numerous functions of each of the primary layers resulting from the diversity of their cell composition have begun to concentrate in separate parts, where the cells performing these functions also concentrate. The concentration of functions is accompanied by segregation of identical cells.[1] On that account new types of structure arise; new tissues, and new organs formed from them. These organs often include portions of several different tissues, often partly of kinetoblastic and partly of phagocytoblastic origin. The integration of organs takes no account of the boundaries of the primary layers. For instance, the tentacles of hydroids are formed of kinetoblastic (ectodermal) epithelium and cellular supporting tissue of phagocytoblastic (endodermal) origin.

Kinetoblast and phagocytoblast form, therefore, the principal body layers and the immediate organs of the animal only in sponge and coelenterate larvae and in hydroids of the simplest structure, such as *Protohydra*. In all other Enterozoa, because of concentration of functions and integration of organs, the primary layers break down into a number of complexly-interwoven derivatives. Consequently in higher Metazoa the primary layers are reduced to the level of embryonic layers. No longer present, as such, in the adult, these embryonic germ layers produce definite cell systems, tissues, and elementary organs of the adult organism. The embryonic layers, however, remain homologous in all Metazoa, retaining everywhere (except in adult sponges) the same basic characteristics of relative location and prospective significance. We may therefore speak of unity of the structural plan of all Enterozoa and of sponge larvae not only in a promorphological sense (homology of primary poles and body axis) but also in an organological sense (homology of primary organs and embryonic layers). Establishment of the unity of the structural plan of Metazoa,

---

[1] Segregation is the grouping together of similar elements, from the Latin *grex* a flock.

which was still unknown in the time of the famous dispute between Cuvier and Geoffroi Saint-Hilaire, is one of the highest achievements of post-Darwinian embryology, linked with the names of A. O. Kovalevskii, E. Mechnikov, O. and R. Hertwig, E. Ray Lankester, B. Hatschek, and a number of other great morphologists of the second half of the 19th century.

Many objections to the theory of embryonic layers are found in the literature. These objections are based on too narrow and literal an interpretation of the theory or on a purely formal understanding of the criteria of the homology of the embryonic layers, and cannot shake the foundations of the theory.

Chapter II

MODIFICATIONS AND PRODUCTS OF KINETOBLAST (ECTODERM)

1. BASIC FUNCTIONS OF KINETOBLAST

Kinetoblast, or ectoderm, forms the outer layer of the body, which divides the body from the external environment. For that reason it is the first independent tissue to develop, becoming distinct from the rest of the parenchyma and forming a regular epithelium. Among modern Enterozoa we observe the conclusion of that process of separation only in a few Turbellaria Acoela. In all other Enterozoa it has been already accomplished.

As the outer layer of the body and its boundary with the external environment, kinetoblast naturally performs functions of interaction with that environment: (*i*) protective, (*ii*) ciliary-locomotor, and (*iii*) neuro-sensory. Kinetoblast shares the respiratory and excretory functions and the function of muscle-contractility equally with the phagocytoblast. In connection with the primary functions of the kinetoblast, new functions arise as their further development in higher or specialised animal forms: the protective function gives rise to the supporting-skeletal; the ciliary-locomotor to the function of creating a current of water (water-moving, or hydrokinetic function); and so on.

In the most primitive Metazoa the division of functions between kinetoblast and phagocytoblast is still very incomplete. In sponges and coelenterates kinetoblast retains traces of a phagocytic function: these are found in the choanocytes of sponges and in the ectoderm of the nematophores of hydroids; in calcareous sponges phagocytes are formed from kinetoblast.

E. Metschnikoff (1883) has described a capacity for phagocytosis in cells of the ectoderm at the tips of the tentacles of *Actinia* and in cells of the ectoderm of the gastrula of the viviparous sea-anemone *Bunodes*, which develops in the gastric cavity of the mother and feeds on particles of her food. On the other hand, the phagocytoblast of coelenterates, Entero-pneusta, echinoderms, and some other Bilateria (see Chaps III and V) retains the neuro-sensory function; the endodermal neural plexus develops from it in coelenterates and some echinoderms, and in echinoderms also the hyponeural and endoneural nerve system. These facts do not controvert the existence of a definite division of functions between kinetoblast and phagocytoblast, but merely show that in lower Metazoa complete division of functions between the layers, their full specialisation, has not yet taken place.

In this chapter we discuss the morphological modifications of kinetoblast and its products, related to fulfilment of its basic functions and also of the respiratory function.

2. CILIARY-LOCOMOTOR AND CILIARY-HYDROKINETIC ADAPTATIONS OF
KINETOBLAST

In good agreement with the theory that Metazoa originated from Zoo-flagellata is the fact that the ciliary function of kinetoblast is performed principally by flagella in the most primitive multicellular groups, in sponges and many Cnidaria, and also in the larvae of Cnidaria, echinoderms, and Acrania. Among adult Bilateria, exclusively flagellate epithelium is possessed by brachiopods (D. Atkins, 1958), and flagellate epithelium is widespread among adult echinoderms (R. Hovasse, 1958) and in the endostyle of ascidians. In most Bilateria the ciliary function is performed by cilia, which result from the polymerisation and simplification of flagella. Among the groups of Metazoa that possess cilia, Turbellaria Acoela are one of the most primitive, and accordingly the basal apparatuses of their cilia are more complex and more like those of flagella than in most Metazoa; in Acoela and Polyclada we sometimes find such a flagellate feature as attachment of rhizoplasts to the nucleus (Fig. 10, E).

In many cases the whole body of a multicellular animal is entirely covered with ciliated epithelium, as we see in the flagellate blastulae of various animals, in coelenterate planulae, in adult turbellarians and nemertines, etc. In other cases ciliated epithelium is concentrated in certain parts of the body and forms a complexly-differentiated ciliary apparatus. Both the uses and the types of that apparatus vary widely in different groups of animals.

Sponges, which stand near the origin of Metazoa in the primitive nature of many of their organisational features, differ considerably from other Metazoa in their use of kinetoblast. All sponge larvae swim by means of flagella on the kinetoblast, which covers the whole or almost the whole body of the larva (in the parenchymula) or forms the animal half of the larva (in the amphiblastula of Calcarea). Unlike the larva, the adult sponge is sessile. Although it is attached to the substrate and cannot move,[1] the flagella continue to work throughout its active life. Once the sponge ceases to travel, the flagellate activity causes water to move. The ciliary-locomotor function is thus converted into a ciliary-hydrokinetic function. From being a locomotor organ, the kinetoblast becomes an organ for creating a current of water and so bringing food to the animal.

Adaptation of sponges to a sessile mode of life causes the most profound distortion in the organisation found in any sessile animals. More intensive introduction of food into the organism is achieved by extension of the kinetoblast within the body, where it forms flagellate chambers. Consequently there is inversion of the embryonic layers (Y. Delage and E. Hérouard, 1899); the kinetoblast ends up within the phagocytoblast; the phagocytoblast forms on its surface a layer of epithelially-arranged

---

[1] Young sponges can crawl from one place to another, apparently, by means of the amoeboid movement of their *dermacytes*, without any connection with the work of their flagellate cells (M. Burton, 1949).

pinacocytes, which serve as an outer covering but are not homologous with the outer covering of other Metazoa, being a formation *sui generis*.

If, as is usually done, we regard sponges as the first descending branch leaving the main stem of Metazoa, the existence of inversion of the embryonic layers in them is very instructive. It provides a clear example of the rule that the lower the organisation of any group of animals, the more profound are the changes produced by any adaptation. In solving a relatively-simple problem, creating an adaptation to bring water, sponges have literally turned themselves inside out. More complex animals (echinoderms, annelids, Bryozoa, barnacles, etc.) adapt themselves to fulfilment of the same function with much less change in their organisation. One of the principal characteristics of progressive evolution is the increasing emancipation of an organism from the direct and immediate action of individual factors of the environment, which can be seen in the realms of both comparative physiology and evolutionary morphology.

The only instance of inversion of embryonic layers known outside of sponges also pertains to an animal very low in the system. It is *Polypodium hydriforme* (an aberrant hydromedusa?). Its parasitic stage, living in sturgeon ovaries, is turned inside out, so that the intestinal epithelium (phagocytoblast) is turned outward towards the surrounding food and the kinetoblast is inside. When individuals destined for a free life are budded off the contrary inversion takes place, and the free-living forms have the layers in the relative positions normal for coelenterates (Lipin, 1910).

The ciliary-hydrokinetic or irrigation apparatus of sponges, developed as a result of inversion of the embryonic layers, is the principal apparatus of their body. In a typical case it is composed of numerous flagellate chambers lined with choanocytes and of a system of afferent and efferent canals lined with unciliated covering cells. The afferent canals begin with numerous pores on the surface of the sponge and communicate with the flagellate chambers by means of porocytes; the efferent canals lead into the atrial cavity, which opens to the exterior by the osculum. Various modifications of this apparatus, together with variations in the composition and structure of the skeleton, constitute the chief source of diversification in the sponge phylum.

The typical coelenterate larva, the planula, is entirely covered with ciliated epithelium and swims by means of flagella or cilia. The only coelenterates that retain the power (although modified) of swimming by means of cilia in the adult state are ctenophores.

Ctenophores are comparatively large animals. Usually only very small animals can swim by means of cilia because cilia are weak organs, and for them to sustain the animal's body in the water there must be a definite relation between the weight of that body and the number of cilia, i.e. between the mass of the animal and the area of its surface. With increase in the animal's size not accompanied by corresponding changes in its shape and specific gravity, the volume and mass increase more rapidly than the surface and the number of cilia; the load on the cilia thus increases. and very large animals can no longer swim by means of cilia.

Ctenophores are a very conservative group. They retain a purely ciliary mode of locomotion, with almost no transition to muscular swimming, and only a few of them have changed to a sessile or a crawling mode of life. How do ctenophores sustain their bodies in the water by ciliary movement? That is achieved chiefly by perfection of the cilia themselves. Instead of separate cilia ctenophores have much more efficient rowing organs: ctenes, each formed by fusion of two rows of cilia. Then a reduction of specific weight is attained, first by a high percentage of water content in the body (resulting in a great loss of mechanical strength), and secondly, perhaps, by regulation of the ionic composition of the tissue fluid, lowering then concentration of heavy ions such as SO_4 (as has been demonstrated in other planktonic animals, e.g. *Aurelia*, *Salpa*, and *Loligo*, see J. Robertson, 1957).

Ctenes are not formed all over a ctenophore's body, but are arranged in eight meridional rows. The simple cilia retained by the animal are also located only on certain parts of the body. Therefore ctenophores exhibit not only integration of the cilia into ctenes but also differentiation of their locations, a phenomenon widespread in all higher forms. This differentiation is one of the simplest examples of the concentration of separate functions of the kinetoblast in separate regions of it, which we have mentioned above as one of the main paths of evolution of the primary layers in Metazoa.

Most Cnidaria lead a sessile mode of life, and the flagella and cilia of the kinetoblast, when retained, no longer play a locomotor role in them but, as with sponge flagella, serve to put water in motion. Generally speaking, the water currents produced by flagella or cilia may perform three main tasks in different animals: first, by renewing the water around the body surface, they facilitate respiration and removal of excreta; secondly, they wash away from the body particles deposited on its surface by the water—in many animals that function is of vital importance, e.g. in Madreporaria species living in calm lagoon waters (C. Yonge, 1956); and thirdly, these currents can bring food particles to the mouth. Coelenterates are primarily predatory animals and seize relatively large prey by means of tentacles and the mobile edges of the mouth. All Hydrozoa and most members of other classes, including all their most primitive members, feed in this way. But a number of coelenterates have secondarily changed to feeding on more or less finely-divided, diffused food brought to the mouth by cilia. Among Scyphomedusae this has been demonstrated for adult *Aurelia* (A. Southward, 1949) and Rhizostomeae, and among ctenophores for *Mnemiopsis* (R. Main, 1928), and it is very probably for some other Lobifera and for Cestidea. Among sea-anemones, besides predatory forms such as *Urticina* with well-developed tentacles and slightly-developed ciliary covering, there are a number of microphages with relatively slightly-developed tentacles and well-developed ciliary covering, which brings food particles to the mouth from the surface of the disc and the tentacles. All Madreporaria are specialised predators: they overcome prey with the aid of nematocysts in their tentacles and bring it to the mouth either by means of the tentacles or (if the latter are short) by means of the cilia on the disc (many species of *Fungia*) (C. Yonge, 1956). The cilia then play a transporting, not a food-catching, role.

From the point of view of the mobility of primary Metazoa one might imagine that Cnidaria passed to a sessile mode of life at that stage of development of the main stem of Metazoa when musculature and a neural system already existed, and the seizure of food was accomplished by active swallowing. Therefore, unlike sponges, they did not need inversion of the principal layers. However profound the regressive changes in their organisation might be as a result of transition to a sessile mode of life, they were much less radical than in sponges.

When some Cnidaria (medusae) returned to a free-swimming mode of life the ciliary method of locomotion, once lost, was not recovered by them, and new locomotor organs (bell, velum) worked by muscles arose. If that was actually the history of medusae, it is a good illustration of Dollo's law about the non-recovery of organs that have been lost in the course of evolution.

The kinetoblast of turbellarians consists mainly of ciliated cells (together with glandular, sensory, and other cells); usually continuous ciliated epithelium covers their whole body. In some forms, however (e.g. most terrestrial Triclada), ciliated epithelium is restricted to the ventral side only; and *Desmote* (a tissue parasite of crinoids, order Rhabdocoela) is quite devoid of cilia in the adult state, although young specimens still possess them (W. N. Beklemishev, 1916).

Only small turbellarians, not more than 3 mm long, swim by means of cilia. Larger ones can still use them to move on the substrate; but the larger the turbellarian, the smaller is the role played by cilia in its locomotion and the greater that played by its musculature. In the turbellarians that use cilia for swimming the ciliary movement is governed by the nerve system and may cease or change direction as an effect of neural impulses (F. Alverdes, 1923). In all large turbellarians the cilia have a hydrokinetic function (changing the water around the body). In this case they beat constantly in one direction, from front to rear, and are no longer subject to regulation by the nerve system. Their movement is autonomous.

Nemertines, like most turbellarians, are entirely covered with cilia, which, however, in them also play a locomotor role only in very small specimens.

Among parasitic flatworms Temnocephala bear cilia only on their tentacles; adult flukes and tapeworms have no cilia. The larvae of flukes (the miracidia of Digenea and many larvae of Monogenea) and of tapeworms (the decacanth larvae and coracidia of primitive families) are covered with ciliated epithelium and swim by means of it.

Among other Scolecida, Rotatoria and Gastrotricha, which are distinguished by their very small size, retain the power of ciliary locomotion. Instead of a continuous ciliary covering, however, both these groups (like some turbellarians) retain only a band of ciliated epithelium on their ventral side. In Rotatoria that original type of ciliary arrangement, which is possessed by benthic Notommatidae (Fig. 6), undergoes very complex modification (see Vol. 1, Chap. IV), mainly because of development of a second function by their ciliary apparatus: in many Rotatoria it serves not only for swimming but also for a new task, bringing food to the mouth. Swimming by means of cilia that form a continuous covering, or, more often, are grouped in rings or ciliated bands, is characteristic of most primary larvae of marine animals: Müller's larvae, pilidia, trochophores, veligers, tornariae, echinoderm larvae, actinotrochae, larvae of Bryozoa and brachiopods, etc. The structure of the ciliary apparatus of the various

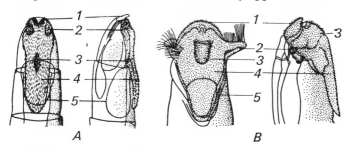

Fig. 6. *Primitive types of structure of the ciliary apparatus of Rotatoria of the family Notommatidae.*

A—*Diglena forcipata.* B—*Copeus cerberus.* 1—front margin of head; 2—lateral processes of the ciliated area; 3—mouth; 4—ciliated area on ventral side of front end of body; 5—mastax (masticatory pharynx) (after Beauchamp).

planktonic larvae is much diversified and permits homologisation of parts only within comparatively restricted groups, e.g. within the group of dipleurula-type larvae or within that of trochophore-type larvae, etc. In many cases, e.g. in several actinotrochae and tornariae, the ciliary apparatus assumes considerable complexity and peculiar configurations.

Among adult animals standing higher than Scolecida in the system ciliary swimming is retained in a few small forms: *Dinophilus* and *Nerilla* among polychaetes, *Aeolosoma* among oligochaetes, some small Nudibranchia (*Rhodope*), and some others.

Most polychaetes retain a more-or-less-well-developed ciliary apparatus even in the adult state, but it is used for swimming only in the above minute forms; its main function is hydrokinetic. Both the structure of the ciliary apparatus and its purpose vary greatly in different families of polychaetes. In many cases it is composed of separate groups or bands of ciliated cells on segments; in others there is continuous ciliated epithelium on the cephalic lobe and on various appendages. Rows of ciliated cells located on the body segments and parapodia of Aphroditidae create water currents towards the gills, which are protected by elytra; cilia also have respiratory significance in a number of other forms. Serpulimorpha have a ciliated groove running from the anus to the front part of the body and serving to remove excrement from the dwelling-tube. Cilia on the cephalic appendages of most sessile forms serve to drive food particles towards the mouth, but when the particles are brought by a strong water current the latter is used for both respiration and removal of excreta.

Ciliary apparatuses used for changing the water around the body are

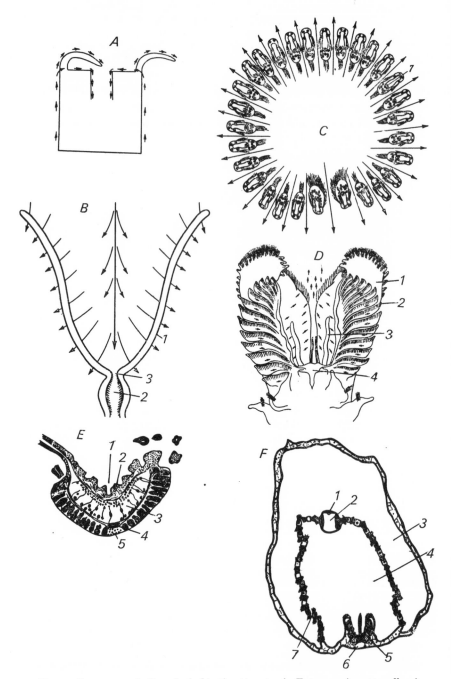

Fig. 7. *Some types of ciliary hydrokinetic apparatus in Enterozoa (except molluscs).*
A—the sea-anemone *Protanthea simplex*, diagram of water currents produced on the
body surface by the activity of the cilia of the ectoderm (from Pax). *B*—Bryozoa
Gymnolaemata, diagrammatic longitudinal section through a ring of tentacles,
showing directions of water currents: 1—tentacles; 2—pharynx; 3—mouth (after
Borg). *C*—*Flustrella* (Bryozoa Gymnolaemata), cross-section through ring of
tentacles near their base (after D. Atkins, from Hyman). *D*—*Sabellaria spinulosa*
(Polychaeta Spiomorpha Hermellidae), front end of body from ventral side.

also widespread among molluscs. They are particularly important when they ensure water circulation in the mantle cavity, e.g. in most Lamellibranchia and Gastropoda Prosobranchia; in these cases special current-directing adaptations often arise, e.g. siphons formed by the edge of the mantle, or the epipodial groove of Aspidobranchia (Risbec, 1955). In Solenogastres ciliated epithelium covers the vestigial foot that lies like a ridge along the ventral groove of Neomenioidea; in Loricata and Gastropoda it covers the sole of the foot, the gills, and the mantle cavity, and in Gastropoda often other parts of the body; Nudibranchia are usually entirely covered with ciliated epithelium. Lamellibranchia also are covered by almost-continuous ciliated epithelium, except for the outer surface of the mantle (which secretes the shell), the epithelium of the siphons, etc. In Cephalopoda ciliated epithelium is greatly reduced, and the water around the gills is changed by contractions of the musculature of the mantle.

The ciliary covering is continuous in Enteropneusta and is well-developed in most echinoderms (T. Gislén, 1924), but it is entirely absent in Holothurioidea and greatly reduced in Ophiuroidea.

In many of the above cases the work of the cilia is important chiefly as a means of hastening change of water near the body surface, which is necessary for the animal's respiration. As stated above, many animals also use ciliary activity to bring food particles to the mouth or the filtering apparatus. This function is secondary for the ciliary apparatus of Enteropneusta, which as a result of it often acquires great complexity and efficiency.

Five main types of ciliary apparatus used for bringing food to the mouth have arisen independently in various groups of Bilateria.

1. Cilia are located over the whole surface of the body or considerable areas of it, and move water, mucus, and food particles adhering to it towards the mouth (Lobifera, many sea-anemones) or towards secondary oral orifices formed when the mouth is grown over (adult *Aurelia*). Probably Amphoroidea also possessed a hydrokinetic apparatus of this type (see Vol. 1, Chap. X) (Fig. 7).

2. The cilia are located around the mouth, usually on a ring of tentacles protruding freely into the water, and together they create a current of water directed towards the mouth. The tentacles may be arranged in a simple circle, as in Sabellidae and in the Bryozoa Stenolaemata and Gymnolaemata, or may form a more complex lophophore, as in *Phoronis*, Phylactolaemata, and some others; sometimes the tentacles are branched pinnately, as in most Serpulimorpha. Apparatuses of this type arise

Arrows show directions of water currents due to activity of cilia: 1—forward-projecting lobes, formed by segments of the prothorax and serving as a food-catching apparatus; 2—rows of tentacles arising from these lobes and covered with ciliated epithelium; 3—palps; 4—mouth (after K. E. Johansson). E—*Cephalodiscus nigrescens* (Graptolithoidea), cross-section through arm: 1—ciliated groove; 2—ectoneural cord running along its bottom; 3—canal of the collar coelom; 4—blood-vessel; 5—dorsal nerve (after T. Gislén). F—*Fragaroides aurantiaca* (Ascidiae), cross-section through gill pouch: 1—dorsal cord; 2—dorsal aorta; 3—peribranchial cavity; 4—branchial gut; 5—endostyle; 6—subendostyle vessel (ventral aorta); 7—gill slits (from Grassé).

convergently in different groups: in Rotatoria (best shown in *Stephanoceros*), Podaxonia, Kamptozoa, and Polychaeta, repeatedly in the latter. For instance, the ciliary apparatus of Serpulimorpha is formed by the branches of the cephalic appendages (palps), whereas the food-catching apparatus with a similar function in Hermellidae (Spiomorpha) is formed by comb-like rows of 'oral filaments' on the dorso-lateral lobes, or on the collar formed by fusion of the first four thoracic segments (a description of the work of the tentacular apparatus of Sabellidae is given by E. A. T. Nicol, 1931, and of that of Hermellidae by K. Johansson, 1927) (Fig. 7).

3. The cilia lie in afferent grooves, which during activity are turned towards the substrate; the ciliary movement drives food particles along the grooves towards the mouth; food is obtained not from the water but from the bottom. The afferent grooves may be located on the body surface or on the food-catching tentacles. The first type is exemplified by the ambulacral grooves that bring food to the mouth in some starfishes of the order Phanerozonia, e.g. some *Ctenodiscus* (W. Spencer, 1951). The second type includes, for example, the appendages of the labial tentacles of Nuculidae and Nuculanidae (Protobranchia) (Fig. 8, *A*); it also includes the food-catching apparatus of polychaetes of the family Terebellidae, which is formed of long thin tentacles arranged like a horse-shoe in front of the mouth. The morphological nature of the tentacles of Terebellidae is debatable. They are often regarded as palps, but in their innervation they correspond rather to the papillae of the front lip of Amphinomidae (Nereimorpha). The arms of *Rhabdopleura* also bear ciliated grooves and, according to the description by T. Gislén (1930), the *Rhabdopleura* zooids climb out of their tubes and creep over the substrate on their cephalic shields, clinging to the substrate by the grooves on their arms and so gathering food particles.

4. The cilia lie in afferent grooves that are turned towards the water mass, catching food particles and bringing them towards the mouth. Ciliary apparatus of this type is very characteristic of Pelmatozoa; as we have seen above, its appearance and evolution largely affected the course of the first stages of the evolution of all echinoderms. Crinoids, having found the most perfect way to construct an apparatus of afferent grooves, thereby acquired predominance over all other Pelmatozoa and alone survived to our times. We may remark, however, that in crinoids the ambulacral feet (tentacles) also take part in the seizure of food (W. Spencer, 1951).

5. A suctorial ciliary-hydrokinetic apparatus is possessed by sponges, brachiopods, bivalve molluscs, some Prosobranchia, and a number of lower chordates (ascidians, *Amphioxus*, Ammocoetes [1]). In all these cases the ciliated epithelia are located within canals that are current-directing adaptations, and create therein water currents that bring food particles to one or another food-catching apparatus; the existence of the canals, which give a definite direction to the afferent and efferent currents, greatly increases the efficiency of the work of the ciliated epithelium, so that the above type of hydrokinetic apparatus is the most perfect.

[1] Name given to the larvae of lampreys (Petromyzontes).

We have already discussed the hydrokinetic apparatus of sponges. Its peculiarity (apart from possession of flagella instead of cilia) consists in having many afferent canals and one efferent canal. In this respect sponges resemble only some compound ascidians, in whose colonies several zooids with independent oral orifices are grouped around a cloacal orifice common to all of them.

The most primitive bivalves are benthic sediment-feeders using as their collecting apparatus long appendages of the oral lobes furnished with ciliated grooves (e.g. *Yoldia* and other Protobranchia). Their gills retain only the respiratory function and have the structure of typical ctenidia with short triangular lamellae, and the ciliated covering of the gills serves only to renew the water at their surface, creating only slight currents of water within the mantle cavity for that purpose.

A major factor in the evolution of bivalves was transition from sediment feeding to feeding on nanoplankton, obtained by filtration; the current created by the cilia of the gill epithelium was then used to bring food particles (C. Yonge, 1941). The ctenidia consequently became hypertrophied, and their lamellae extended into long filaments and often wholly or partly lost their respiratory function.

The hydrokinetic apparatus of lamellibranchs in its most perfect form, in Eulamellibranchia, is constructed as follows. The chief moving force is ciliated epithelium, mainly that of the gills and secondarily that of the oral lobes. Both are located in the mantle cavity; in a typical case the edges of the mantle are either closely pressed together or fused. The mantle cavity actually opens to the exterior only by two siphon orifices lying side by side. A third opening, through which the foot protrudes, is often turned towards the ground and, for one reason or another, has no effect from the hydraulic point of view. The outer gill lamina fuses with the mantle by its margin, and the inner lamina with the foot; behind the foot both inner gill laminae grow together. In that way the whole mantle cavity, except for its anterior, circumoral part, is divided into two sections, dorsal and ventral. The cilia drive water forward, towards the mouth, along the ventral part of the mantle cavity; thence the water moves rearward along the dorsal part, past the urinary, genital, and anal orifices, to the exhalant siphon. Thus the mantle cavity is a U-shaped canal, both ends of which—inhalant and exhalant—lie at the rear end of the animal, which in fossorial forms protrudes from the soil. During its whole journey along the gills the water is gradually filtered through the gill filaments and then passes into the dorsal, exhalant section of the mantle cavity. Food particles are retained on the gills and are gradually moved forward along these to the labial tentacles and thence to the mouth, and partly directly to the mouth. Because of the extensive area of the ciliated epithelium on the gills and the regularity of the currents created by it, the activity of the cilia provides considerable suctorial power, and lamellibranchs are among the most efficient filter-feeders. Both the structure of the gills and the structural details and arrangement of the cilia reach a high level of perfection in 𝚓he higher lamellibranchs.

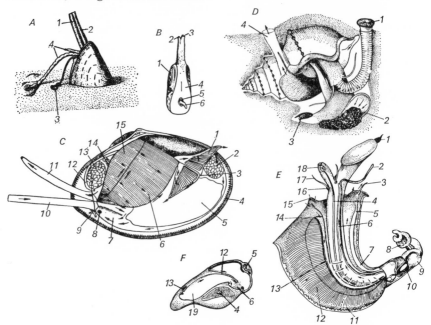

Fig. 8. *Some types of ciliary-hydrokinetic apparatus in molluscs.*

A—Yoldia limatula (Bivalvia Protobranchia), picking up sediment by means of the appendages of the oral lobes: 1—inhalant siphon; 2—exhalant siphon; 3—siphonal tentacles; 4—tentacles of oral lobes, furnished with collecting grooves (after Lang). *B—Saxicava arctica* (Eulamellibranchia), ventral aspect: 1—shell; 2—exhalant siphon; 3—inhalant siphon; 4—fused mantle; 5—anterior opening of mantle; 6—foot (from Pelseneer). *C—Tellina tenuis* (Eulamellibranchia), view from right side after removing right valve of shell and right lobe of mantle: 1—oral lobes; 2—anterior adductor muscle; 3—mouth; 4—edge of mantle; 5—foot; 6—internal lamella of right gill; 7—groove of mantle, which carries to the rear particles rejected by oral lobes; 8—a mass of these particles (later to be carried to the exterior by peristaltic movements of the inhalant siphon); 9—crossed muscles of the mantle; 10—inhalant siphon; 11—exhalant siphon; 12—hind-gut; 13—posterior adductor muscle; 14—posterior retractor muscle; 15—external gill lamella, with arrows showing directions of water currents (after Yonge). *D—Struthiolaria papulosa* (Prosobranchia Taenioglossa Strombacea), in normal position when buried in the soil, having made an outlet for the exhalant current and making an inlet with its proboscis for the inhalant current: 1—fully-extended proboscis; 2—foot; 3—operculum; 4—canal made in soil for exhalant current. *E—Struthiolaria* removed from shell, mantle cavity opened along its right side. *F—*the same, diagrammatic cross-section through mantle cavity: 1—operculum; 2—penis; 3—pallial tentacles; 4—oesophagus; 5—hind-gut; 6—groove bringing food to base of head; 7—mid-gut; 8—liver; 9—stomach; 10—style sac; 11—hypobranchial gland; 12—ctenidium; 13—axis of ctenidium; 14—osphradium; 15—siphonal lobe of mantle; 16—eye; 17—cephalic tentacles; 18—proboscis; 19—inhalant section of mantle cavity (after Morton).

The oral lobes, or palps, of most lamellibranchs have the function of sorting out the particles brought to their surface: small particles are directed into the mouth, and larger ones thrown into the mantle cavity and later ejected through the exhalant siphon by the peristaltic movement of the latter (see Fig. 8, *C*) or, when that is absent, by quick opening and closing of the shell valves, as in Pectinidae and many other Monomyaria (C. Yonge, 1954).

The superfamily Tellinacea (Eulamellibranchia) has secondarily resumed sediment feeding after having perfected a hydrokinetic apparatus and well-developed siphons (Fig. 8, *C*). These animals appose their inhalant siphon to the substrate in order to collect particles lying there, and the upper layer of mud (pelogen) is drawn into the mantle cavity by the current created by the gill cilia (C. Yonge, 1949). In other words, when higher Lamellibranchia returned to benthic scavenging the organs formerly used for that purpose were not restored (in accordance with the law of the irreversibility of evolution) and the former function is now performed by new means.

Among snails, use of the ciliary-hydrokinetic apparatus for bringing food has developed independently in at least five different families of Prosobranchia: in sessile Calyptraeidae (J. Orton, 1912, 1914); Vermetidae and Capulidae (C. Yonge, 1932, 1938; J. Morton, 1955); and in free-moving Turritellidae (C. Yonge, 1946); and Struthiolariidae (J. Morton, 1951) (Fig. 8, *D–F*). In all these cases an exchange of functions takes place, and the ciliary covering of the ctenidia begins to be used for bringing food, having originally served only to renew the water around the gills. In extreme cases (in Calyptraeidae) the ctenidia have largely lost the respiratory function, which has been transferred to the walls of the mantle cavity, as happens in many Lamellibranchia. As in Bivalvia, the lamellae of the ctenidia are extended into long filaments, which increases the efficiency of their work in creating currents of water.

The hydrokinetic apparatus of brachiopods is formed by the ciliated epithelium of their arms, which anatomically represent a modification of the lophophore of the *Phoronis* type and, in the opinion of many authors, are homologous with it. The arms of brachiopods, however, are located in the mantle cavity, inside a bivalvular shell, which (as in Lamellibranchia) regulates the water current created by the ciliated epithelium. In *Lingula* the water enters the shell at the front corners and is ejected at the centre of the front edge (Helmke, 1938); the same is true for *Crania* (J. Orton, 1914) and apparently for most brachiopods.

The hydrokinetic apparatus of brachiopods never attains the perfection it displays in most lamellibranchs, and that is probably one of the reasons why brachiopods have been gradually supplanted by lamellibranchs, as we see in the course of the history of the seas.[1]

The differentiation of the ciliary covering of the gill filaments of filter-feeders among snails and bivalves, and also on the food-catching tentacles

---

[1] The abundance of brachiopods in the Palaeozoic, when many families of lamellibranchs already existed, is perhaps explained by the fact that primitive lamellibranchs, like modern Protobranchia, probably still collected food from the uppermost mud layer and therefore were not yet in competition with brachiopods.

of Sabellidae, *Phoronis,* and several others, convergently attains an unusual
degree of similarity (Orton, 1914, etc.) (Fig. 9).

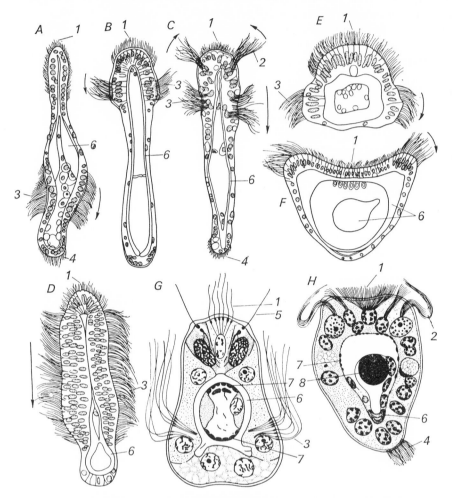

Fig. 9. *Cross-section of filaments or trabeculae of the hydrokinetic apparatus of
various filter-feeders.*

Gill filaments: *A—Crepidula fornicata* (Prosobranchia Ctenobranchia); *B—
Glycemeris glycemeris* (Bivalvia); *C—Mytilus edilis* (Bivalvia). *D—Amphioxus lanceolatus* (Acrania), one of the secondary arches. *E—Crania anomala* (Brachiopoda),
one of the tentacles of the ventral row of the lophophore. *F—Terebratula vitrea*
(Brachiopoda), one of the tentacles of the dorsal row of the lophophore. *G—Sabella
pavonina* (Polychaeta), one of the lateral branches of a tentacle. *H—Hippodiplosia
foliacea* (Bryozoa Eurystomata Ascophora), tentacle. (Drawings *A–F* after J. H. Orton;
G after E. A. Nicol; *H* after G. Lutaud.) 1—frontal cilia; 2—latero-frontal cilia;
3—lateral cilia; 4—abfrontal cilia; 5—sensory setae; 6—supporting skeleton; 7—
longitudinal muscle fibres; 8—blood vessel; arrows show directions of water
currents created by the groups of cilia beside which they are placed.

The hydrokinetic apparatus of lower chordates (Fig. 7) is formed of the
ciliated epithelium of their fore-gut, i.e. it belongs to the phagocytoblast;

it is closely connected with their respiratory organs and is discussed below together with these (see Chap. V). Here we shall merely note the considerable convergence between the hydrokinetic apparatuses of solitary ascidians and of lamellibranchs, with total absence of homology: in each case water enters and leaves through two siphons turned away from the substrate, passing through a U-shaped canal, the afferent part of which is formed of a cribriform gill or gills covered with ciliated epithelium. These analogies, although very superficial from the morphological point of view, have in their time led to attempts to derive ascidians from lamellibranchs, particularly from the Upper Mesozoic Rudistes (P. Steinmann, 1907). Actually ascidians and lamellibranchs belong to very far-separated branches of the animal kingdom, which have developed similar methods of utilising environmental conditions and consequently similar life-forms.

In other tunicates—salps and appendicularians—the hydrokinetic function has been transferred entirely from the ciliated epithelium to the musculature. Fulfilment of the hydrokinetic function by the muscular apparatus is also widespread among other animal groups (annelids, arthropods, etc.) and is discussed in Chap. VI.

In groups of animals that are covered by a thick cuticle the ciliated epithelium of the external covering is much reduced, remaining on small areas of the body (Gastrotricha, Rotatoria, Sipunculoidea, Solenogastres, Ophiuroidea, Synaptida, Pogonophora, etc.), or totally disappears (parasitic flatworms, Acanthocephala, Kinorhyncha, Priapuloidea, most oligochaetes, all leeches, Onychophora, Tunicata, most holothurians, etc.); in extreme cases, as in nematodes, Nematomorpha, arthropods, Tardigrada, etc., ciliated cells have disappeared not only from the skin but from the entire body. Onychophora retain ciliated epithelium only in the coelomoducts, and Kinorhyncha, Priapuloidea, and some Acanthocephala in the nephridia.

3. PROTECTIVE ADAPTATIONS OF KINETOBLAST

Kinetoblast, which forms the external layer of an animal's body and the boundary between its internal and external environments, protects the organism from the multiform harmful effects of the external environment. Above all, it protects the internal environment from penetration by various harmful substances and prevents loss by the body of substances required by it. It also protects the organism from penetration by various animate and inanimate particles, from active assault by enemies, and from damage by rough mechanical contacts.

Morphologically all these functions are expressed in the regular epithelial arrangement of the kinetoblast cells in the lower Metazoa; and in that of the cells of the epidermis or external epithelium of the skin (one of the products of the kinetoblast) in higher forms. The protective function of the kinetoblast is further expressed in the formation of glands, cuticle, and basal membranes. The solidified secretions of the glands, cuticle, and basal membrane may in certain cases assume a supporting (as well as a protective) function, form an external skeleton, and sometimes, with increasing complexity of organisation, even form parts of an internal skeleton. The protective function of the epidermis is often strengthened by its integration with underlying layers, mostly of phagocytoblastic origin, forming the cutis, a connective-tissue layer of the skin or sometimes a musculo-cutaneous sac. The degree of development of such a body covering of composite origin is inversely proportional to the development of cuticle.

Epithelial arrangement of the cells of the epidermis [1] is a general rule; it is the minimum of functional structure required to fulfil the principal, integumental role of an external covering. In some Acoela, as we have seen (Fig. 10), there is still little distinction between the epidermis and the rest of the parenchyma. Its cells are not separated from the underlying parenchyma by a basal membrane or a network of cutaneous muscles; in Acoela a basal membrane is generally absent, and in many Acoela the fibres of the cutaneous musculature lie intra-epithelially. Often the basal ends of the epithelial cells do not form a smooth surface, and project between the parenchyma cells (Fig. 10, *A*). It is easy to derive the four principal types of epidermis in Bilateria from this primitive type.[2]

1. Sunken epithelium (Fig. 10, *B*). The apical parts of the cells fuse into a smooth epithelial layer, in which cell boundaries are sometimes present, sometimes absent. The basal ends of the cells, on the other hand, remain mutually independent and are embedded in the underlying tissue, not forming a surface separate from it. The cutaneous muscles remain epithelial, lying in the epithelial layer, or become distinct from the epidermis and lie at the lower surface of the epithelial layer.

This type of epithelium is found in many turbellarians of the orders Acoela, Alloeocoela, and Triclada, in all parasitic flatworms, and in isolated forms in other groups (e.g. *Protomyzostomum* in the order Myzostomida, according to D. M. Fedotov, 1915). In all these forms, except Turbellaria Acoela, the epithelial layer is separated from the underlying tissues by a basal membrane. The sunken epithelium of parasitic flatworms is distinguished by a high degree of specialisation.

2. Nemertine-type epithelium, found in nemertines and, in a somewhat different form, in some Acoela and other primitive turbellarians (Fig. 10,

---

[1] In very many invertebrates the epidermis is an epithelial layer without cell boundaries, i.e. of syncytial character. Every type of epidermis exists in two versions: cellular and syncytial.

[2] Many histologists, being familiar with the types of epithelium found in higher animals and distinguished by a clear separation between the epithelium and the underlying tissues, are inclined to regard all other types (unseparated, sunken, nemertine-type epithelia) as aberrant, specialised types, as mere curiosities. A correct morphological and evolutionary perspective is obtained, however, only if one starts from lower, not higher, forms. The types of epithelium with incomplete separation from underlying tissues found in lower Metazoa are actually not aberrant but primitive.

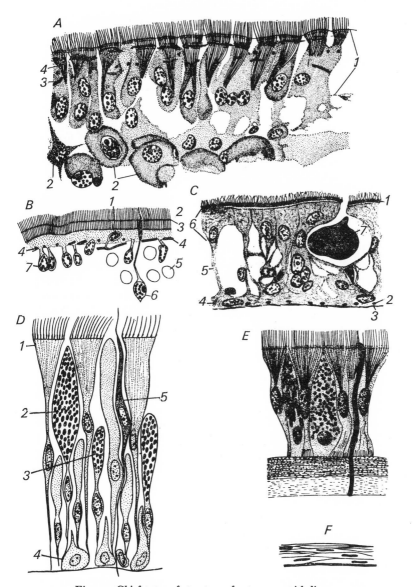

Fig. 10. *Chief types of structure of cutaneous epithelium.*

A—epithelium not separated from underlying tissue (parenchyma)—*Oligochoerus erythrophthalmus* (Turbellaria Acoela): 1—epithelial-muscle cells; 2—parenchyma cells; 3—muscle fibres in epithelial-muscle cells; 4—basal apparatus of cilia (after Beklemishev). *B*—sunken epithelium of *Oligochoerus bakuensis*: 1—nucleated part of muscle cell, remaining in epithelial location as an exception; 2—cilia; 3—epithelial plate; 4—fibre of cutaneous muscle; 5—outlines of nuclei of parenchyma cells; 6—sensory cell; 7—epithelial nuclei sunken in parenchyma (original). *C*—nemertine-type epithelium of *Nemertoderma bathycola* (Turbellaria Acoela): 1—basal apparatus of cilia; 2—annular cutaneous muscles; 3—longitudinal cutaneous muscles; 4—basal layer of epithelial plasma; 5—stems of cells; 6—bodies of epithelial cells; 7—cutaneous gland (after Steinböck). *D*—diagram of structure of epithelium of a nemertine: 1—covering ciliated cells; 2 and 3—glandular cells; 4—basal reserve cells; 5—sensory cells. *E*—external epithelium of usual type, *Cryptocelides loveni* (Polyclada); besides covering ciliated cells, two epithelial glandular cells are visible (including one rhabdite cell), and also the duct of an underlying (adenal) unicellular gland (after S. Bock). *F*—multi-layered epithelium of *Sagitta* (Chaetognatha) (original).

C, D) and also in Enteropneusta. In this type the epithelial cells consist of broadened apical parts combined into a smooth epithelial layer and narrow, sometimes branched, basal outgrowths, which reach the basal membrane or the network of cutaneous muscles. The space between these outgrowths is filled with glandular cells, cells and fibres of supporting type, nerve cells, etc. Here also, then, the epidermis is poorly separated from the underlying tissues on its basal side, but it forms part of a unique cutaneous covering, in the formation of which elements of the peripheral phagocytoblast also participate.

3. Differentiated single-layered epithelium, found in most invertebrates (Fig. 10, *E*). In this type the cells of the epidermis assume a regular prismatic shape and their bases are evened out as much as their already-smooth tops. In Coelenterata and Acoela it consists of epithelial-muscle cells. In higher forms its cells are differentiated into epidermal cells that retain the epithelial linkage among them and muscle cells that have lost their connection with the epithelium and are located beneath its basal surface. Intermediate stages of this differentiation may be observed in Coelenterata and Acoela.

The presence in Acoela of all the first three types of epidermis, as well as the more primitive epidermis of *Oligochoerus*, and the constant absence of a basal membrane in them show that this group stands very close to the level of organisation at which the original separation of kinetoblast took place and the chief types of epidermis were developed.

4. The last of the chief types of epidermis is multilayered epithelium, found in vertebrates and only in Chaetognatha among invertebrates (Fig. 10, *F*). This type of epidermis is a modified and perfected form of the preceding type, 'normal' epithelium.

The external epithelium may be separated from the underlying tissues by a basal membrane (a boundary layer formed of the intercellular matter), which is absent only in Turbellaria Acoela and, according to F. Papi and B. Swedmark (1959), in some Macrostomida. The basal membrane, which is often quite substantial, may be the chief support of the skin. In many turbellarians and Monogenea spines, stylets, and other hard parts of the copulative apparatus, usually erroneously regarded as derived from the cuticle, are formed from the basal membrane of the epithelium of the genital ducts.

A substantial role in the protective function of the kinetoblast is played by cutaneous mucous and protein glands (Fig. 11), which are very numerous in coelenterates, turbellarians, nemertines, annelids, molluscs and many other invertebrates. Usually they are unicellular goblet or pyriform glands, scattered over the whole surface of the skin or grouped in complexes. In lower worms the cephalic complex is particularly constant, being possessed by flatworms, nemertines, and many annelid larvae. With sunken epithelium all glandular cells lie deep in the parenchyma, and only their ducts pierce through the epithelial layer. With other types of epidermis some glandular cells form part of the epithelium (dermal glands), while others are buried under the main layer and only their ducts pierce through the epithelium (adenal glands). Besides the protective function, cutaneous glands assume many others: they participate in the locomotor function (foot glands in terrestrial turbellarians and gastropod molluscs), in adherence to the substrate (adhesive glands in many polyps and worms),

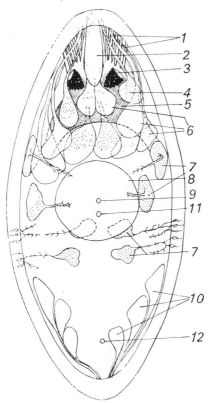

Fig. 11. *Cutaneous glands of* Maehrenthalia agilis (*Turbellaria Rhabdocoela*). *The number and arrangement of glandular cells, as often happens in small forms, are fixed and constant.*

1—sensory nerves of front end of body; 2—'rhabdite paths'—ducts of rhabdite glands; 3—eye; 4—head glands; 5—brain; 6—rhabdite glands of front end of body; 7—glands of centre of body; 8—pharynx; 9—mouth; 10—rhabdite glands of rear end; 11—male genital orifice; 12—female genital orifice (after Luther).

in attacking prey (head glands in turbellarians), in secreting luminous matter (e.g. in *Chaetopterus* among Spiomorpha) (J. A. C. Nicol, 1952), etc.

In turbellarians some slime glands secrete a substance in the form of concrete grains, spheres or rods, known as *hyaloids;* in their most perfect form they are called *rhabdites* (Fig. 11; Fig. 12, *B* and *C*). When ejected from the cells on an enemy or a prey hyaloids adhere to it, swell up, become sticky, and prevent free movement. As well as in turbellarians, rhabdite cells are known in the epithelium of the proboscis of nemertines, e.g. in *Cerebratulus* (O. Bürger, 1897–1907), but U. Pierantoni (1908) suggests that they also include the 'bacillar' glands of polychaetes, e.g. *Protodrilus* (W. Salensky, 1907; G. Jägersten, 1952). A still more highly-developed product of secretion of epidermal glands is the nematocysts of Cnidaria (Fig. 12, *A*).

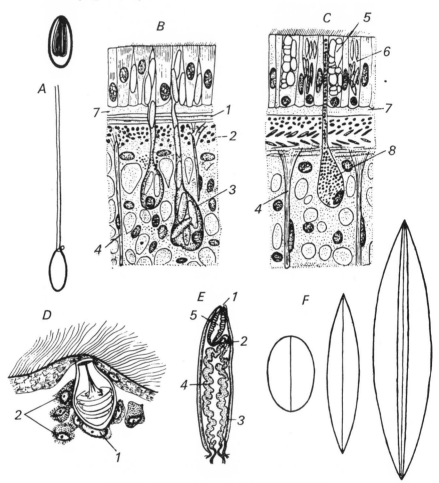

Fig. 12. *Active defensive adaptations of epidermis.*

A—stinging capsules (glutinants) of the hydra *Pelmatohydra oligactis* before and after ejection of filament (from Kükenthal). *B* and *C*—arrangement of rhabdites in the skin of Triclada (*B*) and Polyclada (*C*): 1—annular muscles of skin; 2—longitudinal muscles of skin; 3—adenal rhabdites; 4—dorso-ventral muscles; 5—cutaneous gland; 6—dermal rhabdite cells with rhabdites in protoplasm; 7—basal membrane of epidermis; 8—adenal gland (from Benham). *D*—kleptocnidae: a small penetrant of a hydra (*Hydra vulgaris*) in the skin of the turbellarian *Microstomum lineare* (Macrostomida): 1—cnidoblast; 2—mesenchymal cells of the turbellarian (after J. Meixner). *E*—longitudinal section of dorsal papilla of *Aeolis* (Opisthobranchia Nudibranchia) with cnidosac: 1—external orifice of cnidosac; 2—connecting duct between branch of liver and cnidosac; 3—epidermis; 4—branch of liver; 5—cnidosac (after Pelseneer). *F*—*Convoluta psammophila* (Acoela), three stages of development of sagittocyst (after W. N. Beklemishev).

Nematocysts are not under the control of the neural system and explode automatically under the influence of stimuli originating in the prey. The use of nematocysts by several animals that feed on polyps is based on that automatic action. For instance, in a number of turbellarians (*Paraphanostoma dubium* among Acoela, according to E. Westblad, 1942; *Microstomum* spp. among Macrostomida; *Anonymus* among Polyclada) and nudibranchiate molluscs (*Aeolis*, etc.), some of the nematocysts (cnidocysts) from the polyps eaten are carried by amoebocytes from the gut to the epidermis, where they take up their normal position with the cnidocil outwards (Fig. 12, *D*). It is possible to observe the discharge of cnidocysts located in the skin of *Microstomum* and the killing of other turbellarians by means of them. In Aeolididae cnidocysts are located in the epithelium of special pouches of ectodermal nature (cnidosacs), which are usually in direct communication with branches of the gut (liver) (Fig. 12, *E*). Nematocysts, probably kleptocnidae, are also observed in the epithelium of the proboscis of some nemertines (*Cerebratulus urticans*, etc.) (O. Bürger, 1897–1907).

The only probable homologues of the cnidocysts of Cnidaria are the sagittocysts of Turbellaria Acoela (Fig. 12, *F*). In this case also a fluid-filled capsule develops within a cell (a sagittocyst), and within it is a slender rod attached by both ends to the walls of the capsule. The cnidocysts of *Eudendrium* (Mergner, 1957) also pass through a similar stage in their development. In both cases the rod becomes hollow. Later on, in a cnidocyst one end of the rod separates from the capsule wall while the other remains attached, and the rod is converted into a stinging filament. In the sagittocysts of *Convoluta* both ends of the rod separate from the capsule wall and it becomes a dart, sharp at both ends. If the homology of the sagittocysts of Acoela and cnidocysts is confirmed, that will be one more proof of the close relationship of Acoela to Cnidaria and of their derivation from planula-like larvae of Cnidaria or of some Praecnidaria.

Another refinement of protective glands is shown in the stylet-bearing vesicular glands of many Acoela and Polyclada, which are organs of attack and in several forms are used as part of the male genital apparatus (see Chap. IX).

Cutaneous glands acquire special morphological significance if the protective secretion created by them solidifies and forms permanent protective structures. In some cases (many polychaetes, etc.) these protective formations are not organically united to the body of the animal and have the nature of dwelling-places. We can trace their evolution, beginning with the simplest use of cutaneous gland secretion to lubricate the substrate on which the animal moves. While crawling many turbellarians leave mucous threads behind them, and sometimes turbellarians may hang from such a thread attached to some higher object. When moving through the soil nemertines, many polychaetes, and terrestrial oligochaetes line the walls of their burrows with mucus and so facilitate their movement, lessening friction and protecting their skin from damage by soil particles;

in fact, a slimy tube is constructed within the burrow. Many forms (e.g. *Nereis virens*) convert such a burrow into a real dwelling-place, a tube lined with mucus and having a hydrokinetic as well as a protective function, playing the role of a current-directing adaptation.

In general, the shelters of polychaetes show every stage of transition between temporary and flimsy slimy tubes to firm, permanent tubes using foreign matter cemented together (*Onuphis conchilega* among Eunice-morpha, etc.) or secretion alone, as in the calcareous tubes of Serpulidae, which attain great solidity and (in *Spirorbis*) great regularity in form. Terebellomorpha and Serpulimorpha use the secretion of the glands of the so-called ventral shields of the thorax in building their tubes. Serpulidae and some other polychaetes never leave their tubes, and die when removed from them. We can observe similar developmental series in other groups of marine animals.

In Pterobranchia the tube walls are formed of the solidified secretion of the cutaneous glands of the cephalic shield and are composed of character-istic fusiform pieces (Vol. 1, Fig. 212). Pogonophora have a large number of multicellular flask-shaped glands along the whole length of the trunk, protruding deeply into the coelom and secreting a substance of which these animals' tubes are made (K. Johansson, 1939; A. Ivanov, 1952b).

In other cases the animal remains closely united to the secreted protective formations, which become part of the body. That particularly applies to the perisarc of hydroids, and also to the skeleton of Hexacorallia and the shells of molluscs and brachiopods.

In some solitary Athecata, e.g. *Euphysa* and *Hypolytus*, a gelatinous perisarc forms a fairly loose sheath around the hydranth stem; sometimes it may be shed and replaced by a new one (W. Rees, 1957). In most hydroids the perisarc, or periderm, is a permanent formation covering the whole body of the animal, except the heads of the hydranths in Athecata, and forming bells around these heads in Thecaphora (Vol. 1, Fig. 18). In *Eudendrium* the periderm continues to grow throughout the animal's life and so becomes thicker at the basal part of the stem, meeting the greater mechanical stress there during water movement (H. Mergner, 1957). The perisarc is totally absent only in a few hydropolyps (*Hydra, Protohydra*, etc.).

The skeleton of Hexacorallia is formed of calcium carbonate (aragonite) and is a secretion of the epidermis of the basal disc of the polyp (Fig. 13). In ontogeny, therefore, the first part of it to be formed is the plate covering the basal disc. Later the disc forms a number of radial folds projecting into the chamber of the polyp, and between the two layers of each fold a calcareous plate (a *scleroseptum*) is deposited. The number and arrange-ment of the sclerosepta depend on the number and arrangement of the chambers. An annular fold of the disc, projecting into the body of the polyp from below and concentric with its external wall, similarly secretes the theca. Other parts also, on which we shall not dwell, are deposited in the same way. It is important that the whole skeleton, which topographic-ally is largely internal, is morphologically external: all of its elements are secreted exclusively by the external surface of the epidermis, which projects into the body in various folds.

In just the same way the horny skeleton of octoradiate corals of the suborder Holaxonia, order Gorgonaria, is morphologically external and

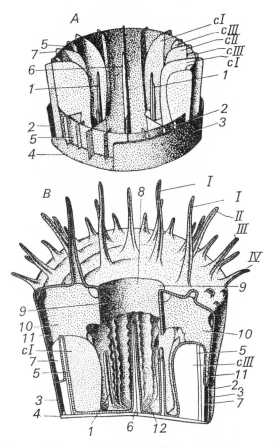

Fig. 13. *Diagram of interrelationships between skeleton and soft parts of the body in Hexacorallia.*

A—skeleton. *B*—polyp with skeleton. 1—columella; 2—perimural cavity; 3—epitheca; 4—foot; 5—wall (theca); 6—central column; 7—costa; 8—mouth; 9—siphonoglyphs; 10—sarcosepta; 11—gastral cavity of exosarc; 12—gastral cavity; I-IV—tentacles of first to fourth orders; cI-cIII—sclerosepta of first to third orders. (From Delage and Hérouard.)

topographically internal. In this case also the skeleton is first secreted by the basal disc and closely attached to a hard substrate, on which the colony grows. Later the skeletogenous epithelium of the basal disc is invaginated into a pouch, grows within the colony, and branches together with it. The skeleton of these Gorgonaria is thus a stem attached to the substrate, running along the axes of the colony and of all its branches.

The skeleton of Pennatularia (Vol. 1, Fig. 48), which are closely related to Gorgonaria and have changed to a life on soft soils, has become modified. In *Veretillum* the ectoderm of the basal disc of the founder-polyp, invaginating, forms a skeletogenous pouch, which later closes, separates from the epidermis, and secretes an axial skeletal shaft; in the less primitive *Funiculina* a thin epithelial layer (instead of a pouch) grows from the basal disc and later develops into skeletogenous epithelium, which secretes the axial skeleton (S. Berg, 1941). The skeletal shaft of Pennatularia, therefore, unlike that of Gorgonaria, is topographically a purely internal skeleton, supporting the colony and not connected with the substrate; the skeletogenous epithelium secreting it, however, is also of epidermal origin.

W. Kükenthal (1925), on the contrary, holds that the skeletogenous epithelium of Gorgonaria is formed not by growth of ectodermal epithelium but by scleroblast cells scattered through the mesogloea. In Alcyonaria and some Gorgonaria (suborder Scleraxonia) the whole skeleton consists of separate calcareous sclerites secreted by similar cells. Even Kükenthal, however, admits the ectodermal origin of all the scleroblasts of Octocorallia.

The most primitive type of protective adaptations of the covering found in any molluscs is apparently that observed in Solenogastres. The whole dorsal side of their body is covered with cuticle; in some species it is thin and has calcareous spicules on its surface, each secreted by a single epidermal cell. In other species the cuticle is thick, and numerous spicules lie within it. The ventral groove and the vestigial foot of Neomenioidea are always devoid of cuticle and covered with ciliated epithelium.

Loricata represent the next stage in shell evolution: in definite areas of the dorsal side, which are differentiated as segments of a shell gland, all the cells begin to secrete calcium and give rise to shell lamellae. That calcareous layer takes over the main protective function, the cuticle remaining on the shell surface as the periostracum. The periostracum and the tegmentum are permeated by sense organs (aesthetes) in which one may see traces of the origin of the shell from accumulation of separate spicules. The shell consists of four layers: periostracum, tegmentum, articulamentum, and hypostracum. Its plates are movably articulated together. On the periphery of the dorsal side, in the perinotum region, the cuticle of Loricata retains approximately the same structure as the cuticle of Solenogastres (see Fig. 134).

In Conchifera the shell covers the whole dorsal side of the mantle and is no longer permeated by aesthetes; it consists of only three layers, periostracum, a prismatic layer, and a nacreous layer, and is a single monolithic formation. The most primitive form of shell known to us in Conchifera is the spoon-shaped or cap-shaped shell of Tryblidiida. Metameric muscle impressions, often eight pairs in number, perhaps indicate that such a shell arose from fusion of the eight plates of primitive Loricata (C. Boettger, 1955). (Loricata are known only since the Upper Cambrian, but their remains are badly preserved, and it is very probable that this group already existed in Precambrian times.) We must admit that in the common prototype of Tryblidiida and Loricata the shell structure was simpler than that in modern Loricata.

The shell of primitive Prosobranchia (Rhipidoglossa) is close to that of Tryblidiida, with a notch at the morphologically-posterior, topographically-anterior edge, facilitating the ingress of water to the gills. From it are derived the spirally-curved shells of Gastropoda, the tubular shells of Scaphopoda, the straight or spiral shells (divided by transverse septa) of Cephalopoda Tetrabranchiata, and the bivalvular shells of Lamellibranchia.

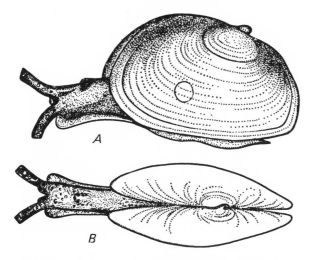

Fig. 14. Tamanovalva limax (*Opisthobranchia Saccoglossa*) *with bivalvular shell; rhinophores, eyes, and protoconch are visible.*
A—view from left side. B—dorsal view (from Cox and Rees).

The latter are derived from dome-shaped shells of the same type as in lower Prosobranchia, which have a single calcification centre and a notch at the rear edge. In Lamellibranchia the shell is broadened, flattened at the sides, and divided into two valves, each with its own calcification centre. This division took place as the result of a second, anterior, notch in the mantle, which caused its two halves to remain united dorsally by only a narrow neck; besides, along the length of that neck the shell is not calcified and forms an elastic ligament joining the two valves (see Chap. VI regarding its function). For the morphology of the ligament within the class see G. Owen, E. R. Trueman, and C. M. Yonge (1953), G. Owen (1958).

A notable instance of convergence is provided by the bivalvular shell of the snail *Tamanovalva limax* (Opisthobranchia Saccoglossa) (Fig. 14), which has a single adductor muscle; the left valve has a spirally-curved larval shell (S. Kawaguti and K. Baba, 1959, quoted from C. Cox and W. Rees, 1960).

The mollusc shell has a protective function and also serves as a place of attachment for muscles. In a number of Gastropoda of different orders, including most Opisthobranchia, and in all modern Cephalopoda Dibranchiata the shell is overgrown by the mantle and becomes internal. It thereby loses its protective function, but may retain the supporting (many Decapoda Dibranchiata) and hydrostatic functions (*Sepia*, E. Denton and J. Gilpin-Brown, 1959). In other cases it becomes vestigial and ultimately disappears (Octopoda, many Gastropoda). In some series of Gastropoda the shell is reduced and does not pass through the internal-shell stage (Taenioglossa Heteropoda and others). There are also cases of shell-reduction among Loricata.

We see numerous examples of disappearance of the shell or its transformation into an internal skeleton in those mollusc classes where its chief function is protective. In fact the whole animal kingdom displays— side by side with the tendency to create shells, armour, and other passive-protection adaptations—an equally-widespread opposite tendency to lighten the body weight, often resulting from a transition to active defence linked with refinement of the neuro-sensory and locomotor apparatuses. In Lamellibranchia the protective function of the shell is combined with a second, not less important function—that of current-directing, i.e. taking part in regulation of the current of water flowing to the mouth and the gills; for that reason reduction of the shell in Lamellibranchia is extremely rare, taking place, for instance, when the current-directing function is transferred to the walls of a burrow or to a calcareous tube lining it (Teredinidae, Clavagellidae) or when the shell is overgrown by an external fold of the mantle (*Chlamydoconcha*, etc.).

The mollusc foot also sometimes acquires the faculty of creating formations similar to shells. In most Prosobranchia the posterior dorsal surface of the foot secretes an operculum—a formation of the same structure and nature as a shell. Among Opisthobranchia *Actaeon* and some others, and among Pulmonata *Amphibola* and the larvae of some other marine species have an operculum (B. Hubendick, 1947). In females of the cephalopod *Argonauta* a shell very similar in form and structure to the true shells of some ammonites is entirely secreted by two broadened arms, i.e. it is a product of the foot.

Another probable product of epithelial secretion (and not a cuticular formation) is the brachiopod shell, with its striking resemblance to a lamellibranch shell and its profound differences from it both in structure and in structural plan (brachiopods have dorsal and ventral valves, as compared with the right and left valves of lamellibranchs).

As in Hexacorallia and Gorgonaria, so in molluscs and brachiopods maximum development of the glandular-protective function of the kinetoblast is accompanied by concentration of that formation in separate, specialised regions of it. We see here another example of the above-mentioned path of development, consisting in concentration of functions and formation of differentiated organs (shell gland, etc.) from the primary layers.

A second type of protective formation (besides secretions of cutaneous glands) is cuticle. Cuticle is the name given to a protective layer covering the surface of epithelium, and is not a product of secretion but the result of degeneration of the external layers of protoplasm in the epithelial cells. Often, however, it is difficult to distinguish cuticle from a secretion, and some protective formations arise as a result of the combined action of the two processes, secretion and cuticle-development.

Indisputable cuticle first appears in parasitic flatworms, and develops in lower worms more or less in parallel with reduction of ciliated epithelium and limitation of its area. Therefore it is totally absent in turbellarians, and in Gastrotricha and Rotatoria it is well developed in those areas of the body where ciliated epithelium is absent. In Gastrotricha cuticle forms a large number of appendages, sometimes very peculiar, in the shape of hairs, spines, and scales. Among Scolecida it attains greatest development in Kinorhyncha and nematodes. The cuticle of Kinorhyncha forms an external skeleton divided into rings, partly resembling that of arthropods (Vol. 1, Fig. 64).

The cuticle of nematodes, in spite of its often-considerable thickness, is flexible and remains unsegmented. The growth of nematodes and Kinorhyncha is often accompanied by the shedding of old cuticle, i.e. by moulting similar to that of arthropods. There are some analogies between the cuticles of nematodes and of arthropods in details: the structure of sensory hairs, etc. Annelids have a thin cuticle of comparatively simple structure. In neither nematodes nor annelids does the cuticle contain chitin.

Cuticle reaches high development on the surface of Bryozoa cystids, forming their external skeleton; it contains chitin, and in Stenolaemata and partly in Gymnolaemata it is calcified. The peak of utilisation of cuticle as an external skeleton, however, is found in arthropods.

The most characteristic feature of arthropods is a continuous cuticle, differentiated into thicker, harder, and inflexible sclerites, which are components of the external skeleton, and thinner, flexible connecting membranes. The cuticle of arthropods consists of chitin and a protein, arthropodin. The latter is condensed (sclerotised) in the sclerites because of tanning by o-quinones, which takes place soon after a moult (M. Pryor, 1940, and a number of subsequent works). In many crustaceans, in Diplopoda, and in some insect larvae a considerable amount of mineral salts is deposited in the cuticle, especially $CaCO_3$, but calcified cuticle is almost devoid of arthropodin. The fine structure of cuticle, especially in sclerites, is generally extremely complex (Fig. 15), providing it with excellent mechanical properties. With regard to use of various methods of studying the fine structure of insect cuticle, including the electron-microscope method, see (for instance) A. G. Richards (1952).

We must point out that the 'cuticle' of arthropods is not cuticle in the full sense of the word. Of the three layers of insect cuticle (endocuticle, exocuticle, and epicuticle) (Fig. 15) the uppermost, epicuticle, is without doubt a secretion of epidermal cells or specialised glands, discharged on the surface of already-formed cuticle (V. Wigglesworth, 1947); the same has been demonstrated for the epicuticle of crustaceans, e.g. Isopoda (H. Gorvett, 1946) and Decapoda (R. Dennel, 1947), Acariformes (A. Lees, 1947), and Myriapoda (G. Blower, 1951). Epicuticle consists of proteins, lipoproteins, and lipoids, and usually also various kinds of wax (regarding the chemistry of arthropod cuticle see A. Richards, 1951). The mandibles

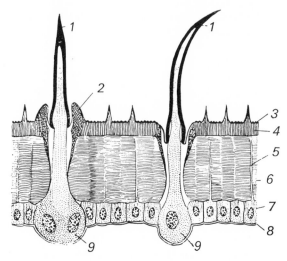

Fig. 15. *Diagram of section through insect skin.*
1—hairs; 2—basal ring of hair; 3—epicuticle; 4—exocuticle; 5—pseudopores; 6—endocuticle; 7—epidermis; 8—basal layer of epidermis; 9—trichogen cells (from Weber).

of Copepoda Calanoidea bear epicuticular teeth (Fig. 16) containing amorphous hydrate of silicon dioxide (opal), which are particularly well developed in forms that feed on diatoms (C. W. Beklemishev, 1954); we may remark that siliceous formations are exceedingly rare in Eumetazoa.[1]

Another type of extreme complexity in external cuticle is shown in the 'mantle' of ascidians and salps. The ascidian mantle, or, more precisely, tunic, is cuticle of external epithelium. It is formed of tunicin, which consists of carbohydrates that are sometimes very close to cellulose, and is the only example of cuticle containing cells. Actually some epidermal cells leave the epidermis and penetrate the tunic, that being made possible by the gelatinous consistency of the latter (A. Kowalevsky, 1892). The result is a structure somewhat resembling connective tissue: gelatinous

---

[1] Regarding the connection between the segmentation of the external skeleton of arthropods and that of their musculature see Chap. VI. Some data on the structure of the external skeleton of arthropods have been presented by us in connection with the question of metamerism in arthropods (see Vol. 1, Chaps VI and VIII). More detailed discussion of this question would take us far beyond the limits of this book.

material occupied by cells (Fig. 17, *A*). Moreover, even blood vessels grow into the substance of the tunic. In the latter respect the ascidian tunic resembles the shell of Rudistes.

The last and most peculiar type of cuticular formation is found in Appendicularia. Their thick tunic is detached from the body by a process similar to moulting and forms a glassy, transparent dwelling-place that encloses the animal (Fig. 17, *B–E*). The mass of the tunic does not contain cells or vessels but has a very complex structure, adapted to filtration of food particles from sea-water (Fig. 17, *F*). The structure of the 'dwelling' is determined by the mosaic epithelium by which it is formed (Fig. 17, *G*).

The passive-protection function of body coverings may be much

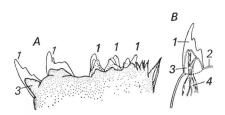

Fig. 16. *Siliceous epicuticular teeth on the mandibles of the last copepodite stage of* Eucalanus bungii (*Copepoda Calanoidea*).
A—chewing edge of mandible (after C. W. Beklemishev). *B*—a single tooth, more highly magnified (original) (optic section, the outline of the tooth base being shown by a broken line). 1—siliceous epicuticular teeth; 2—cuticle; 3—papilla on which epicuticular tooth stands; 4—ducts of cutaneous glands.

intensified by means of all possible types of spines, setae, and hairs. When there is thick cuticle (in Nemathelminthes, arthropods, etc.) formations of this kind are numerous and diversified; morphologically they are divided into three main types: (*i*) microchaetae, which are simple projections on the surface of the cuticle; (*ii*) true hairs, movably connected with the skin and each secreted by a separate trichogen cell; (*iii*) spines or teeth, which are projections of the skin as such, lined inside with hypodermis. All these formations have very diverse functions: defence against enemies or injury, locomotor, prehensile, sensory, etc.

Some turbellarians, most annelids, and some brachiopods have stiff setae although they do not have a continuous chitinous cuticle. In Polyclada and Turbellaria Acoela such setae or spines are agglutinations of separate rods or tubes, each of which is the secretion of a separate epidermal cell; in annelids and brachiopods each seta is secreted by a single large chaetogenous cell. The setae of annelids have principally a locomotor role, but they may also have a protective function, e.g. the setae of the notopodia of Amphinomidae, which are furnished with poison glands, or the paleae of Amphictenidae and other sessile polychaetes.

The spines and pedicellariae of starfishes and sea-urchins are formed not by epidermis but by the whole skin of these animals and contain part of

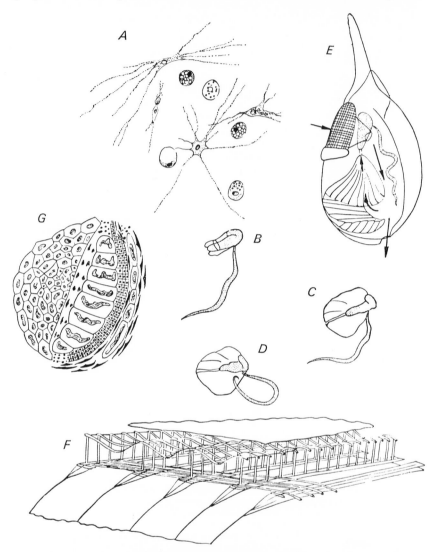

Fig. 17. *Structure of ascidian cuticle.*

A—Morchelium argus (Ascidiae), cells within mass of tunic (after K. K. Saint-Hilaire). *Oikopleura albicans* (Appendicularia): *B, C, D*—during formation of tunic ; *E*—in the completed tunic the arrows show the directions of water currents; *F*—a portion of the filtering network that forms part of the tunic. *G*—part of the oikoplastic epithelium of *Oikopleura albicans*, showing the complex mosaic, formed of epithelial cells and determining the structure of the 'dwelling' (after Lohmann).

the cutaneous skeleton, articulated with the underlying skeletal parts; pedicellariae are possessed by sea-urchins and starfishes and are grasping spines, a curious analogy with the avicularia of Bryozoa, which, however, are modified zooids. The spines of echinoderms may have numerous other functions besides that of protection (see Chaps IV and VI).

Pogonophora have several pairs of bunches of embryonic setae (A. Ivanov, 1957; G. Jägersten, 1957). Unlike the spines of polychaetes, each of them is formed inside a multicellular sac; the adults have setae only in the posterior, metameric part of the metasoma, these being formed in the same way as the embryonic setae (see Vol. 1, Chap. IX). The other attachment organs of adult Pogonophora represent thickenings of the cuticle, not homologous with setae (A. Ivanov, 1957).

4. RESPIRATORY ADAPTATIONS OF KINETOBLAST

Aerobic respiration is characteristic of the great majority of animals, and consequently they require a flow of oxygen from the external environment to their cells and tissues, and removal of carbon dioxide. In small, and even in immobile, organisms these processes may take place by means of diffusion alone. *Hydra* needs no special respiratory adaptations, since almost all its body cells are in immediate contact with the surrounding water, which washes not only the ectoderm but also the endoderm, entering through the mouth and the gastral cavity: here there is only tissue respiration, with no specialised respiratory organs. The same applies to many other lower multicellular animals: hydroids, most Scolecida, and many of the very small forms in higher groups.

When body size increases diffusion alone can no longer ensure adequate respiration, and various mechanisms arise to increase the intensity of gaseous exchange beyond the range provided by diffusion. All such mechanisms are *respiratory adaptations* in the broad sense of the word. They are divided into three groups: (*i*) external respiratory adaptations, increasing the flow of oxygen from the external environment to the respiratory surface of the animal and removing carbon dioxide from that surface; (*ii*) direct respiratory adaptations, facilitating the passage of gas through the bounding epithelium of the animal; (*iii*) internal respiratory adaptations, ensuring the transfer of oxygen from the bounding (respiratory) epithelium to the tissues that require it (A. Lindroth, 1938). Internal respiratory adaptations form part of the distributive apparatus and are discussed in Chap. VIII. External respiratory adaptations create, by means of respiratory movements, currents of water or air that wash the whole body of the animal or specialised respiratory epithelia. Direct respiratory adaptations are expressed, first, in the existence of a specific functional structure in the respiratory epithelia, increasing its permeability by oxygen and carbon dioxide; and secondly, in increase in the surface of the epithelium so modified. Direct respiratory adaptations, as a rule, form part of true respiratory organs and are discussed below.

Sponges and coelenterates possess only external respiratory adaptations, primarily in the form of hydrokinetic organs. These include: (*i*) the irrigation apparatus of sponges; (*ii*) the hydrokinetic adaptations of kinetoblast in sessile coelenterates (see Chap. II); (*iii*) the hydrokinetic adaptations of the phagocytoblast of coelenterates, which ensure exchange of water between the gastrovascular apparatus and the external environment and also circulation of water within the gastrovascular apparatus. During the life of a coelenterate the circulation of water within its gastrovascular apparatus has multifold significance; its respiratory function is particularly obvious in the case of large solitary animals, such as large sea-anemones of massive colonies of Alcyonaria.

External respiratory adaptations also include the subgenital fossae of Scyphomedusae. These are not true respiratory organs, as they do not regulate the gaseous regime of the internal environment of the medusae as a whole but merely bring oxygen to the vicinity of separate organs, particularly the gonads. Apparently the tracheae of Disconantae also are respiratory adaptations, as they carry atmospheric air inside the organism (see Chap. I, Fig. 41).

All these adaptations, however, can ensure respiration and removal of excreta only if the body is not too massive and metabolism is not too intensive. Increases in body size and in metabolism intensity immediately call for additional adaptations. Many lower Metazoa (among sponges, coelenterates, and turbellarians) make use of symbiosis with unicellular algae, green or brown, which dwell in the cells and tissues of the animals. During daylight hours the algae give off oxygen, which is consumed by the tissues, and themselves consume the CO_2 and excreta freed by the tissues. This type of symbiosis is particularly widespread in the littoral communities of warm seas, where the high temperature raises the intensity of metabolism in the animals and the bright light produces intensive photosynthesis in the plants. All reef corals contain zooxanthellae [1] in their endodermal cells, using them for respiration and removal of excreta, but not feeding on them (C. Yonge and A. Nicholls, 1931). On the other hand, small turbellarians such as *Convoluta roskoffensis* (L. von Graff, 1891) and *Achoerus caspius* (Acoela) (W. Beklemishev, 1937) use symbiotic algae primarily as a source of food.

In general, the wide distribution of symbiosis with algae in the lower Metazoa is not accidental but is due to the imperfection of their respiratory and excretory adaptations. Among higher Metazoa only some annelids (Chaetopteridae) and molluscs (Tridacnidae, some Aeolididae) (see P. Büchner, 1953) have symbiotic algae, and, in Tridacnidae at least, the significance of the algae for the host is purely trophic (C. Yonge, 1954).

Among higher animals external respiratory adaptations for bringing water or air to the respiratory organs are much more widespread. Active animals in constant movement can sometimes dispense with these adaptations, since in such cases their role is in fact filled by the locomotor organs. In less mobile aquatic animals external respiratory adaptations are the rule.

---

[1] These are stages in the development of the Dinoflagellata *Gymnodinium* and *Symbiodinium* (Hovasse, 1937; Kawaguti, 1944; Freudenthal, 1962) (Russian Editor's Note).

Most such animals create water currents around themselves by the same means used for their locomotion: movement of cilia, muscular movements of the whole body (e.g. *Nereis*, *Tubifex*, Chironomidae larvae) or of its appendages (e.g. many crustaceans). Echinoderms have highly-developed and diversified ciliary-hydrokinetic adaptations, found in sea-urchins and starfishes among Eleutherozoa (see T. Gilsén, 1924; W. Spencer, 1951). The hydrokinetic apparatus may include current-directing as well as water-moving adaptations, as we have seen in bivalves and ascidians.

In most cases, as we have seen, water currents created by animals have more than one function: they bring oxygen, food (in most filter-feeders and feeders on settling organic particles) or food odours (some polychaetes). Often, however, their function is purely respiratory (e.g. in Tubificidae, which are exclusively mud-feeders).

Respiratory organs in the narrow sense of the word may be absent even in higher animals that possess internal respiratory adaptations, their internal environment doubtless being also a respiratory environment. For instance, we might say in a purely physiological sense that in earthworms, which breathe with their whole skin, the whole skin is a respiratory organ; it is through the skin, in fact, that the worms' internal environment (blood, coelomic fluid) becomes saturated with oxygen and freed from CO_2, and it is only with that internal environment and not directly with the atmosphere that all other tissues of the organism are involved in the exchange. Here, however, we do not find direct respiratory adaptations and morphologically differentiated respiratory organs. These first appear in organisms whose respiratory functions are concentrated in separate areas of one of the principal body layers.

In higher Metazoa separate respiratory organs may be produced by any epithelium in contact with the external respiratory environment, i.e. especially by the skin, but also by the gut; the respiratory epithelium is ectodermal (kinetoblastic) in the first case and endodermal (phagocytoblastic) in the second. In this chapter we shall discuss only the more frequent instances of the respiratory organs being formed by kinetoblast. With regard to respiratory organs of phagocytoblastic origin, see Chap. V.

Among Scolecida we find specific respiratory organs only in *Priapulus*, which is in accordance with the large size and comparatively high organisational level of Priapuloidea. The gills of *Priapulus* are hollow, thin-walled, botryoidal appendages at the rear end of the body. Their lumen constitutes an extension of the animal's coelomic cavity. *P. caudatus* has one and *P. bicaudatus* two gill appendages. Digging into the soil with its front end, *Priapulus* often exposes its caudal appendages in the water, so confirming their respiratory role (L. Zenkevich, 1940).

True respiratory organs are widely distributed only in polychaetes. Polychaete gills are of two main types, blood and coelomic. Both are outgrowths of the body walls, usually branched, which protrude into the water. The most typical blood gills are the pectinate parapodial gills of Eunicidae (Fig. 18, *A*). The role of blood gills may also be played by other appendages, parapodial and cephalic (Vol. 1, Figs 111, 113, and 114). They occur more often in sessile than in errant polychaetes. All blood gills are characterised by abundant branching of capillary blood vessels. In contrast, the skin of coelomic gills contains no capillary blood vessels. These gills are found in animals where the internal respiratory environment is provided not by blood but by coelomic fluid, e.g. Nephthydidae, Glyceridae, Capitellidae, etc. Many polychaetes, especially errant ones,

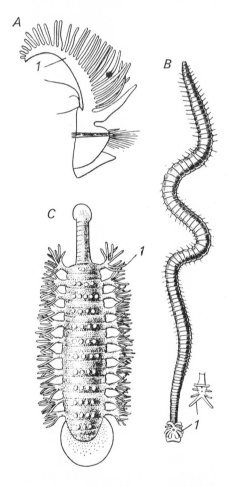

Fig. 18. *Gills of annelids.*
A—parapodial blood gills of *Eunice aphroditoides* (Polychaeta). *B*—anal gills of *Dero incisa* (Oligochaeta Naididae). *C*—lateral gills of *Ozobranchus jantseanus* (Hirudinea Ichthyobdellidae) (after Oka). 1—gills.

generally lack gills (Nereidae, Phyllodocidae, and many others), that lack being sometimes the result of reduction and in other cases perhaps primary (N. Livanov, 1940). In *Nereis* the function of respiratory organs is fulfilled by parapodia and also by special areas of epithelium on the dorsal and ventral sides of the body segments, which are permeated by a dense network of capillary blood vessels (Fig. 163, *C*).

Oligochaetes and leeches lack the parapodial and cephalic gills found in polychaetes, but secondary gills develop in a number of separate forms in both classes (Fig. 18, *B*, *C*), also being outgrowths of the body wall and sometimes provided with a highly-developed system of afferent and efferent vessels (*Branchellion* among Rhynchobdellea). Among leeches many Ichthyobdellidae (Rhynchobdellea), and among oligochaetes separate forms in various families (*Dero* and others) possess gills.

The primary gills of molluscs, or ctenidia (Fig. 19), partly resemble the parapodial gills of Eunicidae in their pectinate form, but have two rows of lamellae each. Like polychaete gills, the ctenidia of Loricata are arranged

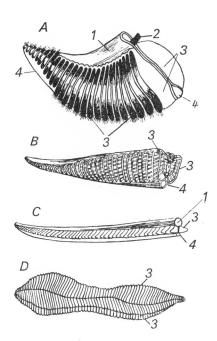

Fig. 19. *Ctenidia of molluscs, semi-diagrammatic* (after Ray Lankester).
A—Chiton. B—Sepia. C—Fissurella. D—Nucula. 1—afferent branchial vessel; 2—longitudinal branchial muscle; 3—paired branchial lamellae; 4—efferent branchial vessel.

metamerically along the sides of the trunk (Vol. 1, Fig. 97, *B*), and the suggestion that they are homologues of polychaete gills is very tempting. *Neopilina* has five pairs of metameric gills, each bearing a single row of well-developed lamellae. Each ctenidium of the first three pairs has also a second row of lamellae, although vestigial (Lemche and K. Wigstrand, 1959). Other molluscs have a single pair of ctenidia, except *Nautilus*, which has two pairs. The ctenidia undergo reduction in most Solenogastres, in a number of specialised groups of Gastropoda, and in some Lamellibranchia (order Septibranchia, Fig. 20, *D*). The evolution of ctenidia has taken a unique course in most Lamellibranchia and in several Prosobranchia (C. M. Yonge, 1938). Among Lamellibranchia only Protobranchia have retained ctenidia in their usual form (Fig. 19, *D*; Fig. 20, *A*). In other orders of this class they grow along the whole length of the mantle cavity, and their lamellae extend into long filaments that are folded in two (order Filibranchia, Fig. 20, *B*). In their further development these filaments fuse together and form laminated, reticular gills, characteristic of the most numerous order, Eulamellibranchia (Fig. 20, *C*). As mentioned above, however, and as pointed out by Orton (1914), the chief functions of gills in the higher Bivalvia are hydrokinetic and filtering, and the conversion of gills from ctenidial to lamellar is itself evidently due to tremendous increase in their hydrokinetic function, consequent on the animals' transition

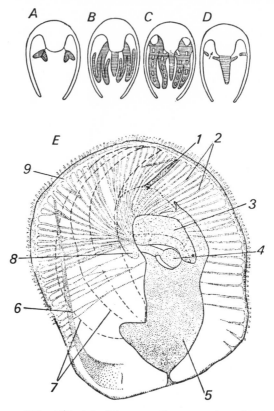

Fig. 20. *Gills of Bivalvia. Diagrammatic cross-sections of body in gill region.*

A—Protobranchia; *B*—Filibranchia; *C*—Eulamellibranchia; *D*—Septibranchia (after Hescheler). *E*—transfer of respiratory function to mantle; great development of blood vessels in mantle of young oyster (*Ostrea lurida*). 1—supplementary heart (contractile part of mantle vessel); 2—radial vessel; 3—shell-adductor muscle; 4—heart; 5—viscera; 6—efferent canal of mantle vessels; 7—outline of gills, visible through mantle; 8—kidney; 9—marginal blood sinus of mantle (after Hopkins).

from sediment-feeding to filter-feeding. A corresponding increase in the respiratory function did not take place, not being required, since the total intensity of metabolism in Eulamellibranchia is scarcely higher than that in Protobranchia. In many cases the respiratory function of the gills even declines, as is seen from their poor supply of blood vessels; in such cases the respiratory function passes partially to the mantle, whose blood supply is correspondingly increased and modified (e.g. in the oyster *Ostrea lurida*, Fig. 20, *E*). The evolution of gills in Lamellibranchia is therefore character- ised by exchange of function, in part or wholly; in the latter case, by substitution of the walls of the mantle cavity for the gills as respiratory organs.

In the order Septibranchia (Poromyacea) the gills are represented by muscular septa that separate the mantle cavity into dorsal and ventral sections, corresponding respectively to the exhalant and inhalant siphons; the septa are pierced by a number of small orifices. Water currents are created in the mantle cavity not by the activity of cilia but by contractions of the septa. The hydrokinetic function has thus been transferred from the ciliary apparatus to the musculature, and been modified: Poromyacea are

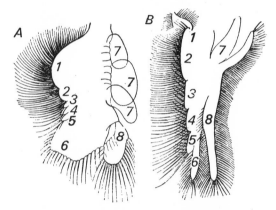

Fig. 21. *Trunk appendages of Phyllopoda.*
A—Chirocephalus grubei (Anostraca). *B—Limnadia lenti-cularis* (Conchostraca). 1—proximal endite; 2-6—other endites; 7—epipodites (respiratory appendages); 8—exopo-dite (after Bening).

not filter-feeders but predators, drawing in small prey by sudden sucking movements of their gill septa (C. Yonge, 1928). As P. Pelseneer (1906) suggests, the gill septa of Poromyacea originated from gills of ordinary type by reduction of downward-hanging lamellae, with retention of their dorsal parts attached to the foot and the mantle; this point of view has a sound comparative-anatomical basis in the existence of transitional forms, but has not yet been confirmed embryologically.

Among Gastropoda, ctenidia have totally disappeared in Opistho-branchia Nudibranchia and in many of them have been functionally replaced by a new type of gills: dorsal (Vol. 1, Fig. 97, *E*) or circumanal (Dorididae, Vol. 1, Fig. 163, *C*). Like ctenidia, they are branched out-growths of the skin and are provided with afferent and efferent vessels. A number of small forms in this group have no gills at all.

Possession of gills in the form of appendages on the outer surface of the limbs, corresponding to the dorsal surface of the parapodia of polychaetes, is characteristic of aquatic arthropods. Such branchial appendages, pre-epipodites and epipodites, are possessed particularly by the unsegmented, leaf-shaped legs of Phyllopoda; they are usually sacciform (Fig. 21). The branchial epipodites of trilobites are described in Vol. 1, Chap. VII; Fig. 117, *A*. One modification of the trilobite prototype is represented by the mesosomatic branchial appendages of aquatic Chelicerata (Vol. 1, Fig. 117, *C*). Most crustaceans also have sacciform or pinnately-branched gills (Vol. 1, Fig. 117, *B*; Vol. 2, Fig. 22), generally representing epipodites

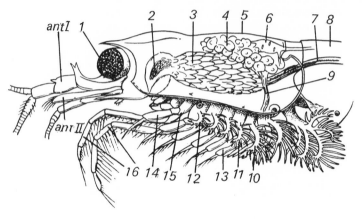

Fig. 22. *Respiratory apparatus of* Euphausida pellucida (*Malacostraca Eucarida Euphausiacea*).

ant I, ant II—antennae I and II; 1—eye (paired); 2—stomach; 3—hepatic appendages of mid-gut; 4—ovaries; 5—carapace; 6—heart; 7—gut; 8—first abdominal segment; 9—oviduct; 10-12—gills of thoracic limbs; 13–14—exopodites of thoracic limbs; 15—epipodite; 16—endopodites of first and second pairs of thoracic limbs (after Sars).

retained on certain pairs of legs, depending on the order. Among crustaceans all members of the superorder Copepodoidea are totally devoid of gills; among other aquatic arthropods the whole class of Pantopoda lack gills, as also do aquatic Acariformes that have secondarily taken to aquatic respiration (the families Hydrachnidae and Halacaridae.

One is greatly tempted to regard the gills of arthropods and the parapodial gills of polychaetes as being homologues. That seems especially plausible if we regard the exopodite and endopodite of the bifurcated limbs of crustaceans as homologues of the dorsal and ventral branches of the parapodia of polychaetes, as many, mostly the old authors, have done. In itself the suggestion that the limbs of arthropods are homologous with the parapodia of polychaetes is plausible, and is supported by the similarity of the extrinsic musculature of the primitive limbs of arthropods and of the parapodia of polychaetes (B. Shvanvich, 1949; E. Becker, 1950). But several authors, e.g. R. Snodgrass (1938), deny even that homology; and although Snodgrass's point of view is extreme and has little foundation, the question of the homology of the gills in the two groups remains open.

Among Atelocerata, many aquatic insects possess gills. These are usually secondary gills, being leaf-shaped or filamentous outgrowths of the body walls, located in the most diverse places. The gills of Diptera larvae are certainly secondary. A. Handlirsch (1925), however, believes that some of the branchial appendages of the larvae of Odonata, Ephemeroptera and Plecoptera are modified abdominal limbs (Fig. 23, *B*, *C*) or epipodites of thoracic limbs (larva of *Taeniopteryx* (Plecoptera) Fig. 23, *A*). In fact, the tracheal gills of the ephemerid *Ephemera vulgata* arise from the rudiments of abdominal limbs in the embryo; the tracheal gills of the larva of *Sialis* and several other aquatic larvae of Neuroptera (Fig. 23, *C*) and Coleoptera, which are segmented and resemble limbs, are also probably products of

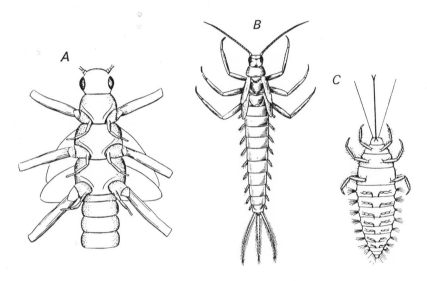

Fig. 23. *Gills of insect larvae.*
A—larva of *Taeniopteryx nebulosa* (Plecoptera) with coxal gills (after Lauterborn).
B—larva of *Phthartus rossicus* (Ephemeroidea) (Permian) (after Handlirsch).
C—larva of *Sisira* (Neuroptera) (after Grube); the two latter with abdominal gills.

limbs (B. N. Shvanvich). If we accept Handlirsch's view on the primary amphibious nature of Odonata, Ephemeroptera, and Plecoptera, the discovery in them of relics of respiratory organs found in other primarily-aquatic arthropods would be quite natural. But even if we take the opposite point of view and regard the aquatic mode of life of their larvae as secondary (M. Gilyarov, 1949), we may still admit that the rudiments of abdominal limbs in the embryos or their vestiges in the nymphs, when meeting new requirements, may develop into gill appendages without violating the law of the irreversibility of evolution (L. Dollo, 1893).

It is characteristic of all the above types of Protostomia gills that they are outgrowths of the body walls, freely washed by the water and mostly pectinate or branched (*evaginated* type of respiratory organs, V. A. Dogiel, 1938). In other words, maximum development of the surface in contact with the external respiratory environment is characteristic of them. A number of observations on *Aëdes* larvae (E. Martini, 1923, etc.) show that with oxygen-poor external environment and other respirational difficulties compensatory enlargement of the gills, and consequently of the respiratory surface, takes place. In small *Anopheles* larvae the relative size of the gills is less than in large ones (W. Beklemishev and J. Mitrofanova, 1926). Hence it is clear that small animals, with their relatively larger surface, are very often satisfied with cutaneous respiration alone, dispensing with the possession of specialised respiratory organs (Copepoda among crustaceans, a number of mites among Arachnoidea, etc.).

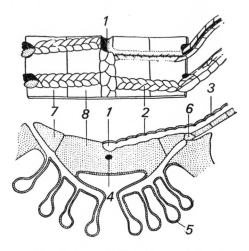

Fig. 24. *Aquatic lungs of Blastoidea: ambulacral area of* Pentremites, *surface and cross-section views.*

1—radial ambulacral canal; 2—covering plates; 3—brachiole; 4—supposed nerve stem; 5—hydropores; 6—point of attachment of brachiole; 7—lateral (porous) plate; 8—lanceolate plate (after Bather).

Among Deuterostomia specialised respiratory organs of kinetoblastic origin occur mainly in echinoderms: they are represented there by the gill appendages of sea-urchins and starfishes, and also of some holothurians; these are thin-walled outgrowths of the body walls, formed of all their layers, including coelomic epithelium. Tube-feet also have a respiratory function. In irregular sea-urchins some of the tube-feet are converted into leaf-shaped respiratory appendages. But the most peculiar respiratory organs were those of Blastoidea: in each of the interradii of the actinal surfaces of their calyces Blastoidea had two rows of slits leading to thin-walled flat sacs, sometimes branched, which projected into the animal's body cavity (Fig. 24). Water evidently circulated in these sacs, thus being enabled to enter into close contact with the coelomic fluid and to exchange gas with the latter in spite of the massive skeleton that covered the whole body of the animal. A similar role is played by the bursae of Ophiuroidea

(W. Spencer, 1951); circulation of water in the bursae is usually achieved by means of the cilia of their epithelium, but sometimes also by rhythmic respiratory movements of the aboral body walls (*Ophiothrix fragilis*, E. MacBride, 1933). The extreme development of the bursae in Euryalae, which are giants among Ophiuroidea, is therefore perhaps an adaptation for better satisfaction of the respiratory needs of their massive bodies.

The arms of Pogonophora and Pterobranchia may also perhaps have respiratory significance as well as being a food-catching apparatus.

5. PROTECTIVE AND RESPIRATORY ADAPTATIONS OF KINETOBLAST ON TRANSITION TO A TERRESTRIAL MODE OF LIFE

Air contains many times more oxygen than water does, even when the water is in equilibrium with the atmosphere. The rate of oxygen diffusion in water is many thousands of times less than that in air; air is therefore a source of oxygen many times better than water is. Consequently many aquatic animals need external respiratory (hydrokinetic) adaptations, which terrestrial animals of the same size and organisational level can largely dispense with. 'The terrestrial animal can count upon its 21 per cent oxygen being available the world over, irrespective of the vagaries of weather. It is no easy matter to change over one's respiratory mechanism from water-breathing to air-breathing, but to do so is to invest in a gilt-edged security' (Y. Ramsay, 1952).

Transition to a terrestrial mode of life by animals probably takes place in different ways. G. Colosi (1927) believes that the general reason for appearance of the higher groups of terrestrial animals is the intensification of metabolism accompanying a higher organisational level. Intensification of metabolism involves higher oxygen requirements, and the group of animals concerned has a constantly-increasing urge to leave the water and gradually change to a terrestrial mode of life. In Colosi's opinion very many crabs, for instance, are at present at that stage of evolution.

If a direct transition from the sea to dry land is possible, however, it is so mainly for higher organisms such as vertebrates and higher crustaceans, which possess sufficient osmotic independence of internal environment to free them from the menace of death from the hypotonic effects of atmospheric precipitation. Primitive lung-breathing molluscs (Pulmonata subclass), particularly Siphonariidae, are also marine littoral forms, and perhaps the protection afforded by a shell aided their emergence on dry land. On the whole, the lower Metazoa move to dry land more easily from fresh water than from the sea. Most terrestrial turbellarians, nematodes, and annelids are related to freshwater and not to marine members of these groups; and the highway for their passage from water to land lay, as M. S. Gilyarov (1949) truly states, in the soil, in forest litter, in moss— in moist habitats representing a series of intermediate conditions between aquatic conditions and open terrestrial habitats. The gradual conquest of these intermediate stages by any group of animals led them, step by step, from aquatic to terrestrial life, and from an organisation adapted to living in water to that of a true terrestrial animal. Many groups of animals went no farther than moist, concealed habitats, and are, in effect, semi-terrestrial forms.

On transition to a terrestrial mode of life the necessity of a number of adaptations faced the animals, especially the problem of protection from evaporation, which breaks down into two partial problems: retention of a supply of water in the body and protection from drying-out of the respiratory surfaces. A number of terrestrial animals are 'imperfectly' adapted to terrestrial life: they are not protected from evaporation, and therefore can live only in an environment where the air is saturated with water vapour. These include terrestrial turbellarians, nemertines, and shell-less molluscs, which have no cuticle, are protected only by cutaneous glands, and live only in a constantly-moist environment. They also include 'terrestrial' Rotatoria, Nematodes, and Tardigrada, which have a thin cuticle that does not protect them from desiccation and which are active only in periods of high humidity, but are to a great extent capable of surviving desiccation in a state of anabiosis. Terrestrial oligochaetes and leeches are somewhat better protected. Shell-bearing molluscs also are active mainly when the air is humid, but some of them can inhabit biotopes with sharp fluctuations in humidity, passing dry periods in a passive state within their shells (e.g. *Xerophila* among Pulmonata).

Invertebrates 'perfectly' adapted to terrestrial life are found only among arthropods, whose cuticle, possessing several adaptations, can provide fully-satisfactory protection against desiccation. Onychophora, which stand morphologically between arthropods and annelids, are also intermediate in resistance to desiccation: in similar conditions *Peripatus* loses only half as much water as earthworms do, but twice as much as *Scolopendra* of the same size (P. Morrison, 1946). In insects the principal adaptation against desiccation, besides the thickness of the cuticle, is possession of an epicuticle formed by a thin layer of waxy matter impervious to aqueous vapour.

The second problem, protection against drying-out of the respiratory surfaces, is linked with the fact that apparently gas-exchange is possible only through a moist surface (G. Colosi, 1928). In all animals the respiratory epithelium is always moist. There are no dry respiratory organs. Colosi calls this situation 'the principle of constancy of respiratory environment' and formulates it in the following (at first glance paradoxical) manner: all animals, regardless of their environment, always breathe in water. By that he means that a water film of molecular thinness, which covers the respiratory epithelium of terrestrial animals, is the intermediary between them and the atmosphere.

Because of the principle of constancy of respiratory environment, external or evaginated respiratory organs become impossible on transition to respiration in a dry environment. 'Imperfectly'-adapted terrestrial animals living in air, but in air saturated with moisture, can still breathe with the whole surface of the body. At the same time the oxygen-rich nature of atmospheric air permits animals living on dry land to dispense with specialised respiratory organs and external respiratory adaptations even when they belong to groups whose aquatic members often possess both of these (e.g. annelids). With transition to life in open habitats and respiration of dry air, however, the development of internal, or *invaginated*, respiratory organs, whose epithelium is protected from drying-out, becomes necessary. Respiratory organs of this type have in fact developed in all 'perfectly'-adapted terrestrial animals, whereas typical invaginated respiratory organs occur relatively rarely in aquatic animals (in chordates, Stomochorda, Odonata larvae, Ophiuroidea, Blastoidea, etc.); but very often the evaginated gills of aquatic animals are covered by folds that protect them from fouling by mud and at the same time act as current-directing adaptations, changing the water in contact with the gills more effectively (Mollusca, Crustacea, etc.).

Among molluscs only lung-breathing snails (Pulmonata) are true terrestrial animals. They have developed a hollow respiratory organ by reducing the ctenidia, transferring the respiratory function to the entire epithelium of the mantle cavity, and reducing its external opening to a small spiracle. The mantle cavity is converted into a lung. Its wall is densely covered with branched blood vessels, whose blood takes part in the gas-exchange. Among all Pulmonata true ctenidia are retained only by primitive marine Siphonariidae (B. Hubendick, 1947) (Fig. 136, *B*).

Comparatively few crustaceans live on dry land, and these mostly have only primitive adaptations to air-breathing. Decapoda, especially crabs, provide a good example of utilisation of a whole series of adaptations, already developed in the water, when moving to dry land. While still in the water their legs became adapted to running on the bottom. Their gills are protected from mechanical damage and fouling by folds of the shell descending on the sides, and are thus located in special gill chambers; when they leave water these folds also protect them from drying-out; the thick shell protects the body from desiccation. Consequently many crabs have taken to a semi-, and sometimes a fully-terrestrial mode of life (*Ocypoda* among Brachyura, *Birgus* among Anomura). In extreme cases the gills become vestigial or even disappear and the respiratory function is transferred to the walls of the gill cavity, which is thus converted into a lung (*Birgus*).

In terrestrial Amphipoda and Isopoda the anatomical adaptations to air-breathing also consist largely in changes in the shape of the gill-limbs and in increased protection for the gills. In some woodlice (Oniscoidea) the broadened, roof-like exopodites of the anterior abdominal limbs are furnished with invaginations in the form of slender, bushy, branched tubes resembling the tracheae of terrestrial arthropods and forming tracheal lungs (Fig. 25). Their respiratory function (absorption of oxygen) has been demonstrated experimentally (Reinders).

Terrestrial Chelicerata, which are combined under the name of

Arachnoidea, possess two types of respiratory organs, lungs and tracheae. The nature of the lung-books of Arachnoidea was discovered by Ray Lankester (1881, 1884, and later), who demonstrated their homology with the podobranchiae of aquatic Chelicerata. Before him the lung-books were regarded as modified tracheae. Strangely enough, the latter view is still held by several authors.

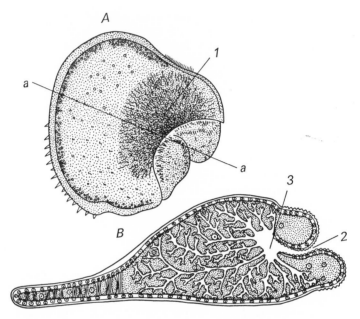

Fig. 25. *Respiratory pleopods* (*abdominal limbs*) *of the wood-louse* Porcellio scaber (*Malacostraca Peracarida Isopoda*).

A—entire limb. *B*—cross-section of limb along line a-a. 1—bundle of tracheae showing through cover; 2—spiracle leading to air-chamber (3), giving rise to branched tracheae (from Dogiel).

The lungs of Arachnoidea (Fig. 26, *C*, *D*) are sacs that open by transverse slits on the ventral surface of the mesosoma; their cavities are filled with a large number of thin leaves rising from the front wall of the lung and almost reaching its rear wall. These leaves in scorpions and spiders are very like the gill-books of *Limulus*, while in Telyphones they are more like those of trilobites. During embryonic development the lungs of Arachnoidea are first formed in the appropriate places in the form of the papillae of rudimentary podobranchiae. Later a pit (the rudiment of the lung cavity) forms behind each papilla, and folds (the rudiments of the first lung-books) appear on the posterior surface of the papilla (Fig. 26, *A*, *B*). The limb does not grow any more in height, but the pit at its foot deepens. The whole front edge of the pit corresponds to the rear surface of the limb (which consists almost solely of an epipodite) and consequently is covered with folds that give rise to the leaves of the lung-books. Embryologists, however, sometimes deny the precise homology of these leaves with the gill-books of *Limulus*, with the curious argument that they do not grow in the lung cavity but, on the contrary, the cavity protrudes into the mass of the limb in folds separating the leaves, as if that difference had any meaning. The lungs of Arachnoidea therefore, both in their method of development and in their structure, as well as in their situation (see Vol. I,

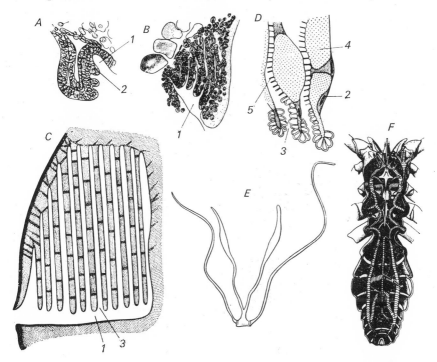

Fig. 26. *Respiratory organs of Arachnoidea.*

A and *B*—two stages in development of lungs in *Lycosa* (Araneina) (from Lang). *C*—diagram of structure of lung of a spider (Araneina) (after MacLeod). *D*—cross-section through free margins of three lung-book leaves of *Euscorpius flavicaudis* (Scorpiones) (from Kästner). *E*—tracheae of spider *Aranea diadematus* (Araneina) (from Lang). *F*—tracheal apparatus of *Galeodes araneoides* (Solifugae), dorsal side of the animal dissected (after Kästner). 1—lung cavity; 2—hypodermis of lung sacs; 3—adits from lung vestibule to area between leaves; 4—blood lacunae within leaves; 5—chitinous villi preventing adhesion of leaves, interwoven in 'lacework' at leaf edges.

Chap. VII), represent modifications of the podobranchiae of aquatic Chelicerata, whose edges have grown to the integument and whose lower surface has sunk into the depth of the body. This is a far-reaching adaptation for protecting the respiratory epithelium from desiccation, although there are no new formations in the composition of the lung and only modifications of the existing organ have taken place.

At the same time what is perhaps the most peculiar feature of cuticular covering—its reaction, by formation of tracheae, to transition to a terrestrial mode of life—appears extremely frequently in Arachnoidea. We find independent formation of tracheae in the most diverse orders of Arachnoidea, in different parts of the body and arising from different sources. Among Acariformes primitive, slightly-sclerotised forms, living in a constantly-moist environment (forest litter, etc.), have no tracheae; in higher forms tracheae develop (together with sclerotisation of the skin) by different methods in different groups. The fact that tracheae are not in any way inherited from the common ancestors of terrestrial arthropods, as was once thought, but arose independently in different groups, long ago became clear with regard to spiders (Araneina). Primitive spiders (Liphistiomorpha, Mygalomorpha) and some families of Araneomorpha (e.g. Pholcidae) have no tracheae. Most Dipneumones (syn. Araneomorpha) have two pairs of tracheal stems in the place of the posterior pair of lungs (Fig. 26, E). In some families of spiders, however (Caponiidae, Telemidae, etc.), tracheae also form in the place of the anterior pair of lungs; all such families are only slightly related and do not form any natural group, so that loss of the anterior pair of lungs and their replacement by tracheae in different groups of Dipneumones has taken place frequently and independently (L. Fage and A. Barras Machado, 1951).

As the history of development shows, the lateral pair of the two posterior pairs of tracheal stems in Dipneumones has developed from lung rudiments and the medial pair represents a modification of simple muscular apophyses. Muscular apophyses are outgrowths of the cuticle arising within the body and used as points of attachment for muscles. Becoming hollow and thin-walled and beginning to branch, the apophyses of one pair are converted into tracheae. Instances of formation of tracheae from intersegmental membranes occur in other Arachnoidea. The origins of tracheae are therefore very diversified.

The places of formation of tracheae in Arachnoidea are equally diversified. Most often they develop on the ventral side of the mesosoma, but in mites and ticks they open in the most varied places: at the base of the chelicerae in Prostigmata (Acariformes, Fig. 27, B); in the joints between the coxal plates and the trochanters of all four pairs of legs in adult Oribatei (Acariformes, Fig. 27, C); on the sides, dorsally to the bases of the third and fourth pairs of legs in Mesostigmata (Parasitiformes, Fig. 27, A). In all these cases there is no possibility of their being formed from lungs. In the opinion of A. B. Lange (1947), the tracheae of Mesostigmata are formed from expansions of a single pair of the slit-like sensory organs widely distributed in mites and ticks. Opiliones have stigmata on the first segments of the legs, etc., besides the mesosomatic stigmata.

The respiratory apparatus of Ricinulei is well-developed and very peculiar. It opens to the exterior by a single pair of stigmata located above the coxae of the fourth pair of legs; the stigmata lead to a respiratory atrium from whose walls dense bunches of fine unbranched tracheoles arise, covering the internal organs (Millot, 1949).

From the above data we draw the following important conclusion:

evidently we cannot speak of a monophyletic origin for tracheae in the development of arthropod classes, since in related forms and even in the same animal they often arise by several independent methods.

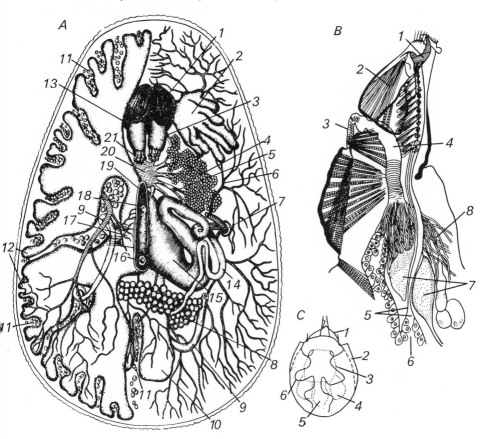

Fig. 27. *Respiratory organs of Acariformes and Parasitiformes.*

A—Argas persicus (Parasitiformes Ixodoidea), dissected dorsally; the branching of the tracheae is well seen: 1—Géné's organ (used to coat deposited ova with a protective secretion); 2—gland of Géné's organ; 3—duct of salivary glands; 4—salivary gland; 5—nerves; 6—tracheae; 7—stigma; 8—ovary; 9—Malpighian duct; 10—rectal diverticula; 11—dorso-ventral muscle bundles (cut across); 12—diverticula of mid-gut; 13—entrance to capitulum; 14—oviduct; 15—uterus; 16—entrance to hind-gut; 17—heart; 18—mid-gut; 19—oesophagus; 20—brain; 21—retractors of chelicerae (after Robinson and Davidson). *B—Allotrombidium fuliginosum* (Acariformes Prostigmata), sagittal section through front end of body with one of the chelicerae; its musculature is seen: 1 and 2—segments of chelicerae; 3—stigmatic plate; 4—tracheal stem arising from it; 5—oesophagus; 6—stomach; 7—central section of neural apparatus; 8—tracheoles (after Berlese). *C—Euzetes seminulum* (Acariformes Oribatei), dorsal view (natural size 1.12 mm); legs not shown: 1—propodosoma; 2—hysterosoma; 3—stomach; 4—its diverticula; 5—hind-gut; 6—tracheal stems of dorsal side (after Willmann).

In some Arachnoidea (Solifugae, Fig. 26, *F*; Parasitiformes Ixodoidea, Fig. 27; Ricinulei) tracheae attain a great size.

As well as Arachnoidea, Myriapoda and Insecta have typical tracheae;

Onychophora have very primitive tracheae; and we find incipient trachea-formation in woodlice. In a word, whenever any group of cuticle-covered animals changes to air-breathing a tendency to development of tracheae appears in it. In Onychophora (Fig. 28, *A*) tracheae are represented by bunches of very thin unbranched tubules (1 to 3 μ in diameter), which open to the exterior by a common orifice (a stigma) and penetrate deeply into the body. Innumerable stigmata cover the whole surface of the body, including the head, but are most numerous on the back. They are irregularly arranged, although sometimes a certain disposition in longitudinal rows is observed.

In the prototype of insects and myriapods the tracheae also form bunches of stems, which are usually branched, but the bunches are much smaller than in Onychophora. They are always arranged strictly metamerically, one pair in each segment. The stigmata lie dorsally to the points of attachment of the limbs.

In most Diplopoda (Fig. 28, *B*) the tracheae are not branched; they arise in bunches from tracheal pouches, which perhaps represent modifications of muscular apophyses. Their arrangement is very homonomous, with one pair of stigmata on each segment of the trunk (two pairs of stigmata on each diplosegment). There are no stigmata on the heads of Diplopoda. Among all Atelocerata only Symphyla and some Collembola possess functional head stigmata. The arrangement of stigmata in Chilopoda is described in the chapter on metamerism in that group (see Vol. 1, Chap. VII); there also the peculiar modification of the tracheal system of Scutigeromorpha is mentioned (Vol. 1, Fig. 140).

The tracheal system of insects is distinguished by increasing heteronomy and integration. In the prototype there is one pair of stigmata in each segment, from the labial to the ninth abdominal segment inclusive. In the labial segment, however, stigmata are known only in the *Apis* embryo and in adult *Sminthurus* (Collembola). The *Leptinotarsa* (Coleoptera) embryo has three thoracic and nine abdominal pairs of stigmata; adult *Heterojapix* (Campodeoidea), three thoracic and eight abdominal; Lepismatidae and many adult Pterygota, two thoracic and eight abdominal; and Machilidae, two thoracic and seven abdominal (Fig. 28, *C*). Further oligomerisation occurs also in Apterygota (three pairs of thoracic stigmata in *Campodea*, two pairs in *Eosentomon*, one pair of head stigmata in Sminthuridae) and is the rule in Holometabola. In the prototype the separate bunches of tracheae, or tracheomeres, are independent of each other. That structure is retained fully in *Campodea* and *Eosentomon* and almost fully in Machilidae. Formation of anastomoses between tracheomeres, however, occurs as an exception in some Diplopoda and as a rule in Chilopoda Epimorpha. In Japigidae all the tracheomeres on each side are joined together in longitudinal trunks, but there is only one transverse commissure. Lepismatidae also have similar longitudinal trunks and one transverse commissure in each segment. Pterygota are generally similar to Lepismatidae in the structure of their tracheal system. In many of them

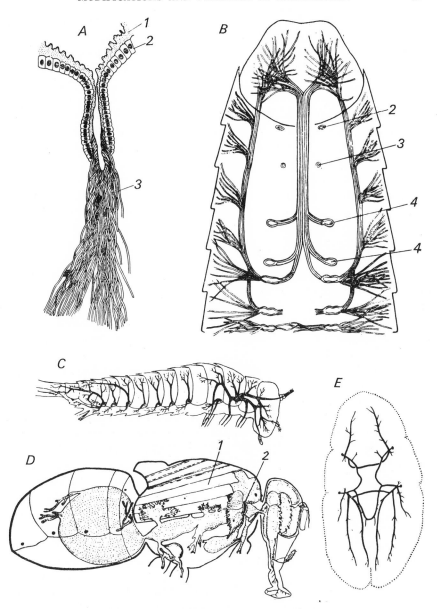

Fig. 28. *Tracheae of Onychophora and Atelocerata.*
A—section through stigma and bunch of tracheae in *Peripatopsis capensis:* 1—
cuticle; 2—epidermis; 3—tracheae (from Lang). *B*—tracheae of front part of
body of *Julus mediterraneus* (Diplopoda): the bunches of tracheae arise from
tracheal pouches (4), commencing with segment IV; in segments II and III there
are undeveloped tracheal pouches (2 and 3) (after Krug). *C—Machilis* (Thysanura)
—primitive type of insect tracheal apparatus (from Lang). *D—Musca domestica*
(Diptera), highly-differentiated tracheal apparatus of higher insects; lateral view to
show air-conducting stems and air-sacs: 1—flight muscles; 2—anterior thoracic
stigma (from Berlese). *E—Lecanium hesperidum* (Homoptera), female, simplified
tracheal apparatus of a small, but not primitive, insect (from Weber).

further reduction of the number of stigmata takes place, and in an extreme case only one pair remains, e.g. in mosquito larvae (Culicidae), which, however, retain several pairs of vestigial stigmata.

The fine tracheoles, unlike the large tracheal trunks, are never anastomosed. In many flying insects (Diptera, Fig. 28, *D*; Hymenoptera) the main trunks form large air-sacs, or a large number of small vesicles form on the small trunks, these also serving to decrease the specific weight of the body (Coleoptera). Spiracles or stigmata become extraordinarily complex and diversified in insects and some Arachnoidea because of the formation of various protective and closing mechanisms (Fig. 29).

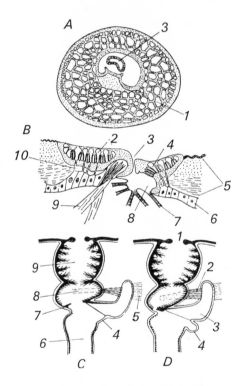

Fig. 29. *Stigmata of terrestrial arthropods (comparatively complex types shown).*

A and *B—Ixodes hexagonus* (Parasitiformes Ixodoidea), stigma and peritreme: *A*—external view; *B*—section: 1—opening in cribriform plate; 2—internal passages in cribriform plate, separated by chitinous branched columns; 3—principal external orifice of stigma; 4—passages leading to post-stigmatic chamber; 5—cuticle; 6—hypodermis; 7—tracheae; 8—anterior post-stigmatic chamber; 9—opening muscle; 10—pores with sensory tips at the bottom of the cribriform plate (after Bonnet, modified). *C* and *D*—sections through open and closed abdominal stigma of ant: 1—orifice of stigma; 2—closing muscle; 3—lever; 4—flexible membrane; 5—opening muscle; 6—trachea; 7—posterior chamber; 8—attachment of closing muscle to wall of chamber; 9—anterior post-stigmatic chamber (after Janet).

Among Arachnoidea, Solifugae possess a highly-integrated tracheal apparatus (Fig. 26, *F*), in this respect developing partly parallel to insects. In most Arachnoidea the tracheae do not anastomose, and even in the prototype, as we have seen, they do not have the regular metameric arrangement we see in Atelocerata.

Among Atelocerata tracheae are absent only in a few groups: Pauropoda, some Protura, and Collembola,[1] and their absence there is probably not historically primary; it is rather due to the small size of these animals and to their life in a humid environment, which facilitates cutaneous respiration.

[1] Among all Collembola only Sminthuridae (Symphytopleona) possess tracheae.

From the physiological point of view tracheae constitute a respiratory apparatus of a unique type. With a high degree of development of the tracheal system, fine tracheal stems and tracheoles directly cover all the internal organs with innumerable ramifications, penetrate into some cells, and ensure direct gaseous exchange in the tissues. Haemolymph largely loses the role of a respiratory environment. Whereas the lungs of a scorpion or the almost-unbranched tracheae of a spider represent a centralised respiratory system, whose function consists in saturating the haemolymph with oxygen and freeing it from CO_2, the tracheae of insects or Solifugae represent a decentralised respiratory apparatus, permeating the whole body, penetrating everywhere, and making tissue respiration possible without the intermediation of haemolymph. This feature of the tracheal system has a great effect on the whole organisation of insects and, in all probability, on the chief trends of evolution in this class (see Chap. VIII).

In the respiratory organs of Myriapoda, most Arachnoidea, and many insects exchange of air takes place by diffusion or, at most, by constriction and straightening of the tracheae with the bending of the animal's body. Many insects, however, and Solifugae and Ricinulei among Arachnoidea, perform actual respiratory movements. Spiders also make some respiratory movements. In Coleoptera the soft dorsal surface of the abdomen rises and falls rhythmically; in Odonata there are similar movements of their narrow sternites; in Tettigonidae the tergites of the abdomen approach and move away from the sternites; and so on. With each increase in intra-abdominal pressure the air-sacs or the thick, soft-walled tracheal stems collapse, and with decrease in pressure they refill with air. In this way the air is changed in the main tracheal stems; in the branched stems it is also changed, by diffusion. The change of air in the main tracheal stems increases the flow of oxygen to the respiratory epithelium of the tracheoles and therefore the skeleto-neuro-muscular mechanisms that make the respiratory movements possible must be regarded as external respiratory adaptations. Their development is clearly connected with increase in the intensity of metabolism in the higher terrestrial arthropods. Apart from the latter, among all terrestrial animals only vertebrates and some lung-breathing snails (Pulmonata) make respiratory movements. With regard to the role of flying muscles in ventilating the tracheal system of higher insects during flight, see Chap. VI.

1. PRELIMINARY REMARKS

Excitability and conductivity are among the basic properties of living protoplasm. In all Enterozoa these properties reach the highest level of development in a special class of cells, nerve cells, or *neurons*. Their function consists in more sensitive and acute reception of action by the external, and later by the internal, environment, and transmission of the stimuli so arising to other neurons, and through the latter to those parts of the organism (effectors) that are capable of responding to a given stimulus by a reaction that is useful to the whole.

Neurons are in contact with each other by the ends of their processes; the points of contact are called synapses. Neurons produce active substances, mediators (e.g. acetylcholine or noradrenalin), which flow along the nerve fibre, the axon, and accumulate in its presynaptic extremities. On stimulation of the neuron these substances flow out and act on the postsynaptic extremities of another neuron, bringing it in turn into a stimulated condition (see J. Welsh, 1957). If the nerve fibre ends in an effector, a muscle or a gland, the secretion of the mediator brings the effector into action. In both cases the active substance secreted enters the reacting cell directly and produces a rapid and brief reaction. In many cases, however, the end of the axon is free, and the active substance produced by it on stimulation of the neuron is discharged into the organism's internal environment, into the blood or the haemolymph. Dispersed by the blood, it exerts slower but more prolonged action on effectors located at any distance from the extremity that discharged it. In that case the active substances secreted by the nerve cell are called neurohormones, and the nerve cell that secretes them is called a neuro-endocrine cell. Carlisle and Knowles (1953) have proposed the name of neurohaemal organs for the nerve ends that innervate nothing but discharge neurohormones into the blood. Besides neuro-endocrine cells and neurohaemal organs, arthropods and vertebrates possess internal-secretion glands, whose cells are not nerve cells. In arthropods, however, these cells are both anatomically and physiologically closely linked with the neural apparatus, especially with its neuro-endocrine organs, and therefore we shall discuss them together with the latter.

When we regard all the nerve cells, or neurons, of the whole body from the tectological point of view we say that they form the *neural system*. Regarding them from the architectonical point of view we find that they form a single *neural apparatus*. Actually, even in the most primitive Enterozoa all the nerve cells of the body are connected with one another, being in contact by means of their processes; and through the intermediation of one another, and often also through that of the body's internal environment, they are connected with all the effectors in the body.

The neural system becomes more complicated with increase in the diversity of the nerve cells and in the complexity of their structure. The neural apparatus becomes more complicated with increasing complexity in the mutual arrangement and especially in the mutual connections of these cells, and also in their connections with the organs served by them (effectors), i.e. muscles, glands, and in fact all the body tissues. Moreover, complexity of the neural apparatus is accompanied by creation of protective, supporting, and other adaptations serving it; these are often formed from elements not of neural origin. The sensory part of the neural apparatus, which forms various sense organs, also becomes extremely complex. Because of their complexity we discuss the individual structure of sense organs in a separate chapter (see Chap. IV); in the present chapter, which is devoted to the neural apparatus as a whole, we shall discuss sense organs only as component parts of that apparatus.

Every neural apparatus, simple or complex, is a single whole both morphologically and functionally; uniting all parts of the body, it intensifies body integrity to the highest degree. In this way the neural apparatus is the most important integrating factor in an animal's body; that integrating role increases immeasurably with increase in the degree of its own centralisation. On the other hand, the more complex and perfect the neural apparatus, the more complex and diversified are the reactions of the organism to external stimuli; consequently, the more complex and perfect the neural apparatus, the greater is the role of neuro-muscular activity in the animal's life and the greater are the advantages given it by its neuromuscular activity in the struggle for existence.

Among the many directions taken by the evolution of different stems of the organic world, we consider that the direction most characteristic of animals as such and most progressive is that leading to perfection of the neural apparatus, to its centralisation and complexity.

Comparing the data of comparative anatomy and comparative ecology, we see that the directions of development of separate stems of Metazoa depend mostly on the character of the activity of the animals. Since the chief aim of animals' activity, especially that of lower animals, is procurement of food, the methods of procuring food have the strongest influence and affect the direction of evolution, particularly in its earlier stages.

From the point of view that interests us here, we may divide all invertebrates into the following groups according to their methods of food procurement.

1. *Grazing animals.* They feed on immobile and almost-immobile food (not necessarily vegetable), each unit of which does not require a special catching action and does not require pursuit.

These include filter-feeders and feeders on settling organic particles that feed on small particles, alive or dead, suspended in water; mud- and soil-feeders, substrate-scrapers, and most phytophages; they also should include permanent parasites. From this list it is seen that grazing animals may be both mobile (scrapers, soil-feeders, some members of the other groups) and sessile (many filter-feeders and feeders on settling organic particles, many parasites).

2. *Hunting animals.* They feed on relatively large and mostly mobile individualised prey, each unit of which requires a separate catching action.

(*a*) *Ambush-hunters.* They are incapable of locomotion (sessile animals) or have imperfect locomotion ability, which they cannot use to take prey (hovering forms such as medusae, slightly-mobile forms such as sea-anemones).

(*b*) *Active hunters.* They are capable of locomotion and use it to seek and capture prey.

As we shall see below, the first appearance of neural apparatus and musculature in primitive Metazoa probably occurred as a result of transition to predation, whether of active or ambush type. The next step in perfecting a neuro-sensory apparatus—the appearance of the first complex sense-organs and, under their influence, the first signs of centralisation of the neural apparatus—was made with transition to an actively-moving mode of life and is seen in medusae, ctenophores, and lower turbellarians, i.e. in both active hunters and moving ambush-hunters. The third step in progressive development, development of distant-stimulus receptors (eyes and olfactory organs for hunting), was stimulated only by active hunting. Active hunting calls for increased mobility, which in turn leads to improvement of the locomotor apparatus. The search for prey and rapid movements require improvement of sense-organs, as we have seen in nemertines (olfactory organ), errant polychaetes (eyes, palps, and olfactory organs), arthropods (eyes and antennae), etc. The development of distant-stimulus receptors, which assume the role of analysers (I. P. Pavlov), and the improvement of the locomotor apparatus lead to increased strength and complexity in the neural centres. On the other hand, the increased neuro-muscular activity of an active hunter produces strengthening of all the ancillary apparatuses: digestive, respiratory, excretory, circulatory. Their improvement includes greater subjection of them to the control of the neural apparatus, i.e. improvement of the sensory and motor innervation of the internal organs and strengthening of the centres governing them. Improvement of the organs of vegetative life leads in turn to intensification of metabolism and of all activities. At the same time the parallel improvement of the neuro-sensory and locomotor apparatuses allows greater complexity of the animal's behaviour.

At definite stages in evolution (the lower Bilateria), therefore, only an active hunting mode of life ensures harmonious progress and a general raising of the organisational level (L. Zenkevich, 1944; N. Livanov, 1945). Every return to other feeding methods results in some kind of regression, first in the structure of sense-organs and then in the structure of the locomotor and neural apparatuses (sessile polychaetes, lamellibranch molluscs) and, in extreme cases, in the organs of vegetative life (many parasites). In animals that have attained a high level in development of the central section of the neural apparatus, however, activities take a number of new directions besides procurement of food: defence against enemies, active distribution of population, care for offspring, search for and attraction of sexual partners, construction of shelters, etc. The existence of these types of activity, often with considerable complexity, in higher animals lessens the detrimental effect of giving up active predation, and we see (e.g. among insects) many phytophagous forms that not only have not degenerated in comparison with their predatory ancestors but have continued to make progress. These include dung-beetles (*Gymnopleurus*, *Scarabaeus*, etc.) and burying beetles (*Necrophorus*) with their complex instincts of brood care, and termites, which represent one of the peaks in development of neural activity among invertebrates. Even among insects, however, when any phase of the life cycle becomes completely inactive the result is marked regression in organisation, as we see in male butterflies of the family Psychidae (D. Fedotov, 1940) or in the family Coccidae in the order Rhynchota, and in other cases.

The most important stimulus to progressive evolution is therefore the animal's own activity. At the lower stages of Metazoan evolution the chief form of activity causing general progress in organisation, including progress in the neural apparatus, is any form of predation; at the next stages, only active hunting activity. With further perfecting of the neural apparatus, however, a number of new forms of activity arise, making further progressive development of a given group of animals more or less independent of their method of procuring food. Only when such a moment arrives can we consider the progressive direction of evolution in that branch to be more or less firmly established.

Three stems of the animal kingdom contain progressive groups that have, in the course of hundreds of millions of years, raised their organisational level: these are vertebrates, Articulata, and molluscs (Cephalopoda!). In these groups, so far as the data of palaeontology and comparative anatomy enable us to judge, there has been prolonged progress in the neural apparatus, particularly its central part; a conclusion first formulated by Dana in the middle of the nineteenth century. The gradual and uninterrupted growth and improvement of the mammalian brain, well traced through the Tertiary epoch, led at the end of it to a colossal leap forward, to the appearance of the human brain, whose complexity and efficiency have made possible social life with its specific and qualitatively–new characteristics.

The tremendous effects of progressive evolution of the neural apparatus of animals on the earth as a whole have been well described by V. I. Vernadskii (1944; Vernadsky, 1945).

The essential content and meaning of progressive evolution in the animal kingdom has therefore been improvement of the central part of the neural apparatus, and the history of that process presents one of the central problems in biology.

2. COELENTERATES AND THE FIRST APPEARANCE OF NERVE CELLS AND OF A NEURAL APPARATUS

As we have seen, sponges have no neural elements;[1] their incipient contractile and glandular cells are independent effectors (G. Parker, 1916), reacting directly to external stimuli. At the same time sponge tissues possess an incipient capacity for conducting stimuli: if one pricks a sponge at some distance from one of its oscula, after some time the latter contracts by the action of its myocytes. That conduction of stimuli, however, which takes place without the help of specialised neural elements, is exceedingly slow, at the rate of 1 cm per minute (C. Prosser, 1950). The function of conduction of stimuli therefore appears in Metazoa before the development of a special apparatus to perform it. A neural apparatus first appears in coelenterates.

The absence of neural and muscular apparatuses in sponges and possession of these by coelenterates is no accident: the difference has the closest connection with the types of feeding and the methods of pro-curing food in the two groups. Sponges feed on small particles suspended in sea-water—ultra- and nanoseston; food is brought by the activity of the flagella in the choanocytes, and the food particles are absorbed by separate body cells, choanocytes and amoebocytes. Co-ordination in the activity of the separate cells is not required in either the bringing or the absorbing of food. Sponges require a definite degree of integration of all cells in the body into a single organism only in the creation and maintenance of a specific structure, with the proper relationship of parts, to ensure the regular functioning of the irrigation apparatus. Although in many respects sponges are very far from other Metazoa, their method of feeding is clearly primary: in fact, the colonies of Flagellata from which Metazoa are most probably derived must originally have lived on food that could be entrapped and digested by separate members of a colony, by separate cells.

In contrast to sponges most coelenterates, including all the primitive ones, feed on large prey, which they take by means of the mouth.[2] Here the seizure and swallowing of prey requires harmonised action by all the epithelial-muscle and muscle cells of the body; that is necessary if the animal is to act as a whole, as a single organism, and to act quickly. Such integration and rapidity of action can be attained only by possession of a neural apparatus and musculature. I think it almost beyond question that in the evolution of Metazoa both the neural and the muscular apparatuses

---

[1] The description of supposed neural elements in sponges given by M. P. Ceccaty is unconvincing.

[2] Except for a number of forms among Anthozoa, Scyphozoa, and Ctenophora, which have secondarily returned to feeding on small particles taken by means of hydrokinetic adaptations (see Chap. II).

arose because of transition from primary feeding on fine dispersed food to the taking of large prey, i.e. because of the appearance of predation. In other words, the transition to predation served as a stimulus also for development of the organisation of Enterozoa, and among the latter predation was the primary, original method of feeding.

In coelenterates both kinetoblast and phagocytoblast contain neurosensory as well as other types of cells.

N. Kleinenberg (1872) has suggested that the epithelial-muscle and neuro-sensory cells of coelenterates were formed by splitting of original epithelial-muscle elements that were independent effectors: a muscular outgrowth of the original cell formed a muscle cell, and its epithelial part formed a neurosensory cell. A. A. Zavarzin (1941) also holds similar views.

From the structural-morphological point of view, of course, the epithelial-muscle and neurosensory cells are the result, not of splitting, but of differentiation of original multifunctional ectodermal cells: some of them lost sensitivity and became dependent effectors, epithelial-muscle cells, while others lost their contractility, retaining and increasing their sensitivity, and became neurosensory cells. Their phylogeny also must have been of that kind. N. G. Khlopin (1946) also believes divergent differentiation to have been the most probable method of original differentiation of neural and muscular elements in Metazoa, and substantiates that view in detail.

The components of the neural system of coelenterates are of three types: sensory, associative, and motor cells. The first retain their epithelial location, but the others have lost it and lie subepithelially, on the same level as the muscular processes of the myoepithelium. They all form numerous outgrowths, which make contact with other nerve cells (neurons), and the outgrowths of the motor cells also make contact with muscle cells. Unlike the neurons of higher forms, the nerve cells of coelenterates (at least their associative cells) are not polarised; all their outgrowths are uniform, and they are not divided into a neurite (axon) and dendrites. All the nerve cells are combined into a single plexus that covers the whole body of the animal and constitutes its neural apparatus. In hydroid polyps this plexus is diffuse and forms only a rather vague concentration around the mouth, and in *Hydra* also on the foot. We have already mentioned a remarkable feature in coelenterates, that their neural plexus is developed not only from the kinetoblast but also from the phagocytoblast, the epithelium of the gut. The two plexuses, ectodermal and endodermal, interpenetrate each other in the region of the mouth.

According to the view expressed in their time by O. and R. Hertwig (1878, 1879, 1880) and accepted by most authors, the primary form of nerve cells was sensory cells lying in the epithelium, from which both associative and motor cells are derived. It is most natural to regard as the prototype the neural system of coelenterates (Fig. 30, *A*) in the form of a network of cells in the epithelium, each having a receptive device at its apex, and at its base outgrowths whereby it made contact both with cells similar to itself and with muscle cells. The direct link between sensory and muscle cells would be the simplest conceivable reflex arc—monomeric. This type of link has actually been described in the tentacles of Actiniaria. Mutual linkage of all the cells in the neural plexus results in reaction to purely local stimuli by wider areas of the body. Let us now imagine (Fig. 30, *B*) that some of these universal cells lose their direct connection with the muscle cells, but retain their epithelial location and their connection with other nerve cells; that others lose their direct connection with both the surface of the body and the musculature, but retain their connection with other nerve cells; and, finally, that a third group retain their connection with the musculature and other nerve cells but lose direct connection with the external world; in such a case the first cells would become sensory,

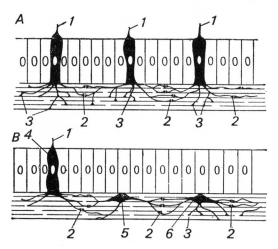

Fig. 30. *Diagram of origin of sensory, associative, and motor nerve cells in coelenterates (B) from universal nerve cells of the prototype (A).*

1—sensory ending; 2—synapses at places of contact of different neurons; 3—motor endings on muscle cells; cells: 4—sensory; 5—associative; 6—motor (original).

the second associative, and the third motor cells, and their assemblage takes the form of the neural apparatus of modern coelenterates. This picture of the course of differentiation of types of nerve cells in coelenterates (which differs somewhat from the usual formulation of O. and R. Hertwig's theory) also explains the absence of polarisation in their associative cells, since from the above point of view all the outgrowths of these cells are equivalent in nature.

The diffuse neural apparatus of coelenterates (Fig. 31) is a very primitive type. It is not divided into central and peripheral sections, there is no specialisation among the associative cells, and there are no long conducting routes consisting of single processes. The network conducts stimuli in all directions, the stimuli being passed from neuron to neuron. At the same time each of them is also connected with motor cells. Therefore a wave of excitation spreading through the neural plexus from any point is accompanied throughout its journey by a wave of muscle contraction. A polyp is incapable, as a rule, of reacting to a stimulus arising at any point of its body by contraction of muscles at a distance from that point without previous reaction by all the muscles between the two points. Stimuli are

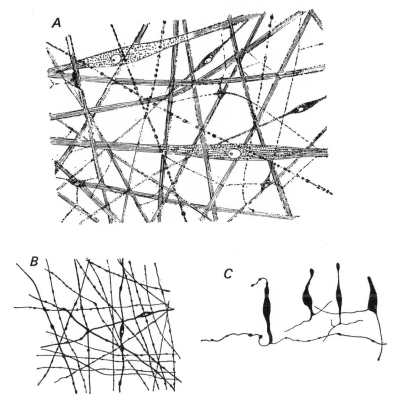

Fig. 31. *Neural apparatus of Cnidaria.*
Neural plexus of the medusa *Rhizostoma* (Scyphozoa): *A*—ectodermal, *B*—endo-dermal. Diagram *A* also shows muscle fibres (after Bozler). *C*—group of endo-dermal sensory cells of the sea-anemone *Bunodes gemmaceus* (after Grošel).

transmitted through the neural plexus far more rapidly than in sponges, where neural elements are absent (in Actiniaria, at a rate of 4 to 15 cm/sec; C. Prosser, 1950), but far more slowly than in higher Metazoa,[1] where they do not have to surmount synapses (i.e. points of contact between the pro-cesses of different neurons) located along their path. Even in Actiniaria, however, there are paths of rapid transmission, which produce, for example, a reaction of defensive contraction by the animal. The rate of transmission in these is rather more than 1 m/sec. Morphologically these paths are

[1] 5 m/sec in the nerves of a crab's leg, and 22 m/sec in the giant nerve fibres of *Loligo*.

represented by a ring on the oral disc and longitudinal strands in the mesenteries (see Fig. 32). Synapses probably exist on these paths, but transmission of stimuli through them is permanently facilitated (J. Ramsay, 1952).

A second peculiarity resulting from the structure of diffuse neural apparatus is that any part cut away from the polyp's body is capable of independent reflexes, of muscular contractions in response to a stimulus, unlike, for instance, the amputated leg of an insect. This is because, with the diffuse, absolutely-uncentralised neural apparatus of polyps, each small part of the body always retains the whole neural mechanism necessary to complete a reflex—sensory, associative, and motor cells and an effector muscle, all linked together.

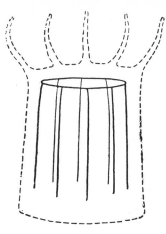

Fig. 32. *Diagram of arrangement of paths of accelerated transmission of stimuli in body of* Actinia (*annular thickening in oral disc and radial nerve cords in mesenteries* (after J. A. Ramsay).

Anthozoa, especially Actiniaria, differ from hydroid polyps in their greater concentration of the neural plexus in the circumoral-disc region; basically, however, their neural apparatus remains diffuse. A second important difference between Anthozoa and hydroids consists in the great development of the parts of the neural apparatus located in the mesenteries of Anthozoa, which are generally lacking in hydroids. The outstanding role of peripheral phagocytoblast in creating the muscular (see Chap. VI) and neural apparatuses of Anthozoa is very characteristic of that group. In this respect they show a very interesting resemblance to the lower Deuterostomia (see below). The endodermal neural plexus of Anthozoa combines with the ectodermal not only in the region of the mouth but also throughout the mesogloea.

All medusae differ from polyps in possessing specialised sense-organs and in making considerably more complex movements. As a result the neural plexus of hydromedusae forms two annular thickenings at the edge of the umbrella, one at the subumbrellar and the other at the exumbrellar side of the annular canal. Possession of a subumbrellar ring has also been demonstrated in several Scyphomedusae (*Aurelia, Rhizostoma*), in which it is joined by radial strands to all the marginal sense-organs. At the base

of each of the latter there is a considerable accumulation of nerve cells—
a marginal ganglion. According to physiological research data, rhythmic
impulses arise in these marginal ganglia, are transmitted by the entire
neural plexus, and produce rhythmic contractions of the bell. The marginal
ganglia of Scyphomedusae represent a first step towards creation of a
central section of the neural apparatus. It is instructive that they originated
because of the formation of the first complex sense-organ, whose origin
was in turn due to the mobile mode of life of medusae.

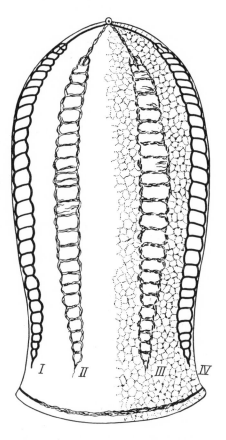

Fig. 33. *Neural apparatus of the ctenophore*
Beroë (*diagrammatic*).

The nerve cords along the ctenes, the
circumoral plexus, and part of the diffuse
cutaneous plexus are shown. In radius II
the course of the nerve fibres, and in
radius III and interradius III-IV the
nuclei of nerve cells, are shown (after
Heider).

The neural apparatus of ctenophores is of exceptional interest to comparative anatomists, but because of the extreme delicacy of these animals
it is difficult to study. According to K. Heider (1927), the general subepithelial plexus of *Beroë* (Fig. 33) forms a thickening (circumoral ring)
around the external mouth. Along all the rows of ctenes the plexus also
thickens, thus forming eight meridional condensations. These meridional
condensations extend still farther aborally, becoming fused in pairs and
converging along the course of the four ciliated bands towards the base of
the 'springs' of the aboral sense-organ. *Coeloplana* has there four ganglion-
like concentrations of neural tissue (T. Komai, 1936). As physiological

experiments indicate, the aboral sense-organ and adjacent parts of the neural plexus certainly exert a tonic and regulating influence on the activity of the ctenes, which are, however, capable of reflex responses to stimuli affecting any other parts of the neural plexus.

Thus a co-ordinating centre linked with the aboral sense-organ, eight meridional strands, and a circumoral ring are developed in the neural plexus of ctenophores, a structural plan very similar to that which we find in polychaete trochophores and partly in Scolecida. B. Hanström (1928) suggests that in ctenophores, as in Scyphomedusae, the appearance of balancing organs (which obviously resulted from a free-swimming mode of life) was the stimulus to the first creation of neural centres.

H. Korn (1959) describes a neural plexus beneath the epithelium of the stomach of *Pleurobrachia pileus;* judging by his data Ctenophora, as well as Cnidaria, possess not only an ectodermal but also an endodermal section of the neural apparatus, a fact of great comparative-anatomical significance.

3. LOWER WORMS, AND GENERAL PRINCIPLES OF THE EVOLUTION OF THE NEURAL APPARATUS

The neural apparatus of lower turbellarians is at the same level as that of coelenterates, but it reaches higher levels in higher turbellarians. Unlike coelenterates, however, turbellarians no longer show any traces of an endodermal neural system; their neural system is entirely of kinetoblastic origin. We deduce the reasons for this by studying the evolution of phago-cytoblast in Protostomia (see Chap. V).

We find the most primitive type of turbellarian neural apparatus in some Acoela and other members of lower orders, e.g. *Xenoturbella* (E. Westblad, 1949) (Fig. 182). This remarkable turbellarian has a diffuse subepithelial neural plexus, without any stems or longitudinal thickenings. The same type of diffuse neural plexus is possessed by *Convoluta stylifera* and *Otocoelis gulmariensis* among Acoela (E. Westblad, 1946) and a number of other lower turbellarians. In all these forms the plexus is more developed at the front end of the body than in the rest of its length, which occurs also in coelenterate planulae (the hydroids *Gonothyrea* and *Clava*). We ascribe the same condensation of the neural plexus at the aboral end to the coelenterate prototype of turbellarians. The high number of neural elements at the aboral pole is quite understandable, since in swimming it is the first part to come into contact with new stimuli.

Like ctenophores, *Xenoturbella* has a balancing organ located at its aboral pole, but (unlike the situation in ctenophores) its statocyst is a closed vesicle lying partly in the external epithelium and partly in the sub-epithelial neural plexus. The central neural apparatus contains no other parts besides the cutaneous plexus, and in this respect *Xenoturbella* stands at the same organisational level as coelenterates. In contrast to *Xeno-turbella*, almost all Acoela and several other lower turbellarians have a statocyst sunk in the parenchyma, which has lost its connection with the epidermis. In *Nemertoderma* (Acoela) a group of nerve cells lies near the statocyst and communicates with the subcutaneous plexus by six or eight radial nerves. *Polychoerus* (Acoela) has four such nerves, lying in the same planes as the four ciliated bands connecting the aboral organ of ctenophores with the ctenes (Löhner, 1910). In the embryo of *Otocoelis chiridotae* (Acoela) four brain rudiments form in the same planes. We may regard the group of nerve cells serving the statocyst in *Nemertoderma* as a brain, even if an incipient one.

It was in consequence of the statocyst—a sense-organ at the aboral pole, sunk beneath the skin—that the brain, or cerebral ganglion, arose in turbellarians; the cerebral ganglion is a part of the cutaneous neural plexus that has separated from the rest and sunk into the body behind the statocyst. We have already seen that in ctenophores and Scyphomedusae the first rudiments of neural centres arise in connection with the service of the balancing organs. In view of the location of the cerebral ganglion of higher turbellarians within the cutaneous plexus, on the body axis, E. Reisinger (1925) calls it the *endon*.

Developing from the above-described prototype, the neural apparatus of flatworms passes through a series of successive stages of improvement. That improvement takes five main directions: (*i*) centralisation of the neural apparatus; (*ii*) migration of its elements into the depth of the body; (*iii*) increase of the role played by the cerebral ganglion (endon); (*iv*) external architectonical simplification of the neural apparatus; (*v*) greater complexity of its internal architectonics, causing greater complexity and plasticity in the animal's behaviour. The latter question is too complex for discussion within the limits of a general course in comparative anatomy. A detailed study of it may be found in the works of B. Hanström (1928) and A. A. Zavarzin (1941); we can touch only on a few points that are relevant here.

Centralisation of the neural apparatus in turbellarians consists in spatial differentiation of the neural plexus and conversion of it from a dense diffuse network into a regular *orthogon* (E. Reisinger's term, 1925) (Fig. 34). Both the associative and the motor cells with all their processes are gathered into several pairs of longitudinal cords or stems, connected by a large number of commissural stems of the same structure. All the intervals between these stems, on the other hand, are devoid of these elements; only sensory cells sometimes remain in the intervals, but they send their central processes to the nearest transverse stem of the orthogon; these spaces also contain peripheral processes of motor cells arising from these stems (Fig. 36). In brief, only elements of the peripheral section of the neural apparatus remain in the spaces between the transverse bars of the orthogon, while the orthogon already represents the first step in the formation of its central section.

Why do we regard the formation of the orthogon as the first step in centralisation of the nervous system? A diffuse neural plexus is characterised by very short peripheral connections and very long connections between

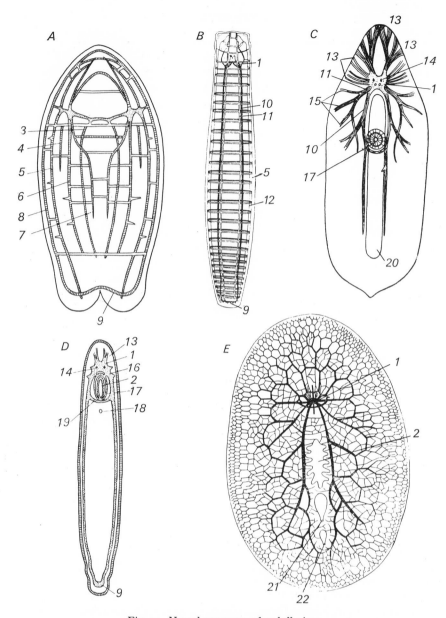

Fig. 34. *Neural apparatus of turbellarians.*

A—Polychoerus caudatus (Acoela) (after Löhner). *B—Bothrioplana semperi* (Alloeocoela) (after Reisinger). *C—Mesostoma ehrenbergii* (Rhabdocoela) (after Bresslau). *D—Macrostomum finnlandense* (Macrostomida) (after Luther). *E—Planocera graffii* (Polyclada) (after Lang). 1—brain; 2—pharynx; 3—internal ventral stems; 4—external ventral stems; 5—lateral stems; 6—external dorsal stems; 7—internal dorsal stems; 8—marginal nerve; 9—caudal commissure; 10—ventral stems; 11—dorsal stems; 12—ventro-lateral stem; 13—sensory nerves arising from the brain; 14—eye; 15—branches of dorsal stem; 16—nerves to pharyngeal plexus; 17—pharyngeal neural plexus; 18—excretory orifice; 19—anterior commissure of ventral longitudinal stems in *Macrostomum*; 20—gut; 21—male genital orifice; 22—female genital orifice.

ganglionic cells (including associative and motor cells in that concept). In fact, with a diffuse plexus any sensory cell lies in the immediate vicinity of associative and motor cells, which constitute the link between it and the nearest effector. On the other hand, with a diffuse plexus the connections between separate ganglionic cells extend throughout the whole body, because the ganglionic cells are dispersed throughout its entire length.

After the ganglionic cells have concentrated in the cross-bars of the orthogon the paths between them are somewhat shortened; but the paths to the muscles and sensory cells lying outside the orthogon are lengthened, and the fewer the cross-bars and the more open the network formed by them the shorter will be the average distances between the ganglionic cells and the greater those separating the sensory cells and the effectors from the ganglionic cells. In other words, the paths within the central section are shortened and those between the centre and the periphery are lengthened.

The general trend of evolution in turbellarians is towards fewer stems and commissures, i.e. towards increased centralisation of the neural apparatus. The smaller the animal, however, the easier is it for centralisation to take place. Among turbellarians, therefore, the higher the systematic position of the group and the smaller the species, the greater is the centralisation of the neural apparatus. For instance, among Acoela, which are at a low level of organisation, large forms such as *Polychoerus* (Fig. 34, *A*) have a very complex plexus with a large number of cross-bars; small forms such as *Convoluta* (Vol. 1, Fig. 51, *E*) have only from three to five pairs of longitudinal stems and a regular orthogon. Among the more highly-organised Triclada the small Maricola have a relatively highly-centralised neural apparatus, while that of the large Terricola is relatively diffuse. Rhabdocoela, which are small forms standing highest in the system, have the most centralised neural apparatus in the whole class (Fig. 34, *C*). The orthogon of Polyclada is only slightly centralised (Fig. 34, *E*), which is in accordance with the large size and comparatively primitive nature of that group; but Polyclada have an excellently-developed and functionally-efficient endon, perhaps originating because of the presence of apical tentacles and brain eyes. We may remark that the orthogon of Polyclada is characterised by the plexus being highly developed on the dorsal and slightly developed on the ventral side (D. Hadenfeldt, 1929; D. Corrêa, 1949) (Fig. 35), for which reason the latter escaped the notice of the old authors (A. Lang, 1884).

A second trend in the development of the neural apparatus is movement of it deeper into the body. In many turbellarians the neural plexus or even the differentiated orthogon lies beneath the epithelium itself, so that the epithelial sensory cells can be directly adjacent to the cross-bars of the orthogon (e.g. in *Pomonotus* in Alloeocoela; W. Beklemischev, 1927).[1] In

---

[1] By the way, it is clear from the location of these cells that those authors are in error who (like N. Livanov, 1955) regard the orthogon as a purely motor section of the neural apparatus of flatworms.

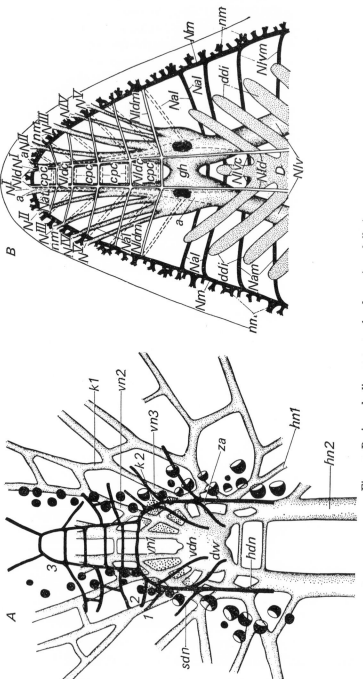

Fig. 35. Brain and adjacent parts of orthogon in turbellarians.

A—*Zygantroplana henriettae* (Polyclada Leptoplanidae): posterior (*dw*) and anterior (*vdn*) commissures between brain and dorsal longitudinal stems; *hdn*—postcerebral parts of these stems; *hn* 1—beginning of lateral stems of ventral plexus; *hn* 2—main stems of ventral plexus; *k* 1—cross-bars in anterior section of ventral plexus; *k* 2—lateral commissural stems of that plexus, bending round the brain; *sdn*—one of the commissures of the dorsal plexus; *vn* 1—anterior stems of central plexus; *vn* 2, *vn* 3—anterolateral stems of that plexus; *za*—one of its cross-bars; 1, 2—trabeculae of dorsal plexus; 3—precerebral parts of dorsal longitudinal stems, joined together in front of the brain by six commissures. The granular masses in front of the brain are accumulations of sensory cells lying outside the brain envelope (after D. D. Corrêa). B—*Bdelloura candida* (Triclada Maricola): *a*—anterior parts of ventral longitudinal stems; *Nal*—commissures between them and lateral stems; *cpc*—commissures between the two stems of the first pair of sensory nerves; *D*—front end of gut; *ddi*—its ramifications; *gh*—brain; *N*I–V—paired sensory nerves of the brain; *Nld*—dorsal longitudinal stems; *Nldm*—commissures between dorsal and lateral longitudinal stems of the brain; *Nldc*—commissures between them; *Nldm*—commissures between them.

many other turbellarians we find the orthogon well sunk within the body; only the epithelial and subepithelial sensory cells remain on the periphery. A similar inward movement takes place very easily in large turbellarians, e.g. in Triclada (Fig. 38) and Polyclada. In contrast, the smaller the turbellarians the more persistently does their neural apparatus keep its subepithelial location, even with a high general organisational level (many Rhabdocoela). We can see, however, that the higher the animal's organisational level, the deeper its central neural apparatus sinks, body sizes being equal.

Different classes of neurons show variations in the tendency to sink from the epithelium into the depth of the body. As a rule motor neurons sink first, then associative, while sensory neurons remain longest near the surfaces that are related to their functioning. We find this law first in flatworms. Some Rhabdocoela Dalyellioidea (E. Marcus, 1946; A. Luther, 1955) and their close relations Digenea (*Opisthorchis felineus*, E. Kolmogorova, 1956, 1959) have a pair of deep longitudinal nerve stems as well as an orthogon and an endonal brain. The three outer pairs of stems in the orthogon are all connected with one another by annular commissures, and

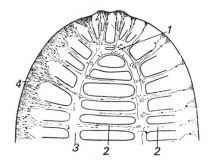

Fig. 36. *Front end of body of* Bdelloura candida *(Turbellaria Triclada) with brain and ventral part of neural apparatus.*

1—brain; 2—transverse commissures; 3—longitudinal stems; 4—'sensory border' with great number of sensory cells. Neurons of several types are visible (after Hanström).

each is connected with the brain by a commissure; the deep stems are connected with the lateral stems and the brain by commissures (Fig. 37, *A*). In *Opisthorchis* (the fine structure of whose neural apparatus has been studied) the outer stems and the annular commissures of the orthogon contain sensory cells, and the brain and the deep stems contain associative and motor cells.

We may believe that the deep stems of *Dalyellia* and *Opisthorchis* have separated from the lateral stems of the orthogon. Deep stems of this type are probably possessed by other Digenea and by many other forms of the Dalyelliida group. Probably the deep intraparenchymal plexus of Triclada Terricola (Fig. 37, *B*) is a similar formation (O. Steinböck, 1925).

We have already seen the first appearance of a cerebral ganglion (endon), related to possession of statocysts, in primitive Acoela. The endon is further developed, however, only in a few Acoela (*Proporus brochi, Childia*, and some others; E. Westblad, 1942, 1945). In its place many Acoela (e.g. *Aphanostoma pallidum*, Fig. 38) have in the front part of the body a thickening of all the longitudinal stems of the orthogon and the annular commissures joining them (W. Beklemishev, 1937). That is evidently a result of the great number of cutaneous sensory cells in the head section and the consequent increased size of the *neuropile* (the plexus of processes of ganglionic cells, which in most invertebrates occupies the central part

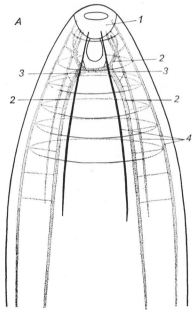

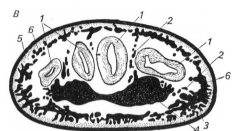

Fig. 37. *Part of the flatworm central neural apparatus withdrawing from the body surface.*
A—*Opisthorchis felineus* (Digenea), location of neural apparatus in front part of body: 1—oral sucker; 2—nerve stems sunk within body; 3—brain; 4—superficial nerve stems (after E. Ya. Kolmogorova). B—*Geoplana whartoni* (Triclada Terricola), cross-section of front part of body: 1—branches of gut; 2—nerves connecting deep neural plexus with internal cutaneous plexus; 3—nerves connecting internal with external cutaneous plexus; 4—deep intraparenchymal plexus; 5—external cutaneous plexus; 6—internal cutaneous plexus (after Steinböck, slightly modified).

of a ganglion, is called the neuropile). Such Acoela, although they possess the rudiment of an endon, have an annular orthogonal brain formed from the orthogon, and not an endonal brain.

Among other turbellarians Catenulidae (order Notandropora) and all Crossocoela (order Alloeocoela) possess a statocyst; in both these the statocyst lies in or near the bulky endonal brain; in these forms and in all higher turbellarians (unlike most Acoela) all the sensory cells in the front end of the body send their processes into the brain, not into the orthogon. Their brain therefore is a large ganglion; it is connected by several pairs of sensory nerves to the sense-organs (olfactory fossae, eyes) and to the cutaneous sensory cells in the front part of the body, and by radial connectives to all the longitudinal stems of the orthogon. The brain of most turbellarians is therefore constructed of elements directly or phylogenetically linked with the statocyst and of elements linked with other sense-organs in the front part of the body.

Originally the brain has little influence, either physiologically or morphologically, on the orthogon. In *Bothrioplana* (Reisinger, 1925) and *Geocentrophora* (Alloeocoela) (Steinböck, 1927) the stems of the orthogon pass by the brain, being joined to it only by connectives. Later, however, since the large number of sensory cells sending their processes into the brain must affect the activity of the musculature of the whole body, the brain begins to take a dominant role. Its associative cells send out long processes extending far along the longitudinal stems of the orthogon (Fig. 36) and thus establish direct nerve connections with distant parts of the body. The parts of the longitudinal stems of the orthogon running rearward from the brain connectives then become much thickened by the addition of a large number of fibres coming from the brain. On the other hand, the parts of the stems lying in front of the brain connectives remain small, and in some Rhabdocoela (especially if the brain lies near the front end, as in *Solenopharynx*) they disappear entirely (W. Beklemischev, 1929).

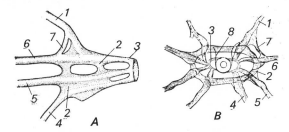

Fig. 38. *Front end of the orthogon of* Aphanostoma pallidum (*Acoela*), *semi-diagrammatic.*

A—view from right side. *B*—view from behind. 1—inner dorsal stems; 2—two annular commissures lying on the extension of the thickened part of the stems that takes the place of a brain; 3—frontal commissure; 4—inner ventral stems; 5—outer ventral stems; 6—outer dorsal stems; 7—first postcerebral transverse commissure between the dorsal stems; 8—nerves to statocyst (after W. N. Beklemishev).

In any case, the forms with a highly-developed brain give the impression that the posterior stems of their neural apparatus arise from the brain and that the radial connectives are merely the roots of these stems, whereas the anterior parts of the longitudinal stems are merely small forward-pointing branches. Thus the increase in the brain and in its influence on the orthogon leads to reconstruction of the latter and to more perfect integration of the whole neural apparatus.

The result of internal architectonical simplification of the neural apparatus is further development of the above two processes, further increase in centralisation and integration. The result of further concentration of the neural apparatus is decrease in the numbers of stems and commissures, a process we have already discussed in our survey of the architectonics of turbellarians (Vol. 1, Chap. IV). Maximum concentration is attained in *Macrostomum* (Fig. 34, *D*), which has only one pair of longitudinal stems, with one transverse commissure behind the pharynx. This structure already represents a high degree of centralisation of the neural apparatus. The whole ventral and dorsal surface of the body is devoid of ganglionic cells. A transverse slice cut from the body is capable of reflex action because it contains parts of the longitudinal stems; but if we were to cut off the edge of the body containing these stems the central part of the slice would be incapable of reflex action, since it would contain only peripheral elements of the neural apparatus. At the same time the role of the brain increases greatly. The precerebral parts of the longitudinal stems are reduced. Numerous sensory nerves run forward from the brain, and a pair of lateral stems run rearward from it. The first phase of integration of the central nervous system has been completed.

The principles we have observed in the evolution of the neural apparatus in turbellarians—centralisation, inward movement, and creation of higher centres resulting from the appearance of complex sense-organs—are among the general principles of evolution of the neural apparatus, and we shall see further development of them in higher types of invertebrates.

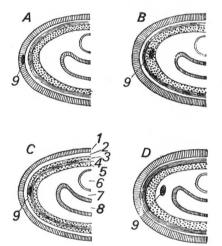

Fig. 39. *Location of longitudinal nerve stems in different nemertines.*
Palaeonemertini: A—*Carinina*,   B—*Cephalothrix.* C—Heteronemertini. D—Metanemertini. 1—epidermis; 2—mesenchyma; 3—outer longitudinal muscles; 4—annular muscles; 5—inner longitudinal muscles of the body; 6—proboscis sheath; 7—internal mesenchyma; 8—gut; 9—lateral nerve stem (after Bürger).

The promorphology of the neural apparatus of the remaining Scolecida has already been discussed in Vol. 1, Chap. IV, in connection with the general structural plan of the body, and we shall add only a little here. Nematodes possess a typical orthogon, without a cerebral ganglion (Vol. 1, Fig. 61, *A*); what is called the brain in nematodes is merely the slightly-thickened front commissure of the orthogon; in flatworms only the orthogonal 'brain' of the *Aphanostoma pallidum* (Acoela) type, formed by thickening of the front ends of the longitudinal stems of the orthogon, partly corresponds to it. Neither are there grounds for speaking of an endonal brain in Nematomorpha, Kinorhyncha, or Priapuloidea; the circumoesophageal ring characteristic of all these groups is rather the homologue of the same feature in nematodes. On the other hand, the cerebral ganglion of Acanthocephala and Rotatoria and the cephalic ganglion of Gastrotricha are, perhaps, homologous with the endonal brain of higher turbellarians; that is almost obvious in the case of Acanthocephala (Vol. 1, Fig. 66).

The neural apparatus of nematodes, like all of their organisation, is characterised by a constant and very small number of cells. According to R. Goldschmidt (1908–10) the circumoral ring of *Ascaris* contains only 162 cells, of which 90 are motor, 50 sensory, and 22 associative.

We have seen that the longitudinal nerve stems of nemertines correspond to the orthogon of flatworms. The brain of nemertines, which is formed of a pair of dorsal and a pair of ventral ganglia joined by commissures, is a ring-like organ like the orthogonal 'brain' of some Acoela and of all

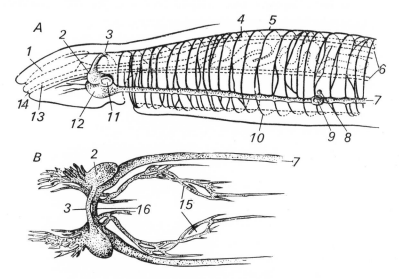

Fig. 40. *Neural apparatus of* Carinella (*Palaeonemertini*).

Diagram of structure of its front part. *A*—view from left side. *B*—dorsal view. 1—cephalic gland; 2—dorsal cephalic ganglia; 3—dorsal brain commissure; 4, 5—internal and external dorsal unpaired nerves; 6—rhynchocoel; 7—lateral nerve stem; 8—external pore of nephridium; 9—lateral sense-organ; 10—annular nerves of trunk; 11—cerebral (sense) organ; 12—ventral cephalic ganglia; 13—rhynchodaeum; 14—rhynchostome (opening for proboscis); 15—pharyngeal nerve plexus; 16—ventral brain commissure (after Bürger).

nematodes, and therein differs sharply from the endonal brain of higher turbellarians, which lies on the main body axis. No modern nemertines have an aboral statocyst, and even if their ancestors had one, no trace of it remains in their brain structure.

Nemertines provide the classic example of gradual movement of the neural apparatus into the depth of the body (Fig. 39). In *Carinina* (Palaeonemertini) the neural apparatus of the trunk is represented by an epithelial plexus, in which the ring-like brain and the longitudinal stems are mere condensations. In other Palaeonemertini the longitudinal stems are better individualised, and in *Tubulanus* they lie between the subepithelial layer of the mesenchyma and the annular musculature. In Heteronemertini also they lie outside the annular muscles, but as that group possesses a thick layer of outer longitudinal muscles the nerve stems lie to the inside of that layer. In *Cephalothrix* (Palaeonemertini) the nerve stems have passed through the annular musculature and sunk into the layer of inner longitudinal muscles. Finally, in Metanemertini the nerve stems, having passed through the whole thickness of the musculature, lie deep within the internal mesenchyma of the body (O. Bürger, 1897–1907).

The brain of nemertines has a much more complex internal structure than the brain of other Scolecida, that circumstance being linked with the great development of their olfactory organs. It is interesting to note that their brain is the first to exhibit accumulations of associative cells resembling glomeruli, the higher associative centres of Articulata.

Like many turbellarians, nemertines have a well-developed neural plexus in the pharynx (Fig. 40, *B*). The pharyngeal plexus is connected with the ventral ganglia of the brain by a pair of connectives. A pair of nerves runs from the ventral commissure of the brain to the proboscis, where a complex network is also developed.

Neuro-endocrine cells already occur in Scolecida, having been described in the brain of Polyclada (R. Turner, 1946) and also in the cells adjacent to the brain in the papillary nerve of *Ascaris* (a single pair of cells) (M. Gersch and H. Scheffel, 1958).

4. MOLLUSCS

In passing to the neural apparatus of trochophore animals we must remember, in the first place, that we take as its prototype the structure of a neural apparatus similar to that of the *Lopadorhynchus* larva (see Vol. 1, Fig. 72): an orthogon, whose longitudinal stems are arranged with radial symmetry around the primary body axis and enter the circumblastopore plexus at their oral ends; cerebral ganglia develop near the aboral pole from the apical plate, from four radial brain rudiments, and from several rudiments connected with various sense-organs in the upper hemisphere. All that remains of that larval neural apparatus in adult annelids and molluscs is the cerebral ganglia, one of the pairs of longitudinal stems of the orthogon, which forms the circumoesophageal connectives, and part of the circumblastopore plexus, which forms the longitudinal nerve stems of the trunk. Hence homologies are possible between parts of the brain of Scolecida and the brain of trochophore animals (although up to the present these have not been traced), but any homologies between the longitudinal stems of the two groups are quite impossible.

The molluscan neural apparatus is found in its most primitive form in Amphineura (Fig. 41). We have already discussed its general structural plan in Vol. 1, Chap. V, and compared it with the neural apparatus of turbellarians, trochophore larvae, and adult annelids. The brain of Loricata (Fig. 41, *A*) consists of a long semi-annular commissure without differentiated ganglia. The slight brain development of Loricata is no doubt

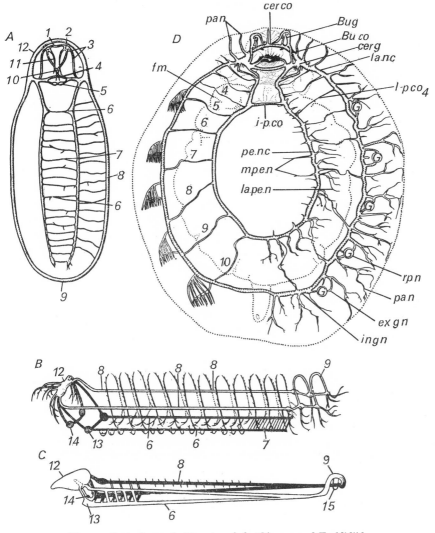

Fig. 41. *Central neural apparatus of Amphineura and Tryblidiida.*

A—Acanthochiton discrepans (Loricata) (after Pelseneer). *B—Proneomenia sluiteri* (Solenogastres). *C—Chaetoderma nitidulum* (Solenogastres). 1—buccal commissure; 2—upper buccal ganglion; 3—stomatogastric commissure; 4—labial commissure; 5—subradular ganglion; 6—pedal stems; 7—pedal commissures; 8—pleuro-visceral stems; 9—commissure of the pleuro-visceral stems, or common commissure of the pedal and pleuro-visceral stems, passing on the dorsal side of the hind-gut; 10—stomatogastric ganglion—subcerebral commissure; 11—nerves to the gullet; 12—cerebral cord or cerebral ganglia; 13—anterior pedal ganglia; 14—sublingual ganglia; 15—commissure passing on ventral side of hind-gut. *D—Neopilina galatheae* (Tryblidiida) (after H. Lemche and K. E. Wingstrand): *cer.co.*—cerebral commissure; *bu.g.*—buccal ganglion; *cer.g.*—cerebral ganglion; *la.n.c.*—pleuro-visceral stem; *l-pco₄*—fourth commissure between pleuro-visceral and pedal stems; *rp.n.*—one of the nerves of the renal orifices; *ex.g.n.* and *in.g.n.*—gill nerves; *la.pe.n.*—lateral pedal nerves; *m.pe.n.*—medial pedal nerves; *pe.n.c.*—pedal stems; *i-p.co.*—single interpedal commissure; *fm*—edge of foot; *pa.n.*—pallial nerves; 4–10—lateral pedal commissures from fourth to tenth; *G*—positions of removed left-side ctenidia.

due to the slight development of sense-organs in the head, and the latter to their sluggish mode of life and their vegetarian diet. Pedal and pleuro-visceral stems arise from the posterior ends of the cerebral commissure and are connected by a large number of transverse commissures. The first of the pedal commissures passes behind the mouth and is called the subcerebral commissure. From it arises the visceral section of the neural apparatus, which consists of two arches each bearing a pair of ganglia: the anterior arch bears the buccal ganglia, which innervate the dorsal side of the oral cavity and, perhaps, the gut; the posterior arch bears the subradular ganglia, which innervate the ventral side of the oral cavity. The buccal neural apparatus of molluscs is not homologous with the pharyngeal section of the neural apparatus of turbellarians, since the mouth and pharynx of turbellarians are not homologous with those of molluscs and annelids (see Chap. V). The pleuro-visceral stems pass into each other above the anal orifice. In Loricata the pedal stems lie within the musculature of the foot, which they innervate, and the visceral stems lie somewhat on the dorsal side of the pallial groove. Even in the most primitive molluscs, therefore, the central section of the neural apparatus has already moved a considerable distance from the skin.

The neural apparatus of *Neopilina galatheae* (Fig. 41) approaches that of Loricata in organisational level, but is less primitive. The number of pedal commissures in *Neopilina* is reduced to two (one anterior and one posterior), and the number of commissures between the pedal and pleuro-visceral stems to ten pairs. There are well-marked cerebral ganglia. In contrast to Loricata, however, the pleuro-visceral stems of *Neopilina* unite in front of the anal orifice, on the ventral side of the hind-gut; in this respect *Neopilina* resembles all other Conchifera.

Besides the central neural apparatus sunk under the skin, all molluscs have a peripheral neural plexus (Fig. 42) consisting of all types of nerve cells and capable of independent reflexes. There are also deeper peripheral plexuses, e.g. in the adductors and other muscles of Lamellibranchia (J. Lowy and J. Bowden, 1955). Thus in molluscs not all the neural elements, but only some of them, have concentrated and moved inwards. It is that part that formed the central section of the neural apparatus, with the peripheral nerves arising from it. The rest have remained beneath the skin in the form of a diffuse network, very like the original network in coelenterates. We have seen a similar process in a more primitive form in some flatworms (Fig. 39). The internal organs of molluscs also contain a complex neural plexus; it is connected with the central section of the neural apparatus through the buccal and visceral ganglia.

The discovery of sensory nerve cells in the gut epithelium of *Anodonta* (F. Gilev, 1952) is very important, as it enables us to surmise that the endoderm of molluscs, like the central phagocytoblast of coelenterates, forms neural elements.

With regard to the degree of centralisation of the neural apparatus we may compare Loricata to such turbellarians as *Mesostoma* (Rhabdocoela), which also have a brain and four longitudinal nerve stems (although they are not homologous with those of Loricata). The levels of development of the neural apparatus are generally fairly close in Loricata and in higher turbellarians, in spite of the radical differences in the structural plans of the two groups. The organisational levels of two apparatuses can therefore be compared even when the architectonical structural plans of the two apparatuses in the groups compared are different, and even when the parts composing these apparatuses are not homologous. Comparison of the fine structure of the central neural apparatuses of annelids, arthropods,

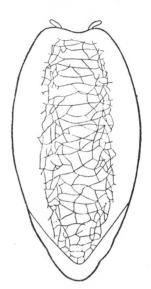

Fig. 42.
Diffuse peripheral neural plexus in the sole of the foot of *Helix pomatia* (Gastropoda Pulmonata) (from Hanström).

molluscs, and chordates has been made on a large scale by A. A. Zavarzin (Zawarzin, 1924, 1941), and it led him to establish 'the principle of parallelism of structure as one of the basic principles of morphology'; we must point out, however, that parallelism, like divergence, occurs in the evolution of all features of organisms, and not only in the evolution of their structure.

The neural apparatus of Solenogastres (Fig. 41, *B*, *C*) is very close to that of Loricata but presents some interesting differences, some of them already described in our discussion of the architectonics of molluscs. Here we shall merely remark that the nerve cells are not arranged regularly along the stems but are concentrated at the points of origin of the commissures, where ganglia are formed, and the intervening parts of the stems have the nature of simple conducting paths, i.e. they are converted into connectives. In Solenogastres, however, this process of ganglionisation has not been completed, and a certain number of nerve cells still remain in the parts of the stems lying between the ganglia.

In most higher molluscs all the cells of the central section of the neural apparatus accumulate in compact, sharply-outlined ganglia, within which a plexus of processes (*neuropile*) is formed and contact between separate neurons takes place. On the other hand, the intervening parts of the stems between two ganglia are totally devoid of nerve cells. They consist of parallel fibres, which begin to branch as soon as they enter a ganglion and there make contact with the processes of other neurons. The ganglia are, as it were, telephone exchanges in which switching is possible, and the intervening parts of the stems are simple multiwire conductors.

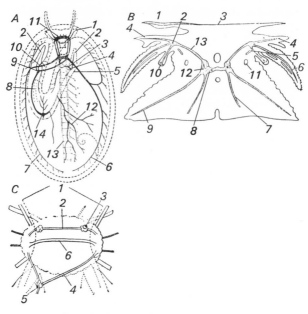

Fig. 43. *Neural apparatus of Gastropoda.*

A—Haliotis (Rhipidoglossa), dorsal view, body contours shown by broken lines: 1—eye; 2—ctenidia; 3—right asymmetrical pallial nerve; 4—right parietal ganglion; 5—margin of mantle cavity; 6—right symmetrical pallial nerve; 7—left symmetrical pallial nerve; 8—rectum; 9—left parietal ganglion; 10—left asymmetrical pallial nerve; 11—cerebral cord; 12—gonadic nerve; 13—pedal stems; 14—visceral ganglion. *B—Gibbula cineraria* (Rhipidoglossa, Trochidae), cross-section through rear part of foot: 1—cells secreting operculum; 2—epipodial ganglion; 3—location of operculum; 4—epipodium; 5—nerve of lateral organ; 6—epipodial tentacle; 7—plantar nerves; 8—pedal commissures; 9—lateral pedal nerve; 10—lateral sense-organ; 11—pedal artery; 12—pedal stems; 13—epipodial nerve. *C—Tethys leporina* (Opisthobranchia Nudibranchia), central neural apparatus, ventral view; outline of gut (oesophagus) shown by broken line; buccal (2) and pedal (6) commissures are seen, also visceral connectives (4), spanning ventral side of gut (1); 3—buccal ganglion; 5—abdominal ganglion (from Hanström).

Ganglionisation is a further stage in centralisation of the neural apparatus. Here we again see shortening of the intra-centre paths, if not as a whole, at least within separate parts of the neural apparatus.

The neural apparatus of the most primitive Prosobranchia shows a certain resemblance to that of Loricata. For instance, *Haliotis* (Aspidobranchia Rhipidoglossa) (Fig. 43, *A*) still has a broad and little-differentiated cerebral cord from which nerves run to the eyes and the cephalic tentacles and fibres run to the statocyst. In all molluscs these organs are innervated from the cerebral section of the neural apparatus, although the fibres to the statocyst usually form part of the cerebro-pedal connectives. Connectives also run from the cerebral cord to the buccal section of the neural apparatus. Two pedal stems run along the foot of *Haliotis*, joined by metameric commissures, the first of which is the subcerebral commissure. At the points where the subcerebral commissure meets the pedal stems pleuro-parietal connectives arise, crossing over each other (chiastoneury; see Vol. 1, Chap. VIII) and running to the parietal ganglia. The latter lie at the bases of the ctenidia, which they innervate. Pleuro-visceral connectives run from them to the visceral (abdominal) ganglion. The arch formed by these three ganglia evidently corresponds to the pleuro-visceral arch in Amphineura. Two pairs of connectives run from the brain to the points where the arch meets the pedal stems: they are obviously cerebro-pedal and cerebro-pleural connectives. Sometimes objections are raised to the homology of the pleuro-visceral arch of higher molluscs with that of Amphineura on the ground that at its posterior end that arch passes beneath the hind-gut in molluscs and above it in Amphineura. We may point out, however, that in *Chaetoderma* (Solenogastres) (Fig. 41, *C*) there is not only a commissure above the gut but also one beneath it, which in fact represents the last of the usual transverse commissures between the stems. The difference between the two mollusc groups in the location of the commissures is therefore most simply explained as reduction of the supra-intestinal commissure in *Neopilina*, Gastropoda, and other higher classes, and retention of the subintestinal commissure, i.e. of the last of the pre-anal commissures. Only one pair of the transverse commissures that connect the pedal to the pleuro-visceral stems in Amphineura remains in *Haliotis*, namely, at the level of the subcerebral commissure. That pleuro-pedal commissure is much shortened in *Haliotis*, so that it seems as if the pedal and pleuro-visceral stems were fused together; but in many cases, e.g. in *Viviparus* (Architaenioglossa), it is easily seen.

The neural apparatus of *Haliotis* therefore differs from that of Loricata in its chiastoneury (a secondary feature), in its abundance of cerebral sense-organs (the absence of which in Loricata may be secondary), and in its complete ganglionisation of the pleuro-visceral arch, while the pedal stems remain at almost the same stage of development as in Loricata. In the pleuro-visceral arch region three ganglia, two parietal and one visceral, have developed. In most Gastropoda additional pleural ganglia form between the parietal and the cerebral ganglia. In *Haliotis* they are fused with the beginning of the pedal stems, and in other forms they are connected with them by a pair of pleuro-pedal commissures. The latter correspond to one of the numerous commissures that connect the pedal stems of Amphineura with the pleuro-visceral stems.

A marked difference is seen in the structure of the peripheral nerves arising from the pleuro-visceral stems in Amphineura and in Gastropoda.

Whereas in the former we see a large number of small metamerically-arranged nerves, *Haliotis* has only a few large stems showing almost no traces of metamerism. We may regard as serially-homologous only the two pairs of pallial nerves: the so-called symmetrical pallial nerves arising from the uncrossed pleural ganglia, and the so-called asymmetrical pallial nerves arising from the crossed parietal ganglia. A curious result of the chiastoneury is the formation of commissures between the symmetrical and asymmetrical pallial nerves on each side (*dialyneury*) (Fig. 44). The displacement of these commissures to the points of origin of the two pallial nerves leads to a condition called *zygoneury*. In this case the commissures directly connect the pleural and parietal nerves lying on the same side, i.e. those actually belonging to different sides of the body. Zygoneury is usually unilateral. An extreme case of zygoneury has been described in *Lamellaria* (Taenioglossa).

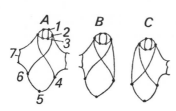

Fig. 44. *Diagram of dialyneural and zygoneural types of structure of the central neural apparatus in Prosobranchia.*

A—dialyneural type. *B*—right-handed zygoneury. *C*—left-handed zygoneury. 1—cerebral ganglion; 2—pleural ganglion; 3—pedal ganglion; 4—subintestinal ganglion; 5—visceral ganglion; 6—supra-intestinal ganglion; 7—pallial anastomosis between symmetrical and asymmetrical pallial nerves (from Lang).

In contrast to the above, metamerism of the nerves arising from the pedal stems is actually better developed in many Rhipidoglossa than in Loricata. The corresponding sections of the body follow the neural apparatus in this respect: the foot generally shows metameric structure more clearly in Rhipidoglossa than in Loricata, whereas the metamerism of the trunk has completely disappeared because of the formation of the visceral hump. The epipodium with its tentacles and other sense-organs is innervated by a whole row of small ganglia, which are metamerically arranged along the line of its attachment and are connected with the pedal stems by the epipodial nerves, which also are metameric (Fig. 43, *B*).

The progressive changes observed in the higher Prosobranchia consist mainly in further ganglionisation of the neural apparatus. The brain of *Haliotis* and of most Aspidobranchia, like that of Loricata, is only slightly developed and is, as it were, a transverse portion of the stem with nerve cells located throughout its length; the slight development of the brain is due to the limited mobility of the lower Aspidobranchia and the slight development of their cephalic sense-organs. The brain of most other snails is differentiated into two cerebral ganglia and a commissure between them (cerebral commissure). Long pedal stems are possessed only by Aspidobranchia and by separate members of Architaenioglossa (*Viviparus*); the structure of the pedal ganglia of *Cypraea* (Taenioglossa) reminds one of the pedal stems of lower snails only in external form (B. Hubendick, 1947).

In all other snails the pedal stems are shortened, concentrated, and converted into a pair of pedal ganglia, connected with each other by a single pedal commissure, with the cerebral ganglia by connectives, and with the pleural ganglia by commissures. The formation of pedal ganglia completes the ganglionisation of the neural apparatus of Gastropoda and is a substantial step forward in the process of its centralisation. The epipodium, an ancient legacy in Rhipidoglossa, no longer occurs in higher forms.

The most primitive of Opisthobranchia Tectibranchia, *Actaeon* among Bullomorpha and *Acera* among Aplysiomorpha (J. Guiart, 1901), are still at the Prosobranchia level in the organisation of their neural apparatus (long connectives, chiastoneury). Among Pulmonata some traces of chiastoneury remain in *Chilina* and some other primitive Basommatophora. In both subclasses (Opisthobranchia and Pulmonata) the processes of untwisting and shortening of the connectives are taking place, leading to liquidation of chiastoneury. But the shortening of connectives observed in these groups, which leads to concentration of all the ganglia around the gullet, has another, more general meaning. It represents a further important step in centralisation of the neural apparatus. Such forms as *Tethys leporina* (Nudibranchia), in which all ganglia except the small buccal and visceral ones are fused into a single mass, attain one of the highest levels of centralisation of the neural apparatus in the whole of the animal kingdom (Fig. 43, *C*). We must remember, however, that Opisthobranchia, like all molluscs, also retain a diffuse subcutaneous neural plexus.

It was on one of the Opisthobranchia, *Aplysia* (Tectibranchia), that S. Baglioni (1910) made experiments that clearly illustrated the difference between the work of a centralised and that of an uncentralised neural apparatus. If one cuts across the long bundle of nerves that stretches from the circumoesophageal ganglionic mass along the whole body, thus disconnecting the nerve centres, and then irritates any part of the body, the reaction obtained is very similar to that observed in *Hydra:* a slight prick causes contraction of the nearest muscles; the stronger the irritation, the wider spreads the wave of muscular contraction. One may make the opposite experiment and cut the animal in halves, leaving only the nerve bundle entire. In that case irritation of one end of the animal produces a reaction in both halves. In an uninjured mollusc the neural centres inhibit and co-ordinate reflexes arising in the peripheral network: a decerebrated *Aplysia* is in a state of constant unrest and aimless motion, swimming and crawling, whereas possession of the centres inhibits these movements and directs them in accordance with the biological significance of the irritants that affect the sense-organs.

Apparently the central neural apparatus of all Gastropoda and Scaphopoda includes neuro-endocrine cells whose secretions affect the animals' annual sexual cycles. In Rhipidoglossa these are scattered in the trunk ganglia and the pedal stems as well as in the brain, whereas in higher snails they are more or less concentrated in the cerebral ganglia. In

Dentalium they are located in the cerebral, buccal, and pleural ganglia (B. Scharrer, 1936, 1937; M. Gabe, 1951, 1953e, 1953f, 1954c).

The neural apparatus of Lamellibranchia (Fig. 45) is constructed on the same plan as that of Gastropoda. The chief difference consists, of course, in the absence of chiastoneury and also in simplification due to considerable reduction in cephalic sense-organs and simplification of the

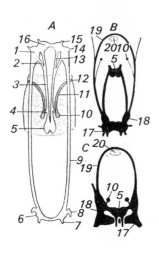

Fig. 45. *Central section of neural apparatus of Lamellibranchia.*

A—*Nucula nucleus* (Protobranchia), one of the most primitive types of the central section of the neural apparatus in Bivalvia (after Pelseneer). B—*Spondylus* (Filibranchia), beginning of concentration of neural ganglia. C—*Lima squamosa* (Filibranchia), extreme degree of concentration of neural apparatus in Bivalvia (from C. M. Yonge). 1—pleural ganglion; 2—pleuro-pedal commissure; 3—parts of pleuro-pedal commissure, united with the cerebro-pedal connectives; 4—nerves to statocysts; 5—pedal ganglia; 6—viscero-branchial ganglion; 7—posterior pallial nerve; 8—osphradium; 9—pleuro-visceral connective; 10—statocyst; 11—its canal; 12—external opening of statocyst canal; 13—cerebro-pedal connective; 14—anterior pallial nerve; 15—nerve of oral lobes; 16—cerebral ganglion; 17—visceral ganglia; 18—cerebro-pleural ganglia; 19—cerebral commissure; 20—mouth (shown by dotted line).

oral apparatus. In the degree of centralisation of their neural apparatus Lamellibranchia stand at approximately the same level as the higher Prosobranchia: it is completely ganglionised but not concentrated, and the connectives are long. Concentration of ganglia is observed even there, namely, in some monomyarian families. It reaches its highest level in *Lima* (P. Pelseneer, 1911). These, however, differ from Gastropoda in that all the ganglia are attracted, not to the cerebral ganglia (which have lost their significance as the principal centre), but to the visceral ganglia innervating the mantle, whose edges form the chief sensory zone in Lamellibranchia. There also, therefore, all ganglia are concentrated around the physiologically-dominant centres (C. Yonge, 1954).

In the structure of their neural apparatus Scaphopoda approach the most primitive of Lamellibranchia, such as *Nucula* (Protobranchia), but their buccal section is better developed.

We have seen that in Gastropoda centralisation of the neural apparatus consists of two processes, ganglionisation of the stems and shortening of the commissures and connectives. These two processes are essentially independent of each other, as may be seen by comparing Gastropoda and Cephalopoda. In Gastropoda ganglionisation of the stems takes place first, followed by the shortening. In Cephalopoda the processes take place in the reverse order.

We do not find the lowest level in development of the central section of the neural apparatus—long undifferentiated stems—in any modern

Cephalopoda. Shortened nerve stems, however, not differentiated into ganglia and connectives, are found in *Nautilus*.

The central section of the neural apparatus of *Nautilus* (Fig. 46, *A*) consists of three arches formed by broad cords that are completely covered with nerve cells. All the cords are shortened and located in the head. One of them, the cerebral cord, spans the gullet on the dorsal side, and the other two span the gullet on the ventral side. The anterior of these corresponds

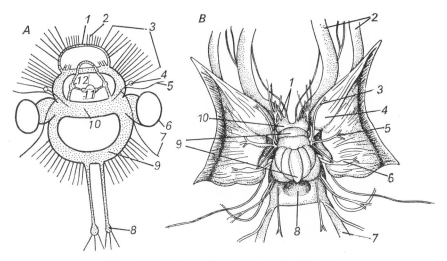

Fig. 46. *Central neural apparatus of Cephalopoda.*

A—Nautilus (Tetrabranchiata), view from above, diagrammatic, tentacular nerves partly shown: 1—prepedal nerve ring; 2—nerves of the lamellar organ; 3—nerves to tentacles; 4—statocyst; 5 and 7—nerves to anterior and posterior eye-tentacles; 6—eye; 8—abdominal ganglia; 9—pleuro-visceral cord; 10—cerebral cord; 11—buccal ganglia; 12—labial ganglia. *B—Eledone moschata* (Dibranchiata Octopoda), dorsal view: 1—labial nerves; 2—brachial nerves; 3—preoptic nerve; 4—optic ganglia; 5—medial optic nerve; 6—upper optic ganglia; 7—pallial nerves; 8—visceral ganglia; 9—cerebral ganglia, divided into several lobes; 10—upper buccal ganglia (from Hanström).

to the pedal stem of Loricata, and the posterior to the pleuro-visceral stem. In addition a buccal commissure arises from the cerebral cord and a prepedal ring from the pedal cords. The prepedal ring innervates some of the pedal tentacles. The pedal cords directly innervate the other tentacles and the infundibulum, which also is a product of the foot and is, according to some authors, the homologue of the epipodium of Rhipidoglossa. The pleuro-visceral cord sends out (besides numerous nerves to the mantle and other organs) two connectives to the abdominal ganglia, innervating the gills and the internal organs.

The forward movement of the pedal ring observed in all Cephalopoda is due to the forward movement of a great part of the foot itself, which is characteristic of that group.

The central section of the neural apparatus of all Dibranchiata (Fig. 46, *B*), unlike that of *Nautilus*, consists of ganglia, not of undifferentiated cords. The connectives and commissures are much shortened, so that the principal ganglia (cerebral, pedal, and visceral) are fused around the gullet into a single mass, externally divided only by slight constrictions.

At the same time additional ganglia, not found in other molluscs, separate out from the peripheral plexus in Cephalopoda. The largest of these are: brachial ganglia, innervating the arms and connected with the pedal ganglia and also directly with the cerebral ganglia; labial or upper buccal ganglia, connected with the cerebral ganglia and also, by commissures, with the brachial ganglia; and ganglia stellata, innervating the mantle and connected by pallial nerves with the visceral ganglia.

In Decapoda the brachial and labial ganglia are connected with the brain by long connectives, which are especially long in Oegopsida. In higher forms the connectives gradually shorten, and in Octopoda all these ganglia are completely fused with the corresponding sections of the brain (Fig. 47).

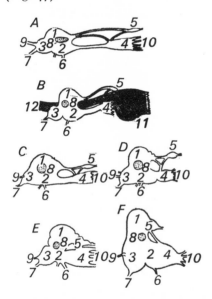

Fig. 47. *Diagrams of central neural apparatuses of different Cephalopoda Dibranchiata, lateral view.*

Decapoda: *A—Ommastrephes; B—Sepiola; C—Loligo; D—Sepia.* Octopoda: *E—Octopus; F—Argonauta.* 1—cerebral ganglion; 2—pedal ganglion; 3—visceral ganglion; 4—brachial ganglion; 5—labial or upper buccal ganglion; 6—infundibular nerve; 7— visceral nerve; 8—optic nerve, cut across; 9—pallial nerve; 10—brachial nerves; 11—pharynx; 12—gullet (after Pelseneer).

Besides the above, a number of subsidiary centres separate out from the diffuse neural plexus in Cephalopoda. The so-called brachial nerves, which run along the arms from the brachial ganglia, are true nerve cords with a coating of ganglionic cells, with a neuropile, and with large longitudinal conducting nerve-bundles.

In addition small ganglia are scattered through the musculature of the arms and at the bases of the suckers. For that reason severed arms retain the power of making complex and specific reactions to external stimuli.

The processes of ganglionisation of the peripheral plexus and of bringing it into the central neural apparatus have gone so far in Cephalopoda that diffuse parts of the plexus, if any remain, no longer play an important role either in the animals' locomotion or in the activity of the cutaneous chromatophores (C. Prosser, 1950). The diffuse neural plexus is retained in the internal organs (the walls of the gut, etc.), and according to some data (Mikhailov, 1921) also in that part of it linked with the musculature of the head.

The neural apparatus of Cephalopoda contains giant motor fibres (J. Young, 1936). They are distinguished by large diameter (700 μ in *Loligo*) and great rapidity of conduction, almost the same as in the myelinised fibres of vertebrates. Giant fibres are found in many animals and usually serve for rapid reactions of flight or concealment in a tube, but in Cephalopoda they assist in the normal function of reactive swimming (see Chap. VI). The latter, however, probably arose phylogenetically from a reaction originally used in flight (J. Ramsay, 1952).

All Cephalopoda are active predators: the sense-organs and the locomotor apparatus of Dibranchiata reach a high level of perfection. Consequently the brain of Cephalopoda is characterised by extreme development of the optic centres, by a very large total brain volume, by complex internal differentiation, and by highly-perfected microarchitectonics. The cephalopod brain is protected by a cartilaginous skull, underneath in *Nautilus* and almost on all sides in Dibranchiata. The brain of Dibranchiata, and particularly of Octopoda, represents the peak of development of the central neural apparatus in molluscs, and as neural equipment it is among the most perfect possessed by any invertebrates.

5. ANNELIDS

The neural apparatus of annelids, like that of molluscs, consists of a brain (originally representing an accumulation of nerve centres in the upper hemisphere of the trochophore) and circumblastopore nerve stems connected with the brain by circumoesophageal connectives. The circumblastopore system of annelids consists of a pair of ventral stems homologous, it is generally thought, with the pedal stems of Loricata. Some annelids also have paired podial ganglia lying at the bases of the parapodia and connected with the ventral ganglia of their segment. Annelids also have a pharyngeal section of the neural apparatus. They do not have the diffuse neural plexus found in all molluscs: that is one of the chief differences between the neural apparatuses of the two groups. On the periphery, outside the central parts of the neural apparatus, there remain a few sensory cells, from whose long central processes sensory nerves are constructed (Fig. 76). As in molluscs, the sensory cells of annelids are formed partly from the endoderm of the mid-gut and not only from the ectoderm, as has been demonstrated by G. A. Nevmyvaka (1947) in the earthworm *Allolobophora* (Fig. 48).

Unlike molluscs, a number of annelids retain the original epithelial location of their neural apparatus. These include small forms: *Polygordius*, *Protodrilus* (Eunicemorpha), several Spiomorpha, Nereimorpha, a number of other polychaetes, and *Aeolosoma* among oligochaetes. In the most primitive polychaete families (Phyllodocidae, Aphroditidae, and Glyceridae among Phyllodocemorpha, Nephthydidae among Nereimorpha) there are even large forms with subepithelial location of the ventral chain. Other polychaetes show all the successive stages in movement of the ventral stems deep into the body: in *Nereis virens* they lie within the annular muscles, in *Eunice punctata* between the annular and the longitudinal

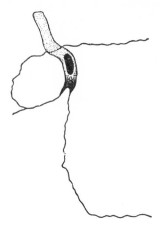

Fig. 48. *Sensory nerve cell from the epithelium of the mid-gut of the earthworm* Allolobophora calliginosa (after G. A. Nevmyvaka).

muscles, and in many Serpulidae within the longitudinal muscles. In most polychaetes they have left the body walls and lie in the coelom. Even in such forms, however, the brain usually remains in contact with the epidermis of the dorsal side of the head.

In some primitive forms the ventral stems are entirely covered with ganglionic cells and are undifferentiated cords, as in Loricata (e.g. in *Polygordius, Protodrilus,* and *Saccocirrus* among Eunicemorpha); in *Aeolosoma* the individualisation of ganglia is scarcely noticeable (E. Brace, 1901). The nerve stems are slightly ganglionised in many Phyllodocidae and Aphroditidae; but in most annelids the ventral stems are fully ganglionised, a single pair of ganglia being formed in each body segment in a typical case. In some Hermellidae (Spiomorpha) (Fig. 49, *B*) and Serpulidae, however, there are two pairs of ganglia in each segment, and in *Pectinaria* (Terebellomorpha) some segments contain as many as three pairs each; in higher Articulata this type of structure of the ventral stems is not widespread, although in Phyllopoda, for instance, the two ganglia in each neurosomite in their ventral nerve chain are joined together by two transverse commissures, not one (Fig. 53, *A*; Fig. 143).[1]

[1] Moreover, each of the posterior thoracic segments of Notostraca bears several pairs of legs and the same number of nerve ganglia; but, as we have seen above, the polypody of Notostraca is probably due to incomplete division of segments, and consequently the presence of two or several ganglia in each segment in that group is secondary, and has no direct relation to the question of the number of ganglia per segment during incipient ganglionisation of the ventral nerve cords in Articulata.

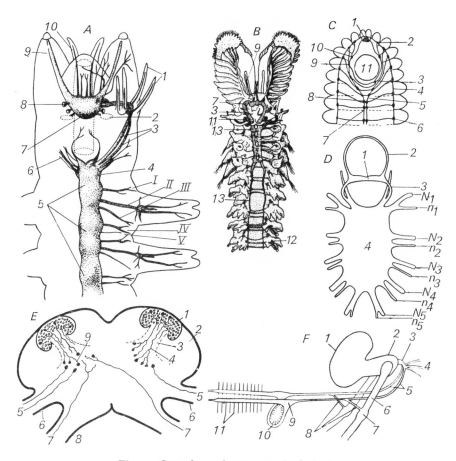

Fig. 49. *Central neural apparatus of polychaetes.*

A—Nereis virens (Polychaeta), front end of body (from Dogiel). *B—Sabellaria alveolata* (Polychaeta Hermellidae), front end of body (after K. E. Johansson): 1—peristomial cirri; 2—olfactory organs; 3—circumoesophageal connectives; 4—ganglia of peristome; 5—ventral nerve chain; 6—peristomial nerve; 7—brain; 8—eye; 9—palps; 10—antennae; 11—excretory nephromixia in thorax; 12—genital nephromixia; 13—ventral nerve ladder, with connectives close together in thorax; I-V—five pairs of nerves in each neurosomite, of which II bears parapodial ganglion. *C—Hermodice* (Polychaeta Amphinomidae), diagram of neural apparatus of front part of body: 1—brain; 2—parapodial ganglia of first trunk segment; 3—commissures of pairs I and II of ventral ganglia, partly encircling the pharynx; 4—lateral nerve; 5 and 7—connectives of ventral stems; 6—parapodial ganglia of eighth segment; ventral ganglia: 8—of sixth segment, 9—of second segment, 10—of first segment; 11—pharynx (after Gustafson). *D*—central section of neural apparatus of female *Pulvinomyzostomum pulvinar* (Myzostomida), an example of concentration of a ventral brain in annelids: 1—brain; 2—pharyngeal ring (stomatogastric section); 3—circumoesophageal connectives; 4—ventral brain; N_1-N_5—main metameric nerves; n_1-n_5—supplementary metameric nerves (after Jägersten). *E*—diagrammatic frontal section through brain of *Stenelais picta* (Polychaeta Aphroditidae): 1—globulus of corpus pedunculatum; 2—suprapharyngeal ganglion; 3—stem of corpus pedunculatum; 4—neurites of cells of globulus; 5—sensory fibres from the palpal nerves; 6—palps; 7—fibre from the circumoesophageal connectives; 8—circumoesophageal connectives; 9—glomerulus of palpal nerve (after Hanström). *F*—diagram of anterior part of central neural apparatus of polychaetes of the family Amphinomidae: 1—metencephalon; 2—prosencephalon; 3—stomatogastric ganglion; 4—labial nerves; 5—cerebral roots of stomatogastric section; 6—circumoesophageal connectives; 7—first ganglion of ventral chain; 8—ventral-brain roots of stomatogastric section; 9—pharyngeal ganglia; 10—their commissure; 11—pharyngeal nerves (after Gustafson).

Each pair of ventral-cord ganglia innervates its own segment; but it is known that in earthworms (Lumbricidae) each pair of ventral-cord ganglia also innervates, although only slightly, the two adjoining segments (C. Prosser, 1950).

In a few primitive annelids, e.g. *Dinophilus* and *Aeolosoma*, the ventral stems are wide apart and are joined by long commissures, so that a 'ladder nerve system' is formed. In the majority the stems approach each other and the commissures are shortened, leading to some fusion of the ganglia and connectives of the right and left sides, so that the ladder becomes a chain. In typical annelids the internal structure of the chain always reveals its paired origin. Fusion of the paired ventral stems into a chain is one of the processes of centralisation of the neural apparatus. Further centralisation, which would be reflected in shortening of the connectives and in approach and fusion of successive pairs of ganglia, is almost absent both in typical polychaetes and in oligochaetes, except for fusion of some of the anterior ganglia into a suboesophageal ganglion (*Nereis*, Fig. 49, *A*; *Hermione* among Phyllodocemorpha, *Myxicola* among Serpulimorpha, etc.). In leeches, besides fusion of a large part of the first four ganglia into a suboesophageal mass, the last seven (in *Acanthobdella*, four) ganglia fuse into the ganglionic mass of the posterior sucker.

Thus within the whole subphylum of Annelides centralisation of the neural apparatus advances little beyond the level of the most primitive members of that group. Only in the higher members of the aberrant, oligomerous order Myzostomida is the whole ventral chain fused into a single continuous neural mass.

Three pairs of peripheral nerves arise, as a rule, from each ganglion in the ventral chain of polychaetes; but when the length of the separate segments increases considerably, as, for instance, in the family Maldanidae (Drilomorpha), the number of peripheral nerves in each neurosomite increases (N. A. Livanov, 1940). In a typical case the front and rear pairs of nerves are mainly motor and the central pair mainly sensory. The two nerves in each motor pair are joined together on the dorsal side, so forming annular nerves. The errant polychaetes, as stated above, have a small podial ganglion at the base of each parapodium, connected with the nerve of the central pair and through it with the ventral chain. In the family Amphinomidae (Nereimorpha) all the podial ganglia on each side are connected directly with one another by longitudinal connectives called lateral nerves (Fig. 49, *C*); the podial ganglia of the first trunk segment are connected with the brain by extensions of the lateral nerves, which meet the brain at the points of entry into it of the dorsal roots of the circumpharyngeal connectives. Something similar is described by M. Prenant (1927) in Sphaerodoridae, which are distantly related to Amphinomidae.

Apart from Amphinomidae and Sphaerodoridae, no other errant polychaete is known to have longitudinal connectives between all the podial ganglia; for that reason O. Storch (1912) gave the name of Tetraneura to Amphinomidae and that of Dineura [1] to all other polychaetes. Some errant polychaetes, however, have direct connections between the podial ganglia, but only along some of the first segments and not in the rest; in *Nereis*, for instance, the podial ganglia of the first four segments, which bear tentacular cirri, are connected with one another; G. Gustafson (1930) considers that connection to be the result of cephalisation. Most sessile polychaetes have no podial ganglia, or only vestiges of them.

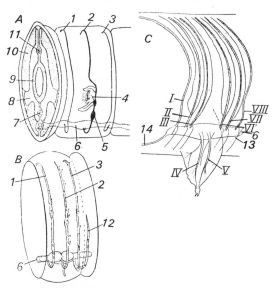

Fig. 50. *Diagrams of structure of neurosomites.*

A—Notocirrus (Polychaeta, Eunicidae) (from Livanov). *B—Branchiobdella parasitica* (Oligochaeta) (from Livanov). *C—Peripatus tholloni* (Onychophora) (after Fedorov), motor nerves shown by stippling, sensory by single lines. 1—anterior annular nerve of somite; 2—medial sensory nerve of somite; 3—posterior annular nerve of somite; 4—parapodia; 5—parapodial ganglion; 6—ventral nerve stems; 7—ventral vessel; 8—coelom; 9—gut; 10—longitudinal musculature; 11—dorsal vessel; 12—fourth nerve of somite; 13—ventral nerve of somite; 14—one of the commissures between the ventral stems; I, II, III—anterior group of nerves of somite in *Peripatus*; IV and V—pedal nerves; VI, VII, VIII—posterior group of nerves of somite.

Oligochaetes also have fine lateral nerves running along the sides of the body and (in contrast to Amphinomidae) connecting all three (or four) peripheral nerves in each segment; at the points of junction there are small ganglia. Connection of the lateral nerves with the brain has not been proved in oligochaetes (V. V. Izosimov). Among leeches only *Acanthobdella* has a lateral nerve (N. Livanov, 1905). In its most general form it may be regarded with some probability as the homologue of the lateral nerve of Amphinomidae (Fig. 50).

[1] That separation is not well founded, since Storch greatly exaggerated the taxonomic significance of the presence or absence of lateral stems.

Storch (1912), who has described the lateral stems of Amphinomidae in detail, regards them as being homologous with the pleuro-visceral stems of Amphineura; but if one regards the epipodium of Gastropoda as a partial homologue of polychaete parapodia (see Vol. 1, Chap. VI, and Vol. 2, Chap. VI), the podial ganglia of polychaetes are homologues of the epipodial ganglia of Gastropoda (Fig. 43, B), and the lateral nerve joining the podial ganglia of Amphinomidae can in no way be homologous with pleuro-visceral stems. Its homologue in molluscs is the plexus in the form of a longitudinal cord that connects the epipodial ganglia in some Rhipidoglossa. From this point of view, therefore, annelids possess no homologues of pleuro-visceral stems.

The annelid brain, which in primitive forms develops from a large number of independent rudiments (in *Lopadorhynchus*, according to E. A. Meyer, 1898), in many other polychaetes develops from a single rudiment, and in oligochaetes even from a common rudiment with the ventral stems. The latter method of development is undoubtedly secondary. Regardless of the method of formation, the brain of the adult annelid is a monolithic organ, and only with great development of the brain itself and of the sense-organs innervated by it (in errant polychaetes) do three sections appear in it: the fore-brain (prosencephalon), which innervates the palps; the mid-brain (mesencephalon), which innervates the eyes and the tentacles; and the hind-brain (metencephalon), which innervates the olfactory organs. In view of the fact that the rudiments of the palps are formed at the sides of the apical plate of the trochophore, we may assume that the prosencephalon arose from rudiments lying around the aboral pole, and the metencephalon, like the olfactory organs served by it, in the dorsal (D) quadrant of the upper hemisphere; in this way the hind-brain fully justifies its name.

The more highly developed any sense-organs are, the more highly developed is the corresponding section of the brain; Amphinomidae, with their huge olfactory organs, have a large metencephalon with well-developed associative centres (glomeruli). In Alciopidae (Phyllodocemorpha), which have large and relatively perfect eyes, the mesencephalon and the optic centres are well developed. In Nereidae and Aphroditidae, which have well-developed palps, the prosencephalon is highly developed. Eunicidae have a pharynx of more complex structure than that of any other polychaetes, with a ventral blind sac and a very complex pharyngeal section of the neural apparatus (see K. Haffner, 1959), and that circumstance stimulates the development of the mesencephalon, which sends nerves to the pharyngeal section in polychaetes.

Sessile polychaetes, which have become soil-feeders or feeders on nanoplankton, have acquired many specialised features and have lost the improved sense-organs and the predatory pharynx of errant families; consequently their brain also is much simplified (Fig. 51). The brain of Sabellidae, Hermellidae, and Chlorhaemidae (K. Johansson, 1927) shows practically no division into sections; in Terebellomorpha the brain is reduced to a transverse stria above the pharynx (see C. Hessle, 1917).

The circumoral connectives of errant polychaetes arise from the ventral surface of the mesencephalon. As a rule they have two roots in the brain, dorsal and ventral; in many polychaetes the retractor muscle and the blood vessel of the palps pass between the two roots and the brain. The suggestion that the polychaete brain contains one of the ventral-chain ganglia, made on the basis of analogy with arthropods (F. Hempelmann, 1911) is not borne out by facts.

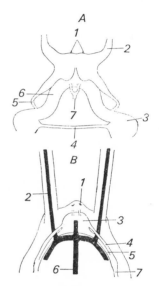

Fig. 51. *Brain of sessile polychaetes.*
A—diagram of structure of brain in *Serpula vermicularis* (dorsal view): 1—anterior medial nerves; 2—palpal nerves; 3—first pair of ganglia in ventral chain; 4—their commissure; 5—dorsal roots of circumoesophageal connectives; 6—ventral roots of circumoesophageal connectives; 7—nerves to dorsal vessel. *B*—diagram of inter-relationship of brain and head vessels in *Stylarioides* (family Chlorhaemidae): 1—eye; 2—palpal vessels; 3—brain; 4—dorsal roots of circumoesophageal connectives; 5—ventral roots of circumoesophageal connectives; 6—dorsal vessel; 7—circumoesophageal connectives (after K. E. Johansson).

A notable feature in the internal structure of the polychaete brain is the development of corpora pedunculata, which are the highest associative centres in the brain of Articulata (Fig. 49, *E*). They are reduced in sessile polychaetes, and absent in oligochaetes and leeches. As N. Holmgren (1916) and B. Hanström (1927, 1928) have shown, the corpora pedunculata of polychaetes have developed in connection with the nerves of the palps, providing their associative links with other parts of the brain. They are most highly developed in the compact, active, mobile Aphroditidae.

The motor centres of annelids that hide quickly in burrows contain 'giant' cells, from which arise 'giant' fibres that extend along the ventral chain. We have seen that other long-bodied invertebrates that make rapid movements also possess giant fibres (P. Kirtisinghe, 1952), e.g. burrow-hiding Enteropneusta and *Lingula*, jumping decapod crustaceans (W. Holmes, 1942) and grasshoppers. *Amphioxus* also has them. In the larva of the dragon-fly *Anax* giant fibres in the ventral chain take part in the reflex of ejection of water from the rectum (G. Hughes, 1953). In *Anax* and *Sepia* giant fibres are merely an extreme 'plus-variant' in a series of changes in the diameter of motor fibres.

Polychaetes have a well-developed pharyngeal neural apparatus (H. Stannius, 1831), whose ganglia are formed from the epithelium of the pharynx and which is connected with the brain, with the circumoral connectives, or with the nerves of the ventral chain, sometimes by several pairs of roots (Fig. 49, *F*) (G. Gustafson, 1930). It is probably homologous with the buccal neural apparatus of molluscs, although perhaps in a very general way, just as it might be with the stomatogastric apparatus of arthropods.

In some cases the nerve cells of annelids assume an endocrine function. The ventral-chain ganglia of many polychaetes, leeches, and earthworms contain motor cells that send processes to the blood-vessel muscles and simultaneously secrete adrenalin (J. Gaskell, 1919); in other words, these cells have a dual action on the circulatory apparatus of the animal, neural and humoral. In addition, by its presence the polychaete brain apparently delays until the proper time the epitokous transformation of the rear segments (regarding epitoky, see Vol. 1, Chap. VI, and Vol. 2, Chap. IX); removal of the central part of the prostomium of *Nereis* produces epitokous reconstruction even of immature specimens; by analogy with insects we may suggest that some brain cells secrete a special hormone that inhibits 'metamorphosis' (M. Durchon, 1948, 1949).

In recent years the existence of neuro-endocrine cells has been demonstrated in the brain and ventral chain of Nereidae and a number of sessile polychaetes (L. Arvy, 1954), of *Nephthys* (R. Clark, 1959), of oligochaetes (H. Herlant-Meewis, 1956; H. Hubl, 1953; M. Gersch, 1959), and of Sipunculoidea (M. Gabe, 1953d). Coincidence of their secretory cycles with sex cycles has been demonstrated in Nereidae. In *Nephthys*, Nereidae, and earthworms many of the axons of the neuro-endocrine cells of the brain end in a haemal sinus or in a plexus of blood vessels covering the supra-pharyngeal ganglion. This cerebro-vascular complex (G. Bobin and M. Durchon, 1952; R. Clark, 1959) is a typical neurohaemal organ (see section 1 of this chapter).

A very important but a very difficult task is the homologisation of the various sections of the neural apparatus of polychaetes with those of Scolecida and molluscs. In Vol. 1, Chap. V, we have seen that only trochophores can be directly compared with Scolecida. Speaking in the most general terms, the neural apparatus of higher turbellarians consists of a cutaneous plexus (orthogon) and a cerebral ganglion (endon), and the neural apparatus of the *Lopadorhynchus* trochophore consists of a neural plexus corresponding to the turbellarian orthogon and of a circumblasto-pore plexus that is absent in turbellarians. Part of the orthogon of the larva, lying in its upper hemisphere, gives rise to the circumpharyngeal connectives, and the ventral stems are formed from the circumblastopore plexus. Thus the polychaete brain contains the part of the neural apparatus of the larva that corresponds to the anterior part of the orthogon of turbellarians. But does the polychaete brain contain anything homologous with the endon of turbellarians? As we have seen, the latter arises in connection with the aboral sense-organ of lower turbellarians, the statocyst, and accordingly lies on the morphological axis of the body. In many polychaetes, such as *Polygordius*, the rudiments giving rise to the brain include elements originally connected with the apical plate, which is the aboral sense-organ of the trochophore. As is easily seen, these elements are to some extent homologues of the endon of turbellarians; but they play only a negligible role in the polychaete brain. Nevertheless the apical plate shows great constancy as one of the sources of development of the brain, and has been traced as such in molluscs and even in lower crustaceans (Vol. 1, Fig. 102, *A*) as well as in polychaetes.

The homology of the polychaete brain with the cerebral section of the neural apparatus of molluscs is beyond question. It is equally clear that the pedal and pleurovisceral stems of lower molluscs, as well as the ventral stems of annelids, are products of the circumblastopore plexus of their common prototype. At the same time we see in the *Lopadorhynchus* larva considerable complexity of that plexus, no fewer than three pairs of longitudinal stems, while there are two pairs in molluscs and only one pair in adult polychaetes. It is usually believed that the pedal stems of molluscs are homologous with the ventral stems of polychaetes, and this view is supported by the similar relation of both to the podial (corresponding to epipodial) ganglia. But whether the pleurovisceral stems of molluscs are homologous with any of the stems of the circumblastopore plexus of the *Lopadorhynchus* larva still remains an open question.

6. ONYCHOPHORA AND ARTHROPODA

Like annelids and unlike molluscs, Onychophora and Arthropoda are distinguished by complete reduction of the subcutaneous neural plexus. Apart from the central section of the neural apparatus and a few visceral ganglia connected with it, they retain no neural mechanism capable of independent reflex action. A good illustration of the resulting situation is provided by a natural physiological experiment, namely, the behaviour of paralysing insects (J. Fabre, 1879). When the solitary wasp *Cerceris* paralyses a weevil it pierces its ventral brain and injects poison into it; the weevil remains alive but becomes immobile. It would be impossible to paralyse a snail by the same method: after all ganglia were paralysed the snail would retain reflexes and mobility because of its possession of a diffuse plexus.

The central section of the neural apparatus of Onychophora and arthropods consists, like that of annelids, of a brain, circumoesophageal connectives, and ventral stems. As we have seen, however (Vol. 1, Chaps VI and VII), the brain of members of both these subphyla includes at least one pair of ventral ganglia, which form the tritocerebrum. The brain of higher Articulata is therefore not completely homologous with that of annelids, and the circumoesophageal connectives of higher Articulata, being connectives between the tritocerebrum and the mandibular segments, are absolutely not homologous with the circumoesophageal connectives of annelids.

The central section of the neural apparatus of higher Articulata always lies deep within the body and never occupies the subepithelial position that it does in some annelids. The ventral nerve 'ladder' of Notostraca,

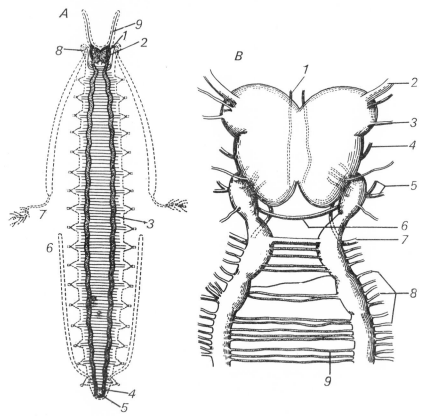

Fig. 52. *Neural apparatus of Onychophora.*

A—Peripatopsis capensis, diagram of structure of neural apparatus: 1—eyes; 2—brain; 3—ventral nerve stems; 4—genital orifice; 5—posterior duct; 6—crural glands of last pair of legs; 7—mucus glands; 8—oral papillae; 9—antennae (from Hanström). *B—Peripatus jamaicensis,* brain and beginning of ventral stems: 1—posterior brain nerves; 2—antennal nerves; 3—optic nerves; 4—labial nerves; 5—mandibular nerves; 6—commissure of mandibular segment; 7—nerves of oral papillae; 8—nerves of first pedipherous segment; 9—commissures between ventral stems (after Bouvier).

however, lies outside their well-developed longitudinal muscles, i.e. it has a more superficial location in them than it does in many polychaetes.

The ventral stems of Onychophora (Fig. 52, *A*) are extremely primitive in some respects: they are nerve cords without definite ganglia, well separated and connected by numerous transverse commissures, the number of which may reach ten in a single body segment. In addition, as stated above, the stems fuse with each other above the anal orifice like the pedal stems of *Chaetoderma* (Solenogastres).

In contrast, the ventral stems of all arthropods never unite above the anal orifice and are almost always fully ganglionised, one pair of ganglia usually

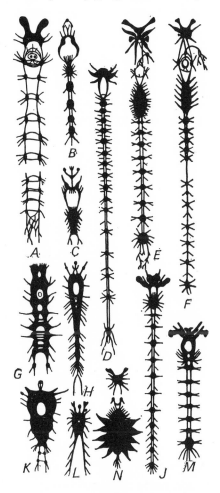

Fig. 53. *Types of structure of central neural apparatus in Crustacea.*

A—Phyllopoda; *B*—Cirripedia Pedunculata; *C*—Cirripedia Operculata; *D*—Euphausiacea; *E*—Stomatopoda; *F*—Decapoda Macrura; *G*—Cladocera; *H*—Copepoda; *J*—Amphipoda Gammaroidea; *K*—Ostracoda; *L*—Copepoda; *M*—Isopoda Oniscoidea; *N*—Decapoda Brachyura (from Dogiel).

being formed in each segment. In Scutigeromorpha, however, the ventral-chain ganglia are not fully separated from the longitudinal connectives, which are also covered with cells (K. Fahlander, 1938). Among all modern arthropods a ventral nerve ladder with widely-separated stems and long, double transverse commissures is found only in Crustacea Phyllopoda (Fig. 53, *A*) and Cephalocarida (H. Sanders, 1957). In all other arthropods the two ganglia of one segment approach into actual contact, so that the transverse commissure is externally invisible, although the connectives often remain separate (most Myriapoda and Crustacea, some insects). In many crustaceans and insects and in all terrestrial Chelicerata there is (at least externally) complete fusion not only of the ganglia but also of the connectives into a single unpaired chain. In this respect the processes of centralisation run in parallel courses in annelids and arthropods, and independently in different groups of arthropods.

In arthropods (unlike annelids), however, concentration of the ventral chain by shortening of the longitudinal connectives and fusion of separate metameric ganglia is widespread. That phenomenon is fully analogous to the similar concentration of ganglia in molluscs, but takes place somewhat differently in arthropods because of the different general plan of the neural apparatus. In separate groups of arthropods the character and degree of concentration of the ventral chain are closely linked with the character and degree of heteronomy of their metamerism, with the extent and degree of integration of their tagmata, and with the length and degree of integration of the whole body.

The neural apparatus of Phyllopoda represents the prototype of a homonomous ventral ladder consisting of mutually-independent ganglia, including the tritocerebral ganglia, which are independent of the brain (Vol. 1, Fig. 102, *B*).[1] In all other arthropods, because of the formation of a head, the tritocerebral ganglia fuse with the brain; but only in a few crustaceans do all ganglia except the tritocerebral remain mutually independent. In the great majority of crustaceans and Atelocerata, at least the ganglia of the mandibles and of the two pairs of maxillae fuse into a single suboesophageal ganglion, but in Chilopoda Anamorpha the ganglia of maxillae II have not yet become so fused (Fig. 54, *B*). In all insects and Chilopoda the ganglia of the last body segments (the ninth to eleventh abdominal segments in insects, the genital segments in Chilopoda) fuse with the ganglion of the preceding segment (the eighth abdominal segment in insects, the segment of the last walking legs in Chilopoda) (Fig. 54, *D*).

Further fusion takes place in somewhat different ways. In Crustacea Malacostraca first the maxillipede ganglia (Decapoda Macrura,' Fig. 53, *F*) and then the other thoracic ganglia fuse with the suboesophageal ganglion; and in crabs (Brachyura, Fig. 53, *N*) all the ganglia of the ventral chain fuse into a single mass.

Equally high concentration of the ventral chain is observed in several Entomostraca (some Copepoda and Cirripedia), in which, however, it is facilitated by decrease in the number of body segments. In all these cases centralisation of the neural apparatus reaches almost the same level as in Opisthobranchia or Cephalopoda. Whereas in polychaetes and Phyllopoda each ganglion lies in the body segment that it innervates, in crabs

---

[1] Mystacocarida also have independent tritocerebral ganglia (E. Dahl, 1952).

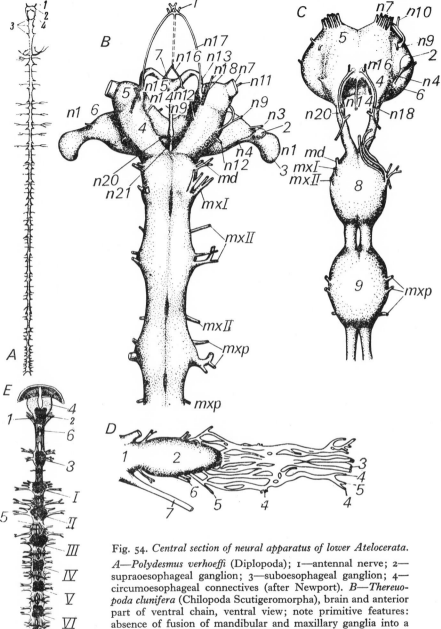

Fig. 54. *Central section of neural apparatus of lower Atelocerata.*
A—Polydesmus verhoeffi (Diplopoda); 1—antennal nerve; 2—
supraoesophageal ganglion; 3—suboesophageal ganglion; 4—
circumoesophageal connectives (after Newport). *B—Thereuopoda clunifera* (Chilopoda Scutigeromorpha), brain and anterior
part of ventral chain, ventral view; note primitive features:
absence of fusion of mandibular and maxillary ganglia into a
single suboesophageal ganglion and poor ganglionisation of the
ventral chain; *C—Scolioplanes hirtipes* (Chilopoda, Geophilomorpha), brain and anterior part of chain, ventral view; note
reduction of optic lobes and of entire protocerebrum because of
blindness due to fossorial mode of life, with a general high
structural level of the ventral chain: 1—prefrontal plexus,
sending nerves to the clypeus and the upper lip; 2—cerebral
gland; 3—optic lobes; 4—tritocerebrum; 5—deutocerebrum;
6—protocerebrum; 7—frontal ganglion; 8—suboesophageal
ganglion; 9—ganglion of maxillipede segment; *md*—mandibular nerves; *mxI*—nerves of maxillae *I*; *mxII*—nerves of

all the ganglia have moved to the front segments of the gnathothorax, but retain connection with their own areas of innervation, each continuing to innervate its own segment. Thus the connectives are shortened to the same extent to which the peripheral nerves are lengthened, and the main task of centralisation of the neural apparatus is fulfilled: shortening of paths between centres.

The same process of concentration of the ventral chain by an independent method (or, more correctly, by a number of independent methods) takes place in other classes of arthropods, insects, Chelicerata, and Pantopoda. The only exception is Myriapoda, in which, because of their slight trunk integration, concentration of the ventral chain also stands at one of the original stages of development mentioned above (Fig. 54, *A*).

Among insects even the most primitive forms, such as *Machilis* (Thysanura, Fig. 54, *E*) have only eight free abdominal ganglia in the adult state. In most cases only the larvae possess that number. Almost all adult insects show a higher degree of concentration of the ventral chain. Most often the first abdominal fuses with the third thoracic ganglion, and that not only in the insects in which the corresponding segments are fused. In some Lepidoptera and Orthoptera not only the first abdominal ganglion but the first two fuse with the third thoracic ganglion (e.g. in *Telea*, family Saturniidae, H. Nuesch, 1957), and in *Locusta* the first three (Roonwall, 1937). Further fusion of the thoracic ganglia with one another and with the suboesophageal ganglion, and of the abdominal ganglia with one another, takes place. Finally, in some forms there is complete fusion of all the ganglia of the ventral chain into a single suboesophageal mass, as in crabs. There is great variety in combinations of these types in separate cases. The total number of variations is further increased by the existence of sexual dimorphism in some forms, and in Holometabola by differences between winged and larval insects.

Using mainly the data of E. Brandt (1879), who studied the anatomical structure of the neural apparatus of very many insects, and also other sources, we may compile the table shown on page 115.

---

maxillae II; *mxp*—nerves of maxillipedes. Nerves arising from protocerebrum: n_1—to compound eyes; n_3—to Tomeshvar's organ; n_4—to cerebral gland. Nerves arising from deutocerebrum: n_7—sensory antennal nerve; n_9-n_{11}—motor antennal nerves; n_{12}—motor nerve to some head muscles; n_{13}—connective to frontal ganglion; n_{14}—nervus recurrens (swelling upon it is the stomatogastric ganglion); n_{15}—motor nerve to pharynx; n_{16}—roots of unpaired nerve following the aorta; n_{17}—labial nerve to prefrontal plexus; n_{18}—sensory nerves to sides of oral cavity; n_{20}—tritocerebral commissure; n_{21}—nerve to tongue retractor. D—*Thereuopoda clunifera*, rear end of ventral nerve chain: 1—ganglion of 21st segment; 2—terminal ganglion, product of fusion of ganglia of genital and pregenital segments; nerves: 3—to rectal plexus, 4—to genital duct, 5—to parietal musculature, 6—to gonopodia, 7—to last pair of walking legs (after K. Fahlander). E—*Machilis* (Insecta, Thysanura), supraoesophageal ganglion and ventral chain: 1—nerves to antennae; 2—supraoesophageal ganglion; 3—suboesophageal ganglion; 4—optic lobes; 5—unpaired medial (Zavarzin's) nerve; 6—oesophagus; I–III—thoracic ganglia; IV–XI—abdominal ganglia (the last of these consists of three fused ganglia) (after Oudemans).

VARIOUS STAGES IN CONCENTRATION OF VENTRAL NERVE CHAIN IN INSECTS

(relation between number of free ganglia in the abdomen and in the head and thorax)

Stage	Head and thorax	Abdomen								
		0	1	2	3	4	5	6	7	8
I	1+1	Hydrometra Rhizotrogus	Stylops	—	—	—	—	—	—	—
II	2+1	Pentatoma Nepa Serica Musca Sarcophaga Lucilia	Conops Ortalis Myopa	Syrphus Volucella	Cyrtus Oncodes	Chrysozona Stratiomya	Tabanus Chrysops	Pangonia	—	—
III	2+2	Lygaeus Cetonia Melolontha Acylius Bostrichus	Gyrinus Phora	Curculioni- dae	Agrilus Eucera Crabro	Vanessa Argynnis Apis ♀+♂	Necrophorus Apis ♀ Bombus ♂ Vespa ♀	Acerentomon Bombus ♀♂	—	
IV	2+3	Podura Psocus All Mallo- phaga Phthirus Geotrupes Aphodius Ateuchus	Hister Scaphidium Coccinella quinque- punctata	Melasoma Chrysomela Harpalus Coccinella septem- punctata	Cassida Mutilla rufipes ♂	Callidium Donacia Meloë Cossus Mutilla europaea ♂	Creophilus Silpha Hepialus Locusta	Forficula Blatta Cicindella Chironomus Culex Tipula Cloëon	Eosentomon Aeschna Elater Telephorus Pulex ♀ Mantodea Most Ephem- eroptera	Machilis Dictyopterus Corydalis Pulex ♂

Note. The table shows the number of free ganglia in the head, thorax, and abdomen of various insects. The figures in the second column signify: 1+1 = one ganglion (suprapharyngeal) in the head and one in the thorax (the latter includes the subpharyngeal ganglion and the entire ventral chain); 2+1 = suprapharyngeal, subpharyngeal, and a single combined thoracic; 2+2 = suprapharyngeal, subpharyngeal, and two thoracic; 2+3 = suprapharyngeal, subpharyngeal, and three ndependent thoracic.

From this table we may deduce some general laws.

1. Maximum concentration of ganglia in the head and thorax always corresponds to high concentration of ganglia in the abdomen. For that reason the forms with the least-centralised ventral nerve chain are absent from the first row in the table.

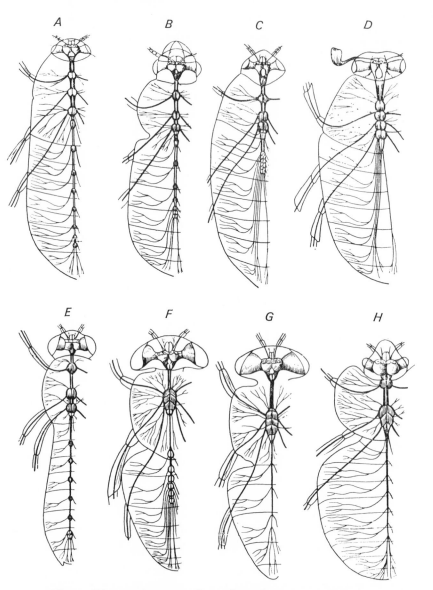

Fig. 55. *Diagram of structure of central neural apparatus in insects.*
A—Dictyopterus sanguineus. B—Cicindella sylvatica ♂. C—Gyrinus natator ♀. D—Rhizotrogus solsticialis ♀ (A–D are all Coleoptera). E—Chironomus plumosus ♂. F—Tabanus bovinus. G—Sarcophaga carnaria (E–G are all Diptera). H—Cimex lectularius (Rhynchota Heteroptera) (after E. K. Brandt).

2. Most Diptera are in the second row and very few of them remain outside it, i.e. Diptera are characterised by great constancy in the number of ganglia in the head and thorax, while the numbers of free ganglia in the abdomen are successively reduced.

3. All bugs are in the zero column, i.e. in all Heteroptera the abdominal ganglia are fused with the last thoracic ganglia (Fig. 55, *H*), while the ganglia in the head and thorax show a number of successive stages of ·fusion with one another. Palaeontologically bugs are a fairly ancient group, but in many respects they are highly specialised. A feature of specialisation in them is also complete concentration of the ventral chain. A high degree of concentration of the ventral chain is also found in Homoptera and other orders of the group Paraneoptera, established by A. V. Martynov (Thysanoptera, Psocoptera, Anoplura).

4. The highest members of Coleoptera (Lamellicornia) and Diptera (Muscoidea) have the most concentrated nerve chain, and the most primitive members (Nematocera among Diptera, and *Cicindella* and *Dictyopterus* among Coleoptera) have the least concentrated nerve chain (Fig. 55, *A–G*).

5. Lepidoptera are almost all concentrated in the third and fourth columns of the fourth row, i.e. they have four abdominal ganglia, and only *Hepialus* (a form very primitive also in other respects) falls into the fifth column, having five abdominal ganglia; the female of the casebearer moth *Pachytelia* retains the larval number of abdominal ganglia (D. Fedotov, 1940), but that form is generally much simplified, with a number of neotenic features.

6. Members of primitive orders (Protura, Thysanura, Odonata, orthopteroid orders, Neuroptera) have, as a rule, a very low level of concentration of the ventral chain, but specialised forms among the lower orders may fall into the zero column, e.g. *Podura* (Collembola).

The table therefore shows clearly that the process of concentration of the ventral chain takes place independently in different orders of insects, and sometimes even in different families of the same order (Coleoptera).

Now a few words on the development of the ventral chain. All its ganglia are mutually independent in insect embryos. In larvae, as a general rule, the chain is more concentrated than in embryos, and in the adults it is still more so (Fig. 56, *A*, *B*). But there are several interesting exceptions to this rule, especially among Diptera larvae. In *Syrphus* and *Volucella* larvae, according to Brandt, the ventral chain is completely concentrated, whereas adult syrphids have two free ganglia in the abdomen (Fig. 56, *C*, *D*). In other words, decentralisation takes place together with development. This is an example of a biogenetic 'law' reversed: in the structure of the ventral chain the *Syrphus* larva runs too far forward, as it were, in the general direction of the evolution of the group, and the imaginal form falls back.

The separate metameric components of the dipterous larva's concentrated ventral chain send out peripheral nerves to their own segments (law of retention of region of innervation), but they also receive tracheal branches from the metameric branches of their own segments (Fig. 56, E) (J. Buck and M. Keister, 1953)—an instance of retention of region of tracheation, similar to retention of region of blood supply by the arteries of

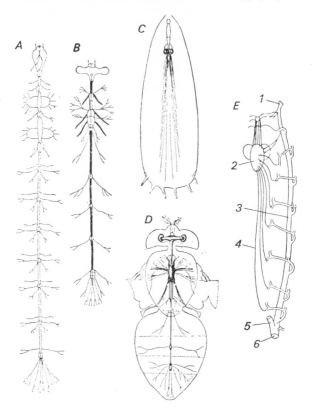

Fig. 56. *Development of ventral neural chain in insects.*

Neural apparatus of: *A*—caterpillar of the cabbage white butterfly *Pieris brassicae* (Lepidoptera); *B*—adult butterfly of that species (after Brandt); *C*—larva of *Volucella zonaria* (Diptera); *D*—adult of same species (after Künckel d'Herculais); *E*—larva of *Phormia regina* (Diptera), metameric nerves shown running from abdominal ganglia lying in thorax to their own segments, and metameric branches of tracheae running to the same ganglia, which are displaced forward and fused into a single mass: 1—anterior spiracle; 2—brain; 3—dorsal tracheal stem; 4—tracheae to nerve mass; 5—visceral stem; 6—posterior spiracle (after Buck and Keister).

vertebrates (E. Geoffroi St. Hilaire, 1822). Retention of region of innervation and retention of region of blood supply are in a way, forms of inertia. Eventual reconstructions are quite possible in the course of further evolution.

Concentration of the ventral chain also takes place in Pantopoda (V. Dogiel, 1954, see Fig. 57).

The class Chelicerata also shows gradual concentration of the ventral chain but no early stages of that process. Xiphosura have the least concentrated ventral chain, but all the ganglia of their prosoma are fused together near the oesophagus; all the mesosomatic ganglia remain independent, and a separate ganglion represents the three fused metasomatic ganglia. Besides, the two ventral stems of the Xiphosura larva are quite widely separated, in this respect resembling the situation in Phyllopoda.

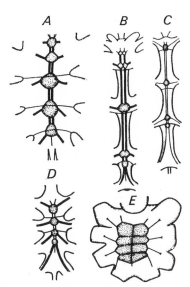

Fig. 57. *Concentration of ventral nerve chain in Pantopoda.*

A—Hannonia, showing suboesophageal ganglion (product of fusion of ganglia of palps and oviducts) and four independent pairs of ganglia of limb segments. *B—Rhopalorhynchus. C—Pipetta. D—Anoplodactylus. E—Halosoma,* all ganglia of ventral chain fused into a single mass (from V. A. Dogiel).

In scorpions (Fig. 58, *A*) all the prosomatic ganglia and four mesosomatic ganglia are fused into a single suboesophageal mass, behind which runs a chain of seven ganglia that innervate two mesosomatic and six metasomatic segments. The last metasomatic ganglion, lying in the fifth metasomatic segment, innervates also the sixth segment and the telson.

In Telyphones (Fig. 58, *B*) all the ganglia of the ventral chain are fused together, and only one of them, of triple origin, is moved far to the rear (as in *Stylops* in the order Strepsiptera among insects) and innervates the last three segments of the body. A similar structure is retained in Solifugae. In Araneina the whole central section of the neural apparatus, including the brain and the ventral stems, is fused into one compact mass traversed by the oesophagus. In Mesothelae (= Liphistiomorpha), however, that mass contains 17 well-developed neuromeres (Fig. 58, *D*), corresponding to the full number of post-cheliceral segments. In other Araneina the same number of neuromeres are included, but the five posterior ones degenerate and only 12 are included in the ventral nerve chain. In Pseudoscorpionoidea (Fig. 58, *C*), Opiliones (harvest spiders), and all Acariformes that have been studied in this connection, all the ganglia of the body are similarly fused into a single mass.

From the above it is clear that in Chelicerata, as in other classes, there

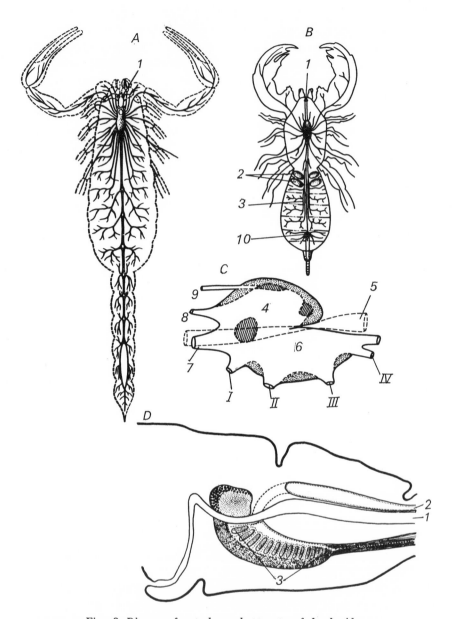

Fig. 58. *Diagram of central neural apparatus of Arachnoidea.*
A—Androctonus (Scorpionoidea); *B—Telyphonus caudatus* (both dorsal view).
C—brain and ventral nerve mass of *Chelifer* (Pseudoscorpionoidea), view from
left side. 1—eye; 2—lung-sacs; 3—abdominal nerves; 4—brain; 5—gullet;
6—ventral nerve mass; 7—nerves of pedipalps; 8—nerves of chelicerae; 9—optic
nerves; 10—compound ganglion of last three body segments; I–IV—nerves of
walking legs (from Hanström). *D—Liphistius desultor* (Araneina), medial cross-
section through prosoma showing front part of gut (1), front part of aorta (2),
brain, and ventral nerve mass, in which are seen sections of 17 commissures
between ganglia of successive metameric pairs (3) (from Millot).

is a correlation between the shortening and integration of the body on the one hand and the degree of concentration of the ventral chain on the other. We also see that the ganglia of the prosoma are the first to fuse, then the ganglia of the mesosomatic segments; but the ganglia of the metasoma, almost everywhere that they are retained, remain partly independent, and the ventral chain usually becomes completely concentrated only after the metasoma has been eliminated. With a highly-integrated body, however, a completely-concentrated ventral chain is found even when traces of the metasoma are present, e.g. in Pseudoscorpionoidea.

Many orders of Chelicerata therefore attain the same concentration of the neural apparatus as (or even higher concentration than) that attained by crabs among crustaceans, by some highly-specialised groups of Entomostraca, by *Hydrometra* or *Rhizotrogus* among insects, or by *Myzostomum* among annelids. The process of shortening of connectives and commissures, leading to centralisation of the neural apparatus, has thus taken place independently in various classes, orders, and even families of arthropods, just as it has taken place independently in Cephalopoda, Lamellibranchia, and various orders of Gastropoda.

The composition of the brain in arthropods and Onychophora and the origin of its separate parts have been discussed in Vol. 1, Chaps VI and VII. We shall now note only a few features in the internal structure of the brain.

Most arthropods are characterised by great development of, and a major role is played by, the optic centres, which is doubtless due to the development of compound eyes. The optic centres form part of the protocerebrum, where the corpora pedunculata (which also are highly developed in arthropods) are located. The corpora pedunculata of arthropods are connected with both the optic and the olfactory (antennal glomeruli) centres, which are located in the deutocerebrum. We may say that the arthropod brain originally develops under the influence of two principal sense-organs, eyes and antennae. Chelicerata have lost the deutocerebrum as well as the antennae, and with it the antennal glomeruli, which have persisted only in Xiphosura (see Vol. 1, Fig. 123). The antennal glomeruli are also much reduced in the arthropods of other classes that have very slightly-developed antennae, e.g. Ephemeroptera (B. Hanström, 1928). In Chilopoda Epimorpha reduction of the optic lobes of the fore-brain keeps pace with reduction of the eyes, being most complete in the eyeless Geophilomorpha (Fig. 54, *C*). Generally speaking, wherever there is considerable reduction of sense-organs in arthropods it produces considerable simplification in brain structure, as we see in most Arachnoidea, in Pantopoda, in parasitic crustaceans, etc.

In its internal structure the brain of all arthropods shows modifications of a single general plan, crustaceans and Atelocerata being closer to each other in this respect than either is to Chelicerata. Detailed comparison of the brain structure of insects with that of crustaceans and myriapods enables us to conclude that in its principal organisational features the brain of insects is somewhat closer to that of myriapods, whereas in the degree of differentiation of their brain higher insects are very close to Crustacea Decapoda, with myriapods and lower insects (e.g. Campodeoidea) being approximately at the level of Phyllopoda. The directions of development in the two stems, crustaceans and Atelocerata, are very similar; in particular, the structure of the compound eye and the optic centres in insects closely resembles that in Crustacea Decapoda (A. Zawarzin, 1924; A. Zavarzin, 1941).

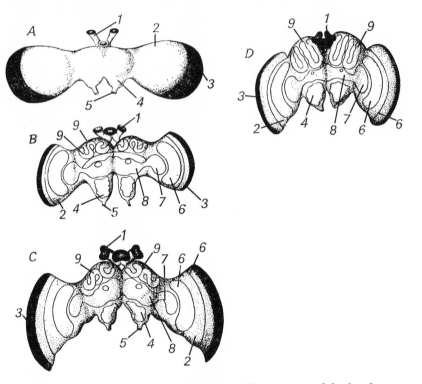

Fig. 59. *Development of corpora pedunculata in Hymenoptera as behaviour becomes more complex.*

A—brain of *Tenthredo flava* (sawfly), front view. *B*—*Ichneumon obsessor* (ichneumon fly). *C*—*Andraena albicans* (solitary bee). *D*—*Vespa vulgaris* (social wasp). 1—frontal eyes; 2—optic lobes; 3—retina of compound eye; 4—deutocerebrum; 5—antennal nerves; 6 and 7—optic centres; 8—protocerebrum; 9—corpora pedunculata (after Alten).

The degree of development of the corpora pedunculata in different arthropods shows some remarkable features of regularity, obliging us to conclude that these are higher centres that make possible the most complex forms of behaviour found in arthropods. In fact, a number of data indicate that the qualitative and quantitative development of these organs is directly correlated with the complexity of the animals' behaviour. H. Alten (1910) found that among Hymenoptera the corpora pedunculata are least developed in Tenthredinidae (sawflies, Fig. 59, *A*) and Siricidae (woodwasps), which have the simplest maternal instincts; better developed in Ichneumonidae (Fig. 59, *B*); still better in solitary wasps (*Andraena*, Fig. 59, *C*; *Anthophora*); and best of all in social wasps (*Vespa*, Fig. 59, *D*), which have the most complex instincts among all the insects studied by him. Among bees the social forms *Apis* and *Bombus* have the most highly-developed corpora pedunculata. In accordance with their role in the life of the species, female Bombidae have corpora pedunculata as well developed as those of workers, whereas in hive bees the workers certainly take first place. In the ant *Camponotus*, according to H. Pietschker (1911), the optic centres are developed best in males, then in females, and least in workers, while the olfactory centres and the corpora pedunculata stand in the reverse order: the volumes of the corpora pedunculata in workers, females, and males of that ant are in the ratio 8 : 4 : 1. In the relative sizes of their corpora pedunculata worker ants take first place among insects. Besides the above-mentioned correlation there is another: the larger the body of any arthropod, the more highly developed are its corpora pedunculata, other conditions being equal. Among all arthropods examined the corpora pedunculata are most highly developed in *Limulus*. They are better developed in the large *Vespa crabro* than in the smaller *V. vulgaris*. They reach great size in Crustacea Decapoda and in some Telyphones; but neither *Limulus* nor Telyphones are distinguished by complexity of behaviour.[1] On the basis of these data some authors have objected to assigning significance to the corpora pedunculata as centres of higher neural activity. These objections, however, are hardly justified, as is easily seen from analogy with mammals, in which the degree of development of the cortical convolutions of the cerebrum is linked not only with the complexity of an animal's behaviour but also with the size of its body.

Let us pass to the peripheral section of the neural apparatus of higher Articulata. The peripheral trunk nerves of Onychophora have a strictly metameric arrangement; there are eight pairs of them in each body segment, some of these nerves uniting dorsally with their partners from the opposite side of the body to form rings (B. Fedorow, 1926, 1929; Fig. 50, *C*). In this respect Onychophora resemble annelids, especially those polychaetes that have a large number of peripheral nerves in each segment. Solenogastres also show a tendency to an annular course for the metameric peripheral nerves. Annular nerves of this type are not found in arthropods. In primitive arthropods the number of peripheral nerves arising from each ganglion in the ventral chain does not exceed three or four. With the development of heteronomy that number greatly increases in some segments and falls in others. The water-beetle *Dytiscus* has six pairs of peripheral nerves in the prothorax and seven pairs each in the mesothorax and metathorax, and only one pair in each abdominal segment (G. Holste, 1910).

---

[1] Goossen (1949) also demonstrates a positive correlation between body size and the development of the corpora pedunculata within the orders Hymenoptera and Coleoptera.

The internal organs of arthropods, like those of polychaetes, are innervated first by the stomatogastric section of the neural apparatus and secondly by the ventral chain.

It has been demonstrated that the hearts of Malacostraca (Alexandrowicz, 1932, 1934, 1952), Chilopoda, Xiphosura, and some other arthropods have a longitudinal nerve cord capable, as Alexandrowicz asserts, of independent reflexes; in Stomatopoda three pairs of regulating nerves, which branch within the neuropile of the cardiac cord, come to it from the ganglia of the ventral chain, and 15 pairs come to the valves of the 15 pairs of arteries leaving the heart. The hearts of many insects (cockroaches, bees, Chinese silkworms, etc.) contain two longitudinal nerve cords connected with the stomatogastric section of the neural apparatus; regulating nerves come to them from most of the ganglia of the ventral chain. Nerves running from the brain to the dorsal vessel are found also in polychaetes, e.g. in *Serpula*, but there they are connected directly with the posterior surface of the brain, not with the stomatogastric section (K. Johansson, 1927).

In some insects, e.g. mosquitoes, the heart has no nerve cells of its own; its work depends on the autonomous rhythm of the musculature, but is regulated by metameric branches of the unpaired, so-called sympathetic nerve (Fig. 54, *E*, *5*; Fig. 60, *A*), which runs along the ventral chain (L. Yaguzhinskaya, 1954). Other branches of that nerve innervate the muscles of the spiracle-closing apparatus in insects (A. Zawarzin, 1924) and part of the musculature of the mid-gut (S. Plotnikova, 1949). According to J. S. Alexandrowicz (1952), the pterygoid heart muscles of *Lygia* (Isopoda), like those of *Anopheles* (according to L. V. Yaguzhinskaya), are innervated by branches of the unpaired nerve. The muscles of the spiracle-closing apparatus of *Platysamia* (Lepidoptera) are innervated both by the unpaired nerve of the preceding segment and by the paired nerves of their own segment (H. Schneiderman, 1956).

The hind-gut of both polychaetes and arthropods is innervated by paired nerves arising from the ganglion of the last body segment, since the anal lobe has no nerve centres. In Chilopoda the ganglia of both the genital and the anal segments [1] are fused into one compound ganglion, which sends nerves directly to the gonopodia and through a plexus to the genital apparatus, to the parietal musculature, and (the rearmost pair of nerves) to the rectum (Fig. 54, *D*). In insects the ganglia of the posterior segments (i.e. at least the genital and postgenital segments) are also always fused into a compound ganglion, which sends out the rearmost nerves to the rectum and (one or two pairs) to the genital apparatus and the body walls (*Dytiscus*, G. Holste, 1910; *Apis*, E. Rehm, 1939; *Tenebrio*, E. Jösting, 1942; *Anopheles*, V. Polovodova, 1953) (Fig. 60, *B*). Apparently, therefore, the genital apparatus of arthropods is as a rule innervated by the ganglia of the segment on which the genital orifice is situated.

The fore-gut, and in some groups the mid-gut, and also the salivary

---

[1] The product of fusion of the anal lobe of an arthropod with the last body segment is called the anal segment.

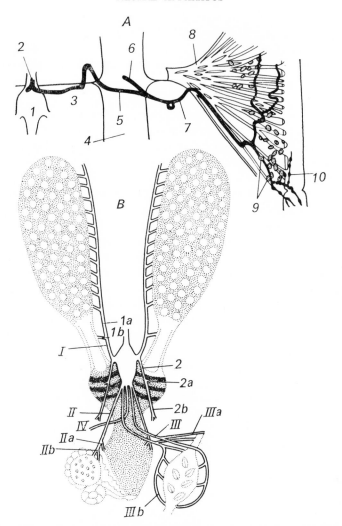

Fig. 60. *Innervation of internal organs in lower Diptera (Insecta Diptera Orthorrhapha).*
A—innervation of pterygoid muscles and heart of the crane-fly *Pachyrrhina cornicina* (family Tipulidae): 1—ventral ganglion; 2—unpaired nerve; 3—sternite; 4—pleural membrane; 5—branch of unpaired nerve; 6—branch of paired segmental nerve; 7—nerve to heart and pterygoid muscle; 8—pterygoid muscle; 9—pericardial cells; 10—nerve ends on heart (after L. V. Yagunzhinskaya). *B*—last ganglion in the ventral chain of *Anopheles maculipennis* and organs innervated by it: I—nerve of seventh segment of abdomen (n. septimi segmenti abdominis); 1—nervus ovarialis; 1a—its branch to ovary (ramus ovarialis); 1b—its branch to oviduct (r. oviducticus); 2—n. ampullo-parietalis; 2a—its branch to ampullae of oviduct (r. ampullaris); 2b—its branch to body wall of seventh segment (r. parietalis); II—nerve of eighth segment of abdomen (n. octavi segmenti abdominis); IIa—its branch to vagina (r. vaginalis); IIb—its branch to spermatheca (r. spermathecalis); III—nerve of hind-gut (n. intestinalis ultimi segmenti abdominis); IIIa—its branch to hind-gut (r. colicus); IIIb—its branch to rectum (r. rectalis); IV—unpaired nerve of last segment of abdomen (n. impar ultimi segmenti abdominis) (after V. P. Polovodova).

glands are innervated by the stomatogastric section of the neural apparatus, which is possessed by all arthropods and Onychophora.

We have seen that in polychaetes the stomatogastric section is connected with both the brain (its anterior part) and with the first ganglia of the ventral chain. In arthropods the stomatogastric section is connected by frontal connectives with the tritocerebrum, which, as has been stated in Vol. 1, Chap. VI, is probably the first ganglion of the ventral chain, which has become part of the brain. Chilopoda and many insects (*Aeschna*, termites, *Machilis*) also have an unpaired connective nerve joining the stomatogastric section of the neural apparatus to the protocerebrum. Among crustaceans this second connection has been described in Cumacea. In arthropods, therefore, at least in the prototype, the stomatogastric section retains connections with both the brain and the first ganglion of the ventral chain (the tritocerebrum) (K. Fahlander, 1938).

The stomatogastric section of the neural apparatus of crayfish (*Astacus*) is joined to the brain by several roots and forms a fairly complex network, in which (Fig. 61) an unpaired nerve (n. recurrens) running back along the dorsal side of the gullet is well individualised; two small unpaired ganglia lie on it, containing motor cells but incapable of independent reflexes (J. Orlov, 1925, 1927). Insects (Fig. 61) have, on the dorsal side of the gullet in front of the brain, a frontal ganglion (ganglion frontale), which is connected to the tritocerebrum by paired connectives and to the proto-cerebrum by one connective nerve; a recurrent nerve bearing from one to three additional small ganglia arises from the frontal ganglion. Among all these only the frontal ganglion is capable of independent reflexes (J. Orlov, 1924). From a purely anatomical point of view this apparatus is a simplification of that of crustaceans.

Among Myriapoda, Scutigeromorpha have a stomatogastric section of the neural apparatus close to that of insects (Fig. 54, *B*); in other Chilopoda, and also in Diplopoda and Pauropoda the frontal ganglion fuses with the tritocerebrum, forming the so-called pharyngeal bridge, which is often erroneously taken to be the pre-oral commissure of the tritocerebrum. The inclusion of the frontal ganglion in the brain presents a certain analogy with the inclusion of the labial ganglion in the brain of Octopoda. This process also takes place in all Chelicerata. In *Limulus* (Fig. 65) the pharyngeal bridge still has the form of a free commissure, which encircles the gullet in front and from which nerves run to the fore-gut. In Arachnoidea the pharyngeal bridge is completely fused with the brain, although it retains some independence in its internal structure. In Arachnoidea, as in insects, the n. recurrens runs from the frontal ganglion, and in both groups bears a small oesophageal ganglion.

The stomatogastric section of the neural apparatus develops most peculiarly in Pantopoda (Fig. 61, *C*), in which it lies in the proboscis. Near the end of the proboscis lies a nerve ring connected with the brain by three roots, like the stomatogastric plexus in crustaceans and insects. The ring contains three ganglionic thickenings and corresponds to an expanded frontal ganglion. A plexus runs rearward from it along the gullet, corresponding to the recurrent nerve in other arthropods. The independence and great development of the stomatogastric section in Pantopoda are due to their possession of a proboscis, and constitute one of the significant features of difference between them and Chelicerata.

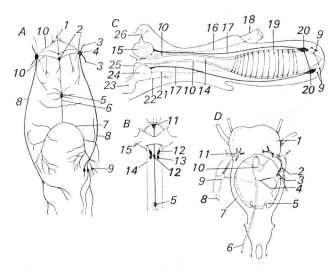

Fig. 61. *Diagram of structure of stomatogastric section of neural apparatus of arthropods.*

A—Astacus leptodactylus (Crustacea Decapoda) (after Orlov). *B—Periplaneta* (Insecta Blattoidea). *C—Nymphon brevirostre* (Pantopoda) (from Hanström). 1—motor plexus (labial); 2—oesophageal ganglion; 3—oesophageal connectives; 4—ganglia along them; 5—stomach ganglion; 6—dorsal unpaired gastric nerve; 7—paired sensory gastric nerves; 8—neurites of sensory cells; 9—groups of sensory cells; 10—oesophageal nerves; 11—frontal ganglion; 12—corpora pharyngaea; 13—corpora allata; 14—gullet; 15—brain; 16—nerve of chelophores; 17—dorsal and ventral roots of neural apparatus of proboscis; 18—chelophores; 19—oesophageal plexus; 20—ganglia of terminal nerve ring of proboscis; 21—palpal nerves; 22—ovigeral nerves; 23—ganglia of first pair of legs; 24—suboesophageal ganglion; 25—mid-gut; 26—optic nerve. *D—Bombyx mori* (Insecta Lepidoptera), last-instar larva: 1—supraoesophageal ganglion; 2—nervi corporis cardiaci II; 3—n. corp. cardiaci I; 4—anastomosis of n. recurrens and corp. cardiaca; 5—tritocerebral commissure; 6—n. recurrens; 7—circumoesophageal connectives; 8—corpora allata; 9—c. pharyngaea; 10—ganglion frontale; 11—tracheae (after Bounhiol, Gabe, and Arvy).

The endocrine apparatus of arthropods develops in close connection with the brain and the stomatogastric section of the neural apparatus.

The endocrine apparatus of Crustacea Decapoda consists of neurosecretory cells located in both the brain and the ventral chain, and also of a pair of separate plexuses lying in the eye-stalks and known as the X-organs or Hanström's organs (Fig. 62, *A*). In Isopoda also Hanström's organs lie near the optic ganglia. The secretion of all these cells travels along their axons (on which it may be observed in the form of small drops) and collects in two sinus glands, which also lie in the eye-stalks. According to Bliss these are not glands, but bunches of flask-shaped swollen ends of the axons of neurosecretory cells. These 'glands' are places where hormones accumulate and enter the haemolymph (D. Bliss and J. Welsh, 1952) and therefore are neurohaemal organs, like the cerebrovascular

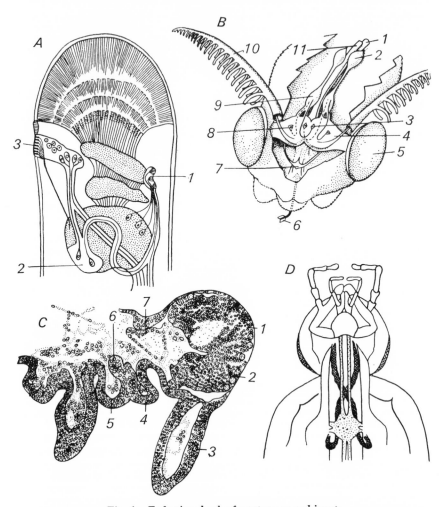

Fig. 62. *Endocrine glands of crustaceans and insects.*

A—eye-stalk of *Lysmata seticaudata* (Decapoda); 1—sinus gland; 2—ganglionic part of X-organ (Hanström's organ); 3—distal part of that organ. *B*—Chinese silkworm (*Bombyx mori*), head of adult, dissected on dorsal side: 1—oesophagus; 2—corpora allata; 3—neuro-secretory cells of central part of protocerebrum; 4— brain; 5—compound eyes; 6—arrow leading to oral cavity; 7—ganglion frontale; 8—lateral groups of neuro-secretory cells of protocerebrum; 9—corpora cardiaca or pharyngaea; 10—antennae (boundaries of their segments not shown); 11—aorta (after Bounhiol, Gabe, and Arvy). *C—Carausius* (Phasmatodea), paramedial section of embryo: 1—protocerebrum; 2—deutocerebrum; 3—antenna; 4— mandible; 5—maxilla; 6—rudiment of corpora allata; 7—coelom. *D*—nymph of *Leucophaea* (Blattoidea), head and prothorax, dorsal view, prothoracic glands shown in black (after Pflugfelder).

complex of annelids (see above). Usually the sinus glands also contain glandular cells secreting their own hormone. The X-organ produces moult-inhibiting substances (D. Bliss, 1953; L. Passano). Sinus glands also contain accumulations of hormones that affect metabolism, growth, sexual activity (G. Stephens, 1952), the activity of chromatophores in the skin, adaptation of the retina of the eyes to changes in lighting, etc.

The X-organs of higher Crustacea (Malacostraca) are, in the opinion of B. Hanström (1953), vestigial sense-organs, and in some forms partly retain sense-organ structure. The most primitive crustaceans, Phyllopoda Anostraca, have neurosecretory cells in both the brain and the sub-oesophageal ganglion, but no sinus glands (J. Lochhead and R. Resner, 1958).

Other neurohaemal organs besides the sinus glands have been described in higher crustaceans, e.g. the pericardial organs of Decapoda (J. S. Alexandrowicz, 1953), which discharge a heart-accelerating hormone into the blood (J. Alexandrowicz and D. Carlisle, 1953). M. Gabe (1953g, 1954c) describes Y-organs in Malacostraca, located in the maxillary and antennal segments and having glandular structure. Their secretion apparently accelerates moulting.

In Chilopoda neurosecretory cells lie in the protocerebrum, their axons ending in the so-called cerebral glands (Fig. 54), between the cells of which their secretion is discharged. These glands are perhaps homologous with the corpora pharyngaea of insects (see below) and perhaps also with the sinus gland of crustaceans. Similar interrelationships occur also in Diplopoda (M. Gabe, 1954b).

In Pterygota neurosecretory cells are concentrated mainly in the protocerebrum (B. Hanström, 1940; M. Gabe, 1954c), although there are paired metameric groups of such cells also in the ganglia of the ventral chain (M. Gersch, 1959). In all insects the neurosecretory cells of the protocerebrum send their processes through the brain to the ganglia pharyngaea, or corpora cardiaca, which lie behind the brain along the sides of the aorta. The corpora cardiaca (Fig. 62, B) also contain nerve cells and broadened ends of the axons of the neurosecretory cells of the protocerebrum. In relation to the latter cells the corpora cardiaca therefore play the role of a neurohaemal organ. Like sinus glands, they have their own hormones (according to M. Gabe, 1951). Embryologically the corpora cardiaca develop from the ectoderm of the dorsal wall of the fore-gut (O. Pflugfelder, 1952).

In Hemiptera the neurosecretory cells of the brain secrete a hormone affecting the moult. In Diptera (*Calliphora*) the secretion of these cells activates the corpora allata, and in Lepidoptera it activates the prothoracic glands (see below).

In all Pterygota a corpus allatum is connected with each of the corpora cardiaca (Fig. 62, B). The corpora allata arise from the ectoderm at the boundary between the mandibular and maxillary segments (R. Heymons, 1895) (Fig. 62, C). Like the corpora cardiaca, they are innervated by the neurosecretory cells of the protocerebrum. Their endocrine function was first suggested by A. Nabert (1913). Their secretion inhibits the onset of metamorphosis in larvae and nymphs of younger instars, activates the ovaries of adult females of most insects and the accessory glands of males, and increases metabolism (V. Wigglesworth, 1936; T. Detinova, 1945, and many others; cf. O. Pflugfelder, 1952, and V. Novak, 1960).

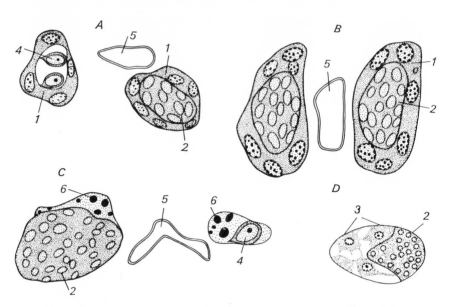

Fig. 63. *Postembryonic development of retrocerebral endocrine glands in* Anopheles
(Insecta, Diptera) (cross-sections, except D).

A—larva of last (fourth) instar (oblique section, on one side cutting corpus allatum, and on the other side cutting corpus pharyngaeum lying in front of it). *B*—pupa. *C*—young female (oblique section, like *A*) (after M. V. Mednikova). *D*—adult female, longitudinal section through c. pharyngaeum, including c. allatum (after T. S. Detinova). 1—pericardial or peritracheal glands; 2—corpora allata; 3—c. pharyngaea; 4—nerve cells; 5—aorta; 6—degenerating cells of pericardial glands.

Most Hemimetabola have ventral glands in the head, formed by the ectoderm of the second maxillary segment. They resemble the corpora allata in structure and are perhaps serial homologues of these. The ventral glands degenerate after the last moult. They secrete a hormone affecting metamorphosis. Lepidoptera, Hymenoptera, and Blattoidea have prothoracic glands (Fig. 62, *D*), also formed by the ectoderm of the second maxillary segment, but they are located in the prothorax and are perhaps homologous with the ventral glands of lower insects. In Lepidoptera they secrete hormones affecting moulting and metamorphosis. After the last moult they degenerate (C. Williams, 1948, 1949). In addition, pericardial or peritracheal glands have been described in Phasmatodea and Diptera (Fig. 63), lying near the corpora cardiaca.

Like the ventral and prothoracic glands they secrete a moulting hormone and undergo involution after metamorphosis (see O. Pflugfelder, 1950; M. Mednikova, 1952). The pericardial glands in Phasmatodea develop from the mesoderm (O. Pflugfelder, 1952) and consequently are not homologous with the ventral glands.

In higher Diptera (Cyclorrhapha) all the retrocerebral glands (corpora allata, corpora pharyngaea, and pericardial glands) grow together into a glandular ring encircling the aorta immediately behind the brain (Weissmann's ring, so named after A. Weissmann, 1864, who was the first to describe it).

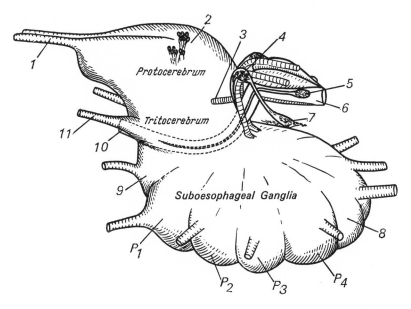

Fig. 64. *Central section of neural apparatus and retrocerebral endocrine glands in* Pardosa lugubris *(Araneina, Dipneumones).*

1—optic nerves; 2—neuro-secretory cells; 3—arteries: 4—first pair of Schneider's organs; 5—second pair of Schneider's organs; 6—gullet; 7—haemal part of first pair of Schneider's organs; 8—opisthosomatic section of ventral nerve mass; 9—ganglion of pedipalps; 10—location of mouth; 11—nerve of chelicera; p_1–p_4—ganglia of walking legs (after H. Kühne).

As was first stated by B. Hanström (1941) and later by B. and E. Scharrer (1944), the endocrine apparatus of insects shows surprising convergence with that of vertebrates. In both cases the neurosecretory cells of the brain (lying in the protocerebrum in insects and in the hypothalamus in mammals) send a hormone along their axons into an organ of neural origin (corpora cardiaca in insects, posterior lobe of the hypophysis in vertebrates), which accumulates that hormone and discharges it into the blood. A third link in both cases is an internal-secretion gland, formed from the ectoderm of the oral cavity and secreting a gonadotropic hormone (corpora allata in insects and anterior lobe of hypophysis in vertebrates). Finally, the ventral, prothoracic, and pericardial glands of insects present, in the Scharrers' opinion, a certain analogy to the thymus of mammals.

In Chelicerata the neurosecretory cells are arranged relatively homonomously: as a rule they lie in all the ganglia of the ventral brain and their

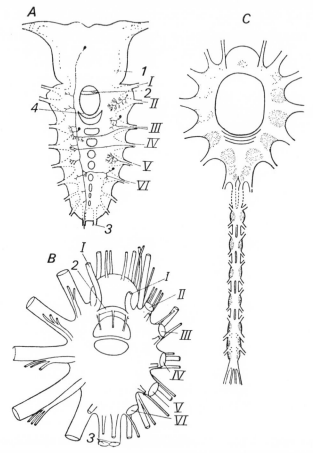

Fig. 65. *Central neural apparatus of Xiphosura* (Limulus polyphemus).
A—brain and prosomatic section of ventral chain of larva, dorsal view; fine structure partly shown. *B*—the same parts of the neural apparatus of adult *Limulus*, ventral view. 1—protocerebrum; 2—stomodaeal commissure; 3—connectives to mesosomatic part of ventral chain; 4—tritocerebral commissure; I–VI—nerves to prosomatic appendages (from Hanström). *C*—location of neurosecretory cells in central section of neural apparatus of *Limulus*, denoted by stippling (after B. Scharrer).

accumulation in the protocerebrum is less dominant than in insects (Fig. 65, *C*). Araneina have two pairs of retrocerebral glands (Schneider's organs), representing an accumulation of neurosecretory cells. Each pair receives its own pair of nerves from the tritocerebral, i.e. cheliceral ganglion (Fig. 64). The lateral pair has a separate neurohaemal section, which discharges secretion into the haemolymph (R. Legendre, 1953a, 1953b, 1954, 1956a, 1956b; M. Gabe, 1955; H. Kühne, 1959). To judge by some data, there are perhaps homologues of Schneider's organs in scorpions, Telyphones, and Solifugae (Kühne, loc. cit.). The homology of Schneider's organs with the retrocerebral organs of Atelocerata has not yet been proved. Perhaps one pair, or both together, may be homologous with the corpora pharyngaea of insects.

7. Deuterostomia

The neural apparatus of the lower Deuterostomia (Hemichorda) stands at a very low level of development, which is one of the proofs of their direct connection with coelenterate ancestors (see Vol. 1, Chap. IX).

A second characteristic feature possessed in common by the lower Deuterostomia and coelenterates is possession of neural plexuses not only in the ectoderm but also in the epithelia of the gut and the coelom.

The larvae of Enteropneusta and of almost all echinoderms possess an apical plate (Vol. 1, Fig. 166). In crinoid larvae it is connected with a ganglion from which two nerves arise (*Antedon*, *Notocrinus*). On metamorphosis it completely disappears, and no traces of it remain in the organisation of the definitive neural apparatus. Its reduction in echinoderms is evidently due to the larva's attachment to the substrate by its aboral pole. The reasons for its disappearance in Enteropneusta are not clear. Enteropneusta have no complex sense-organs.

Like coelenterates and lower turbellarians, Enteropneusta have a continuous subepithelial neural plexus (Fig. 68, *A*), which extends into the oral cavity (C. Dawydoff, 1948). Its fine structure has been described by T. H. Bullock (1945) and L. Silén (1950). The epidermis contains a large number of sensory cells. Their neural processes form a plexus lying between the bases of the epithelial cells and the boundary membrane of the epithelium. Associative and motor cells lie in that plexus. As in coelenterates, both of these are unpolarised nerve cells. The peripheral processes of the motor cells, according to Bullock's data, penetrate the boundary membrane of the epithelium and go inside the body, to the muscles. Synapses also are scattered throughout the epithelium. The presence everywhere in the skin of sensory cells, synapses, ganglionic cells, and links with the internal organs explains why the smallest separate part of a worm is capable of independent reflexes. There is no individualised central section of the neural apparatus. As in Anthozoa and Ctenophora, however, the cutaneous plexus of Enteropneusta forms condensations representing the first rudiments of centralisation: but the plan of arrangement of these condensations after the apical plate disappears has no longer anything in common with their arrangement in coelenterates and turbellarians (Fig. 66, *A*). It is quite unique. The trunk of Enteropneusta contains two longitudinal cords, dorsal and ventral; at the edge of the collar they are connected by an annular cord, and only the dorsal cord continues into the collar. A second fine annular cord is formed along the front edge of the collar. On entering the proboscis the dorsal cord divides into two branches, which form a ring round the base of the proboscis. Along the collar the external epithelium, from which the nerve stem is formed,

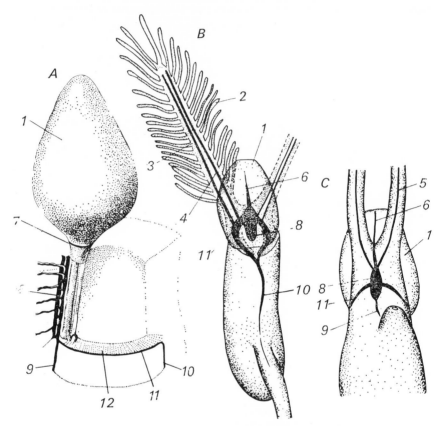

Fig. 66. *Diagram of location of the principal nerve cords in Hemichorda.*
A—Enteropneusta, front end of body, sagittal section of collar (after Delage and Hérouard). *B*—*Rhabdopleura*, ventral view. *C*—the same, dorsal view (after A. A. Shchepot'ev). 1—proboscis, or shield; 2—tentacle; 3—right arm of lophophore; 4—ventral cords of right arm; 5—dorsal cord of right arm; 6—dorsal cord of cephalic shield; 7—annular cord of proboscis; 8—dorsal cord of collar (tubular in Enteropneusta); 9—dorsal cord of trunk; 10—ventral cord of trunk; 11—annular cord of collar; 12—diaphragm (dissepiment at boundary of collar and trunk).

condenses into a narrow band beneath the rest of the skin, forming a neural tube. In *Ptychodera* and some other species the tube retains a lumen; in *Balanoglossus minutus* it retains the lumen only in the juvenile state; but in the majority it forms a solid cord. Nerve cells, of course, line the lumen of the tube, so that the latter resembles the dorsal brain of chordates. In chordates, however, the dorsal branch extends not only into that part of the body corresponding to the collar but also along the whole trunk, whereas in Enteropneusta it is developed only in the collar.

Histologically the neural system of Enteropneusta, unlike that of Protostomia and like those of echinoderms and chordates, is characterised by
absence of unipolar neurons; the only exception is provided by the giant
nerve cells with their giant fibres, which play the same role in them as in
other animals (see section 5).

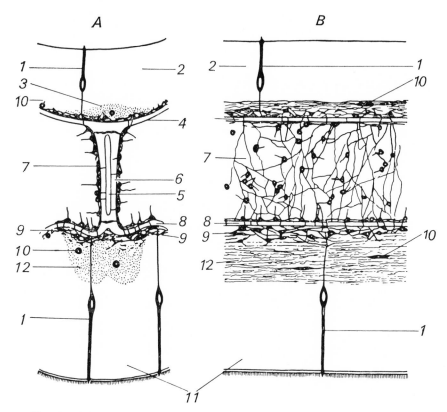

Fig. 67. *Diagram of structure of neural system of* Glossobalanus marginatus
(*Enteropneusta*).

A—part of cross-section of ventral part of trunk. B—parasagittal section in same
part of body, in the plane of one of the mesenterial neural plexuses. 1—sensory
cells; 2—gut epithelium; 3—ventral cord of intestinal part of neural apparatus;
4—boundary layer of gut; 5—ventral bloodvessel; 6—boundary layer of mesentery;
7—neural plexus of coelomic epithelium; 8—boundary layer of skin; 9—plexus of
associative cells; 10—bipolar cells; 11—epidermis; 12—ventral nerve cord of
ectoneural section of neural apparatus (after L. Silén).

A remarkable feature of Enteropneusta is the possession of a neural
plexus over the whole length of the gut (except the stomochord). This
intestinal section of the neural apparatus is fully similar in structure to
the ectodermal section, and also forms two longitudinal medial thickenings
or cords, dorsal and ventral; the ventral cord in the gut, like that in the
skin, is thicker than the dorsal (Fig. 67).

Moreover, the two leaves of the ventral mesentery of *Glossobalanus* contain neural plexuses connected both to each other and to the cutaneous and intestinal plexuses (L. Silén, 1950). In Silén's opinion, the coelomic plexuses are formed only of motor cells. The absence of primary sensory cells in the coelomic epithelium, together with lack of embryological data, leaves the origin of the coelomic section of the neural apparatus an open question: it is impossible as yet to say definitely whether the cells constituting it were formed from the coelomic epithelium or whether they are sunken motor cells belonging to the cutaneous and intestinal sections of the neural apparatus. As we shall see below, this question is one of great morphological significance.

The neural apparatus of Pterobranchia (Fig. 66, *B, C*) is constructed on the same general plan as that of Enteropneusta, but is still less separate from the external epithelium. The collar section of the dorsal cord is greatly developed, but is not sunken beneath the skin; cords run from it to the arms of the lophophore and along their dorsal sides, and cords also run from it to the ventral side of the body and form the nerve ring of the collar. Uniting on the ventral side they give rise to a ventral cord that runs along the front part of the metasoma and on into the stalk. From the annular cord arise the ventral cords of the arms, which run along afferent grooves like the ectoneural cords of starfishes and crinoids (Fig. 69, *E*).

Because of the absence of a 'cerebrospinal tube' in the collar, the neural apparatus of Pterobranchia is more primitive than that of Enteropneusta. E. W. MacBride (1910) suggests that possession of a lophophore was the reason for the formation of the principal condensation of the neural plexus on the dorsal side of the collar. A connection between the location of that condensation and the location of the tentacles is also seen in Pogonophora (see below). MacBride further suggests that Enteropneusta, on transition to a fossorial mode of life, lost the lophophore but kept the neural tube in the collar, which by that time had already acquired independent significance as the incipient nerve centre of the body. In discussing the architectonics of Hemichorda (see Vol. 1, Chap. IX) we have stated that the absence of a lophophore in Enteropneusta is secondary. Analysis of the structure of the neural apparatus of Hemichorda also supports that view.

The neural apparatus of Pogonophora (A. Ivanov, 1958b) is a diffuse plexus lying entirely in the external epithelium. It is apparently absent in the rear part of the body and in the arms. A dorsal cord that runs along the whole body is differentiated in the cutaneous plexus (see Vol. 1, Chap. IX, regarding the anatomy of Pogonophora). In the prosoma a widening of that cord, called the brain, is formed. Paired cords arise from the brain and unite on the ventral side to form a ring. Nerves run from them to the tentacles. In the mesosoma of the large *Spirobrachia* the epithelium of the dorsal side forms a deep groove, at the bottom of which lies the dorsal cord. That groove represents, as it were, the beginning of a collar neural tube of the same kind as in Enteropneusta. In the front part of the metasoma is a mediodorsal band of ciliated epithelium, perhaps a chemical sense-organ. It is underlaid along its whole length by the dorsal cord, which is there widened. The dorsal cord contains giant nerve fibres. There are tentacle nerves only in the epithelium on the dorsal side of the arms, whereas in the arms of Enteropneusta there are also nerve cords in the epithelium of the ciliated grooves.

The epidermal neural plexus, the dorsal cord, and the rudimentary cerebrospinal tube in the collar section bring Pogonophora close to Enteropneusta; possession of a dorsal cord only, with no ventral cord, brings them close to Chordata. The development of a brain in the prosoma, and the associated presence of tentacles there also, are specific features of Pogonophora distinguishing them from all other Deuterostomia.

From our survey of Hemichorda and Pogonophora we see that the neural apparatus of Deuterostomia is derived directly from an almost-diffuse neural plexus. Hence it is clear that all attempts to derive the anatomical structure of the neural apparatus of vertebrates from that of any group of Protostomia possessing a differentiated neural apparatus (whether annelids, nemertines, or any other) were foredoomed to failure.

Because of the level of its organisation the neural apparatus of Chordata cannot be compared with that of lower Deuterostomia, but architectonically it somewhat resembles that of Enteropneusta and Pogonophora. The central section of the neural apparatus of *Amphioxus* is a cerebrospinal tube: it extends along the whole body, like the dorsal cord of Enteropneusta,[1] and in its method of formation resembles the collar neural tube of the latter. In *Amphioxus*, however, no trace of the diffuse neural plexus of Enteropneusta remains; all the motor and associative cells in its body (except those in the vegetative section of its neural apparatus) are located in the cerebrospinal tube, and even some of the sensory cells have left the periphery and lie along the sensory nerves, far from the tissues innervated by them (A. Dogiel, 1903). In this respect *Amphioxus* stands higher than most invertebrates, although lower than vertebrates, since its sensory cells are not yet grouped into cerebrospinal ganglia. The walls of the hepatic process and of the gut in *Amphioxus* contain a vegetative nervous system, evidently homologous with that of vertebrates, but of simpler structure; it is connected with the central section of the neural apparatus by visceral branches of the dorsal nerves (J. Boeke, 1935; P. Kirtisinghe, 1940).

Ascidian larvae stand higher than *Amphioxus* in that their cerebrospinal tube forms a swelling at its front end, which apparently corresponds to the cerebrum of vertebrates and is provided with intracerebral sense-organs (Vol. 1, Fig. 176). On metamorphosis all the caudal part of the neural tube disappears. The cerebral vesicle, together with the larval sense-organs, breaks down, but the single ganglion of the adult ascidian develops from an expansion of its dorsal wall. From the ventral wall of the vesicle a

---

[1] The dorsal cord of Enteropneusta does not extend into the proboscis, but in Chordata the section of the body homologous with the proboscis is very slightly developed, for which reason it appears as if their brain reaches almost to the front extremity of their body.

subneural gland is formed, which opens by its own duct into the ciliated organ of the dorsal side of the oral cavity. The posterior part of the neural tube persists in the form of a dorsal band extending along the dorsal side of the pharynx as far as the visceral mass. It does not have neural structure, and it is connected with the gland but not with the ganglion.

From two to five pairs of nerves arise from the ganglion: to the oral siphon and the pharynx, to the cloacal siphon and the peribranchial cavity, etc., and one unpaired nerve to the viscera. The latter runs above the dorsal band. In the internal organs (gut, genital organs, heart) there is a neural plexus; it is also found in salps (Fedele, 1927). A general plexus has not been observed morphologically in the body walls of tunicates (Fig. 68, C), but E. Florey (1951) postulates its existence on the basis of physiological data.

Nevertheless the neural apparatus of tunicates, which is impoverished and simplified because of their sessile mode of life, is not really very primitive. Its structure is one of the best proofs of the origin of tunicates from highly-organised mobile animals.

It is interesting to observe in lower chordates the first rudiments of some of the endocrine organs that play an important role in vertebrates.

The endostyle of *Amphioxus* and ascidians, which functions as part of their food-procuring apparatus and still has the same role in Ammocoetes, is known to be the homologue of the thyroid gland in vertebrates. In *Amphioxus* it has been found to secrete an iodine-containing substance, evidently close to the hormone of the thyroid gland; the implantation of a few *Amphioxus* endostyles in an axolotl produces metamorphosis in the latter (K. Sembrat, 1958). Organic compounds of iodine are also found in the endostyle of *Ciona intestinalis* (E. Barrington and L. Franchi, 1956).

The cerebral gland of ascidians (Vol. 1, Fig. 178) was regarded by C. Julin (1881) as the homologue of the neurohypophysis (posterior lobe of the hypophysis) of vertebrates, on the basis of morphological data. It was later reported that substances had been found in that gland that acted on vertebrates like oxytocin and other hormones of the posterior lobe of the hypophysis (E. Butcher, 1930; Z. Bacq and M. Florkin, 1935); that, of course, would strongly support Julin's hypothesis, but the physiological aspect of the question, unfortunately, still remains unclear (P. Brien, 1948).

The neural apparatus of echinoderms, judging by the level of its organisation in the most primitive cases (starfishes), stands no higher than that of Enteropneusta and consists of the same three sections: cutaneous, intestinal, and coelomic. In accordance with the complex structural plan of echinoderms, however, it presents considerable anatomical complexity. In other classes the organisational level of the neural apparatus has been raised somewhat, although that process has not advanced very far in any echinoderms.

The ectodermal, or ectoneural, section of the neural apparatus of echinoderms is fully homologous with that of Enteropneusta, and in the

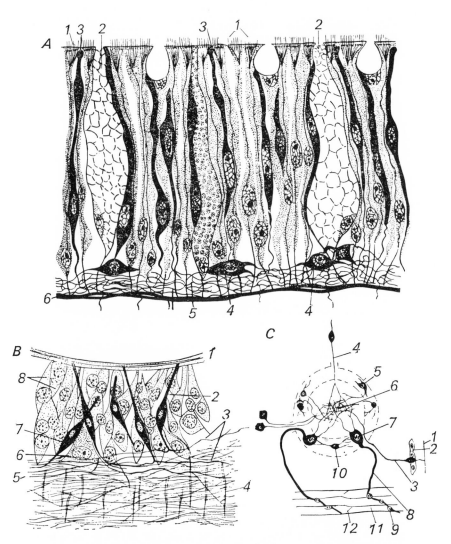

Fig. 68. *Histological structure of the ectoneural nervous system of Deuterostomia.*
A—diagram of structure of cutaneous neural plexus of *Saccoglossus pusillus*
(Enteropneusta): 1—ciliated (covering) cells; 2—glandular cells; 3—sensory cells;
4—nerve cells; 5—plexus of their fibres; 6—basal membrane of epithelium (after
Bullock). *B*—longitudinal section through radial ectoneural cord of the starfish
Asterias: 1—cuticle; 2—receptive processes of sensory cells; 3—basal processes
of sensory cells; 4—supporting fibres; 5—nerve fibres; 6—ganglionic cells; 7—
sensory cells; 8—supporting cells (after Meyer). *C*—diagram of structure of neural
apparatus of salps (interrelationship of central and peripheral sections): 1—tunic;
2—epidermis; 3—central processes of sensory cells of skin; 4—central processes
of epithelium of pharynx; 5—ganglionic cells; 6—central neuropile (plexus of processes of nerve cells); 7—motor cells; 8—their neurites; 9—motor endings on
muscles; 10—commissural cells; 11—muscle fibres; 12—boundaries between
muscle fibres (after Fedele).

most primitive cases is also a diffuse subepithelial plexus (Fig. 68), with local condensations in the form of cords and ganglia. In the starfish *Martasterias glacialis* sensory cells of a primitive type are distributed all through the skin, with 4,000 or more per mm², being especially numerous on the arms and around the spines and pedicellariae. A neural plexus lying in the basal part of the epithelium is formed by the neurites of sensory cells and by the associative cells situated there. The motor cells of the plexus

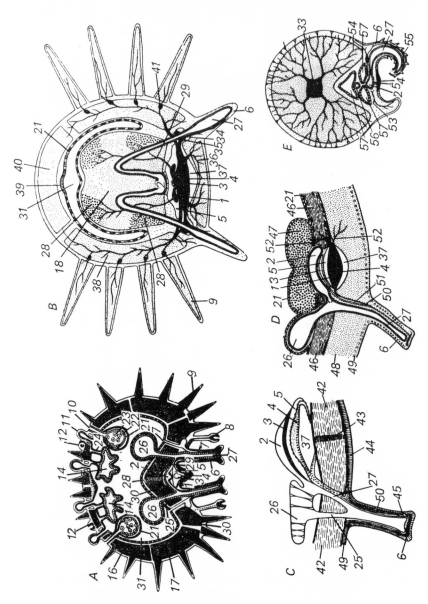

Fig. 69.—See caption opposite.

lie deeper than it does, and are separated from it by an interlayer of connective tissue permeated by a large number of nerve fibres (J. Smith, 1937).

We thus find in starfishes the first step in the inward movement of neural elements, motor cells being the first to sink inward; in most animals associative cells follow them, while in all invertebrates sensory cells almost always remain on the periphery.

The formation of cords and condensations in the ectoneural apparatus of echinoderms is due to the development of afferent ambulacral grooves. Therefore the ectoneural apparatus of starfishes and crinoids contains five radial cords running along the ambulacral grooves and formed by the epithelium of the bottom of these (Fig. 69, *A*, *E*).

In starfishes the radial ectoneural cords are connected with one another by a circumoral annular cord; in crinoids the latter is poorly developed— many authors believe it to be absent.

The ectoneural section of the neural apparatus of starfishes is sensory-motor. In particular, bunches of fibres arising metamerically from the ectoneural cords innervate the longitudinal muscles that connect the adambulacral plates. The activity of tube-feet, spines, and pedicellariae is also governed entirely by the ectoneural plexus.

In holothurians (Fig. 69, *D*), sea-urchins (Fig. 69, *C*), and the great majority of Ophiuroidea (Fig. 69, *B*), which have covered ambulacral grooves, the latter are converted into closed radial epineural canals connected by a circumoral epineural ring. In these groups the ectoneural stems are formed from the epithelium of the bottom of the epineural

Fig. 69. *Structure and arrangement of radial nerve stems in echinoderms.*
Diagrammatic cross-sections: *A*—through arm of starfish (from Sedgwick); *B*—through arm of brittle star (Ophiuroidea) from Hanström; *C*—through ambulacrum of regular sea-urchin (from Bütschli). *D*—semi-diagrammatic cross-section through ambulacrum of holothurian (from Hanström). *E*—diagrammatic cross-section through arm of crinoid (from Hanström). 1—hyponeural stem; 2—radial ambulacral canal; 3—radial "bloodvessel" of lacunar system; 4—radial ectoneural cord; 5—hyponeural sinus; 6—tube-foot; 7—its coelomic canals; 8—pedunculate pedicellaria; 9—spine; 10—genital orifice; 11—cutaneous gills; 12—sessile pedicellariae; 14—hepatic appendages (of gut); 16—supramarginal plate; 17—inframarginal plate; 18—vertebra (arm ossicle) of Ophiuroidea; 21—coelomic epithelium; 22—genital sinus; 23—gonad cavity; 24—mesenteries of hepatic appendages; 25—canal of ampulla; 26—ampullae; 27—canals of tube-feet; 28—muscles between ambulacral plates in starfishes, or intervertebral muscles in Ophiuroidea; 29—nerves of the hyponeural system innervating these muscles; 30—ambulacral plates of starfishes; 31—coelom; 33—radial stem of apical neural system; 34—epineural canal around bases of tube-feet in Ophiuroidea; 35—annular nerve cord in same place; 36—ventral shield in Ophiuroidea; 37—radial epineural canal; 38—lateral shields in Ophiuroidea; 39—ciliated groove along coelom of arm in Ophiuroidea; 40—dorsal shields; 41—ganglion at base of spine; 42—ambulacral plate in sea-urchin; 43—suture between ambulacral plates; 44—diffuse ectoneural plexus; 45—ganglion of tube-foot; 46—annular musculature of body in holothurians; 47—longitudinal musculature of the body in holothurians; 48—cutis; 49—epidermis; 50—nerve of tube-foot; 51—blood vessel of tube-foot; 52—cutaneous nerves; 53—paired hyponeural nerves of crinoids; 54—junction between endoneural (apical) and hyponeural sections of neural apparatus; 55—sensory papillae; 56—genital sinus with genital cord; 57—coelomic canals of arm.

canals. In echinoderms, therefore, as in other phyla, we observe a move-ment of the nervous system into the depth of the body, that movement taking place in them by fusion of the margins of the grooves and the forma-tion of a subcutaneous neuro-epithelial tube; a method very different from all others serving the same purpose in Protostomia and most nearly resembling the formation of the cerebrospinal tube in chordates.[1] Its chief difference from the latter consists in the fact that in chordates nerve tissue develops from all the walls of the tube and in echinoderms from only one wall, the others remaining purely epithelial (V. Dogiel, 1940; Livanov, 1945). In holothurians considerable reduction of the epineural canals is observed, the epineural ring is absent, and even the method of embryonic development of the ectoneural apparatus is greatly modified.

In all Eleutherozoa the centralised part of the ectoneural apparatus sends independent nerves to the musculature and at the same time sends connecting branches to the subcutaneous neural plexus and to the hypo-neural and apical (endoneural) sections of the neural apparatus. Together with the latter it plays the role of a central section of the neural apparatus. In sea-urchins the external plexus forms a number of condensations and ganglia connected with the tube-feet, the spines, and the pedicellariae, and, as in starfishes, completely governs the expedient reflexes of all these appen-dages. The most complex behavioural acts, however (turning the body over, taking to flight, etc.), are performed by means of the co-ordinating role of the radial cords and the circumoral ring. In Ophiuroidea the diffuse ectoneural plexus is considerably reduced and mainly consists of condensations, or ganglia, connected with the tube-feet, spines, and hooks. In holothurians it is still more reduced, although the ganglia at the bases of the tube-feet and those of the oral tentacles are retained. In this way a certain degree of centralisation of the ectoneural section undoubtedly takes place in the series: Asteroidea→Echinoidea→Ophiuroidea→Holothuroidea.

In starfishes and sea-urchins part of the ectoneural plexus connected with the circumoral ring passes directly into the epithelial neural plexus of the pharynx and the gut, at the edge of the mouth. The latter plexus is evidently formed from the endodermal gut-epithelium, i.e. the central phagocytoblast, in the same way as in coelenterates, in which the two plexuses pass into each other at the edge of the intestinal mouth. In star-fishes the neural plexus envelops the whole gut, and in sea-urchins at least the first part of the gut (L. Cuénot, 1948).

At the same time the relatively highly-differentiated sections of the neural apparatus in echinoderms are also connected with the epithelium of the coelomic cavities. The hyponeural section is connected with the epithelium of the hyponeural, or perihaemal, sinuses, and the apical section with the epithelium of the general coelomic cavity (see Chap. V).

Almost all echinoderms have a hyponeural section. As far as is known,

---

[1] Formation of ganglia by means of invagination of ectoderm has been described in Phylactolaemata (A. Gerwerzhagen, 1913), but we can hardly compare that process with the formation of neural tubes in Deuterostomia.

it has a purely motor function. In starfishes and Ophiuroidea its radial cords closely adjoin the radial ectoneural cords, being separated from them by two basal membranes of external epithelium and of hyponeural canals. These membranes are penetrated by nerve fibres joining the ectoneural and hyponeural cords. The annular hyponeural cord stands in the same relation to the annular ectoneural cord.

The hyponeural cords of starfishes and Ophiuroidea innervate part of the musculature of the arms and of the skeletal oral ring. In view of the slight mobility of the latter in starfishes, their hyponeural annular cord also is slightly developed. Sea-urchins, on the other hand, with their immovable shell and well-developed masticatory apparatus, lack radial hyponeural cords but do have a hyponeural ring with five ganglia. Spatangoidea, because of the reduction of Aristotle's lantern, have lost even these ganglia. In holothurians, which have strong body musculature and an immovable oral skeleton, the radial hyponeural cords are developed but the circumoral hyponeural ring is absent.

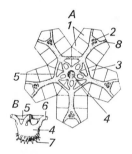

Fig. 70. *Apical section of the neural apparatus of the crinoid* Antedon bifidum.

A—projection on horizontal plane. *B*—profile view of neural cup, freed from surrounding skeleton. 1—first brachialia (segments of skeleton of arms); 2—commissures between nerve stems of arms (8); 3—radialia; 4—neural cup, containing five-chambered organ; 5—commissures between radial nerve stems; 6—radial stems; 7—nerves to cirri (from Cuénot).

The apical section is extremely well developed in crinoids, in which it plays the principal role in the whole neural apparatus, whereas in most Eleutherozoa it is more or less vestigial and in holothurians it is totally absent.

In crinoids the apical section is developed from the epithelium of a five-chambered organ containing five parallel and radially-arranged tubular processes of the right somatocoel, which extend along the body axis. In its definitive form it is a cup-shaped neural mass lying in the bottom of the crinoid calyx and sending out processes into the stalk and its ramifications on one side and into the arms on the other side. The stems directed towards the arms lie within the skeleton of the theca and divide dichotomously every time they pass from one plate to another, anastomosing together and forming a complex and geometrically-regular network (Fig. 70). In the arms the stems of the apical section run along the abactinal side. Both physiologically and anatomically the apical section of the neural apparatus is the central neural apparatus of crinoids, and governs the entire locomotor activity of free-living forms. It is connected by means of its branches with both the hyponeural and the ectoneural sections.

In starfishes the apical section of the neural apparatus is represented by five fine radial cords lying in the coelomic epithelium on the aboral side of the body; they all unite at the apical pole. Echinoidea and Ophiuroidea have a thin nerve ring in the walls of the annular aboral sinus, sending out nerves to the gonads and connected with the branches of the ectoneural section. This ring is usually regarded as the homologue of the apical section of the neural apparatus of crinoids and starfishes, which innervates their genital organs.

The structure of the neural apparatus in Enteropneusta and in the most primitive echinoderms is therefore comparable to its structure in the higher coelenterates, e.g. Anthozoa or Ctenophora. In all these cases parts of the neural apparatus are linked with all the principal epithelia of the body: (*i*) cutaneous epithelium (kinetoblast); (*ii*) intestinal epithelium (central phagocytoblast); and (*iii*) the epithelium of the peripheral parts of the gastrovascular apparatus in coelenterates and the coelomic epithelium in the lower Deuterostomia. In coelenterates the nerve cells located in any epithelium belong also to that epithelium by their origin, ontogenetically and phylogenetically. In the lower Deuterostomia the same is evidently true not only for the ectoneural plexus but also for the endodermal plexus with its numerous sensory cells and with its structure closely resembling that of the ectoneural plexus. The problem regarding the coelomic sections of the neural apparatus is somewhat more complex. The hyponeural cords of echinoderms and the mesenterial cords of Enteropneusta, so far as is known, consist only of motor cells. Until embryological data are obtained, therefore, it is very difficult to decide whether they represent parts of the coelomic-epithelial section of the neural apparatus or, taken together, an accumulation of motor neurons that have migrated from other plexuses, as Silén (1950) is inclined to believe. The apical section of the neural apparatus of Eleutherozoa has been too little studied for definitive conclusions about its nature and origin to be drawn, although formally it belongs to the coelomic epithelium. The situation is clearest regarding the apical section in crinoids, which originates from the five-chambered organ, far from the ectoderm and the endoderm, and is a fully-developed neural apparatus; there can therefore be no doubt of its coelomic-epithelial nature. In any case the problem of the nature of the coelomic sections of the neural apparatus of the lower Deuterostomia is worthy of the most diligent further study. It is also important to solve this problem in order to understand the origin of Deuterostomia themselves (the extent of their direct connection with Coelenterata) and to solve the problem of the origin of the coelom (see Chap. V).

Returning to the question of the general level of development of the neural apparatus in echinoderms, we may remark that it is very low, which is doubtless due to the slight development in them of the complex sense-organs whose possession by other animals stimulates the formation of physiologically-dominant centres and, through these, further differentiation of the entire neural apparatus. Slight development of sense-organs is in turn due to slight development of locomotor capacity in even the most mobile of echinoderms. In other words, deficient activity is the chief reason for the low level of organisation in echinoderms.

The most remarkable peculiarity of crinoids is the extremely high development of the apical section of their neural apparatus, which is of coelomic-epithelial origin and which in that group has assumed the role of a central section of the neural apparatus, governing (e.g. in Antedonidae) their complex swimming movements.

1. Preliminary Remarks

In their simplest form sense organs are represented by single sensory cells, and thus form part of the nervous system. Nerve tissue continues to be the most important and specific part of even the most complex sense organs; but the latter also include accessory formations not of neural, and partly not even of kinetoblastic, nature. The cephalopod eye does not belong entirely to the nervous system (i.e. to the aggregate of neural elements in the body), since it includes complex accessory parts not of neural nature, but it continues to be a part of the neural sensory apparatus of the animal. Because of the high degree of integration of all their parts, both neural and accessory, the individual sense organs acquire a good deal of independence, and their evolution often has a decisive effect on the evolution of the entire neural apparatus.

In their development sense organs show, perhaps more clearly than any other organs, the principal paths of their advancement, of increase in their efficiency and in the complexity of their functions. These paths are: (*i*) the gathering together of specific cells that were originally scattered through the whole body-layer (segregation); (*ii*) their union into a systematically-constructed organ (integration); (*iii*) inclusion in that organ of other types of cells and tissues, and construction of accessory apparatus from them. It often happens that similar sensory cells, when becoming part of differently-constructed organs possessing different accessory apparatus, serve as receptors of totally different stimuli.

Sense organs are extraordinarily diversified. They are usually classified, according to the location of the stimuli received, into: (*i*) exteroceptors, receiving stimuli from the external environment; (*ii*) proprioceptors, receiving knowledge of the position of the body and the relative positions and movements of its parts; and (*iii*) interoceptors, receiving stimuli from the internal organs. Besides that rather physiological classification, morphological classification is possible, based on the characteristics of the sensory cells of the sense organs, and of the accessory apparatus.

Adult sponges, which have no nervous system, have no sense organs. Sponge larvae, however, which are mobile and are therefore more highly integrated than the adults, possess some kind of sense organ (an eye?) (H. Meewis, 1939).

Among all other Metazoa those having the most primitive sense organs are hydroid polyps and Anthozoa, i.e. all polypoid coelenterates. They have only unco-ordinated epithelial sensory cells. That is the original condition from which the systems of sense organs of all other Metazoa are derived.

The only differentiation observed in polyps is the irregular, but at the same time systematic, arrangement of sensory cells over their body surface, which is linked with the irregular distribution of their entire neural plexus, i.e. with the structural plan of their neuro-sensory apparatus.

The primary sensory cell of coelenterates is an epithelial cell, passing at its basal end into a nerve fibre and possessing a receptive tip at its apical

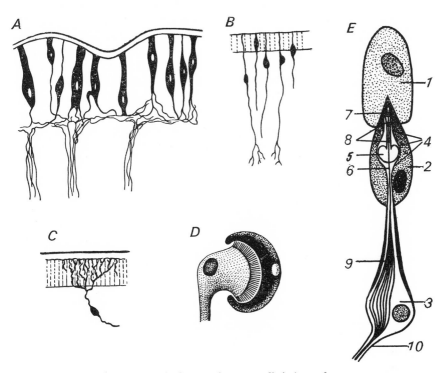

Fig. 71. *Principal types of sensory cells in invertebrates.*
A—epithelial sensory cells in the skin of *Lumbricus* (Oligochaeta) (after Retzius). *B*—movement of sensory cells from epithelium into depth of skin: epithelium of tentacle of *Helix pomatia* (after Hanström). *C*—sensory cells with much-branched peripheral processes (tentacle of *Helix pomatia*) (after Hanström). *D*—visual cell and contiguous pigmented cells of *Polycelis felinea* (Turbellaria Triclada). *E*— scolopal apparatus of an insect: 1—hypodermal covering cell; 2—enveloping cell; 3—sensory cell; 4—its outline within the enveloping cells; 5—vacuole; 6— axial column; 7—its terminal thickening; 8—thickening of vacuole wall; 9— neurofibrillae; 10—neurite (from Stempel).

end. In worms, molluscs, echinoderms, etc., some sensory cells fully retain their epithelial location (Fig. 71, *A*), but at the same time even in turbellarians the sensory cells are subject to the general tendency of the nervous system to sink into the body. In that process we see a number of successive stages (Fig. 71, *B*). In some cases only the part of the cell containing the nucleus sinks, while the apical part remains part of the epithelial layer; such a sensory cell resembles the cells of sunken epithelium. In other cases the peripheral part of the cell, becoming longer and narrower, is converted into a long narrow peripheral process, so that the body of the cell comes to lie deep beneath the epithelium. If then the peripheral process begins to branch, the primary sensory cell is converted into a sensory cell with free extremities (Fig. 71, *C*). Such cells are known in several turbellarians (Triclada). The extremities of such a cell may lie in the epidermis as before, but they may form secondary connections with the deep-lying tissues, producing sensory innervation of the latter. A sensory cell lying deep in the phagocytoblastic tissues, intimately linked with these tissues by its processes, and having lost connection with the epidermis but joined by a neurite to the central section of the neural apparatus, is a clear example of mutual interpenetration of the primary layers, which increases as an animal's structure becomes more complex. Primary sensory cells and cells with free extremities are the only types of sensory cells found in invertebrates. Secondary sensory cells lacking their own nerve processes are widespread in the sense organs of vertebrates, but are totally absent in invertebrates.

Primary sensory cells are the chief component in the structure of complex sense organs in invertebrates. They are of several kinds. The simplest kind consists of primary sensory cells without any receptive apparatus, or provided with simple sensory cilia, flagella, setae, or spines (undifferentiated sensory cells). In other cases we find primary sensory cells with specialised receptive apparatus, making possible the reception of special kinds of stimuli. These include optic cells (Fig. 71, *D*) and the *scolophore*-bearing cells of the chordotonal organs of insects (Fig. 71, *E*). There are also ordinary sensory cells, the receptor ends of which come into contact with the external world only through the medium of apparatus formed by other (not nerve) cells; these include the cells that innervate the bases of the sensory hairs of arthropods (Fig. 79), cells forming part of various balancing organs, etc.

2. LOWER SENSE ORGANS

Lower sense organs are those constructed of undifferentiated sensory cells and lacking apparatus that would transform external influences into a form suitable for reception. According to their function (when that is known) these organs can be divided into tactile organs and *chemoreceptors*. The *hygroreceptors* (organs sensitive to atmospheric humidity) of terrestrial arthropods are not essentially different in structure from chemoreceptors. Rheoreceptors (J. Gelei, 1930), which are sensitive to water movement, have been described in some turbellarians, and are probably close to tactile organs in their nature.

We have seen that the only sense organs in polyps are primary sensory cells, not collected into groups and possessing no accessory apparatuses formed of other types of cells. Such elementary sense organs continue to be found in other invertebrates, although usually more complex organs occur along with them. When sensory cells are diffusely distributed it is difficult to discover precisely what stimuli they receive. Experiment shows that *Hydra* possesses a tactile sense, i.e. it is able to receive mechanical stimuli, and also a chemical sense; but whether both types of stimuli are received by the same or by different nerve endings we cannot say. It is possible to answer that question when sensory cells are distributed in an uneven, regional manner, when separate groups of them are confined to definite parts of the body. In *Euplanaria lugubris* (freshwater Triclada) (Vol. i, Fig. i, *A*) a large number of sensory cells are concentrated at the front edge of the head, and also in the so-called *auricular* organs at the sides of the head. The auricular organs are pits lined with ciliated and sensory cells, which bring water from a distance by ciliary activity. The approach of prey excites the planaria: it crawls, turns its head from side to side, and receives water from different directions in its auricular organs. In this way it determines the direction of its prey and crawls directly towards it. If the margins of the head with the auricular organs are cut off, the planaria loses the faculty of locating prey, but on accidental encounter it recognises it and begins to devour it. On the other hand, if the front edge of the head is cut off without injuring the sides the planaria easily finds prey, but on finding it does not eat it. Evidently in *Euplanaria lugubris* the sensory endings in the front edge of the head serve organs of taste, which enable the animal to recognise prey when it is in immediate contact with it, whereas the auricular organs have an olfactory function.

The above example clearly shows the differences between the three points of view—morphological, physiological, and ecological—regarding sense organs. From the morphological point of view we find both in the auricular organs and on the front edge of the head aggregates of primary sensory cells, with only rather insignificant differences of structure, but differently situated: in one case they are in an open area, in the other they lie in ciliated pits. From the physiological point of view the two groups of sensory cells are almost identical, since both are used to perceive substances dissolved in the water. From that point of view they are all gustatory. But from the ecological point of view the two complexes of sensory cells serve different purposes: one group receives chemical stimuli from a distance and serves to locate prey, and the other receives chemical stimuli from immediate contact and identifies prey when found. From the ecological point of view the former are long-range chemoreceptors, i.e. olfactory organs, and the latter are contact chemoreceptors, i.e. gustatory organs.

For terrestrial animals the physiological and ecological classifications of chemoreceptors coincide. From the physiological point of view olfactory organs receive stimuli from gaseous substances, and gustatory organs those from liquid substances; but from the ecological point of view the former are organs of smell and the latter are organs of taste. For terrestrial animals reception of chemical stimuli from a distance is possible only through the medium of the atmosphere. On the other hand, for aquatic animals the two points of view regarding organs of chemical sensitivity are, as we have just seen, entirely different.

Purely morphologically, then, it is impossible to distinguish tactile, gustatory, and olfactory organs among the elementary organs of soft-skinned invertebrates; physiologically and ecologically that is possible only when they are combined into larger complexes.

In primitive cases, as stated above, elementary sense organs are represented by separate sensory cells. In Turbellaria Acoela larger sense organs, called *sensillae* (Fig. 72), are also formed, by the aggregation of

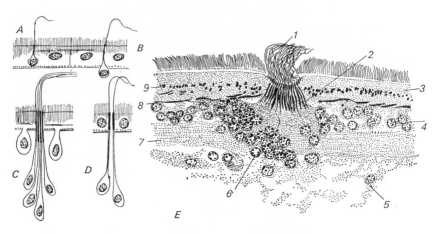

Fig. 72. *Development of sensillae in turbellarians.*
A and *B*—*Anaperus traerminnensis* (Acoela). *C*—*Childia groenlandica* (Acoela). *D*—*Macrostomum* (Macrostomida) (after Luther). *E*—*Convoluta sagittifera* (Acoela)—sensilla in longitudinal section (after A. V. Ivanov). 1—sensory flagella; 2—peripheral processes of sensory cells; 3—epithelial layer; 4—sunken epithelial nuclei; 5—parenchymal cells; 6—nuclei of sensory cells; 7—longitudinal nerve cord; 8—longitudinal skin muscles; 9—annular skin muscles.

a number of similar sensory cells. In many annelids and molluscs sensillae consist of compact aggregates of a fairly large number of primary sensory cells (Fig. 73), whose bodies often retain their epithelial location. Such sensillae often stand out as small papillae on the skin surface. It is very probable that they also have a tactile function, and doubtless also a chemo-receptive function; for instance, when parts of the skin of Lumbricidae are smeared with a quinine solution, the greater the number of sensillae in a given part, the stronger is the reaction.

In arthropods, which are covered with thick cuticle, the undifferentiated sensory cells form organs of chemical sensitivity (of taste and smell) and also hygroreceptors, which perceive the level of atmospheric humidity. The cell body always lies beneath the hypoderm, and only a thin peripheral process approaches the surface of the latter. Where it approaches the epithelium the cuticle is always thinned, so that the end of the process, passing through the epithelium, is separated from the external environment only by a very thin cuticular membrane. It thus becomes possible for the process to register the chemical properties of the environment in spite of the insulation that a thick armour might seem to impose. Elementary chemoreceptors of this type are found in arthropods in several different forms. In some cases the peripheral process of a sensory cell penetrates

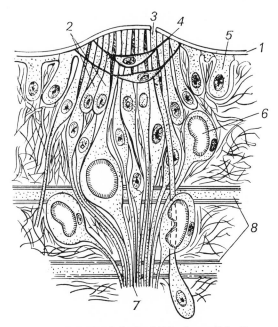

Fig. 73. *Sensilla of the leech* Hirudo medicinalis.
1—cuticle; 2—sensory cells; 3—glandular cell; 4—muscle cells for protrusion of sensilla; 5—covering cell; 6—optic cell; 7—sensillar nerve; 8—annular muscles of skin (from Bütschli).

the cavity of a thin-walled hair and extends through its whole length (a *trichoid* sensilla). That structure occurs in the olfactory hairs on antennae I of many Crustacea (Fig. 74), in the olfactory hairs of Xiphosura, and in many other instances. In the chemoreceptors of terrestrial arthropods the hair is sometimes reduced, but it is often retained (e.g., in a very characteristic form, in the olfactory organs of many Acariformes).

Insects' organs of taste take the form of *basiconic sensillae*, i.e. protruding papillae (Fig. 75, *A*), or of thin-walled hairs (trichoid sensillae), so that the gustatory nerve endings come into contact with the objects investigated.[1] It has been experimentally proved that the oral organs of insects contain gustatory nerve endings capable of detecting the presence of sugar: in insects with masticatory oral parts, they lie at the ends of the

---

[1] The thermoreceptors of insects are also represented by trichoid sensillae (e.g. in the beetle *Dorcus*, H. Gebhardt, 1951).

labial and maxillary palps, of the maxillae, and of the hypopharynx; in insects with suctorial oral parts, at the end of the proboscis; in flies, on the labial palps (labella); and in bees, on the antennae. Flies, butterflies, bees, and some bugs have chemoreceptors also on the tarsal segments (D. Minnich, 1921, 1930; H. and M. Frings, 1949; C. Lewis, 1954), and ichneumon flies have them on the ovipositor (Dethier).

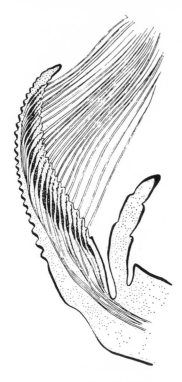

Fig. 74. *Longitudinal section through end of antenna I (antennula) of the deep sea crab* Geryon affinis.

A bundle of peripheral processes of the fusiform sensory cells lying at the base of the fibre projects into each of the thin-walled olfactory hairs; the central processes of these cells combine into a sensory branch of the antennular nerve (after Doflein).

Insects' olfactory organs (distant chemoreceptors) usually have the form of a pit surrounded by a theca (Fig. 75, *B*); at the bottom of the pit there may be, instead of a hair, a small papilla scarcely projecting above the skin, but even that may be lacking. Both the floor of the pit and the papilla are covered with a very thin layer of cuticle, to which the endings of the sensory cells approach. The sensory endings hidden in the pit cannot come into direct contact with objects, but odours have full access to them. Olfactory organs are mainly, but not exclusively, located on the antennae. Their number may fluctuate widely, depending on the role of the sense of smell in the insect's life, reaching 14,000 to 15,000 pits on each antenna in bees and falling to 10 or 12 in the dragonfly *Libellula depressa*.

Some insects, e.g. ants (A. Schmidt, 1938), also possess contact chemoreceptors on the antennae, with a papilla, not a pit. When investigating any object with its antennae an ant applies the sensory endings directly to the odoriferous surface. As a result the sharpness of its 'low perception' becomes greatly intensified, and because of the delicacy and precision of the movements of its antennae the ant can discern the size and shape of the odoriferous surface. Consequently the olfactory sense of such insects is of different, and much greater, significance than that of most other animals.

Generally speaking, organs of touch and chemical sensitivity in the bodies of all animals are distributed not uniformly but according to the structure of their neurosensory apparatus. Lack of uniformity is displayed in:

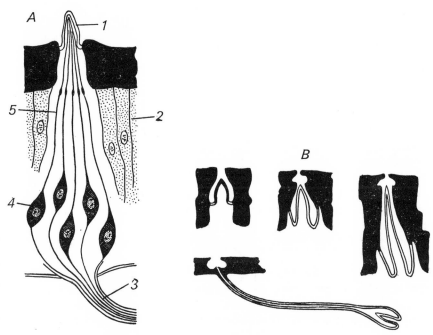

Fig. 75. *Sensillae of the chemical sense of insects.*
A—diagram of structure of basiconic sensilla. *B*—cuticular components of holo-conic sensillae (some examples). 1—sensory papilla; 2—hypoderm; 3—neural processes; 4—sensory cells; 5—their peripheral neural processes (after Snodgrass).

(*i*) uneven distribution of the organs through regions of the body, (*ii*) concentration of them in body appendages, and (*iii*) combination of them into organs of a higher order.

Uneven regional distribution of elementary sense organs is to some extent characteristic of all invertebrates. It varies greatly in separate groups and is difficult to describe concisely. *Hydra*, for instance, has many sensory cells on the tentacles, the oral disc, and the base; other hydroids, on the oral disc; ctenophores, on the edges of the mouth and the tentacles; turbellarians, as well as hydroid planulae, on the front end of the body; and many flat Triclada have an accumulation of sensory cells along the edges of the body ('sensory border'). The suckers of parasitic flatworms are rich in sensory cells. In nematodes and Rotatoria with a stable cellular composition the number of sensory cells is limited and their arrangement becomes constant and regular. In higher groups the diversity is still greater because of the increasing complexity of the neural apparatus.

The tentacles of coelenterates, as we have seen, are rich in sensory cells, but they are appendages of general use, mainly for grasping. Some of their probable or suggested homologues in Protostomia have retained only a sensory function: these include the apical tentacles of Polyclada; the palps, tentacles and cirri of most polychaetes; the cephalic and epi-podial tentacles of molluscs; etc. One glance at Fig. 76, *A* is enough to show that the cephalic appendages of errant polychaetes have been con-verted into sense organs of a higher order, in which the separate sensory cells are merely specific components. The antennae of arthropods are also essentially sense organs, particularly of touch and smell. In insects they may also bear hygroreceptors in the form of basiconic sensillae, e.g. in the beetles *Tenebrio* (Pielou, 1940) and *Tribolium* (L. Roth and E. Willis, 1951). In a number of instances antennae attain a high level of complexity and perfection as sense organs, as we have seen with regard to ants' antennae.

Other appendages of arthropods also may assume a sensory function, e.g. the first pair of legs of Notostraca, Telyphones, Araneina, and some Acariformes,[1] the pedipalps of Opiliones, etc. That commonly takes place in Chelicerata, which have no antennae; Notostraca also have much reduced antennae.

Among molluscs *Neopilina* has one, and Gastropoda one or two, pairs of cephalic tentacles, the posterior pair of which forms specialised olfactory organs (*rhinophores*—see Vol. 1, Fig. 97, *E* and Fig. 163) in Opisthobranchia. The homologues of rhinophores are perhaps the cephalic olfactory pits (or rarely papillae) of Cephalopoda (V. Dogiel, 1940). The olfactory pits of Cephalopoda are certainly organs of scent for hunting. Many molluscs have what are known to be secondary tentacular appendages, also with a sensory function, such as the marginal pallial tentacles of various Lamelli-branchia. The arms of Cephalopoda, which are outgrowths of the foot and fulfil mainly locomotor and grasping functions, are also very rich in sensory nerve endings.

Among the appendages of Deuterostomia, the tube-feet of echinoderms are noteworthy for their abundant content of sensory cells.

Aggregates of sensory cells in body appendages are sometimes diffuse and sometimes well individualised. We find individualised aggregates of sensory cells also in other parts of the body. Aggregates of sensory cells, much exceeding ordinary sensillae in size, are widespread among molluscs

---

[1] The tarsi of the anterior limbs of gamasid mites (Gamasoidea, Parasitiformes) bear some basiconic and thin-walled trichoid sensillae (E. Nelzina, 1951) as well as tactile hairs, and serve as both contact and distant chemoreceptors (Men Yan-Tsun', 1959). In Ixodoidea all these sensillae lie in a deep pit, forming the so-called Haller's organ, which is an organ of both smell and humidity-perception (P. Schultze, 1941; A. Lees, 1948). The anterior tarsi of spiders also bear the so-called tarsal sense organs in the form of several basiconic sensillae, which lie in a deep pit and are hygroreceptors (H. Blumenthal, 1935).

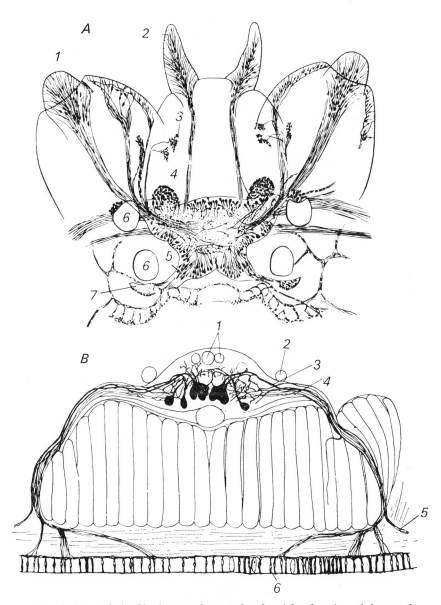

Fig. 76. *Inter-relationships between the central and peripheral sections of the neural apparatus in annelids.*

A—supraoesophageal ganglion and sensory nerve-endings in the head of *Nereis diversicolor*: 1—palps with their sensory cells; 2—antennae; 3—nerve endings on head muscles; 4—pedunculate bodies; 5—sensory cells of olfactory (nuchal) organs, lying deep in the brain; 6—eyes; 7—olfactory (nuchal) organs. *B*—cross-section of ventral wall of body of earthworm (Oligochaeta Lumbricidae): 1—'giant fibres' of ventral nerve chain; 2—blood vessel; 3—motor fibres (thicker); 4—sensory fibres (thinner); 5—annular sensory nerve; 6—sensory skin cells (after G. Retzius).

and polychaetes. Among the former, these are most numerous in the mantle cavity, although they also occur elsewhere, e.g. in the mouth, where they act as organs of taste (the gustatory papillae of Rhipidoglossa, the sub-radular organ of Loricata and its homologues in *Neopilina*, Scaphopoda, and Cephalopoda).

The most constant of the numerous sense organs in the mantle cavity are the *osphradia*, which are paired organs lying near the ctenidia. They are found in Gastropoda (except those that have lost the ctenidia), in Lamelli-branchia, and in *Nautilus* among Cephalopoda; they have homologues among the pallial organs of Loricata. Osphradia occur in Fissurellidae in the primitive form of loose accumulations of sensory cells on the walls of the mantle cavity; in other Gastropoda they are well developed, and in higher Prosobranchia they are usually cylindrical and covered with pinnately-arranged folds (Vol. 1, Fig. 160), whose epithelium contains a large number of sensory (as well as ciliated and glandular) cells. Osphradia are believed to be organs of scent that test the water coming to the ctenidia. Some authors think that the chief function of osphradia is to perceive the amount of suspended matter in water entering the mantle cavity.

The lateral organs of Rhipidoglossa are round areas of sensory epi-thelium, capable of invagination or evagination in the form of papillae. They lie at the bases of the epipodial tentacles of *Fissurella* and Trochidae (Fig. 43, *B*) and also in the epipodia of *Haliotis*. These organs are very similar in structure to the lateral organs of annelids. Among annelids, lateral organs are found in Eunicidae and in many sessile polychaetes, and in Myzostomidae. They usually lie between the two branches of the parapodia. The similarity in structure, location, and innervation of the lateral organs of polychaetes and of Rhipidoglossa is one of the arguments in favour of the view that the epipodia of the latter are composed of elements homologous with the components of the parapodia of poly-chaetes. The view that the lateral organs of polychaetes correspond to the placodes of vertebrates (H. Eisig, 1887) and to the lateral organs of Palaeonemertini and of some Heteronemertini can scarcely be regarded as sound, in view of the scarcity of characteristic features of the organs themselves and the great differences in the groups' structural plans. Nemertines have only one pair of lateral organs.

The cephalic olfactory fossae of Protostomia have widely-distributed homologues among lower sense organs. We meet them first in turbellarians. In their simplest form they are represented by two patches of sensory epithelium, free from rhabdites, situated on the head (e.g. in some Typhlo-planidae among Rhabdocoela and in some Triclada). In *Stenostomum* (Notandropora) (Fig. 77, *A*) they are represented by well-defined cup-shaped depressions of the integument lying at the sides of the head; their epithelium contains sensory cells and also glandular and ciliated cells; the latter create strong currents that bring water to the olfactory fossae. The auricular organs of Triclada and other hunting-olfactory organs (H. Müller, 1936) also belong to this group. All such organs are innervated directly from the anterior part of the brain. Astonishingly similar formations are the *amphids* of nematodes (without cilia, like the whole body of a nematode), the cerebral organs of nemertines, and the *olfactory* organs of polychaetes. The cerebral organs of Palaeonemertini (Fig. 77, *B*) differ only slightly from the olfactory fossae of turbellarians; in other orders of nemertines they attain considerable complexity (Fig. 77, *C*). As we have seen (Vol. 1, Chaps IV and V) there are many data supporting the homology

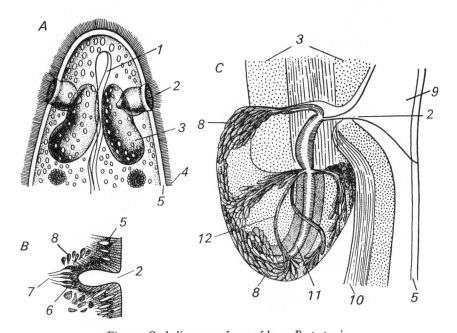

Fig. 77. *Cephalic sensory fossae of lower Protostomia.*
A—Stenostomum leucops (Turbellaria Notandropora) (after von Graff). *B—Tubulanus annulatus* (Palaeonemertini) (from Bütschli). *C—Micrura fasciolata* (Heteronemertini) (from Bütschli). 1—nephridium; 2—aperture of fossa or canal; 3—brain; 4—cilia; 5—epidermis; 6—sensory cells; 7—nerve of sensory fossa; 8—glands of sense organ; 9—cephalic clefts; 10—longitudinal nerve stem; 11—canal of sense organ; 12—duct of glands.

of the front end of the body in Scolecida with that in trochophore animals. Homology between the cephalic olfactory fossae in these two groups is therefore very probable. The nuchal organs of polychaetes (Fig. 76, *A*) are organs that aid in the finding and recognition of prey; when the nuchal organs of *Nereis* are removed, it ceases to take food (F. Rullier, 1950). In some polychaete families (Amphinomidae, Spionidae) the nuchal organs are extraordinarily developed, growing back along the dorsal side of several body segments. The nuchal organs of polychaetes are probably homologous with those of Sipunculoidea; the latter possess canals lined with sensory ciliated epithelium; these canals are deeply sunk in the brain, like the cerebral organs of nemertines.

No certain homologues of cerebral olfactory fossae are known in molluscs or arthropods. Attempts to find homologues of these fossae in coelenterates or Deuterostomia have been made, but can scarcely be regarded as successful.

3. BALANCING ORGANS

In the olfactory fossae of worms ciliated cells play the role of an ancillary mechanism that creates water currents and so improves the functioning of the organs. By strengthening a sense organ and altering its ecological significance for an animal (its ability to discern properties of distant objects), however, such ancillary mechanisms do not serve as means of transforming the energy itself, as means of converting it from a form that cannot be perceived directly into a form that can be so perceived. Mechanisms of the latter type are quite indispensable in organs serving the sense of equilibrium (balancing organs).

No nerve-ending, of course, can directly perceive the direction of the force of gravity. Very many animals, apparently, are generally incapable of orienting themselves with respect to that direction; in taking up a definite position relative to the substrate they are guided only by tactile impressions (*Planaria*, starfishes). On the other hand, many animals that lack special organs of equilibrium, such as Ixodidae in the order Parasitiformes (V. Mironov, 1939), assume a definite 'geotactic' position on a vertical plane, orienting themselves with respect to the direction of the force of gravity by the aid of a muscle sense—discernment of the amount of stretching of the muscles and mechanical tissues of their bodies. The same has been shown to be true of a number of spiders.

A number of other animals have developed special adaptations that enable them to discern the direction of the force of gravity with more sensitivity and precision. Such balancing organs have developed mostly in the large groups of planktonic Metazoa, beginning with ctenophores and medusae. The unenclosed statocyst of ctenophores has been described in Vol. 1, Chap. III, and results from the direct progressive development of an aboral aggregate of sensory cells possessed by the primarily-planktonic larvae of coelenterates. The narcomedusa *Hydroctena salenskii* has two shortened tentacles, which enclose statocysts that have moved to the aboral pole of the umbrella, resembling an aboral sense organ (C. Dawydoff, 1953). The balancing organs of other medusae lie on the margin of the umbrella and are of two different structural types. In Scyphomedusae, Trachymedusae, and Narcomedusae the balancing organs are modified tentacles; in Leptomedusae they are pits or vesicles.

A primitive example of a balancing organ of the first type is found in the narcomedusa *Cunina* (Fig. 78, *A*), in which some of the marginal tentacles are much shortened, thickened, and united to the margin of the umbrella by thin stalks. Calcareous concretions form in the endoderm of the tentacle, weighing it down. Around the place of attachment of the stalks the epithelium of the umbrella margin is thickened into a pad containing a large number of sensory cells provided with long sensory hairs. With every tilt of the body the tentacle bends on the stalk and presses on

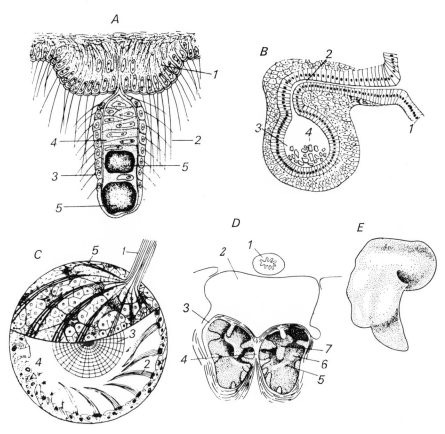

Fig. 78. *Balancing organs.*

A—club-shaped modified tentacle of the medusa *Cunina lativentris* (Narcomedusae): 1—sensory papilla; 2—sensory hairs; 3—ectoderm; 4—endoderm of the club; 5—statolith. *B*—statocyst of *Arenicola marina* (Polychaeta Drilomorpha): 1—epidermis; 2—duct of statocyst; 3—its vesicle; 4—grains of quartz, acting as statoliths. *C*—statocyst of *Pterotrachea* (Gastropoda Prosobranchia); 1—statocyst nerve; 2—ciliated cells; 3—statolith; 4—sensory cells; 5—nerve fibres (from Bütschli). *D*—*Sepia officinalis* (Cephalopoda), front half of statocyst in cross-section through head, view from rear: 1—oesophagus; 2—suboesophageal ganglionic mass; 3—statocyst capsule; 4—septum between statocysts; 5—'auditory' crest; 6—curved papillae on inner surface of statocyst; 7—statolith. *E*—statolith of right statocyst of *Sepia* (from A. V. Ivanov).

the hairs of the corresponding side. In Trachymedusae, such as *Carmaria*, the margin of the umbrella grows around a balancing tentacle forming something like a bell, whose walls also are provided with sensory hairs.

Leptomedusae have balancing organs of a different type. In their simplest form they are small pits lying at the base of the velum and having no connection with the tentacles. In most Leptomedusae the pits become enclosed and form a vesicle, the *statocyst*. Some cells of its epithelium secrete concretions within them and fall into the statocyst cavity, forming *statoliths*. Other cells are sensory, receiving pressure from the statoliths, and the animal is enabled to regulate its movements, depending on which particular sensory cells of the statocyst are pressed upon by the statoliths.

Organs of both these types can, of course, perceive vibrations and all kinds of abrupt movements (particularly contractions of the umbrella of the medusa itself) as well as the direction of the force of gravity, i.e. they are *seismoreceptors* and *proprioceptors;* they therefore take part in regulation of the rhythm of contraction of the umbrella, at least in Scyphomedusae. Sessile Scyphozoa (Lucernariida) have no balancing organs.

Balancing organs of the club-shaped-appendage type are rare in the animal kingdom, and develop in widely different ways in separate instances. They include the intracerebral balancing organ of ascidian larvae, the club-shaped hairs (*trichobothria*) of Acariformes, and the *spheridia* of sea-urchins. The latter are short, club-shaped spines, which turn passively on their articulated papillae and press on the fibres of sensory cells lying at their bases. On the other hand, we find a balancing organ in the form of a statocyst, i.e. a vesicle or a pit (a still-unclosed vesicle) with statoliths, in the most diverse animal groups: ctenophores, turbellarians, nemertines, annelids (Fig. 78, *B*), molluscs (Fig. 78, *C*), Malacostraca, holothurians, tunicates (*Doliolum*), and vertebrates. All these organs resemble one another in many respects, but they differ widely in their location and have arisen independently in the various groups. It is true that the unpaired intracerebral statocyst of turbellarians is homologous with the aboral balancing organ of ctenophores. In *Xenoturbella* it still lies in the subepithelial neural plexus of the front end of the body (Fig. 182). We have seen above the part taken by the statocyst in the development of the cerebral ganglion in turbellarians (Chap. III). A primary aboral statocyst homologous with that of ctenophores no longer occurs outside the class Turbellaria, and the aboral sense organ is represented by a simple apical plate. The statocysts of all other animals are new formations. In the only nemertines possessing statocysts (*Ototyphlonemertes* in the order Metanemertini) they lie in the brain, but are paired. In other Protostomia they appear in the most diverse places.

In polychaetes the statocysts are located above the setigerous sac of the notopodium; in Ariciidae they are metameric, and occur in several of the anterior body segments (e.g. in *Scoloplos* in the fourth to the thirteenth segments); in Arenicolidae they lie in the first, and in Terebellidae and Sabellidae in the second, body segment.

The statocysts of molluscs are also paired, and although they are innervated from the cerebral ganglia they always lie in the region of the pedal ganglia. As we have seen in Vol. 1, Chap. VIII, the right statocyst in Pectinidae is vestigial because of the general asymmetry of their body.

Among molluscs only Amphineura lack statocysts entirely. The balancing organs of Cephalopoda, which are enclosed in special chambers in the cartilaginous skull, are very complex (Fig. 78, D).

The statocysts of Crustacea Decapoda are situated in the basal segments of the antennules, those of Mysidacea in the last pair of pleopods, and those of Amphipoda and Isopoda in various parts of the body. The balancing function of the statocysts in Decapoda has been fully proved experimentally. No doubt they are also able to register vibrations.

In the life of many animals (e.g. all fossorial forms) statocysts are not so much balancing organs as seismoreceptors, organs for sensing vibrations. When you approach the burrow of the littoral polychaete *Arenicola* it does not see you, but it immediately disappears into the burrow, sensing the soil vibrations made by your tread. Statocysts are usually seismoreceptors in other sessile polychaetes also. Statocysts are not organs of hearing, as old authors believed, and their former name of otocysts (auditory vesicles) has now been abandoned.

The independent appearance of statocysts in different groups of animals is proved by differences in their location. In almost every group we find them both in the form of vesicles and in the primitive form of pits. Most molluscs, for instance, have vesicular statocysts, formed ontogenetically by invagination of ectoderm; but in *Neopilina*, in *Nucula* and *Leda* (Protobranchia), and in some other primitive Lamellibranchia each statocyst retains connection with the surface of the foot by means of a long narrow canal. Among polychaetes, *Scoloplos armiger* has statocysts in the form of pits; *Arenicola grubei*, in the form of a closed vesicle; and *A. marina* (Fig. 38, B), in the form of a vesicle still communicating with the external environment by a narrow duct. We could cite similar stages of development in connection with the statocysts of Crustacea Decapoda, not to mention vertebrates.

4. Tactile Hairs in Arthropods

When the cuticle is thin, as in caterpillars or in thin areas of the skin of other arthropods, touch sensations are received by means of sensory cells with free endings; the cells lie at some depth in the body, and their peripheral processes are distributed over a definite area of the skin and register every deformation of the latter, every contact (Fig. 79). The sclerites, however, suffer no deformation with light contact, and there the sensation of touch is registered by tactile hairs (Fig. 79). Each tactile hair of an insect is secreted by a single trichogenous cell. Around it a thecogenous cell secretes a theca, or ring, with which the hair is movably articulated by means of the thin connecting cuticle. The theca surrounds a fine canal that penetrates the cuticle towards the base of the hair and is filled with the protoplasm of the trichogenous cell. Every hair of this kind is provided with one or more sensory cells, which lie beneath the skin (Fig. 79) but which send their unbranched peripheral processes through the canal of the theca to the place where the hair is articulated with the cuticle.[1] When the hair is lightly touched it is moved from its position, and the consequent pressure on the articulation is recorded by the endings of the sensory cells. In this way extreme sensitivity to touch is attained, no matter how thick and tough the cuticle may be (A. Zawarzin, 1912).

Hairs of this type are possessed by all arthropods and cover their bodies in great numbers, serving as elementary touch organs and simultaneously fulfilling a number of other functions. In small arthropods, such as many Acariformes, the number and arrangement of sensory hairs on the surface

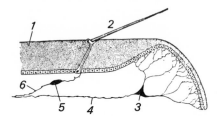

Fig. 79. *Tactile nerve-endings in insects (diagrammatic).*

1—cuticle of sclerite; 2—hair; 3—sensory cell with numerous endings on the thin cuticle of the joint membrane; 4—its central process; 5—sensory cell with ending at base of hair; 6—its central process (after A. A. Zavarzin).

of the body and appendages (chaetotaxy) is often strictly constant, so that the structure of the tactile apparatus (chaetoma) of the animal becomes extremely regular (Vol. 1, Fig. 129). In many cases tactile hairs may also assume special functions. In Crustacea Decapoda, for instance, the sensory hairs arranged in a ring on the distal end of each limb segment lie parallel to the next segment and act as proprioceptors: every time the limb bends it touches several of these hairs, and thus the animal knows the position of its limbs, which is obviously necessary for co-ordination of its movements. A number of insects certainly become aware of air vibrations, i.e. sound waves, by means of sensory hairs, and it has been suggested that spiders possess the same faculty.

The hairs on the cerci of cockroaches register sound waves with frequencies up to 3,000 vibrations per second (J. Ramsay, 1952).

Water-scorpion larvae (family Nepidae, order Rhynchota) have two grooves on the ventral side, joining together all the spiracles on each side. When the animal dives the grooves are full of air, which is retained in them by rows of stiff, immovable spines along their margins. In each segment, however, there is a short section of the groove bordered only by short, flexibly-articulated sensory hairs. When the animal crawls on an inclined surface the air, moving forward or backward along the canal, parts the sensory hairs at one end of the canal and lets them come together at the other: in this way the animal perceives that its body is tilted with

---

[1] All the four cells that form a tactile hair (trichogenous, thecogenous, sensory and neurilemmal) arise through successive division of one epidermal cell, at least in butterflies and caddis-flies (K. Henke and G. Rönsch, 1951). In some cases the sensory cell degenerates, so that a non-innervated hair is the result. In other cases the sensory cell of the hair divides, sometimes repeatedly, and then several sensory cells appear at the base of a single hair.

relation to the force of gravity. Geotaxis is therefore well developed in water-scorpions. In adults the apparatus is somewhat more complex than in larvae (W. Baunacke, 1912). It is interesting to compare that balancing organ with the statocysts of other animals: statocysts contain bodies heavier than water and so operate on the plumb-line principle, whereas the canals of water-scorpions contain air and operate on the spirit-level principle. In both cases, however, the accessory apparatus for perceiving the direction of the force of gravity is a system of two bodies of different specific weight.

The few examples cited above show how varied are the uses of the sensory hairs on the bodies of arthropods.

5. SCOLOPHORES

These formations (Fig. 71, *E*) are also possessed by arthropods, but are much less widespread than sensory hairs, being found only in insects.[1] The principal part of a typical scolophore is a receptive bipolar cell. From one pole a central process arises, and from the other a receptive apparatus consisting of an elastic filament, which either is stretched in a cavity formed by the so-called enveloping cell or has its distal end free in that cavity. The cavity is lined with cuticle. The enveloping cell generally loses its direct connection with the hypoderm and is joined to it only by an elongated cell, the so-called cap.

Scolopal cells are usually arranged in groups in different parts of the insect's body, forming *chordotonal* organs. Primarily they perceive tensions in the parts of the cuticle with which they are associated, and also vibrations in the substrate on or in which the insect is at the time. Very many insects are sensitive to vibrations. Sometimes insects can perceive sound vibrations in the substrate by means of their chordotonal organs, but judging by their structure these organs themselves are scarcely capable of receiving sound waves directly from the air. Usually they display tactile sound reception, resembling to some extent the reception of sound vibrations by a hand laid on the back of a human speaker. In insect larvae chordotonal organs are generally located on the trunk, and in adults on the wings and appendages, including the antennae.

Chordotonal organs, however, served also as the material from which true hearing organs were formed. Of such origin are *tympanal* organs (Figs 80 and 81), which have arisen independently in several orders of insects. Differing in location and in structural details these organs are clearly of independent origin in each group. The tympanal organs of locusts are located on the first abdominal segment; in their close relatives grasshoppers and crickets, on the tibia of the front leg, and in cicadas on the second abdominal segment; among Lepidoptera, in the families Pyralidae, Geometridae, Brephidae, and Cymatophoridae and in females of the family

[1] Recently, however, very similar sensillae have been described in Crustacea Decapoda (M. Whitear, 1960).

Uraniidae they are located on the first abdominal segment, in male Uraniidae on the second abdominal segment (H. Eltringham, 1933), and in Noctuidae on the sides of the metathorax (F. Eggers, 1937). The chief component of the tympanal organ is the tympanic membrane, formed by an area of skin with thin cuticle, to which the cuticular membrane of the large tracheal trunk is closely apposed from within. The external layer of the tympanic membrane is stretched on a frame of thickened cuticle, and usually lies at the bottom of a depression, forming, as it were, an 'external

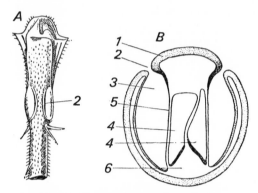

Fig. 80. *Tympanal organ of grasshopper.*
A—fore-tibia of grasshopper from front, showing two openings of tympanal organ (auditory clefts). *B*—cross-section through tibia in region of tympanal organ. 1—external cuticle of leg; 2—auditory cleft; 3—tympanal cavity; 4—tracheae; 5—tympanal membrane; 6—cavity of leg (after Schwabe).

ear'. Sound waves cause the tympanic membrane to vibrate; it transmits the vibrations to the tympanal organ. In grasshoppers the latter is an aggregate of scolophores of different lengths, arranged in order of decreasing length in a single row. This structure suggests that the tympanal organ is an analyser, capable not only of receiving sounds but also of separate reception of sound vibrations of different frequencies. In fact, the tympanal organs of Acrididae receive sound waves of 500–10,000 vibrations per second. The physiological role of tympanal organs as organs of hearing was first positively proved for the grasshopper *Thamnotrizon apertus* (J. Regen, 1901). Later the perception of sound by Orthoptera, Cicadidae, and Lepidoptera was studied in detail by measurement of the action potential in the tympanic nerve resulting from the action of sound waves of various frequencies on the tympanal organ (P. Haskell and P. Belton, 1956).

Most insects that have tympanal organs also have vocal organs; these include locusts, crickets, grasshoppers, and cicadas. In all of these the voice and hearing are a means of co-ordinating the behaviour of members of a single species and, above all, of helping the sexes to find each other. The Carboniferous Protorthoptera did not yet have either vocal or hearing organs, nor did the primitive Jurassic Orthoptera Locustopsidae and

Elcanidae (A. Handlirsch, 1907). It was also in the Jurassic, however, that true grasshoppers and crickets made their appearance (A. Handlirsch, 1925). Not until the Mesozoic was the world enriched with the songs of insects and the natural audience for these songs.

Lepidoptera are, generally speaking, without vocal organs, except death's-head moths (*Acherontia*) and a few others, and in them the tympanal organs serve another purpose: the hearing organs of Noctuidae

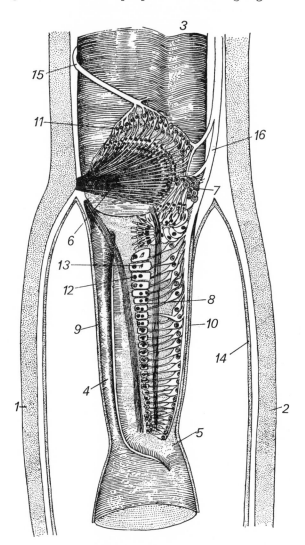

Fig. 81. *Tympanal organ of grasshopper, opened on front side.*

1 and 2—external cuticle; 3, 4, and 5—tracheal trunks; 6—subtibial organ; 7—intermediate organ; 8—crista acustica (auditory crest); 9 and 10—tympanal membranes; 11—sensory cells of subtibial organ; 12—terminal cells of crista acustica; 13—sensory rods; 14—cuticle of auditory concha; 15—subtibial nerve; 16—tympanal nerve (after Schwabe).

are attuned to the voices of bats; some species of Noctuidae react to sounds similar to the voices of bats by taking to flight, and others by immediate landing.

Johnston's organ, which also is formed of a large accumulation of chordotonal organs and is located on the second segment of the antennae of Pterygota and Lepismatidae, is partly an organ of hearing; it is especially intricate in mosquitoes (Diptera Culicidae), in which its significance as an organ of hearing has been proved physiologically (A. M. Mayer, 1874; L. Roth, 1948; H. Tischner, 1953) and is evident from its structure (H. Risler, 1953). The ability of male mosquitoes to distinguish the sounds of females of their own species has now been fully proved. In particular, males of the mosquito *Anopheles albimanus* are trapped in large numbers when the lure is a loud-speaker repeating the previously-recorded sounds of females of that species (M. Kahn, W. Celestin, and W. Offenhauser, 1945).

6. ORGANS OF SIGHT

Sensory cells that perceive light are always provided with special receptive structures in the form of a layer of rods bordering the apical surface of the cells (Fig. 82, *A*). Each rod is apparently prolonged into a neurofibril. In the simplest case the rods lie in the direction of the light-ray to be received. Often, however, the apical surface of the cell together with its layer of rods is curved, assuming a convex (Fig. 82, *C*), concave, or folded shape (the latter in the eyes of the polychaete *Polyophthalmus* of the order Drilomorpha, Fig. 82, *D*). In oligochaetes and leeches the rod layer lines a closed cavity or vacuole within the cell (Fig. 82, *E*); such a structure is most easily explained as resulting from closure of the concave rod-lined surface of a cell. In other cases separate elements of the rod layer become fused into larger structures, as in the formation of *rhabdomeres* in the visual cells of many arthropods (Fig. 82, *H*).

The original location of visual cells, like that of all primary sensory cells, is epithelial. Therefore the simplest type of organ of sight consists of separate visual cells scattered in epithelium. In coelenterates and turbellarians such organs of sight have not yet been described, but they are known in earthworms (Lumbricidae) (R. Hesse, 1896) and in the metameric sensillae of leeches (M. Bhathia, 1956).

Modifications of that primary type take several forms: (*i*) aggregation of a number of visual cells into compact, well-defined groups; (*ii*) migration of visual cells within the body; (*iii*) creation of pigmented screens; (*iv*) creation of dioptric adaptations; (*v*) creation of protective adaptations; (*vi*) creation of accommodation and oculomotor adaptations.

In spite of wide variation in details, the chief structural types of sight organs are few and recur in various branches of the animal kingdom, evolving independently in each. There are two chief types of eyes, parenchymatous and epithelial.

Parenchymatous eyes arise through migration of visual cells from the epithelium into deeper-lying tissues. The visual apparatuses of the leech *Pontobdella muricata* (Rhynchobdellea) and some turbellarians, and also that of *Balanoglossus*, consist of separate visual cells scattered beneath the skin. The separate visual cells themselves can distinguish light from darkness, and variations in lighting intensity, but cannot help the animal to react to the direction of incident light. To discern the latter, a visual cell must be screened on all sides by pigment and exposed to light from only one direction.

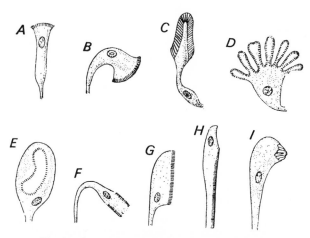

Fig. 82. *Some forms of visual cells in invertebrates.*

A—visual cell from epithelial eye of a medusa. *B*—inverted cell of parenchymatous eye of a turbellarian. *C*—cell from eye of *Limax* with strongly-convex rod layer. *D*—cells from lateral eye of *Polyophthalmus* (Polychaeta Drilomorpha) with folded rod layer. *E*—visual cell of leech with rod layer closed into a vesicle. *F*—cell from medial eye of *Cypris* (Ostracoda), rod layer developed only on the sides. *G*—cell from eye of *Lithobius* (Chilopoda), in which rod layer evenly covers apical surface of cell, but axis of cell is bent at a right angle. *H*—cell from ommatidium of compound eye of *Branchipus* (Phyllopoda), like *G*, but the whole apical surface of the cell is much elongated and narrowed, and the rod layer is fused into a rhabdomere. *I*—cell from simple eye of larva of *Gastropacha* (Lepidoptera), like *G*, but extent of rod layer is much reduced.

A multitude of isolated visual cells, however, when scattered over the animal's skin, represent in the aggregate a visual apparatus with a power not possessed by a single isolated cell: the animal can perceive varied illumination of different parts of its body, and thus can react both to gradients of illumination and to the direction of incident light rays.

In the eyes of many turbellarians the screening of visual cells by pigment is attained by the juxtaposition of these cells and pigment cells, a visual cell being encompassed on one or several sides by a pigment cell. In this case the layer of rods is turned towards the pigment, i.e. away from the light that strikes the rods on passing through the body of the visual cell.

Such eyes are therefore called *inverted*. In *Mesostoma ehrenbergii* (Rhabdo-coela) (Fig. 83, *A*) each eye consists of one visual cell and one cup-shaped pigment cell, and the openings of the pigmented cups of the two eyes point in different directions (A. Luther, 1904). *Planaria torva* (Triclada) has two visual cells and one pigment cell in each eye; in another member of Triclada, *Euplanaria gonocephala*, the pigmented cup of each of its two eyes is formed of several pigment cells and contains several tens of visual cells, arranged irregularly and in several layers. With such an eye structure the animal can orient itself in the direction of incident light even after removal of one eye. In the eyes of Polyclada visual cells line the inner surface of the pigmented cup in a regular epithelioid layer, that being a substantial step towards formation of a true retina. All that is required to receive visual images is that the separate receptive circuits should be isolated from one another by layers of pigment. That further step is taken in the eyes of many two-eyed Triclada in the families Rhynchodemidae and Cotyloplanidae (L. von Graff, 1889). In them we find a pigmented cup formed by an epithelioid layer of pigment cells. The visual cells are narrow and arranged radially around the eye. Their central processes disappear in an optic ganglion, but the peripheral ends pass between the pigment cells; at the level of the latter they form optic rods, and within the eyeball they expand into transparent refractile prisms, which together form a vitreous body (Fig. 83, *C*). Such an eye doubtless enables the animal, to a certain extent, to perceive images of objects.

The visual apparatus of most turbellarians consists of one pair of eyes in the cerebral region; less often there are four eyes (several Alloeocoela, Holocoela, a few Acoela, Rhabdocoela, and Triclada); the possession of paired, symmetrically-placed eyes enables the animal, even with their imperfect structure, to react to the direction of incident light. A single eye occurs as an exception among Alloeocoela, e.g. in *Otomesostoma* and some Monocelididae. Many Triclada and most Polyclada have numerous eyes, arranged in groups in the cerebral region, and sometimes additionally along the edges of the body. A visual apparatus composed of numerous ocelli pointing in various directions must be a more perfect analyser than a single pair of eyes of the same structure. The eyes of nemertines are very similar to those of turbellarians and are mostly at the same level of develop-ment as those of *Euplanaria gonocephala* (Fig. 83, *B*) or of Polyclada; they number two, four, or more, and are situated in the cephalic lobe.

Simply-constructed, inverted eyes, at the level of development of those of *Mesostoma ehrenbergii*, are fairly widespread among sessile polychaetes (Spionidae, Capitellidae, Terebellomorpha, Serpulidae, etc.). We find in leeches all stages of development of parenchymatous eyes, from separate scattered visual cells (*Pontobdella*), through inverted eyes of the *Planaria* or *Euplanaria* type (found in Rhynchobdellea, Fig. 83, *D*), to secondarily-reverted eyes (*Hirudo*, Fig. 83, *E*, *F*) resembling in that respect the eyes of Rhynchodemidae, but less perfect.

Eyes of parenchymatous types are rare except in Scolecida and annelids.

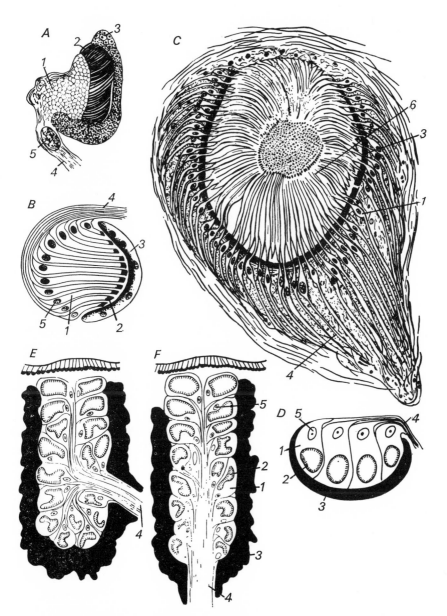

Fig. 83. *Parenchymatous eyes of turbellarians, nemertines, and leeches.*
A—*Mesostoma ehrenbergii* (Rhabdocoela) (after Luther). B—*Baseodiscus delineatus*
(Heteronemertini) (after Hesse). C—*Platydemus grandis* (Triclada Terricola) (after
von Graff, but made diagrammatic). D—Rhynchobdellidae (after Hesse). E and F—
Hirudo (Gnathobdellida) (after Hesse). 1—visual cells; 2—light-sensitive rods;
3—pigment cells; 4—neural processes of visual cells; 5—nuclei of visual cells;
6—vitreous body.

They include, for instance, the intracerebral ocelli of *Amphioxus*, which are of the *Mesostoma* structural type, and the eyes of the *Ceratopogon* larva (Diptera Heleidae), which approach the *Planaria torva* type. The latter type is also approached by the somewhat more complex nauplial eyes of Crustacea. These examples show how easily similar structures arise independently in the most diverse groups of animals.

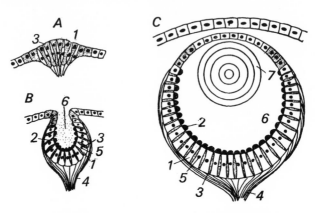

Fig. 84. *Epithelial eyes of hydromedusae (diagrammatic).*
A—eye-spot. *B*—optic pit. *C*—optic vesicle. 7—lens; other numerals denote the same items as in Fig. 83 (from Dogiel).

The second main eye type is *epithelial eyes*. Apart from the above-mentioned separate epithelial visual cells of Lumbricidae, the simplest eyes of this type are the eye-spots of turbellarians and medusae. Among turbellarians they are found in Acoela (*Otocoelis rubropunctata*) and Macrostomida (*Microstomum, Alaurina*), and consist of paired aggregates of epithelial visual cells containing pigment in their basal parts. These eyes lie on the sides of the head, and in Microstomidae are connected with the brain. Unfortunately their fine structure is unknown, but apparently the same cells are both visual and pigment cells. In the eye-spots of Anthomedusae (e.g. *Oceania*) and some Scyphomedusae (*Aurelia*) the visual cells lack pigment and alternate with pigment cells (Fig. 84, *A*); the eye-spots of medusae are thus somewhat better developed than those of turbellarians. In turbellarians the epithelial type of eye has not developed further, whereas in medusae it has developed into optic pits, optic cups, and optic vesicles. Invagination of the epithelium of an eye-spot to form a pit (Fig. 84, *B*) provides it with the best protection and is at the same time the first step towards a chambered eye. When the pit closes to form a vesicle (Fig. 84, *C*) its outer wall loses visual and pigment cells and its epithelium becomes transparent, as well as all the overlying tissues. The vesicle is usually filled with jelly, forming the vitreous body. Sometimes a thickening within the visual vesicle forms a lens. Already in medusae, therefore, the principal dioptric adaptations possessed by the chambered eyes of higher animals have developed. The most perfectly constructed eyes among all medusae are some of those of *Charybdea* (W. Schewiakoff, 1889).

A similar developmental series is shown by the cephalic eyes of polychaetes and molluscs (Gastropoda and Cephalopoda) and many noncephalic eyes of sessile polychaetes, Lamellibranchia, and Gastropoda.

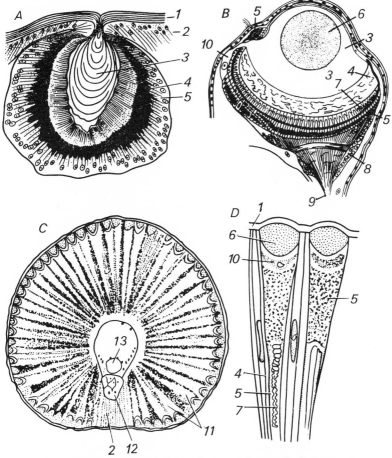

Fig. 85. *Flask-shaped, vesicular, and compound types of polychaete eyes.*
A—section through flask-shaped eye of *Eunice torquata*. *B*—section through eye of *Alciopa cantrainii* (Phyllodocemorpha). *C*—*Branchiomma* (Serpulimorpha), cross-section through gill filament with compound eye. *D*—similar cross-section, two separate eyes, highly magnified. 1—cuticle; 2—epidermis; 3—vitreous body; 4—pigment cells; 5—retina or its separate cells; 6—lens; 7—rods and other light-receiving endings; 8—optic ganglion; 9—optic nerve; 10—cell that secretes lens or vitreous body; 11—ocelli of compound eye; 12—cartilaginous cord of gill filament; 13—blood vessel (from N. A. Livanov).

The cephalic eyes of errant polychaetes are flask-shaped (Eunicidae, Fig. 85, *A*; Amphinomidae) or vesicular; the latter in most cases have a lens. They reach higher developmental levels in the pelagic family Alciopidae and in some Aphroditidae. Alciopidae possess (besides a highly-developed dioptric apparatus, Fig. 85, B) the power of accommodation: under the action of contractile fibres the lens can approach the retina, and it can be moved away from the retina by the elasticity of the eye (Demoll, 1909).

The visual apparatus as a whole differs in principle in errant and in sessile polychaetes. The errant polychaetes have only cephalic eyes, one or several pairs, these often being highly developed. That is the visual apparatus of a predator,[1] which in many cases (Aphroditidae, Alciopidae, etc.) doubtless enables them to see their prey. On the other hand, the visual apparatus of sessile polychaetes is that of prey: we may say roughly that it needs merely to enable its possessors to see the shadow of an approaching predator, so that they can retreat into their tubes. We have seen that their statocyst-seismoreceptors serve the same end, warning of

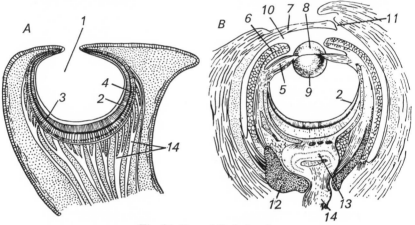

Fig. 86. *Eyes of Cephalopoda.*

Longitudinal sections through eyes of : *A—Nautilus, B—Sepia.* 1—cavity of optic pit; 2—layer of optic rods; 3—pigmented layer; 4—layer of retinal cells; 5—epithelial body; 6—iris; 7—cornea; 8—external half of lens; 9—internal half of lens; 10—anterior eye chamber; 11—its aperture; 12—cartilaginous envelope of eye (sclera); 13—optic ganglion; 14—optic nerve (from Lang).

an approaching enemy. The cephalic eyes of sessile polychaetes are much simplified, sometimes reduced; they also often have similarly-imperfect eyes on segments of the body (Opheliidae), on branches of the palps (*Branchiomma*), on the anal lobe (*Fabricia sabella*), and so on.

Among molluscs flask-shaped eyes are found in *Nautilus* (Fig. 86, A) and in many Prosobranchia Aspidobranchia. In most snails the eyes are vesicular and possess a vitreous body, and sometimes a lens; the best-developed are the 'telescopic' eyes of the predatory planktonic Heteropoda (Taenioglossa), which have a much-elongated eyeball and a very large, spherical lens. They do not, however, have the power of accommodation.

[1] Of course, not all predatory polychaetes use sight to seek prey (e.g. the blind *Glycera*!), and not nearly all errant polychaetes are predators. Hence the variation in the degree of eye development observed among them.

The most perfect eyes among all molluscs and also among all invertebrates are those of higher Cephalopoda (Dibranchiata, Fig. 86, *B*). In them the internal wall of the optic vesicle forms a retina and the outer wall fuses with the external epithelium to form the so-called epithelial body. Its central part secretes the two halves of the spherical lens: one of these is developed from the optic vesicle and the other from the external epithelium. Radial muscle fibres (the ciliary muscle) develop in the peripheral part of the epithelial body. A diaphragm-like fold of pigmented skin over the eye forms the iris; its aperture forms a pupil, capable of widening and narrowing under the action of dilator and sphincter muscles. Around the iris a second fold of skin, concentric with it and absolutely transparent, develops. It fits above the iris, forming the cornea and being the boundary of the

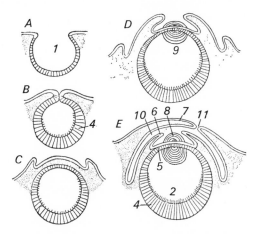

Fig. 87. *Development of eye of Cephalopoda Dibranchiata* (from Lang).

The numerals have the same denotation as in Fig. 86.

anterior chamber of the eye. Sometimes a small aperture remains in the cornea. In some Octopoda additional folds of skin form eyelids. Around the eyeball a cartilaginous protective capsule develops, the sclera, to which the oculomotor muscles are attached. Contraction of the ciliary muscle moves the lens away from the retina, shortening the focal distance of the eye, so that the cephalopod eye is capable of accommodation. The histological structural details are as indicative of the perfection of the cephalopod eye as is its general structural plan. The cephalopod eye shows clearly the unusually high level of organisation of the group. The vision of cephalopods is probably very close to that of vertebrates; the similarity in structure of the eyes of cephalopods and vertebrates is astonishing, in spite of the entirely different method of development of most of the eye parts (Fig. 87) and the entirely different course of evolution taken by the eyes of the two groups. The similarity is one of the most striking instances of convergence in the entire field of the comparative anatomy of animals.

We shall not dwell on the structure of the extremely diversified non-cephalic eyes of molluscs; we shall merely remark that in Lamellibranchia they lie along the edges of the mantle and sometimes on the siphons, and in Loricata and Oncidiidae (Pulmonata) on the back. The visual apparatus consisting of non-cephalic eyes has the same ecological significance in molluscs as in sessile polychaetes; the structural types of the non-cephalic eyes of the two groups are also very similar.

Onychophora have vesicular cephalic eyes, very similar in general structure to the typical eyes of errant polychaetes such as *Lycastis*. Like the latter, Onychophora are predators. As in many polychaetes and molluscs, the eye pigment of Onychophora lies in the basal parts of the visual cells themselves.

In their simplest form the eyes of arthropods, like the cephalic eyes of other trochophore animals, are epithelial flask-shaped ocelli, but their further development diverges widely from that of the cephalic eyes of annelids, Onychophora, and molluscs.

The eyes of Myriapoda such as *Julus* (Diplopoda, Proterandria) or *Lithobius* are nearest to the original type. The eye of *Lithobius* (Fig. 88, *A*) resembles a cup formed of epithelial cells; their bodies contain pigment, and the apical ends of the cells, which are turned towards the axis of the eye, are covered with a layer of photoreceptor rods. The central processes of the visual cells arise from their basal ends, not directly along the cell axis, as is usual in other animals; their point of origin is moved towards the bottom of the optic cup. The outer skin cuticle above the entrance to the optic cup forms a lenticular thickening (the *corneola*) which does not, however, fill the whole cavity of the optic cup. Separate epidermal cells move in beneath its base. The rest of the cup cavity is occupied by a poorly-differentiated vitreous body. Among insects, the larva of the water-beetle *Dytiscus* (Fig. 88, *B*) has very complex eyes. In *Scutigera* numerous ocelli are grouped into a pair of compound eyes. Each of the ocelli or ommatidia, taken separately (Fig. 88, *C*), resembles the eye of *Lithobius*, but has a smaller number of cells; the visual (retinal) cells are arranged in two tiers. The vitreous body is shortened, so that the rod layers of the retinal cells of the basal tier are closely apposed. At the same time the vitreous body assumes a more individualised, differentiated character. The corneola is always underlaid by corneagenous cells.

In the ommatidia of the compound eye of *Lepisma* (Thysanura, Fig. 88, *D*) the upper tier of visual cells also moves together; instead of the vitreous body, a crystalline conus develops, lying entirely between the upper tier of visual cells and the lens and secreted by special cells among the corneagenous cells. The ommatidia of cockroaches such as *Periplaneta* also still have two tiers of visual cells and are distinguished from the ommatidia of *Lepisma* mainly by their more extended and more compact form and the more perfect development of the crystalline conus (Fig. 88, *E*). The two-tier arrangement of visual cells generally predominates in lower insects, and in higher insects (Fig. 88, *F*) only one tier usually remains. The number of elements in the retinula [1] is strictly fixed (in insects it is

---

[1] The aggregate of visual cells in one ommatidium is called 'retinula'.

usually seven); the radial arrangement of visual cells and rhabdomeres belonging to them (Fig. 88, *G*) enables the ommatidium, it is generally believed, to distinguish the direction of a plane of polarised light falling on the compound eye. The number of all other cells in the ommatidium

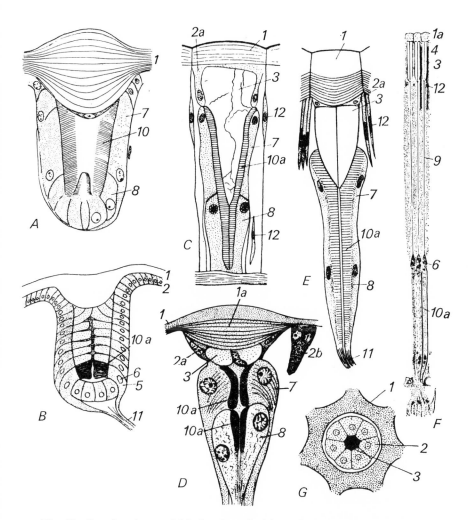

Fig. 88. *Cup-shaped eyes of Myriapoda and of insect larvae (archaeommata) and transition from them to ommatidia of compound eyes.*

A—archaeomma of *Lithobius* (Chilopoda). *B*—archaeomma of larva of *Dytiscus* (Coleoptera). Ommatidia of compound eyes: *C*—*Scutigera* (Chilopoda); *D*—*Lepisma* (Thysanura); *E*—*Periplaneta* (Blattoidea); *F*—*Macroglossa* (Lepidoptera). 1—cuticle; 1a—corneolae; 2—hypoderm; 2a—corneagenous cells; 2b—hypodermal pigment cells; 3—cells of crystalline conus; 4—nuclei of cells of crystalline conus; 5—visual cells; 6—nuclei of visual cells; 7—distal visual cells; 8—proximal visual cells; 9—distal, filiform part of retinula; 10—rod layer; 10a—rhabdome; 11—nerve processes of visual cells; 12—nuclei of pigment cells (drawings taken from Hesse). *G*—diagrammatic cross-section of ommatidium at rhabdome level: 1—pigment cells; 2—visual cells (retinula); 3—rhabdome (from Imms).

is likewise fixed. There is usually a layer of pigment cells between separate ommatidia.

When a large number of simple eyes (archaeommata) are integrated into a single compound eye, the number of elements in them usually decreases, they become simplified and specialised, and they are converted into ommatidia, which at first glance are very unlike the original flask-shaped ocelli. The visual cells of ommatidia are much elongated along the axis of the eye and their central processes arise from their proximal ends. An impression is created that the rhabdomere is formed on the lateral wall of the cell. That is often actually stated. In fact, however, the side of the cell on which the rhabdomere is located is its apical surface; the basal surfaces of the cells are oriented towards the periphery of the ommatidium, and the central processes arise from the proximal sides of the cells (proximal with respect to the bottom of the optic cup from which the ommatidium has developed) (Fig. 88).

We must stress the extreme similarity in the structure of the compound eyes of insects and of crustaceans. Apparently there are no features whereby one may always distinguish the one from the other. To some extent that identity of eye structure has been inherited from common ancestors, but it has largely developed on parallel lines in the two classes.

Besides their compound lateral eyes, many insects also have from one to three simple occipital ocelli; in the larvae of Holometabola the lateral eyes also are usually simple eyes or groups of them. Several simplified groups such as Anoplura, Aphaniptera, etc., also have simple lateral ocelli. Simple ocelli vary widely in structure. Besides the flask-shaped lateral ocelli of the larvae of *Dytiscus* and many other Holometabola, adult insects have occipital ocelli with a thick corneal lens, and beneath it a *retina* in the form of a flattened cup, formed of a very large number of visual cells (Fig. 89, *B*). The visual cells are much elongated; their rod layer lies on their lateral surfaces; and the rods of adjacent cells fuse into rhabdomes. A layer of hypodermal cells lies between the lens and the retina (*Cicada*, etc.). In other cases, e.g. in many caterpillars (Fig. 89, *C*), simple (lateral) ocelli have an oligomerised number of visual cells and resemble the ommatidia of a compound eye in their structure. Such ocelli are doubtless simplified, and they apparently occur only in the most specialised Holometabola larvae; to derive the compound eyes of insects and crustaceans from them, as does A. Berlese (1909–12), would be incorrect.

The eyes of several higher Crustacea (Mysidacea, Euphausiacea, Decapoda), being set on movably-articulated stalks, are movable, as is also the single eye of Cladocera. In the process of ontogeny the two compound eyes of Cladocera move together and fuse into one medial eye. The outer skin forms a fold around it, which closes, producing the external cornea and the anterior chamber of the eye. Oculomotor muscles are attached to the eye, and the eye acquires the power of rotation within the anterior chamber.

Trilobites had a pair of compound eyes, sometimes formed of a large number of ocelli, but still poorly integrated, as is seen from the round (not hexagonal) shape of their corneolae.

The eyes of Chelicerata, which are unique in many respects, have probably been derived from the compound eyes of trilobites by means of far-reaching integration of separate ommatidia into a single whole. The first stage of that integration is shown in the lateral eyes of Xiphosura

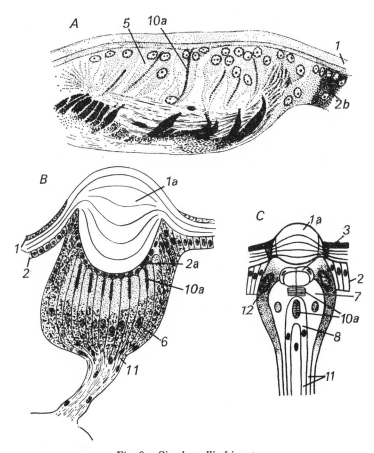

Fig. 89. *Simple ocelli of insects.*

A—*Machilis* (Thysanura) (after Hesse). B—*Perla abdominalis* (Plecoptera) (after Link). C—caterpillar of *Gastropacha* (Lepidoptera) (from Bütschli). Numerals have the same denotation as in Fig. 88.

(R. Demoll, 1914). In Xiphosura the separate corneolae of the compound eye are partly fused together, forming an externally-smooth cornea (Fig. 90, *A*). At the basal surface of the cornea, however, the corneolae remain separate. Beneath each of them is a separate goblet-shaped ocellus. Each of these ocelli (Fig. 90, *B*) is formed of a deep layer of corneagenous epithelium, which forms a continuous lining to the whole optic cup, and of a retinula lying within that epithelium. The retinula consists of from

10 to 15 visual cells, radially arranged around the optic axis of the ocellus, and of one eccentric cell, which sends its processes between the other cells. The presence of that cell is a secondary feature in the Xiphosura eye. Each cell forms a rhabdomere on the side turned towards the axis of the ocellus. On the whole, such an ocellus differs from the eyes of Myriapoda mainly in the large number of its cells and in their differentiation into corneagenous and retinal.

In the medial eyes of Xiphosura (Fig. 90, C) integration has gone further. The cuticle forms a single, strongly-convex lens, covered on the eye side with deep corneagenous epithelium; in Chelicerata that is called the vitreous body. The retina is entirely sunk beneath that epithelium. It possesses preretinal and postretinal membranes, between which the retinulae lie fairly loosely. Each retinula consists of from six to eight cells with irregularly radial arrangement. The rod layer of each cell is fused into a rhabdomere; the separate rhabdomeres in one retinula are turned towards each other. Between the retinulae lie cells filled with bluish pigment.

Scorpions have from six to twelve eyes. Two of these are the principal or medial eyes, the others are lateral or accessory. The principal eyes resemble the medial eyes of Xiphosura in structure. They have a large lens, a well-developed vitreous body, and a retina enclosed between pre-retinal and postretinal membranes. The retina consists of visual and pigment cells. The visual cells are grouped into retinulae, each with five cells. Each cell has a short rhabdomere, the rhabdomeres being combined into a rhabdome. The rhabdomeres lie in the distal halves of the visual cells, and the nuclei in the proximal halves. The eyes of Telyphones, Araneina, and Opiliones (Fig. 90, E) are fairly similar in structure to those of scorpions.

The eyes of Arachnoidea often possess a tapetum, i.e. a light-reflecting layer lying externally to the retina. In such cases the rhabdomes of the visual cells are turned to the tapetum, not to the vitreous body, e.g. in the lateral eyes of Telyphones, in the eyes of Araneina, etc. Eyes of this type are usually called inverted, but they only partly resemble true inverted eyes, the light in both cases passing through a body of visual cells before falling on the perceptive apparatus. In true inverted eyes the apical surface of the cell, which bears the rod layer, is turned away from the light, whereas in the 'inverted' eyes of Arachnoidea (as in uninverted eyes) that surface lies parallel to the incident light, so that the rods, which are there fused into rhabdomeres, are perpendicular to the long axis of the latter, and the long axis of the rhabdomeres always lies along the long axis of the retinula. For that reason the existence of inversion in the eyes of Arachnoidea does not imply rotation of the cells with their basal ends away from the light, but only movement of the nuclei (and sometimes also of the points of origin of the nerve processes) parallel to the apical surface of the cells.

In the medial eyes of Solifugae (Fig. 90, F) the visual cells are not grouped into retinulae and have long rod-like rhabdomes. Apparently

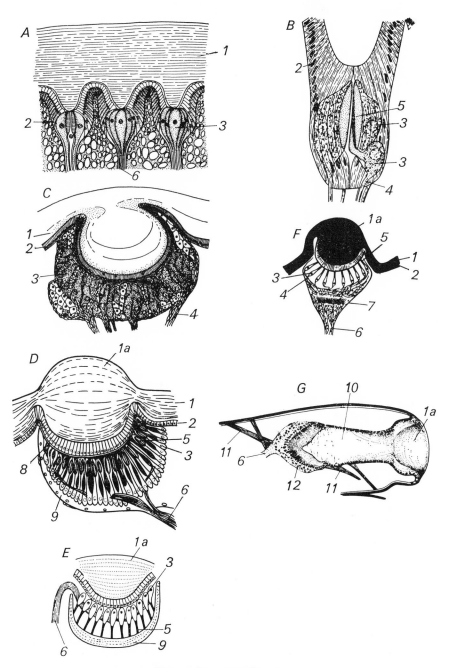

Fig. 90. *Eyes of Chelicerata.*

Limulus: A—part of lateral eye in section (from Lang); *B*—separate ocellus (after Demoll); *C*—medial eye (after Demoll). *D*—one of the principal eyes of *Euscorpius* (after Scheuering). *E*—lateral eye of *Telyphonus*. *F*—medial eye of Solifugae (from Kästner). *G*—*Salticus scenicus* (Araneina), anterior medial eye in longitudinal section (after Scheuering). 1—cuticle; 1a—lens; 2—hypoderm; 3—visual cells; 4—their nerve processes; 5—rhabdome; 6—optic nerve; 7—optic ganglion; 8—preretinal membrane; 9—postretinal membrane; 10—vitreous body; 11—oculomotor muscles; 12—retina.

oligomerisation of the ommatidium cells has here reached an extreme, and each retinula is reduced to a single cell. If this is so, the eye of Solifugae represents the last step in a series of gradual integration of the aggregate of simple ocelli into a compound eye and of further integration of the latter into a simple eye. If it were not for the structure of the rhabdome and for comparison with other Chelicerata, the eyes of Solifugae might be taken to be simple pit-like eyes, and it would be difficult to imagine the complicated course of development that produced them.

Errant spiders are among the few arthropods that possess oculomotor muscles. In particular, the medial eyes of Salticidae (Fig. 90, *G*) contain six muscles: Salticidae can 'follow prey with their eyes' while remaining motionless, and are distinguished by perfect vision. The eyes of Salticidae and other errant spiders are the eyes of active hunters, whereas those of web-making spiders are designed mainly for warning of danger. In general the quiescent mode of life of an ambush-lurking predator has led to simplification of the visual apparatus and the brain in web-making spiders (A. Ivanov, 1946); we have seen the same happen to polychaetes on transition to a sessile mode of life.

The medial eyes of scorpions and spiders are characterised by a unique process of development (Fig. 91). They begin as a pair of ectodermal invaginations, shaped like flattened asymmetrical sacs, and therein resemble the frontal eyes of some insects (e.g. *Hydrophilus*). In scorpions and spiders the outermost of the three layers of the fold forms the vitreous body and secretes the lens, the central layer forms the retina, and the lowest layer forms the postretinal membrane. In all probability that method of eye development is secondary and in no way recapitulates the phylogeny of the medial eyes of scorpions and spiders. On the opposite assumption there would be an impassable gulf between the eyes of these groups and those of other Arachnoidea.

The gradual transformation of a compound into a simple eye by the fusion of lenses is the chief direction of development of the eyes of Chelicerata; in addition, many Chelicerata exhibit breakdown of each of the lateral eyes into a group of smaller eyes, each of which continues to be a highly-concentrated compound eye. That is often followed by a decrease in the number of these eyes: scorpions have from six to twelve eyes; spiders, as a rule, have eight, but some have six, four, two, or none; Solifugae have two normal and one or two pairs of vestigial eyes; Opiliones have only two eyes. Xiphosura, however, already have vestigial as well as functional eyes. In Pseudoscorpiones, many Acariformes, and a number of other Arachnoidea the eyes are much simplified or absent.

Among Mandibulata (Crustacea and Atelocerata) fusion of all the corneolae of an indubitably compound eye into one spherical lens occurs as a very rare exception (*Ampelisca* in the order Amphipoda). The frontal ocelli of several insects, however, with their single lens and their retina consisting of a number of retinulae, have a definite resemblance to the eyes of Arachnoidea. Whether such ocelli arose in insects through complication of primitive flask-shaped eyes of Myriapoda type, or through integration of compound eyes, is difficult to say.

Outside the subphylum of Arthropoda compound eyes consisting of a large number of ommatidia are found sporadically in other trochophore animals, polychaetes and molluscs. Among polychaetes, such eyes are formed on the tentacles of a number of Serpulimorpha, e.g. *Branchiomma* (Fig. 85 C); in different Serpulimorpha we find all transitions between a loose aggregate of separate ocelli and highly-concentrated eyes resembling those of arthropods. Among molluscs, compound eyes occur along the edges of the mantle in some Lamellibranchia of the order Filibranchia (*Arca, Pectunculus*).

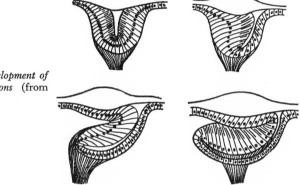

Fig. 91. *Diagram of development of principal eye of scorpions* (from P. P. Ivanov).

Physiologically, the compound eyes of crustaceans and insects are highly-perfected organs, not only enabling the animals to perceive the shape of objects but often also possessing considerable acuity of vision. As is well known, many insects have in addition a well-proved faculty of distinguishing colour (see particularly the experiments of K. Frisch (1914) with bees); but even the most perfect insect eyes certainly fall behind the eyes of vertebrates and higher cephalopods. In particular there are no perfect accommodation adaptations in the eyes of arthropods, and, as we have seen, oculomotor adaptations occur there only rarely.

In surveying the visual apparatuses of various arthropods as a whole (Fig. 92), we may first remark that all arthropods, unlike many annelids and molluscs, have only cephalic eyes. Primitive groups of arthropods have a large number of eyes (the rule of multiple formation of organs, V. Dogiel, 1954), which may be divided (purely topographically) into lateral and dorsal (or medial) eyes.

In most myriapods and in the larvae of insects with complete metamorphosis lateral eyes are represented by groups of simple ocelli, there called *stemmata*: in most arthropods groups of stemmata are fused into compound eyes, which in trilobites are still poorly integrated. All lateral eyes, whether groups of stemmata or compound eyes, are always innervated from the optic lobes of the protocerebrum. In most Chelicerata each of the lateral compound eyes breaks down into several parts, some of which may later disappear. In spite of all their variations, the lateral eyes of all arthropods may be regarded as homologues of one another.

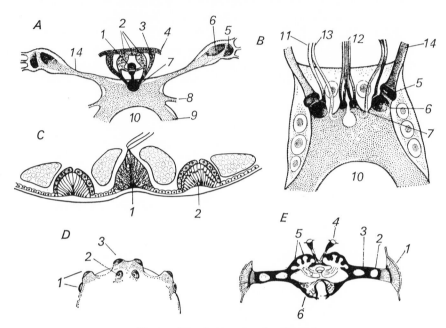

Fig. 92. *Visual apparatus of arthropods.*

A—Artemia salina (Phyllopoda Anostraca), cross-section of brain, showing innervation of the nauplial and compound lateral eyes. *B—Limulus* (Xiphosura), frontal section through supraoesophageal ganglion, showing innervation of eye; 1—ventral frontal organ; 2—nauplial eye; 3—dorsal frontal organ; 4—hypoderm; 5 and 6—nearest optic centres of compound eyes; 7—optic centres of nauplial eye of *Artemia* and of dorsal eyes of *Limulus*; 8—nerve of antennulae; 9—pharyngeal connectives; 10—oesophagus; 11—nerves of ventral eyes; 12—nerves of dorsal eyes; 13—ventral nerve in skin (supposed homologue of antennular nerve of crabs); 14—nerves of lateral compound eyes. *C*—larva of *Limulus*, cross-section through frontal organ (1) and ventral ocelli (2) (after Hanström). *D*—front edge of prosoma of *Aranea* (Araneina): 1—lateral eyes; 2—posterior medial eyes; 3—anterior medial (principal) eyes (from Grassé). *E—Lasius umbratus* (Insecta, Hymenoptera Formicoidea), cross-section through brain of female: 1—lateral (compound) eyes; 2 and 3—optic centres of compound eyes; 4—ocelli; 5—pedunculate bodies (higher associative centres of brain); 6—deutocerebrum (from B. N. Shvanvich).

The medial eyes of primarily-aquatic arthropods are represented by the so-called nauplial eyes possessed by all lower Crustacea, larvae of higher Crustacea, Pantopoda, and perhaps Trilobita. The nauplial eye is usually composed of three united ocelli, but an unpaired ocellus included in these is paired in origin (D. Pedashenko, 1899); the nauplial eye is therefore actually formed of two pairs of ocelli. It is innervated from the front margin of the protocerebrum. In many Copepoda all three ocelli remain independent; Mystacocarida have four independent ocelli.

In typical cases the dorsal eyes of insects (ocelli) are three in number, but there also the unpaired eye is the result of fusion of a pair of ocelli. They are innervated from the intercerebral (mediodorsal) part of the protocerebrum.

Trilobites, Xiphosura, and Eurypteroidea have a pair of simple medial ocelli as well as a pair of lateral compound eyes; trilobites and Xiphosura have also another pair of medial ventral ocelli: in Xiphosura these lie on the epistome, along the sides of the unpaired frontal organ; in adults these ocelli are vestigial, but perhaps they function in larvae; Johansson (1933) ascribes an olfactory function to the frontal organ. B. Hanström (1928a) believes that in Eurypteroidea the ventral ocelli migrate to the dorsal side of the head, and that their homologues are the medial ('principal') eyes of scorpions.

The homology of the nauplial eye with the dorsal ocelli of insects and Xiphosura and with the homologues of these in trilobites has not been proved, but cannot be rejected. If in the future that homology should be confirmed, we may then conclude that the visual apparatuses of all arthropods, in spite of all the diversity of their modifications, are constructed on a single general plan.

Chapter V STRUCTURES DERIVED FROM
PHAGOCYTOBLAST (ENDODERM
AND MESODERM)

1. Phagocytoblast of Lower Metazoa

As we have seen in Chap. I, the most primitive phagocytoblast structure
is found in sponges, coelenterates, and acoelous turbellarians.

Some sponges and coelenterates lack the strict division of labour between
phagocytoblast and kinetoblast that is found in higher animals. In particular, digestion is performed in sponges by various classes of body cells
belonging partly to kinetoblast (choanocytes) and partly to phagocytoblast
(amoebocytes). In calcareous sponges (*Grantia*) choanocytes seize food
particles and digest them (Fig. 93). In siliceous sponges (*Tethya, Suberites*

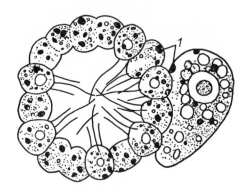

Fig. 93. Grantia (*Calcarea*). *An
amoebocyte receiving grains of carbon* (1) *from choanocytes* (after N.
Pourbaix, from L. H. Hyman).

Polymastia) and horny sponges (*Halichondria*, Spongillidae) the food is
seized by choanocytes, but they pass it on to amoebocytes, which digest it.
In a number of species (*Pellina, Petrosia, Stylotella, Hyppospongia*) food
is both seized and digested by amoebocytes. In the first group the choanocytes are larger than the amoebocytes, in the second group both types of
cells are more or less equal in size, and in the third the amoebocytes are
larger than the choanocytes (N. Pourbaix, 1933). Most sponges, therefore,
have digestive elements of phagocytoblastic origin, but they are scattered
through the whole body and do not form a spatially-delimited whole.
The digestive cells of sponges are free-moving phagocytes. Sponges have
no digestive apparatus in the strict sense of the term.

All sponges feed on very fine particles (nanoseston and ultraseston),
which are suitable in size for phagocytosis by choanocytes and amoebocytes,
and all phases of digestion take place intracellularly. In the vacuoles of
sponge phagocytes, as in the digestive vacuoles of Protozoa, an acid

reaction is observed in the early stages of digestion and an alkaline reaction in the later stages.

Unlike sponges, most coelenterates (including all hydroids) feed on relatively large prey, which they usually seize by means of their tentacles and then swallow.

Among coelenterates the simplest phagocytoblast structure is seen in some hydroid polyps such as *Hydra* (Fig. 94, *A*), in which the phagocytoblast forms a single intestinal layer consisting of epithelially-arranged cells. Among these cells amoeboid-flagellate digestive cells with muscular processes preponderate; the intestinal epithelium contains glandular and sensory cells as well as digestive cells. The chief advance beyond the sponge level consists in the fact that all the phagocytoblast cells have assumed an epithelial arrangement and form a single sacciform gut—the first digestive organ, and at the same time the first digestive apparatus provided with a true mouth. Another substantial progressive feature is the presence of glandular cells in the phagocytoblast.

The gut of *Hydra* contains two types of glandular cells, mucous and digestive. The former lie mainly around the mouth, and their secretion aids swallowing by lubricating the prey. The second type lie mainly in the broader part of the body, and their secretion affects the first stages of digestion: the prey breaks down into portions that are phagocytised by the cells of the gut, and digestion is completed in the vacuoles of the latter. Both these phases, extracellular and intracellular, appear in the digestive process of all coelenterates. In spite of that complex process, however, *Hydra* still stands at the sponge level in one respect, namely, that in both *Hydra* and sponges there is no segregation of similar elements: all types of cells contained in the phagocytoblast of *Hydra* are scattered (although not entirely uniformly) through the whole length of its gut. The phagocytoblast of sponges is amorphous and that of *Hydra* is a polarised tissue, but each is a composite tissue.

Colonial hydroids (Leptolida) differ from *Hydra* in having rather more differentiation of the phagocytoblast (Fig. 94, *B, C, D*). In them the endoderm of the tentacles is converted into a purely supporting tissue. Besides, there are glandular cells only in the intestinal epithelium of the heads of the hydranths, and only in the latter does extracellular digestion take place; on the other hand, intracellular digestion takes place both in the heads and in the endoderm of the perisarc. There are also flagella along the whole length of the gastrovascular apparatus of the colony.

The phagocytoblast of Anthozoa is still more sharply differentiated than that of Leptolida, being divided into two parts, peripheral and central. Radial septa or mesenteries project into the gut cavity (Vol. 1, Fig. 22, and Vol. 2, Fig. 127), being folds of the endoderm (phagocytoblast); within each septum is a boundary layer uniting the two layers of the fold; the muscles of the septum, likewise formed from endoderm, are also there. Thickened areas run along the free margins of all the septa, formed by thickening of the local epithelium, which is actually digestive epithelium

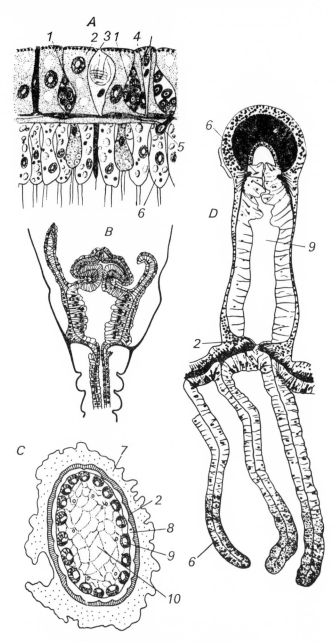

Fig. 94. *Phagocytoblast of Hydrozoa.*

A—diagrammatic section through body wall of *Hydra*; sensory and nerve cells are shown in black (after Schulze). *B*—longitudinal section through hydranth of *Laomedea flexuosa* (after Kühn): note difference in structure of endoderm of head of hydranth, endoderm of coenosarc, and tentacular endoderm (converted into supporting tissue). *C*—cross-section through stalk of *Corymorpha nutans*. *D*—similar section through part of wall of hydranth of *Myriothela penola* with endodermal villi and hollow capitate tentacles (from W. J. Rees). 1—interstitial cells, partly converted into cnidoblasts; 2—ectoderm; 3—cnidoblast (penetrant); 4—mother cell (cnidoblast) of glutinant, enclosed in ectodermal cell; 5—interstitial cell penetrating ectoderm; 6—endoderm; 7—perisarc; 8—mesogloea; 9—endodermal canals; 10—parenchymatous tissue of endodermal origin.

and contains digestive glands (Fig. 95, *A*). When one dissects a sea-anemone these thickenings of the septum margins appear as cords, whence the name 'mesenteric filaments'. From the margin of each septum in a sea-anemone there also arises a thread-like process (*acontium*), like an internal tentacle (Fig. 127, *A*). The epithelium of the radial chambers is not digestive amoeboid epithelium, but simple smooth epithelium resembling that of the body cavity of higher animals. When food is eaten it passes through the pharynx and enters the central part of the gut, which is bounded by the closely-apposed mesenteric filaments, and never passes

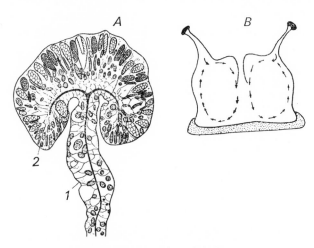

Fig. 95. *Phagocytoblast of Anthozoa.*

A—cross-section through free margin of radial septum of *Parantipathes larix* (Antipatharia): note difference in structure of endoderm of mesenteric filament (2) and of septum (1). *B*—circulation of fluid in gastrovascular cavity of juvenile Madreporaria, caused by activity of flagella of endoderm (from F. Pax).

outside that central part. Undigested food does not pass through into the chambers, at least in some sea-anemones. As in hydroids, digestive breakdown of proteins takes place in the gut cavity; digestion is completed in the epithelial cells of the mesenteric filaments E. Mechinkov). The central part of the gastral cavity is thoroughly differentiated physiologically, although anatomically it is joined to the peripheral part by a large number of longitudinal slits. As soon as food enters it the mesenteric filaments close around the food and the slits disappear. Anthozoa thus have a physiologically-differentiated gut in the true sense of the term, with something like a body cavity around it. We may therefore say that in them the gastral cavity is divided into two sections, central and peripheral. The phagocytoblast of Anthozoa, like that of sponges, contains a large number of non-digestive as well as digestive cells; but whereas the two kinds are mixed in sponges, they are separated in Anthozoa: the digestive cells (glandular and phagocytising) are concentrated in the epithelium of the mesenteric filaments (the central phagocytoblast), while the non-digestive cells predominate in the epithelium of the radial chambers (peripheral phagocytoblast).

In the sea-anemone *Nematostella vectensis* multicellular bodies containing nematocysts, called nematosomes, float in the gastral cavity. Their role, perhaps, is similar to that of acontia in other sea-anemones (T. Stephenson, 1935; E. Robson, 1957). As multicellular bodies swimming freely in the body cavity, they remind one of the mobile urnules in the coelom of Sipunculoidea (see Fig. 160).

In all medusae the gastrovascular apparatus is divided into a central section (gastral cavity) and a peripheral section. The gastral cavity is lined with amoeboid and glandular epithelium, and food is digested there. The extracellular phase of digestion takes place under the influence of a

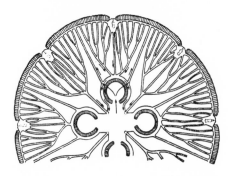

Fig. 96. *Direction of flow of fluid (shown by arrows) in gastrovascular canals of Scyphomedusae* (after J. A. Ramsay).

powerful protease, which acts in an acid medium (C. Yonge, 1931). In higher Scyphomedusae, as in Hydromedusae, the peripheral section consists of a system of radial canals united by an annular canal (see Vol. 1, Chap. III). *Aurelia* (order Discomedusae; Vol. 1, Fig. 20, *B*) has eight branched, and eight unbranched adradial, canals. All the canals are lined with flame-cell epithelium, the flagella of which force fluid in a definite direction, creating regular circulation within the gastrovascular apparatus (Fig. 96). In the unbranched canals fluid moves from the gastral cavity towards the periphery, passes into the annular canal, and returns to the gastral cavity through the branched canals (I. Vetokhin, 1926). In most hydroid medusae the canal system is simpler, but in its main features the structural plan of the gastrovascular apparatus is the same. Circulation of fluid through the activity of ciliated epithelial cells also takes place in the gastrovascular cavity of sea-anemones.

The gastrovascular apparatus of ctenophores includes, after the ectodermal pharynx, an endodermal stomach and a complex canal system, whose structure is described in Vol. 1, Chap. III. Division of the phagocytoblast into central and peripheral sections is therefore found to some extent in all coelenterates except hydropolyps. In ctenophores the extracellular phase of digestion is carried out in the pharynx by the enzymes of its glands, and the intracellular phase takes place in the epithelial cells of the stomach and the canals.

We have already mentioned the high degree of polarisation of all tissues in coelenterates. In hydroids all tissue cells are included in epithelium or myoepithelium. In Anthozoa, Scyphomedusae, and Ctenophora the mesogloea is occupied by cells; in Anthozoa and Scyphozoa these cells arrive there secondarily, in late stages of development, and are derived from both the primary layers, mainly from the kinetoblast. Their formative powers are negligible and almost no organs are formed from them, unless one includes (*i*) the internal skeleton of some Octocorallia, which is formed from scleroblasts within the mesogloea, and (*ii*) the musculature of ctenophores. The scleroblasts of Octocorallia, however, are ontogenetically relatively late immigrants from the ectoderm, and it is rather doubtful whether they belong to the peripheral phagocytoblast.

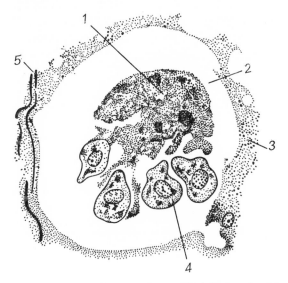

Fig. 97. Oxyposthia praedator (*Turbellaria Acoela*). *Phagocytosis of food by wandering digestive cells.*

1—food; 2—digestive vacuole; 3—syncytial parenchyma; 4—digestive cell; 5—fibre of dorso-ventral muscle (after A. V. Ivanov).

As we have seen in Chapter I, the structure of the lower Turbellaria (Acoela) is at a level somewhat similar (purely formally, it is true) to that of sponges. The phagocytoblast of Acoela, like that of sponges, has a very heterogeneous cellular composition; in some species the digestive cells are not collected together and are represented by scattered phagocytes wandering in the parenchyma (*Proporus venenosus* and *Convoluta sordida*, according to L. von Graff, Fig. 2; *Oxyposthia*, according to A. V. Ivanov, 1952c, Fig. 97). In such forms there is not yet a true digestive apparatus. The mouth is often a simple opening in the epidermis, leading directly into semi-liquid parenchyma (e.g. in *Aphanostoma sanguineum*, W. Beklemishev, 1915).

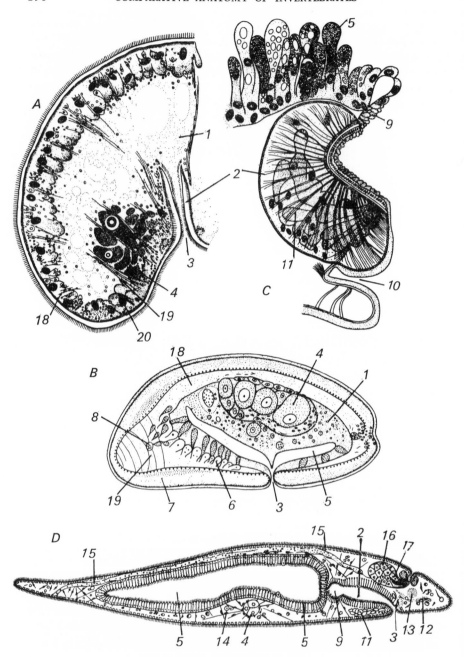

Fig. 98. *Phagocytoblast of Turbellaria.*

A—Convoluta convoluta (Acoela), part of cross-section in pharynx region (after von Graff). *B—Nemertoderma bathycola*, sagittal section (after Westblad). *C—Mesostoma ehrenbergii* (Rhabdocoela), half of pharynx and part of gut wall in section; the gut epithelium consists of amoeboid (digestive) and glandular cells; in gorged specimens the epithelium may blend into the syncytium, the gut cavity

[*continued opposite*

In most Acoela, however, we already see the next step in the development of phagocytoblast: the originally-scattered phagocytes fuse into a single digestive syncytium lying in the central part of the body and communicating with the mouth, closing the mouth, as it were (Fig. 98, A). Swallowed food is digested in the vacuoles of the syncytium. The periphery of the body is occupied by the rest of the phagocytoblast elements—supporting, muscular, etc.—among which lie glandular and nerve cells belonging to the kinetoblast. The degree of separation of the digestive syncytium from that peripheral parenchyma varies greatly. In *Aechmalotus* sustentacular and motor cells are partly mixed with the digestive syncytium in a single frothy mass, and only in the very centre, above the mouth, is there almost pure syncytium. In *Aphanostoma pallidum* and many other species the syncytium by is entirely separate from the supporting elements, but is still permeated bundles of dorsoventral muscles (W. Beklemishev, 1915). From that point we may distinguish in Acoela, as in higher coelenterates, the central phagocytoblast (which is digestive) from the peripheral (which lacks that function, or, more precisely, retains it only to a limited degree). The latter reservation must be made because a certain number of cells capable of phagocytosis remain in the peripheral phagocytoblast of all higher Metazoa from turbellarians upwards (E. Metschnikoff, 1883, and a long series of subsequent publications); in most animals they destroy microbes that enter the organism, remove particles of foreign matter and excreta (see Chap. VII), and break down and consume dead cells of the organism; but simultaneously, because of their mobility, phagocytes of the peripheral phagocytoblast of some animals also take an active part in digestion, penetrating for some time into the intestinal cavity for that purpose; that applies particularly to Echinodermata, Lamellibranchia, and lower Gastropoda.

The entire phagocytoblast (both central and peripheral) of typical Acoela, unlike that of coelenterates, remains completely amorphous and shows no sign of epithelial arrangement of its elements. In all higher turbellarians (beginning with Xenoturbellida, which in other respects are close to Acoela) the peripheral phagocytoblast retains amorphous structure, but instead of a digestive syncytium a sacciform gut develops, formed of a single layer of amoeboid epithelium; its cavity communicates with the oral orifice (Fig. 98, B). In the great majority of turbellarians, however, digestion continues to be intracellular. During digestion, when the gut cells are distended with food particles, the volume of the cells increases and the gut cavity may disappear. In many species the cell boundaries then also disappear and the gut is transformed into a continuous

---

disappearing (after von Graff). D—*Stenostomum* (Notandropora), sagittal section of a solitary individual; the gut epithelium consists of ciliated cells (absorbent), incapable of phagocytosis (Kepner, Carter, Hesse). 1—digestive syncytium; 2—pharynx; 3—mouth; 4—ovary; 5—gut; 6—glandular cells; 7—neural plexus of skin; 8—statocyst; 9—gullet; 10—pharyngeal pouch; 11—pharyngeal glands; 12—anterior loop of protonephridium; 13—brain; 14—phagocyte of peripheral phagocytoblast; 15—protonephridium; 16—testis; 17—penis; 18—peripheral parenchyma; 19—dorso-ventral muscles; 20—zooxanthellae.

syncytium. On conclusion of digestion the cell boundaries, the gut lumen, and the epithelial arrangement of the cells reappear. Often the old epithelial cells then break down and new gut epithelium is formed from reserve, non-functioning cells. The gut of most turbellarians thus has an unstable epithelial structure, appearing and disappearing in accordance with its functional condition.

Among all turbellarians only the orders Notandropora and Macrostomida have extracellular digestion and a stable epithelial structure of the gut. Their gut is lined with ciliated epithelium (Fig. 98, D), the cells of which absorb only already-digested food. Food is coated at the time of swallowing with the secretion of the pharyngeal glands, and is digested in the gut cavity under the influence of enzymes in that secretion (E. Westblad, 1923). This is the highest point in the development of the structure of the central phagocytoblast of any flatworms. Some Polyclada and flukes also apparently have extra-cellular digestion.

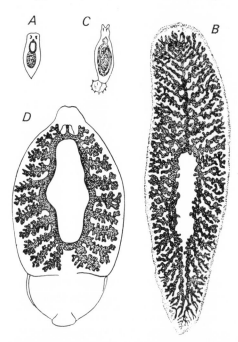

Fig. 99. *Forms of gut in flatworms.* A—*Dalyellia* (Rhabdocoela), length of animal about 1 mm (original). B—*Dendrocoelum lacteum* (Triclada), length about 1 cm (after Gelei). C—*Gyrodactylus elegans* (Monogenea), length 0·5 mm (from Pavlovskii). D—*Epibdella hippoglossi* (Monogenea), length 24 mm (from Pavlovskii).

The shape of the gut in flatworms depends mainly on the body size of the animal (Fig. 99). Small turbellarians and some Monogenea have a simple sacciform gut. In most flukes (Monogenea and Digenea), because of the flattened body and the great development of the genital apparatus, the gut divides into two lateral branches, but in small species it retains a simple tubular form. In large flatworms, on the other hand (Polyclada, Triclada, some Alloeocoela, some flukes), the gut becomes intricately branched. The reason for that correlation between body size and gut form is that the gut of flatworms has a distributive as well as a digestive function. With compact parenchyma and no circulatory apparatus, the nutrition of organs and tissues is possible only if these are all located not too far from the gut. In other words, the gut of large flatworms is physiologically the same kind of digestive apparatus as the gut of coelenterates.

Can we deduce homology between the branched gut of flatworms and the peripheral parts of the gastrovascular apparatus of coelenterates? In other words, can we compare the walls of the branched gut of flatworms to the epithelised peripheral phagocytoblast of coelenterates? That suggestion is negated primarily by comparison of large and small species of flatworms: it is quite clear that the branched gut of the large worms, with all its ramifications, is homologous with the sacciform gut of the small species and consequently, like the latter, is merely the central phagocytoblast. The amorphous peripheral phagocytoblast of flatworms, with its tremendous formative possibilities, is homologous not only with the mesenchyma but also with the peripheral phagocytoblast of coelenterates. That homology is supported, as we shall see, by the fate of the peripheral phagocytoblast in higher Bilateria.[1]

In some sea-sand-dwelling Alloeocoela Crossocoela (Otoplanidae, Polystylophoridae) the front end of the gut is converted into a supporting formation of chordoidal structure (P. Ax, 1957, 1958), resembling the supporting endodermal structure of the tentacles of hydroid polyps or of the chordae of Deuterostomia (Fig. 100).

The most characteristic features of the phagocytoblast of flatworms are therefore (*i*) diversity of its cellular composition and abundance of formative possibilities and (*ii*) its late epithelisation, tenacious retention of amorphous structure. Epithelisation of the central phagocytoblast, and with it the development of a gut, take place before our eyes within the class Turbellaria, and we can observe all stages of that process in its modern representatives, whereas epithelisation of the peripheral phagocytoblast has not even begun in the overwhelming majority of lower worms (Scolecida); the only exception is Priapuloidea, if the existence of coelomic epithelium in them is confirmed.

The independent development of a gut within the class Turbellaria shows that they, and all lower worms, cannot be derived from a prototype at the level of modern adult coelenterates. The organisation of lower worms can be derived only from that of a planula—a view advanced in his time by L. von Graff.

We may also remark that the stages of development of a gut in turbellarians are very similar to the hypothetical stages of phylogeny of the gut of hydroids, assumed by E. Metschnikoff's 'parenchymula theory' (1886). In other words, the stages of individualisation of the gut in turbellarians illustrate the most probable course of its development also in all other Metazoa.

---

[1] B. Hatschek (1888) also believed that the entire gut of flatworms was homologous only with the central section of the coelenterate gut, but based that view on erroneous arguments, assuming that the homologues of the gastrovascular canals of ctenophores were the nephridia and gonads of flatworms.

In several Rhabdocoela and in Nemathelminthes related to them the phagocytoblast undergoes considerable reduction, and schizocoelic cavities (see below) develop within it.

We may recall that cestodes and Acanthocephala are distinguished by complete reduction of the central phagocytoblast and of the entire internal

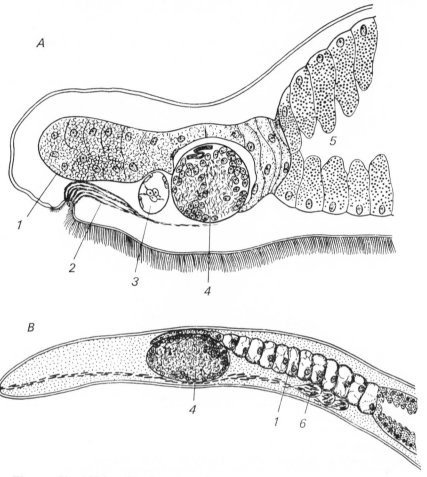

Fig. 100. *Chordoidal modifications of anterior part of gut in turbellarians (Alloeocoela, Proseriata). Sagittal sections through front end of body.*

A—Otoplana intermedia. B—Polystyliphora filum. 1—chordoidal process; 2—frontal glands; 3—statocyst; 4—brain; 5—gut; 6—rhabdite glands (after P. Ax).

digestive apparatus, as a result of their parasitism in the gut of vertebrates. In these forms the absorptive function passes to the external epithelium, to the kinetoblast. The fact that lower worms readily lose the gut is doubtless due to the low organisational level of the whole of that group.

We shall discuss the question of the peripheral phagocytoblast of higher forms in connection with the question of the body cavity.

2. Origin and Development of the Digestive Apparatus

We shall touch only briefly on the chief points in the evolution of the digestive apparatus in invertebrates as a whole. It would not be possible to discuss here its evolution within separate phyla: because of the extreme diversity of the adaptations we would encounter, such discussion would be fruitful only at the expense of much time and space.

The principal part of every digestive apparatus is the digestive section of the gut, formed from the central phagocytoblast.

Sponges have no separate digestive organ and therefore we cannot speak of the digestive apparatus of sponges in the strict sense of the term. Sponges do have an irrigation apparatus—a complex system of canals lined with pinacocytes, which bring water and food to the flagellate chambers and remove water from these chambers. The flagellate chambers, however, are not a gut or part of a gut, and the whole irrigation apparatus is not homologous with any part of the digestive apparatus of other animals. It is a formation *sui generis*, possessed only by sponges. Only from a purely physiological point of view can we say that the irrigation apparatus, which brings food into the body, fulfils in sponges (besides many other functions) the function of a digestive apparatus in which the central link, the gut, is missing.

The digestive apparatus of most hydroid polyps and Narcomedusae consists of a single digestive section, a sac formed of amoeboid epithelium and opening to the exterior by the oral orifice. Complexities in its structure occur in higher polyps, and are evidently due to the need to increase the relative area of endodermal surface with increase in body size. In large species of *Corymorpha* and *Tubularia* the gastral cavity of the stem of a hydranth breaks up into longitudinal anastomosing canals (Fig. 94, *C*) embedded in connective tissue, apparently also of phagocytoblastic origin. The most complex and peculiar system of endodermal canals is that of Disconantae (see Vol. 1, Chap. III). In *Myriothela* the surface of the endoderm is increased in a different way, by formation of intestinal villi (Fig. 94, *D*); these occur also in the gastrozooids of Physalia (Mackie, 1960). Therefore only separate, diversified, and clearly secondary complexities of the digestive apparatus appear in hydropolyps. In contrast, higher classes of coelenterates and hydromedusae show regular series of increasing complexity of the digestive apparatus developing throughout large groups: in all cases we see division of the digestive apparatus into central and peripheral sections, the latter having a distributive as well as a digestive function. The digestive apparatus of coelenterates is therefore usually called the gastrovascular apparatus. Its peripheral sections are the canals of medusae and ctenophores, and the radial chambers of Anthozoa.[1] Another complexity in the structure of the digestive apparatus of higher coelenterates is the formation of a pharynx in Anthozoa and Ctenophora. The pharynx is a fold of the ectoderm that projects like a tube into the

[1] For the plans of arrangement of each of these see Vol. 1, Chap. III.

body, the gut being attached to its inner end. The opening of the gut, or *enterostome*, ceases to be the external mouth and becomes merely the communication between the pharynx and the gut. The external opening of the pharynx becomes the external mouth. The digestive apparatus ceases to be of purely phagocytoblastic origin. Its endodermal (phagocytoblastic) part is thenceforward called the mid-gut; the newly-formed endodermal part, the pharynx, is the fore-gut. The formation of the pharynx, with its powerful musculature and with the pharyngeal section of the neural apparatus, leads to considerable strengthening of the swallowing function.

Among turbellarians, many Acoela (Vol. 1, Fig. 51, *E*, and Vol. 2, Fig. 98, *B*) lack a pharynx entirely, as do hydroids. We find in the order Acoela all stages of development of the tubular pharynx, from an insignificant fold (Fig. 2) to a long epithelial tube furnished with annular, longitudinal, and radial musculature and numerous unicellular glands (Vol. 1, Fig. 51, *B*, *C*, and Vol. 2, Fig. 98, *A*). A pharynx of that type, known as a simple pharynx (*pharynx simplex*: L. von Graff), is found in lower orders of turbellarians (Fig. 98, *D*) as well as in coelenterates and Acoela, and also in all nemertines, and is the prototype from which all kinds of fore-gut have developed.

A compound pharynx is a pharynx divided into two sections, a pharyngeal process and a pharyngeal sheath (Fig. 101). There are two types of compound pharynx, plicate (*pharynx plicatus*) and massive (*pharynx bulbosus*: L. von Graff, 1908–12–17). A plicate pharynx has a large annular process projecting into the invagination, furnished with strong musculature and capable of being protruded from the oral orifice and of grasping, enveloping, and sucking prey. Such a pharynx is possessed by Polyclada, Triclada, and some Alloeocoela; a similar formation is found in Gastropoda (the pleurembolic proboscis of Prosobranchia). A massive pharynx (Fig. 101, *C*; Fig. 185, *C*) has a smaller and less mobile process, but its chief difference lies in its possession of a boundary layer separating the body of the pharynx from the parenchyma. Consequently the pharynx is a massive body consisting of muscles and glands, encircling the fore-gut like a sleeve. The glands open on the pharyngeal process. Because of its possession of retractor and protractor muscules the pharynx is mobile, and its grasping process can be protruded through the external oral orifice. Such a pharynx is also a powerful swallowing and sucking organ. It is found in some Alloeocoela, almost all Rhabdocoela, all Temnocephala, flukes, and (in a somewhat modified form) Nemathelminthes: Gastrotricha, Nematodes,[1] Kinorhyncha; the pharynx of Rotatoria is still more modified, possessing a unique masticatory apparatus.

In a number of Triclada (species of *Crenobia*, *Phagocata*, and *Digonopyla*) there is multiplication of pharynxes (polymerisation). The terrestrial genus *Digonopyla* has more than 100 pharynxes and up to 63 oral orifices.

---

[1] Most nematodologists incorrectly call the pharynx of nematodes the gullet; W. Wieser (1954) was almost the first to call it the pharynx.

In coelenterates and flatworms the mouth serves both to introduce food into the gut and, usually, to expel undigested particles. In many coelenterates, however, it is not the only communication between the gut and the external environment. We have already mentioned the basal pore of *Hydra* and the acrogaster pore-canals of ctenophores. The latter are used by *Mnemiopsis* (Lobifera), which feeds on fine dispersed food, for defecation, although they are in no way homologous with the anal orifice in any member of Bilateria.

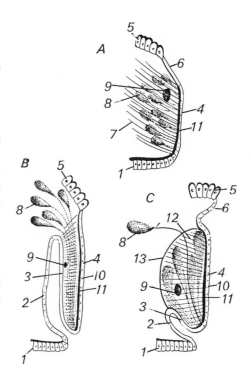

Fig. 101. *Chief types of structure of pharynx in turbellarians.*

Diagrammatic longitudinal sections: *A*—through pharynx simplex; *B*—through pharynx plicatus; *C*—through pharynx bulbosus (after von Graff). 1—epithelium of skin of pharyngeal pouch (2); 3—external epithelium of pharynx; 4—internal epithelium of pharynx; 5—epithelium of mid-gut; 6—gullet; 7—dilators of pharynx simplex; 8—pharyngeal (salivary) glands; 9—pharyngeal neural plexus; 10—internal annular muscles of pharynx; 11—internal longitudinal muscles of pharynx; 12—radial muscles of pharynx; 13—boundary layer separating pharynx bulbosus from parenchyma. In all drawings only half of the pharynx is shown (cf. Fig. 102, 2).

Among other coelenterates, additional communications between the gut cavity and the external environment have been described in some hydromedusae, e.g. *Tima pellucida* and *Zygodactyla rosea*, which have pores leading from the annular canal to the exterior and opening on papillae between the tentacles and the velum (E. Mechnikov and L. Mechnikova, 1870). In *Aequorea* such pores are used to excrete small undigested particles (L. Hyman, 1940). Sea-anemones also have intestinal pores leading from the radial chambers to the exterior. An excretory function is sometimes ascribed to these pores, but they are never used for defecation. In some Pennatularia the two principal medial canals (Vol. 1, Fig. 48, *C*), which are a continuation of the gastral cavity of the first polyp, open to the exterior by fine pores at the basal end of the stalk.

Pores of the above type occur in both coelenterates and flatworms. In several Polyclada many branches of the gut open externally by pores

(*Yungia, Cycloporus*), which do not, however, have an anal function. *Leptoteredra* (Fig. 102) retains a single anal pore, one end of which opens into the posterior extremity of the main gut and the other dorsally to the exterior. The function of that opening is unknown, but anatomically it entirely resembles the anal orifice of higher Scolecida. The same situation occurs in Digenea. In some flukes each of the two main branches of the gut opens externally by a separate orifice (*Schistorchis carneus* and *Diploproctodaeum haustrum*) or even by one common orifice (*Opecoelus sphaericus*). In *Chaunocephalus ferox, Balfouria monogramma*, and members of the family Accacoeliidae the gut opens into the terminal vesicle of the excretory apparatus (K. Skryabin, 1947), as in Rotatoria.

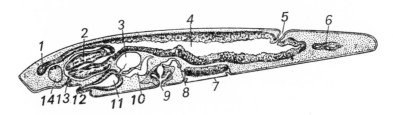

Fig. 102. Leptoteredra maculata (*Polyclada*), sagittal section. *A turbellarian with a single intestinal pore, resembling an anal orifice.*

1—anterior unpaired branch of gut; 2—pharynx; 3—secretion receptacle (vesicula granulorum) of accessory male gland; 4—main gut; 5—anal pore; 6—one of posterior branches of gut; 7—sucker; 8—female genital canal; 9—female genital orifice; 10—seminal vesicle; 11—male copulatory organ; 12—tip of penis; 13—external mouth; 14—brain (from Bresslau).

It is difficult to say whether the anal orifices of nematodes and Gastrotricha (ventral in Macrodasyoidea, dorsal in Chaetonotoidea) are merely new formations or the result of use of intestinal pores of *Leptoteredra* type, since such pores have too few characteristic features. In any case, they show one of the possible methods of formation of an anal orifice in lower worms.

In Rotatoria the gut opens into the terminal section of the excretory apparatus, which thus becomes a *cloaca* (Vol. 1, Fig. 63, *C*). That term denotes a common efferent duct of the gut, the excretory apparatus, and the genital apparatus in all animals in which such a duct occurs. In Rotatoria the anal orifice is cloacal. The communication between the gut and the excretory apparatus is, without doubt, a new formation, and the anal orifice of Rotatoria has no relation to the intestinal pores of Polyclada, which are, however, similar in principle to those of Digenea, in which the gut opens into the urinary bladder (*Chaunocephalus*, etc.); that similarity is the result of parallel development in relatively-distant groups. In any case it is clear that the anal orifice of Rotatoria is not homologous with that of Gastrotricha or of nematodes; not only that, but even within the class Gastrotricha the anal orifices of the two orders Chaetonotoidea and Macrodasyoidea are scarcely homologous with each other. Generally speaking, the anal orifices of the higher classes of Scolecida not only have arisen independently in the various classes, but in many cases are not morphologically homologous with one another. We must stress that conclusion, because many zoologists tacitly assume that the anal orifices of all Bilateria are homologous. Even within the phylum Scolecida that view already proves to be incorrect. It is equally certain that the mouth and anus of all trochophore animals are not homologous with the mouth and any of the types of anal orifices of lower worms, but are products of the differentiation of a single primitive mouth, homologous with the mouth of coelenterates and similar to the slit-like mouth of sea-anemones and ctenophores (see Vol. 1, Chap. V). From all that we know of the development of trochophore animals it follows that they are derived from radial forms that already possessed a slit-like mouth and consequently an epithelial gut. There are, accordingly, sensory cells in the epithelium of the gut of annelids and molluscs (G. Nevmyvaka, 1947; F. Gilev, 1952). Indirect evidence of the primary nature of the epithelial gut of Trochozoa is the fact that movement of food through the gut of the most primitive molluscs is performed entirely by the ciliated epithelium of the gut, without participation by the musculature. Trochophore animals must therefore be derived from animals considerably more highly organised than the planula-like creatures from which we derive the lower worms. Scolecida and Trochozoa are groups related and in many respects parallel to each other, but much farther apart than is often thought.

If the mouth of Trochozoa is not homologous with that of Scolecida, then the ectodermal fore-gut, which in all Trochozoa is formed in association with that mouth, cannot be fully homologous with the fore-gut of coelenterates and Scolecida, which includes the whole of the primary mouth and not only its anterior part.

The diversified pharynxes of Trochozoa, which in structure are sometimes very similar to the massive or plicate pharynxes of Scolecida, are never precise homologues of the latter: they are *homotypic* formations, i.e. constructed on the same general plan, but not homologous, since the concept of homology includes, besides the homotypy of two parts, identity of their topographical and ontogenetic relationships with the other parts of the organism.

In the same way the ectodermal hind-gut of Trochozoa is not homologous with the hind-gut of Scolecida (nematodes, Rotatoria, etc.), which has no relationship with the primary mouth.

In the architectonical section (see Vol. 1, Chap. IX) we have already remarked that in the nature of their oral orifice Deuterostomia stand nearer to Scolecida than to Trochozoa; as far as we can judge, the mouth of lower Deuterostomia is homologous with the primary mouth of coelenterates and Scolecida, and the anal orifice, as in the latter, is a new formation. Although ontogenetically the blastopore of Deuterostomia also is converted into the anal orifice, that method of development is, as we have seen, a secondary modification, and morphologically the anal orifice of these animals is not a derivative of the blastopore but in fact a new formation. Therefore the ectodermal fore-gut of Deuterostomia may be fully homologous with that of coelenterates and Scolecida; in this case, of course, no question can arise of its inheritance from common ancestors of Deuterostomia and Scolecida, of homophyly; when speaking of homology, I have here in mind only morphological correspondence, *homoplasty*. In fact, even if we admit that Deuterostomia have inherited the ectodermal fore-gut from coelenterate ancestors, in Scolecida the ectodermal pharynx is, as we have seen, an acquisition of their own. In the same way the anal orifice of Deuterostomia, although similar in method of formation to that of some Scolecida (e.g. nemertines), has evidently arisen quite independently phylogenetically.

We may point out that only the larvae of echinoderms have a well-developed ectodermal fore-gut. In Enteropneusta larvae the ectodermal part of the fore-gut is only slightly developed. The hind-gut is almost absent in both groups. In adult echinoderms and Enteropneusta almost the whole of the gut is formed of endoderm.

A curious condition is found in some Bivalvia: division of the oral orifice into two symmetrical orifices, each of which receives the afferent grooves of the labial tentacles on its own side. A similar partial overgrowth of the mouth, with formation of several peripheral entrance orifices, occurs in several other animals that feed on finely-dispersed food that they swallow by means of mucous-ciliate adaptations (e.g. among Scyphomedusae, adult *Aurelia;* see Chap. II). Among Bivalvia that structure has developed in some species of *Lima* (family Limidae) and, independently of these, in *Spondylus setosus* (family Pectinidae) (P. Pelseneer, 1931; C. Yonge, 1954). It is obvious that we cannot speak of inheritance of a bifurcate mouth from common ancestors in this case, and we are confronted with an unusually clear example of phylogenetic parallelism.

Even outside the subphylum of Platyhelminthes some Bilateria have retained the intracellular method of digestion that we have seen in sponges and in most turbellarians. Mechnikov was correct in stating that extracellular digestion is a very late acquisition; not only that, but we can now confirm that it was acquired independently by members of the higher stems of Metazoa. We recall that we have already seen in turbellarians a transition from the purely intracellular digestion of most forms to purely extracellular digestion in Microstomidae and Stenostomidae. Among other Scolecida, in Rotatoria the first stage of digestion is performed in the stomach cavity by the secretion of digestive glands, and the second stage in the digestive vacuoles of the syncytial walls of the stomach. Nematodes possess only extracellular digestion.

Most Articulata have purely extracellular digestion, but in leeches, in all Chelicerata studied in this connection (Xiphosura, Scorpiones, Acariformes (M. André, 1927), Opiliones, Araneina, etc.), and in Pantopoda (E. Schlottke, 1933) the final stages of digestion are intracellular; in Chelicerata they take place in the cells of the epithelium of the 'hepatic' processes of the mid-gut.

Purely intracellular digestion is, as a rule, possible only in animals that feed on finely-dispersed food, each particle of which is available for capture by the phagocytic cells of the eater. For that reason we find purely intracellular digestion, or a close approximation to it, in many microphagous animals, i.e. animals that (*i*) feed on nanoplankton, (*ii*) scrape off a growth of unicellular algae from the surface of a hard substrate, (*iii*) select fine food particles from mud at the bottom of lakes, etc. Thus purely intracellular digestion is characteristic of Brachiopoda (demonstrated for *Lingula*), Phoronoidea, and Bryozoa, which feed on nanoplankton. Bivalvia and lower Gastropoda also have almost purely intracellular digestion. Many and various facts indicate that primitive molluscs ('Prorhipidoglossa' of Pelseneer) had a life-form similar to that of modern lower Aspidobranchia, such as Fissurellidae. They must have been slightly-mobile shell-covered animals, with a flat crawling foot, scraping rock surfaces with their radula and feeding on growths of micro-organisms. That food was sorted in the stomach, and the smaller particles were swept by the ciliary movement of the epithelium of the stomach walls into the blind ('hepatic') processes, where they were subjected to intracellular digestion (A. Graham, 1953). I take the following description of the evolution of the digestive apparatus of Gastropoda, in the main, from J. E. Morton (1953).

In *Fissurella* swallowed particles are coated with mucus in the gullet and reach the stomach contained in a slimy cord, from which they are there freed. The stomach has three different functional regions: (*i*) a *sorting area*, covered with cilia and forming a complex system of folds and grooves. Large particles arriving there are forced into the pre-intestinal groove of the stomach, and thence into the gut; small particles remain in suspension in the stomach, being acted upon by the digestive enzymes of the saliva, and are then passed into the liver, where they are phagocytised by the cells of its epithelium and undergo the second, intracellular phase of digestion. In the sorting area numerous amoebocytes from the blood pass into the lumen of the stomach and also phagocytise small food particles; (*ii*) the *cuticular stomach shield*, which covers one wall of the stomach; and (*iii*) the *protostyle sac*, adjoining the pre-intestinal groove and separated from the stomach by longitudinal folds. It contains the protostyle, a rod of solidified mucus with faeces (large particles rejected by the sorting area and undigested residues from the liver) embedded in it. The protostyle is rotated by the cilia of its sac and so draws the slimy cord with food (to which it is united at its anterior end) from the gullet into the stomach. The well-developed narrow gut takes no part in digestion or absorption of food, but forms the faeces and their coating of solidified mucus. That is very important in Gastropoda, whose anus opens into the mantle cavity beside the gills, fouling of which must be avoided (C. Yonge, 1937). In some microphages among Prosobranchia Pectinibranchia and in most Bivalvia the protostyle becomes a real crystalline style, solid, transparent, not containing faeces and not attached to the faecal cord in the gut. The crystalline style, a product of secretion of its own sac, is a gelatinous formation consisting of proteins (globulins) with adsorbed enzymes, mainly amylase, but in many forms also cellulase. The latter occurs not only in *Teredo* (F. Potts, 1923) but also in *Ostrea*, *Mytilus*, and other filter-feeders (B. Newell, 1953) and in some Gastropoda that possess a crystalline style (G. Fish, 1955). Food particles stick to the surface of the style, which is continually rotated by the cilia of its sac. The end of it that lies in the alkaline environment of the stomach is gradually dissolved and liberates enzymes that aid in the digestion of starch and other carbohydrates in the food and (which is especially important) of the cellulose coverings of many algae. There are no free proteolytic enzymes in the gut.

We may ascribe to the common ancestors of Gastropoda and Bivalvia the largest feature—the protostyle, as in Fissurellidae and Protobranchia (G. Owen, 1956); its development into the present crystalline style evidently took place along parallel lines and independently in the two classes; moreover, among Gastropoda that process has taken place independently in several groups. The crystalline style has also evidently arisen independently in *Neopilina* (Tryblidiida).

In many Rhipidoglossa (Haliotidae, Trochidae, etc.) the posterior part of the gut forms a smooth, spirally-twisted process, into which the sorting area extends (Fig. 103, *B*).

Primary phytophagy is retained by a number of Pectinibranchia. These include *Aporrhais pes-pelicani*, which gleans plant remains in the surface layers of the soil; its stomach has a crystalline style and a sorting area, and generally is surprisingly like the stomach of many Lamellibranchia (C. Yonge, 1937).

Some primarily-vegetarian Pectinibranchia have changed over to filter-feeding, drawing in water by the movement of cilia in the mantle cavity, mainly those on the much-enlarged ctenidia (see Chap. II). Forms of this type, such as *Struthiolaria* (Fig. 8) or *Crepidula*, also have crystalline styles and intracellular digestion.

In the predatory Stenoglossa the stomach is simplified: the sorting area, the pre-intestinal groove, the style, and the shield have disappeared or become vestigial, but the buccal mass, the proboscis, and the oesophageal glands are more developed. Among Stenoglossa the crystalline style is known only in *Nassarius*, a soil-feeder (C. Jenner, 1956).

Most Pulmonata and Opisthobranchia have given up microphagy and its associated complex stomach structure. The ciliary-mucous mechanism for movement of food along the gut, which predominates in Bivalvia and lower Gastropoda, is in them replaced by peristaltic intestinal musculature.

Among primitive Pulmonata, however, *Otina* has traces of sorting folds and of a protostyle with its sac; a number of other forms have retained a vestige of the blind process of the stomach, which in *Limnea* has partly retained its sorting role. In many Pulmonata a muscular-cuticular section of the stomach has been formed by expansion of the stomach shield.

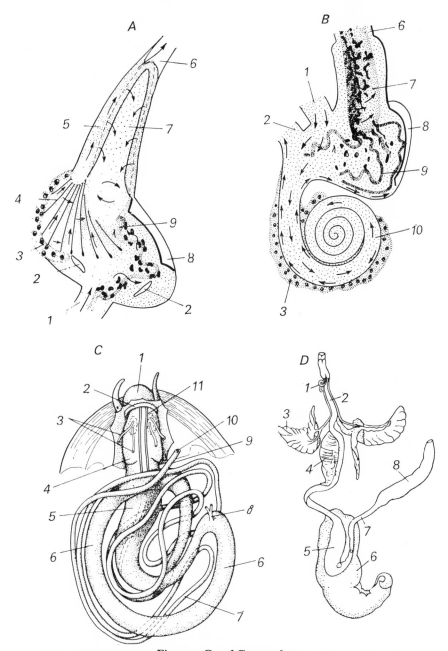

Fig. 103. *Gut of Gastropoda.*

Diagram of stomach structure: *A*—Fissurellidae (Prosobranchia Rhipidoglossa); *B*—Trochidae and Haliotidae (Prosobranchia Rhipidoglossa): 1—gullet; 2—point of entry of liver ducts; 3—phagocytes; 4—sorting area of stomach; 5—groove of gut; 6—mid-gut; 7—protostyle; 8—stomach shield; 9—mucous cord with attached food particles; 10—blind process of stomach (spiral sac) (after J. E. Morton). *C*—*Patella* (Prosobranchia Docoglossa): 1—snout; 2—cerebral commissure; 3—salivary gland ducts; 4—buccal pouch; 5—crop; 6—stomach; 7—gut; 8—liver duct; 9—rectum; 10—anal orifice; 11—pharynx, with gullet arising from its anterior-dorsal part and lying upon it (from A. V. Ivanov). *D*—*Ranella* (Pectinibranchia Taenioglossa): 1—sac of radula; 2—gullet; 3—salivary glands; 4—crop; 5—stomach; 6—liver, the two points where it enters the stomach being denoted by small circles; 7—mid-gut; 8—hind-gut (from Lang).

In Opisthobranchia Tectibranchia the muscular crop, in addition to the cuticle, is furnished with calcareous teeth. It is formed, however, by the posterior part of the gullet and is not homologous with the muscular stomach of Pulmonata. In *Philine* and *Haminea* it grinds detritus; in *Scaphander* it crushes swallowed shells of Bivalvia. The stomach of *Haminea* has vestiges of sorting folds and of a shield. In many Tectibranchia all that is left of the stomach is a narrow place where the gullet, the liver ducts, and the gut meet. *Aplysia* and some Dorididae have a vestige of the style sac, and *Cymbulia* has a true crystalline style as well as a masticatory crop.

On the whole, Gastropoda show all stages of transition from almost purely intracellular to purely extracellular digestion. In carnivorous forms (Muricidae, Buccinidae, etc.) proteolytic enzymes are secreted by the 'salivary' and a number of other glands. The liver of Gastropoda has complex functions: whereas in the most primitive forms its cells, like those of the liver of Bivalvia, phagocytise and digest food particles, in higher forms it is a digestive gland and a place of food absorption. In Pulmonata and Opisthobranchia the blood amoebocytes no longer take part in digestion, but their place is occupied by cells of the digestive epithelium of the liver or fragments of them (which also phagocytise food particles) that have been ejected into the lumen of the liver.

Bivalvia (see C. Yonge, 1923, 1926, 1935, 1941; A. Graham, 1931) resemble lower Gastropoda in the structure of their gut (Fig. 104). Their gut begins with a gullet leading into the stomach. Unlike Gastropoda, they have neither radula nor salivary glands. Their stomach contains the same functional regions as that of Fissurellidae: a sorting area, a pre-intestinal groove, a sac with a crystalline style, and a stomach shield. Paired ducts of the 'liver', which is an accumulation of diverticula, open into the stomach. The small intestine starts from the stomach, describing several loops and opening by an anal orifice into the efferent section of the mantle cavity. In the most primitive forms the sac of the crystalline style is separated from the anal part of the gut by two folds, but in a number of forms it is completely separated from the gut and is an independent blind sac. In Bivalvia the amoebocytes of the peripheral phagocytoblast penetrate not only into the stomach but also into the mantle cavity; in both places they engulf food particles and either return into the tissues with them or break down and enter the liver together with the semi-digested food. The fluid contained in the sections of the mantle cavity lying outside the main tracts of water circulation is practically indistinguishable from the blood in its composition and properties (at least in *Anodonta*, according to A. Jatzenko, 1928).

Cephalopoda Dibranchiata have purely, or almost purely, extracellular digestion. They are all active predators. Squids (Teuthoidea: *Loligo*, etc.) bite their prey (fish) with their jaws, kill it with poison secreted by their salivary glands, bite off pieces, and swallow them. They have purely extracellular digestion: its first phase takes place in the stomach under the action of the liver secretion; the second phase takes place under the action of the pancreas in the blind process of the stomach, and in *Loligo* absorption takes place there also. The digestive gland (hepatopancreas) is homologous with the liver of other molluscs and is divided into liver and pancreas. In *Loligo* food does not enter the cavity of the gland and there is no intracellular digestion (A. Bidder, 1950), but the power of absorption is ascribed to the hepatic epithelium of Octopoda.

In a number of other microphagous filter-feeders the digestive apparatus shows remarkable similarity to that of lower Conchifera. The mucous cord to which food particles adhere, rotated by means of the ciliated epithelium of the narrow posterior part of the stomach, is found in the actinotrochae

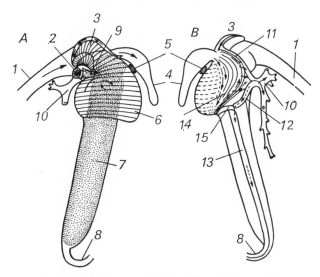

Fig. 104. *Stomach of* Tellina tenuis (*Lamellibranchia*), *semi-diagrammatic.* *A*—from the left, *B*—from the right (after C. M. Yonge). 1—gullet; 2—left liver duct; 3—dorsal (sorting) caecum; 4—postero-dorsal (reserve) caecum; 5—lesser shell adductor; 6—stomach; 7—crystalline style; 8—mid-gut; 9—tooth; 10—right liver duct; 11—food-conducting groove of stomach, leading into hepatic processes; 12—typhlosole carrying refuse from liver to gut; 13—intestinal groove; 14—sorting area; 15—typhlosole removing coarse material from sorting area.

of Phoronoidea (Cori, 1932, 1939), in Kamptozoa (Cori, 1936; G. Becker, 1937), in many Bryozoa (L. Silén, 1944), in Brachiopoda (*Lingula*, S. Chuang, 1959), and in the mid-gut of *Amphioxus* (E. Barrington). All these are analogues of the protostyle of molluscs, but are still more primitive. There are similarities in the structure of the gut in Brachiopoda and in Bivalvia; the stomach of Brachiopoda has no sorting apparatus, but has a ciliated groove that removes undigested residues from the hepatic processes to the small intestine. The whole of the intestinal epithelium is flagellate except the epithelium of the hepatic processes, but peristalsis still plays a larger part in the intestine of Brachiopoda than in that of Bivalvia. The epithelium of both the stomach and the processes is capable of phagocytosis; as in many molluscs, amoebocytes emerging into the intestinal cavity also take part in digestion (S. Chuang, 1959).

In *Phoronis* and Bryozoa, in spite of their lack of multicellular digestive glands, the first phases of digestion take place in the lumen of the stomach and the final phases are intracellular. In Kamptozoa intracellular digestion plays a much smaller role.

Among Deuterostomia, all echinoderms studied in this connection have a mixed type of digestion: the first phase takes place in the intestinal cavity and the final phase is intracellular. In predatory starfishes the stomach walls secrete an active proteolytic enzyme, under the action of which swallowed prey breaks down, and the food particles are carried by ciliary movement into the hepatic processes, where they are phagocytised; the processes have no glandular function. Wandering phagocytes play a large part in absorption in all echinoderms. On the whole, echinoderms stand at a fairly low level in regard to digestion.

In *Amphioxus*, as in Bivalvia, only carbohydrates are subjected to extracellular digestion; fats and proteins are phagocytised by the stomach cells and digested intracellularly (P. van Weel, 1937). In tunicates digestion is entirely extracellular—one more of many indications that they are a simplified and not a primitive group. Tunicates have two digestive glands, a pyloric gland and a liver. The pyloric gland is much branched and usually covers a large part of the intestinal loop; it is found in all tunicates except Appendicularia, and apparently has a double function, excretory and digestive (G. Fouque, 1953; J. Godeaux, 1954). The liver of ascidians is formed of outgrowths of the stomach, and is well developed in Molgulidae and Cynthiidae; it apparently secretes digestive enzymes. Appendicularia and Salpae have one or two blind processes, perhaps homologues of the hepatic processes of ascidians.

We thus see that extracellular digestion has developed from intracellular digestion independently in different stems of the animal kingdom. A diet of finely-dispersed food (nanoplankton) does not stimulate the development of extracellular digestion,. and groups with that diet often retain either fully intracellular (sponges, *Lingula*, *Amphioxus*) or predominantly intracellular digestion (Lamellibranchia). The absence of intracellular digestion in tunicates, which also feed on nanoplankton, indicates their derivation from more active predatory forms, which had already lost the faculty of intracellular digestion.

We have seen that in all animals with mixed or purely extracellular digestion glandular cells are differentiated in the intestinal epithelium and discharge digestive enzymes into the gut cavity. Increased differentiation of glandular cells and the need to increase the area of digestive epithelium lead to the appearance of digestive glands. Often accessory, or even purely digestive, glands develop also in the fore-gut (salivary glands). These include the numerous unicellular salivary glands of flatworms and of Fissurellidae among Gastropoda, and the multicellular salivary glands of Rotatoria, annelids, most molluscs, Atelocerata, Acariformes, etc. Thus the digestive apparatus of the majority of Metazoa consists of an alimentary tract and its associated glands.

A unique component of the digestive apparatus, which only partly fulfils a glandular function, is the liver, which is found in many invertebrates, Mollusca, Crustacea, Chelicerata, Brachiopoda, Enteropneusta, Asteroidea, Ascidiae; among annelids it occurs as an exception in some terrestrial oligochaetes, e.g. *Eutyphoeus* (K. Bahl and M. Lal, 1938). In all these it is a bulky organ with the structure of a complex gland, single or consisting of paired metameric parts, and formed from the epithelium of the mid-gut. The liver has somewhat diversified functions. In primitive forms (Bivalvia, lower Gastropoda, Brachiopoda, Chelicerata, Asteroidea) the main function of the epithelium of the liver is intracellular digestion of food. In higher Mollusca that function is gradually replaced by a glandular function (discharging digestive enzymes into the gut cavity) and usually by the function of absorption of fully-digested food. The liver of Crustacea also has these functions. Besides participating in digestion, the liver almost always has also the function of storage of reserve food supplies, like the liver of vertebrates. As is well known, the vertebrate liver is a very important regulator of the chemistry of the internal environment; it is difficult to say to what extent it assumes that role in different groups of invertebrates.

Can we speak of homology of the livers of different groups of invertebrates? In a certain sense we can: in many cases it is the result of similar differentiation of material of homologous origin, which is adequate basis for admitting the existence of some degree of *homoplasty*. Primitive types of liver in molluscs (Solenogastres), oligochaetes (*Eutyphoeus*), arthropods (trilobites, A. A. Öpik, 1959), and Deuterostomia (Enteropneusta) are paired metameric hepatic processes extending for a considerable length along the mid-gut. In most molluscs the number of hepatic processes has decreased, and no trace of their metameric arrangement remains. Their fate has been similar in Chelicerata. Trilobites had hepatic processes in the head and two pairs in each trunk segment; oligomerisation and heteronomous reconstruction of the system of these processes have taken place in Chelicerata; in this respect scorpions have retained the most primitive type of structure.

In spite of such analogies, there is no possibility of homophyly of the liver even among Trochozoa. The liver is one of the best examples of independent formation of similar organs in far-apart groups of animals.

In one of the most brilliant empirical generalisations of his time, G. Cuvier stated that all animals possessing highly-centralised respiratory organs also have a heart and a liver. The functional relationship between the structure of the respiratory and circulatory apparatuses is clear (see Chap. II), and therefore correlation in their structure is understandable. With regard to the liver, that remains empirical. Nevertheless such correlation certainly exists within large groups of Trochozoa. Generally speaking, Mollusca, Crustacea, and Arachnoidea have centralised respiratory organs, and all of them have a heart and a liver; Annelides and Atelocerata, which have homonomously-metameric respiratory organs, lack a compact, shortened heart and liver. Within separate classes the correlation also holds: for instance, Copepoda, which have no gills, have neither heart nor liver; Phyllopoda, which have gills on all the thoracic segments, have a slightly-differentiated dorsal vessel instead of a heart, and only rudiments of a liver; and so on. In such cases, however, it is usually a question not of definite dependence but of correlation: for instance, polychaetes with oligomerised gills, such as Terebellidae, still lack both a true heart and a liver. On the other hand several Copepoda (e.g. Calanidae, Pontellidae), in spite of loss of gills, have retained the heart but lost the liver. Outside Trochozoa, Cuvier's rule applies well to Brachiopoda, Enteropneusta, and Ascidiae, not to mention higher Chordata. Hepatic processes, however, which are analogous to the liver of other animals, are found in Asteroidea, which lack both a centralised respiratory apparatus and a heart; Siphonophora Disconantae have something in the nature of a liver, i.e. again we see correlation but no definite dependence.

The functioning of the gut may be improved by increase in the length of the intestinal tract, as well as by branching of the gut and formation of specialised digestive glands or of hepatic processes that take part in absorption of food. Within the same group, the forms feeding on less nourishing food are not infrequently distinguished by a longer gut, a fact well known in several groups of vertebrates and insects. A long coiled gut is already observed in. a number of soil-feeding polychaetes, in Sipunculoidea, in insects, in most echinoderms (except Asteroidea and Ophiuroidea), and in most molluscs (except Solenogastres). V. A. Dogiel (1938) presents the following table of the main features of complexity of the gut

Group of animals	Lengthening of gut	Salivary glands	Liver
Platyhelminthes	—	+	—
Nemathelminthes	—	+	—
Annelides (majority)	(+)	+	(+)
Onychophora	—	+	—
Myriapoda	—	+	—
Insecta	+	+	—
Crustacea and Chelicerata (majority)	(+)	(+)	+
Mollusca	+	+	+
Asteroidea	—	—	+
Echinoidea, Holothurioidea, Crinoidea	+	—	—
Ophiuroidea	—	—	—
Enteropneusta	—	—	+
Urochorda (except Appendicularia)	+	—	+

in different groups of Bilateria. I have altered it somewhat to correspond with the system of classification of animals used in this book: a plus sign in parentheses (+) denotes the presence of a given feature in a particular group only in rare instances.

The table shows at a glance the negative correlation between lengthening of the gut and possession of a liver: both adaptations serve to increase the gut surface and are therefore partly interchangeable; their simultaneous presence in tunicates indicates a particularly high level of development of the gut. A second deduction from the table is that all the lower Deutero-stomia lack salivary glands, which is perhaps connected with the poor development of the ectodermal fore-gut in them. In Protostomia, on the other hand, possession of salivary glands is almost universal.

We must point out, however, that 'tabular' generalisations of this kind are graphic but call for use of discretion, since similar morphological features may have different physiological significance in different groups of animals. Thus the lengthening of the gut in insects is generally associated with increase in its digestive function, whereas in Gastropoda it is mainly due to the need to smear faecal masses with mucus to avoid fouling the mantle cavity.

The digestive apparatus of Metazoa has to a certain extent an excretory as well as a digestive function, and in some groups it has also a respiratory function. In some animals that do not have separate excretory organs the gut is the chief organ of excretion. In many turbellarians, for example, metabolic products or pigments artificially introduced into the organism concentrate in the cells of the gut, from which they pass into its lumen and thence to the exterior (E. Westblad, 1923). The 'kidney' of disconant siphonophores doubtless has an excretory function, being a plexus of fine gastrovascular canals containing crystals of guanin (see Vol. 1, Chap. III). In some colonial coelenterates the intestinal endoderm of certain specialised individuals assumes a specific excretory function (cystozooids in Siphon-antae, siphonozooids in Octocorallia, etc.). The colony may thus possess special excretory organs, which the separate coelenterate individuals do not possess. In worms, up to polychaetes and Sipunculoidea, and in molluscs the gut and its separate parts retain an excretory function, largely in spite of the presence of other excretory organs. The gut attains its highest morphological level as an excretory organ in arthropods, in which separate excretory gut appendages, Malpighian tubules, develop (Fig. 105). Formation of the Malpighian tubules by the gut is the same kind of phylogenetic reaction to transition to a terrestrial mode of life as is forma-tion of tracheae by the skin. We find neither of these in aquatic arthropods; both have developed independently in all terrestrial groups, in Atelocerata and Chelicerata. We find neither Malpighian tubules nor tracheae in very small forms (Pauropoda, Aphididae, many Collembola, many Acariformes In small insects without Malpighian tubules, such as lice, the hind-gut itself has an excretory function. In Chelicerata the Malpighian tubules are formed from the posterior part of the mid-gut (Fig. 173), but in

Atelocerata they are mainly formed from the hind-gut: in the first case they are of endodermal, in the second of ectodermal, origin. Malpighian tubules occur in a rudimentary form in some terrestrial Crustacea, such as *Orchestia* (Amphipoda). F. E. Beddard (1895) believed the Malpighian tubules to be homologues of the nephridia of some Megascolecidae (Oligochaeta), which have entered the hind-gut. There are no grounds, however, for accepting that homology. Malpighian tubules develop from the gut itself by (*i*) concentration in specific parts of the gut of one of the functions possessed by the whole gut in less specialised forms, and (*ii*) great expansion of these specific parts.

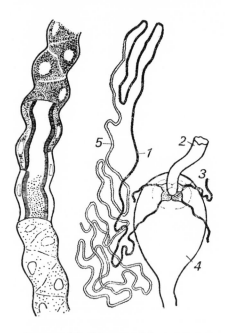

Fig. 105. *Malpighian tubules in* Rhodnius prolixus (*Insecta Rhynchota*).

On the left is part of a tubule, highly magnified, partly in optic section. 1 and 5—two parts of a tubule in different functional conditions; 2—stomach; 3—terminal expansions of tubules before they enter the hind-gut (4) (after Wigglesworth).

As stated in Chapter II, respiratory organs also may arise from the gut. Frequently the fore-gut or the hind-gut may acquire a respiratory function without the development of any anatomical adaptations. Such adaptations first appear in polychaetes. Capitellidae have the so-called siphon along the mid-gut: it is a fine tube, the two ends of which open at the ends of the mid-gut. Water enters it from the hind-gut and is used for respiration: it moves along the siphon independently of food, which moves along the mid-gut at the same time. Such a siphon is found in Echiuroidea, and in Echinoidea among Deuterostomia. Among Echinoidea the more primitive Cidaroidea have the siphon represented by a groove, still united to the gut along its whole length; this is also observed in some polychaetes, and explains the method of origin of the siphons in general. Another respiratory adaptation of the gut is the lungs of holothurians: these are paired, hollow outgrowths of the posterior part of the gut, much branched in large forms (Fig. 175); water is alternately forced into them and pumped out of them by the muscular cloaca. Their respiratory function has been proved beyond doubt.

Respiratory organs formed from the gut become most important in Hemichorda and Chordata, a characteristic feature of which is the possession of gill pouches and gill slits (Fig. 106). The gill slits in these groups are always associated with the fore-gut. Blood vessels run within their walls. In the groups where that apparatus is used only for respiration,

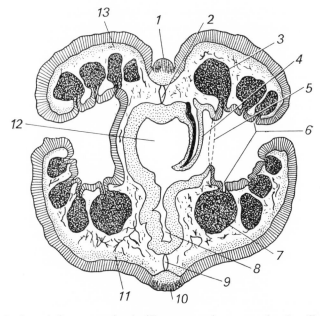

Fig. 106. Stereobalanus canadensis (*Enteropneusta*), *cross-section in gill area of metasoma.*

1—dorsal ectoneural cord; 2—dorsal vessel; 3—gonads (their dorsal sacs); 4—tongue (i.e. a fold descending from the dorsal side of the gill slit); 5—outline of gill slit; 6—gonad ducts; 7—gonads (their lateral sacs); 8—respiratory section of branchial gut; 9—ventral vessel; 10—ventral ectoneural cord; 11—parietal muscles of ventral side; 12—oesophageal section of branchial gut; 13—parietal muscles of dorsal side (after J. W. Spengel).

the gill slits open directly to the exterior (Pterobranchia, Appendicularia) or into grooves on the dorsal side of the body (Enteropneusta). In Ascidiae, *Amphioxus*, and larval cyclostomes the branchial apparatus is used to filter finely-dispersed food from the water and is therefore greatly developed; we have seen such development of the branchial apparatus linked with its use for food procurement also in Lamellibranchia (Chap. II). In both Ascidiae and *Amphioxus* the gill slits open into the peribranchial cavity and not directly to the exterior. In Ascidiae the peribranchial cavity opens to the exterior by a cloacal siphon. The branchial gut and the peribranchial cavity, with the oral and cloacal siphons, form a current-directing adaptation, and the epithelium of the branchial gut creates the current. The current-creating and food-procuring apparatus of *Amphioxus* (J. Orton, 1913) is close to that of Ascidiae; in both cases the endostyle secretes mucus, which is distributed by the frontal cilia of the gill arches throughout the branchial gut in the form of a very fine film or web; that film filters suspended matter from the water passing from the gut to the peribranchial cavity through the gill slits. Gradually moving towards the suprabranchial groove, the film with the attached food becomes a cord that moves rearward along the groove by ciliary action and enters the gullet.

In *Amphioxus* the cord is caused to rotate by the cilia of the anterior part of the mid-gut, and thus moves onward with a screw-like movement, losing the attached food particles in the stomach (E. Barrington).

Some deep-sea ascidians have a much-reduced branchial apparatus, with few gill slits or none. Such forms occur among Molgulidae (e.g. *Aspiracula*) and Ascidiidae (e.g. *Pterygascidia*). Possibly the reduction of the branchial apparatus there is due to its ceasing to participate in the nutrition process when the animals changed from filtration of seston to sediment-feeding.

In Salpae the hydrokinetic (and simultaneously the locomotor) function has been transferred to the annular musculature of the body, and consequently the ciliated part of the branchial gut is also much reduced, becoming a narrow bridge between the cavity of the branchial gut and the cloacal cavity and serving to catch food particles. In Appendicularia the hydrokinetic function is fulfilled by the tail and the current-directing function by the canals of the tunic, food being filtered out by a network stretched in one of these; the animal collects the food from the network with its mouth. The branchial apparatus of Appendicularia, which has only a respiratory function, is therefore much simplified.

We may note that the close similarity in structural details of the gill slits of Enteropneusta and Acrania is evidence in favour of the close relationship of these two groups.

Thus the secondary functions of the digestive apparatus, excretory and respiratory, which are carried out by many animals without special, anatomically-expressed adaptations, attain a high level in some groups. These groups deal with them using special adaptations produced by expansion of the walls of the gut, which vary greatly in different instances.

From the point of view of the structure of the digestive apparatus we must further point out that only in very primitive forms or in very small animals do the mid-gut and other sections of the digestive apparatus consist of epithelium alone. Usually accessory layers, consisting of muscular and supporting elements belonging to the peripheral phagocytoblast, are attached to the intestinal epithelium. The structure of the gut walls consequently becomes very complex, especially if one takes into account the nerves, the blood vessels in many animals, and the tracheae in insects, all of which form part of the gut walls. It is to be noted that the nerves and tracheae are kinetoblastic in origin.

The digestive apparatus (like other apparatuses) in higher invertebrates

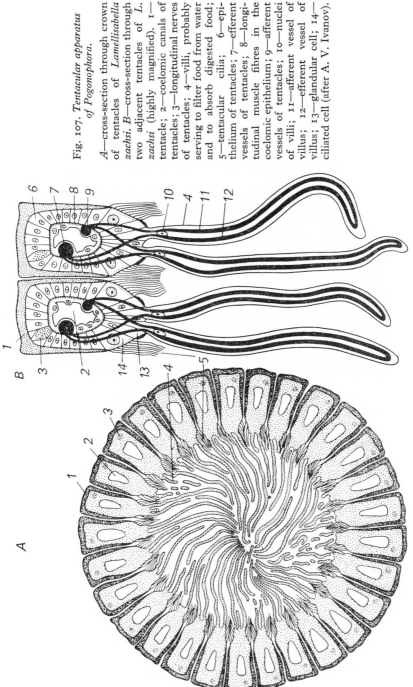

Fig. 107. *Tentacular apparatus of Pogonophora.*

A—cross-section through crown of tentacles of *Lamellisabella zachsi. B*—cross-section through two adjacent tentacles of *L. zachsi* (highly magnified). 1—tentacle; 2—coelomic canals of tentacles; 3—longitudinal nerves of tentacles; 4—villi, probably serving to filter food from water and to absorb digested food; 5—tentacular cilia; 6—epithelium of tentacles; 7—efferent vessels of tentacles; 8—longitudinal muscle fibres in the coelomic epithelium; 9—afferent vessels of tentacles; 10—nuclei of villi; 11—afferent vessel of villus; 12—efferent vessel of villus; 13—glandular cell; 14—ciliated cell (after A. V. Ivanov).

therefore no longer belongs to a single primary layer but is formed from several layers. In each apparatus, however, the products of one layer play a dominant role, being the specific elements that fulfil the main function of that apparatus and around which the accessory elements are grouped. These include the cells of the central phagocytoblast in the digestive apparatus, the kinetoblastic external epithelium in integuments, and kinetoblastic nerve cells in the neurosensory apparatus.

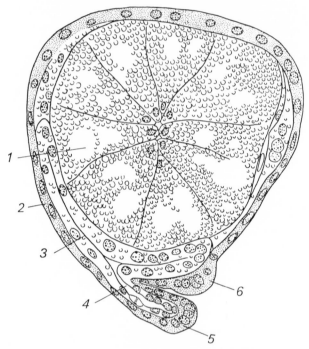

Fig. 108. *Temporary gut of Pogonophora: cross-section through embryo of* Oligo-brachia dogieli.

1—endoderm; 2—ectoderm; 3—first coelom; 4—its continuation into tentacle; 5—right tentacle; 6—left tentacle (after A. V. Ivanov).

As was stated in Vol. 1, Chap. IX, in Pogonophora the gut is totally absent. Digestion of food filtered from the water apparently takes place within the crown of tentacles, which forms a hollow cylinder (A. Ivanov, 1955a). If the suggested digestive role of the tentacles of Pogonophora is confirmed, it will represent one of the most remarkable instances of substitution in the whole animal kingdom (Figs 107, 108).

3. THEORY OF THE COELOM

TYPES OF STRUCTURE OF BODY CAVITY

The history of the peripheral phagocytoblast in Bilateria is closely bound up with the development of the body cavity.

The peripheral phagocytoblast takes part in the formation of the skin and the gut, forming the accessory, mesodermal parts of each. It gives rise to the greater part of the musculature and participates in the structure of other apparatuses and systems of organs, and also to the supporting or connective tissues that fill the interstices between these, as happens in most flatworms, molluscs, etc. In other cases connective tissue does not develop, and spaces filled with liquid, called body cavities, are formed between the organs. Usually a distinction is made between the primary body cavity and the secondary body cavity, or *coelom*. The term 'primary body cavity', however, again denotes two different formations, *blastocoel* and *schizocoel*.

The *blastocoel* is the cavity of a blastula. When there is an invaginated gastrula, as in sea-urchins, the blastocoel persists for some time between the skin and the gut of the gastrula, in addition to the newly-formed gut cavity. The blastocoel is a purely temporary formation. Sometimes it still occurs in early larval stages, but it always disappears later. In fully-polarised animals such as hydroids, Chaetognatha, and *Amphioxus*, the epithelial layers are closely apposed and adhere to each other, and no traces of the blastocoel remain. In animals with a well-developed mesenchyma the blastocoel is filled up by the latter, e.g. in echinoderms and molluscs. There is therefore usually no continuity between the blastocoel and any cavity in the adult animal. If sometimes such a continuity seems to exist, it is only when the method of development is far from primitive (e.g. in the case of development of the heart in some arthropods). In any case the blastocoel plays no part in the organisation of the adult animal, and no products of it exist in adult animals, in spite of the views expressed by some earlier authors.

The *schizocoel* is a cavity that appears in the peripheral phagocytoblast as a result of the moving apart or breakdown of cells. We find such cavities in many Rhabdocoela; in *Rhynchomesostoma* there are such large cleft-like spaces in the parenchyma between the gut and the body wall that I have observed parasitic infusoria (*Ophryoglena* of the order Holotricha) swimming in them.

We also find large schizocoels in Rotatoria, nematodes, and other nematomorphic groups. The schizocoel is especially well developed in Nematomorpha, in which it is characterised by considerable regularity of outlines.

When we call the schizocoel and the blastocoel primary cavities we imply that a common characteristic of them is absence of walls of their own. Every primary cavity is bounded not by walls of its own but by the immediately-adjacent tissues or organs, and thus is merely a space between these. A schizocoel may be bounded on one side by supporting tissue and on the other side by the basal surface of the intestinal epithelium, etc.

A secondary body cavity or *coelom*, on the contrary, is a body cavity lined with its own epithelium. Although the coelom usually occupies all the spaces between organs and its epithelium lines all the crevices between them, the coelom is essentially nothing else than a thin-walled sac. It is not merely a hole, it is an organ. Moreover, the coelom may have its own ducts opening to the exterior, the *coelomoducts*.

A coelom in one form or another is possessed by all trochophore animals except Kamptozoa, and by all Deuterostomia, Podaxonia, Brachiopoda, and Chaetognatha. The origin of the coelom as a separate organ (or even apparatus) presents an important problem. That problem has always called forth a number of divergent views, and even today it has no generally-accepted solution.

THEORIES OF THE ORIGIN OF THE COELOM

What does 'origin of the coelom' imply? From a purely morphological point of view the problem consists in finding homologues of coelomic sacs in lower forms, in coelenterates and Scolecida. From the historical point of view we must produce a satisfactory hypothesis of the probable structure of organs homologous with the coelom in ancestors of coelomate animals, and of the methods whereby they were transformed into the present coelom; that is a very difficult task.

Several theories have been advanced in the search for homologues of the coelom. We list them below, beginning with those less plausible.

The least plausible theory is the so-called *nephrocoelic* theory, advanced by H. Ziegler in 1898 and supported by R. Snodgrass (1938). Ziegler regarded coelomic sacs as expansions of the nephridia of lower worms, in the same way that the air-sacs of Hymenoptera or Diptera are expansions of tracheae. The nephridia of turbellarians and other Scolecida are fine tubes of ectodermal origin ramifying throughout the body. According to Ziegler, the terminal sections of the nephridia expanded into vesicles and the vesicles became coelomic sacs. Later the genital cells became connected with them, and the close association between the genital apparatus and the coelom (observable in all coelomate animals) arose. Expanding, the coelom came into close contact with the musculature, and finally (Ziegler says) all of these organs began to develop from a single rudiment, from the walls of the coelomic sacs.

Not one of the premises of this theory is soundly based or is in any way plausible. As E. Goodrich (1897, 1945) has stated, what were formerly called nephridia are combinations of heterogeneous objects: true ectodermal nephridia, and coelomoducts developing from the walls of the coelom and having the function of genital ducts.[1] It is only through that confusion that the nephrocoelic theory could have arisen. The coelom could not originate from expansion of coelomoducts, since animals without a coelom also lack coelomoducts; the coelom could not originate from expansion of true nephridia, since fully-normal nephridia are retained in typical coelomate animals, particularly in polychaetes. Moreover, it is

[1] The term 'coelomoduct' was suggested by Ray Lankester (1900).

impossible to reconcile the hypothesis of the origin of the coelom from protonephridia with embryological data.

A second theory is the *schizocoelic* theory, advanced by A. Goette (1884) and supported by N. N. Poléjaeff (1893) and J. Thiele (1902). According to this theory, the coelom of annelids and molluscs is homologous with the schizocoel of Scolecida, i.e. it is merely a perfected schizocoel. This theory does not contradict facts, but it gives very little help. A theory of the origin of the coelom should explain why the coelom has a genital function, why genital products develop in its walls. It should explain the origin of coelomoducts; it should explain why in some animals the coelom is formed from the gut, why the musculature of the body is often partly formed from it, and so on. The schizocoelic theory gives a convincing answer to none of these questions, and therefore cannot be fully accepted. It still, however, contains a grain of truth. In the ontogeny of all Protostomia the coelom is formed by separation of cells or, in general, by epithelisation of an originally massive and amorphous rudiment, so that the schizocoelic theory correctly describes what takes place in the individual development of a large proportion of coelomate animals. Therefore we shall see that in the final synthesis something should be borrowed from this theory.

N. A. Livanov (1955) has proposed the *myocoelic* theory of origin of the coelom, according to which the coelom arose as a cavity in the musculature of an animal, which became filled with fluid and therefore played the role of a supporting formation for the surrounding muscles. It is easily seen that this is a modernised version of the schizocoelic theory. The suggestion that the coelom had a supporting role during its evolution is a valuable one.

A third theory is the *gonocoelic* theory, which derives the coelom from the genital glands. It was advanced by the Prague zoologist B. Hatschek (1878), and formulated by the Dane R. Bergh (1885) as follows: the coelom originated from the gonads of lower worms; the cavity of each half-segment of an annelid corresponds to the cavity of a single genital follicle of a turbellarian or a nemertine; the epithelium of the coelom (peritoneal epithelium) corresponds to the wall of a gonad, and the coelomoducts correspond to genital ducts.

From this point of view the best ancestral form from which to derive the metameric coelom of annelids is Nemertini, with their partly-metameric arrangement of sacciform gonads, each of which opens to the exterior by a separate short duct (Fig. 109). Bergh believes that the epithelium lining such a gland was transformed into coelomic epithelium, the formation of genital cells was concentrated in a small area of the wall, and the duct was transformed into a coelomoduct. From this point of view the cells of the coelomic epithelium are modifications of either genital or follicular cells.

There are many facts to support this theory. In all animals that possess a coelom it invariably has a genital function; genital products develop in the walls of the coelom and are conveyed to the exterior by coelomoducts.

We see that process in annelids, molluscs (Solenogastres, Cephalopoda), echinoderms, and vertebrates right up to mammals.

The chief difficulties in the gonocoelic theory are due to its attempt not only to derive the cavity of the coelom from gonad cavities but also to derive the walls of the coelom from elements of a genital gland. To do that we must make the assumption that some of the genital cells lost their original nature and became simple somatic cells, and that the rudiments

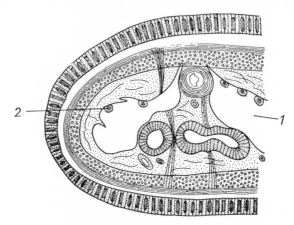

Fig. 109. *Cross-section through trunk of* Drepanophorus crassus (*Hetero-nemertini*) (after O. Bürger).

Empty cavities of the gonads (1), with small oocytes (2), externally resembling coelomic cavities. Presented here to illustrate the basic premise of the gonocoelic theory of the origin of the coelom.

of all organs produced ontogenetically from the walls of the coelom are derived from a genital rudiment. As a result, a number of authors writing on this subject have become thoroughly entangled in the physiology of development.

The conjecture of the old gonocoelic theory that the walls of the coelom were derived from elements of a genital gland was disposed of by W. Faussek (1891, 1897, 1900, 1911), who cited instances of early individualisation of genital rudiments in opposition to the theory. In a number of vertebrates, arthropods, and cephalopods the genital cells are identifiable at very early stages of development, long before formation of the coelom, sometimes at the very beginning of cleavage. Faussek drew the following conclusion: since the genital cells are individualised much earlier than, and independently of, the coelom, these two formations differ in their nature, and the walls of the coelom cannot be derived from genital cells.

That argument was correct, and was later confirmed in the case of polychaetes. P. P. Ivanov (1912) discovered that in the development of Sabellidae and Spionidae genital cells arise in the front part of the body, creep rearward along the gut, and enter separate post-larval segments, where they lie beneath the coelomic epithelium. It is clear, therefore, that in polychaetes also the genital elements and the coelom are formations that were originally mutually independent, and that the walls of the coelom cannot have originated from genital cells. It is true that A. Lang (1903) and others have tried to object, stating that the rudiment of a genital gland could have differentiated into a strictly genital rudiment and a coeloblast, the rudiment of coelomic epithelium, which changed its function. This artificial assumption, however, can hardly save the gonocoelic theory in its classic form.

A fourth theory of the origin of the coelom is the enterocoelic theory, which postulates an intestinal origin for the coelom. It was first advanced by E. Mechnikov (1874), who based it on comparison of echinoderms at early stages of development with ctenophores (Fig. 110), and it was developed by O. and R. Hertwig (1881), working from comparison of sea-anemones with higher animals. Variants of the enterocoelic theory are the theories of A. Sedgwick (1884) (see Vol. 1, Chap. VI) and more recently of A. Remane (1950) and H. Lemche (1960).

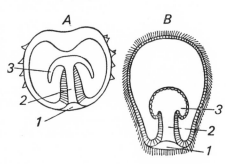

Fig. 110. *Similarity in method of development of coelom in echinoderms and in that of peripheral sections of gastrovascular apparatus in ctenophores; illustrating the basic premise of the enterocoelic theory.*

A—larva of *Cydippe* (Ctenophora). *B*—larva of *Astropecten* (Asteroidea). 1—primary mouth; 2—rudiment of gut; 3—rudiment of coelom in starfish and of gastrovascular canals in ctenophore (after E. Mechnikov).

From the point of view of the enterocoelic theory, the coelom is the homologue of the peripheral section of the gastrovascular apparatus of coelenterates. Mechnikov based this view on the remarkable similarity between the location of the coelomic sacs in a young dipleurula and the location of the gastrovascular canals in a ctenophore, which we have mentioned in our discussion of the promorphology of Deuterostomia (see Vol. 1, Chap. IX). To what was stated there we must add another similarity: the genital products of ctenophores develop in the walls of the gastrovascular canals, and the genital cords and genital products of echinoderms develop in precisely the same way from the walls of the coelom.

In the embryonic stages, when the coelomic sacs of an echinoderm have not yet separated from the gut and are still in the form of pockets, they are not homologues of gastrovascular canals, they are true gastrovascular canals. In later stages the coeloms of echinoderms separate from the gut, but retain all the functions of the peripheral section of the gastrovascular apparatus.

It is easy to see that this theory explains not only the origin of the coelom but also the origin of the coelomoducts, which prove to be homologues of the intestinal pores of coelenterates. At the same time it explains the genital function of the coelom, since the peripheral sections of the gastro-vascular apparatus of higher coelenterates (ctenophores, Anthozoa) also fulfil a genital function. The enterocoelic theory thus includes the gono-coelic theory, but in a somewhat modified form; the homologues of the coelom are not the genital follicles of turbellarians and nemertines, but the peripheral canals or chambers of coelenterates.

APPLICABILITY OF THE ENTEROCOELIC THEORY TO PROTOSTOMIA

The enterocoelic theory would raise no doubts if only Deuterostomia were involved. Deuterostomia are, in fact, naturally derived from animals at the ctenophore level. Important supplementary arguments in favour of the great antiquity of epithelisation of the peripheral phagocytoblast of Deuterostomia are the possession by echinoderms of endoneural and hyponeural sections of the neural apparatus and the presence of neural elements in the coelomic epithelium of Enteropneusta (L. Silén, 1950), if the coelomic origin of their neurons is confirmed. But the question becomes much more complex as soon as we turn to Protostomia, to annelids and molluscs. We recall the heroic attempt of A. Sedgwick (see Vol. 1, Chap. VI) to apply the enterocoelic theory to annelids. From the purely comparative-anatomical point of view, application of the enterocoelic theory to Protostomia could be, to a certain extent, defended; difficulties arise from embryological considerations. There is no enterocoelic formation of the coelom in either annelids or molluscs. Some exceptions have been described, but all such statements have proved to be erroneous. The coelom of annelids is formed from massive paired rudiments, the so-called mesodermal bands, in which the coelomic cavities arise by separation of cells, i.e. by the schizocoelic method.

There is a means of escape from that difficulty. Blastomere 4d belongs to the fourth quartet, and all the cells of that quartet go to form the gut, i.e. are endodermal cells. In many cases 4d itself detaches some small cells, enteroblasts, that form part of the gut. On that basis it is easy to assume that cell 4d also is endodermal, and that formation of the mesodermal bands from blastomere 4d is formation of them from endoderm, from the gut. From this point of view the difference from Deuterostomia is merely that here the coelom is formed, not by evagination of the wall of the gut, but by separation from the endodermal rudiment of a single cell, which by multiplication creates the epithelium of the coelom. That method of development is secondary, and owes its origin to reduction of the number of cells in the embryo. We have here a case of teloblastic establishment of mesoderm.

We should point out that teloblastic establishment of mesoderm occurs not only in Protostomia but also in Deuterostomia, e.g. in Ascidiae, where it masks the enterocoelic formation of the coelom. On account of all these considerations a number of authors, especially V. V. Zalenskii, believe it possible to apply the enterocoelic theory to Protostomia as well as to Deuterostomia, and not only to Trochozoa but even to Scolecida, in particular to nemertines and turbellarians, which have no coelom and whose peripheral phagocytoblast is amorphous. Most supporters of the enterocoelic theory hold the view that if the mesoderm of these animals is formed from blastomere 4d it must be assumed that they have possessed and lost a coelom, that these are simplified animals (V. Zalenskii, A. Remane). As we have seen, it is not possible to agree with that view, or with the theory of a dual origin of mesoderm on which it is based (see Chap. I). It is also contradicted by the development of coelomic mesoderm in *Phoronopsis*, which takes place without participation by D-quadrant cells and without formation of mesodermal bands (see Vol. 1, Chap. V).

Another point of view on the origin of the coelom of Protostomia, based on analogy with the gut, is possible. We may deduce the evolution of the peripheral phagocytoblast from comparison with the evolution of the central phagocytoblast. In this case a deduction from analogy is quite proper, since the two formations are very similar to each other.

We have seen above that epithelisation of the central phagocytoblast, leading to formation of the gut, took place independently in coelenterates and in turbellarians. Nevertheless the gut is certainly homologous in members of the two groups. In both cases a process of epithelisation, qualitatively identical, takes place on an identical substrate—central phagocytoblast. Since basically-similar processes of development take place on a monotypic living substrate, the results cannot be other than equally similar. It is true that homoplasty, and not homophyly, exists between the organs so developed, since they are not inherited from a common ancestor; but their homology is indisputable.

We must use the same reasoning with regard to the peripheral phago-cytoblast. In Actiniaria it is fully epithelised, and is represented by the epithelium of the radial chambers. In ctenophores and medusae most of it is epithelised, but a few cells remain scattered in the mesogloea. In this respect echinoderm larvae stand at the same level as ctenophores. In them the epithelised part of the peripheral phagocytoblast is represented by the coelomic pouches, and the amorphous part by mesenchymatous cells that have migrated into the blastocoel before the beginning of gastrulation.

In Scolecida, including nemertines, the peripheral phagocytoblast generally remains amorphous, and the presence of genital glands does not produce epithelisation in it. We see some hint of such a process only in Nematomorpha. A wide body cavity in Priapuloidea, long believed to be a true coelom, has no coelomic epithelium (K. Lang, 1953), but L. N. Zhinkin (1955) and W. L. Shapeero (1961) have questioned that. If it should be proved that Priapuloidea have a true coelom, that would be an instance of independent development of a coelom in the phylum Scolecida. It is very probable that a coelom arose independently also in Actino-trochozoa, in the first place in Phoronoidea.

Attempts have more than once been made to find a coelom or vestiges of it in nemertines, but as yet without success. In particular, the data of

V. V. Zalenskii regarding the presence of a temporary coelom in that group are apparently erroneous. If, however, a coelom should be found in nemertines, the meaning of that discovery would be quite different from what Zalenskii sought in it: it would not indicate the relationship of nemertines to annelids, but would serve as one more confirmation of the possibility of a coelom arising independently in different stems of Protostomia.

So if a coelom does occur in Scolecida it is as a very rare exception, in the highly-specialised group of Priapuloidea. In contrast, possession of a coelom is in the highest degree characteristic of Trochozoa. We find it, above all, in oligomerous annelids and molluscs. In both molluscs and *Dinophilus* [1] the progeny of cell 4d become part of the amorphous phagocytoblast, in which coelomic cavities lined with epithelium appear only in comparatively late stages of ontogeny. These cavities are formed by the schizocoelic method, according to the schizocoelic theory; they are occupied by genital cells, and the ducts of these coelomic cavities become genital ducts, according to the gonocoelic theory. According to the enterocoelic theory, however, the coelomic cavities of oligomerous annelids and primitive molluscs (being cavities arising from epithelisation of the peripheral phagocytoblast, opening by pores to the exterior, and having a genital function) are homologous with the radial chambers of Actiniaria and the gastrovascular canals of ctenophores.

The post-larval body of polymerous annelids has a coelom that is highly developed, and which therefore appears much earlier in their ontogeny: the first progeny-cells of the mesoblasts are epithelised, and the coelomic pouches are formed from embryonic bands.

The forms with the most primitive type of ontogeny—oligomerous annelids and molluscs—are therefore distinguished by later development of the coelom and less polarisation of the peripheral phagocytoblast. If we regard the development of the coelom as being more primitive in molluscs than in annelids, then it is more probable that the coelom arose independently in Protostomia, but by a process of epithelisation of the peripheral phagocytoblast parallel to that in coelenterates. We have seen that the method of formation of the embryonic layers in animals with spiral cleavage is derived from multipolar immigration and a parenchymatous type of development, as in hydroids (see Chap. I), and not from invagination and the enterocoelic method of formation of mesoderm characteristic of Deuterostomia.

The teloblastic development of Protostomia, unlike the teloblastic development of ascidians, is not derived from the enterocoelic type of development but from the more primitive parenchymatous type of development found in hydroids (see Chap. I). Ontogenetic comparison does not support the derivation of Protostomia from higher coelenterates with epithelised peripheral phagocytoblast. The data of both comparative anatomy and comparative embryology therefore indicate independent,

[1] The development of coelomic cavities in Myzostomidae has not been studied.

homoplastic but not homophyletic, origins of the coelom in Articulata and Mollusca on the one hand and in Deuterostomia on the other.

4. Structure of the Coelom in Annelids and Molluscs

Oligomerous Annelids and Molluscs

The most primitive form of coelom among Trochozoa occurs in the oligomerous annelids Dinophilidae and Myzostomidae. In both of these the coelom is represented only by genital cavities. In the female *Dinophilus* (Vol. 1, Fig. 95) the coelomic cavity lies in the rearmost of the five body segments; the metamerism of oligomerous annelids, as we have seen, mainly involves only the ectodermal (kinetoblastic) organs. The coelomic cavity opens to the exterior by two genital ducts, the coelomoducts. Along its lateral walls lie a pair of ovaries, which develop from primary genital elements originally foreign to the coelom and migrating thither only during embryonic development. The coelom functions as a uterus and a spermatheca. The anterior part of the body of *Dinophilus* contains fairly loose mesenchymatous tissue with irregular schizocoelic areas, without epithelial covering.

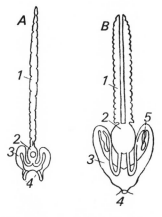

Fig. 111. *Structure of the coelom of Solenogastres.* A—*Chaetoderma nitidulum.* B—*Rhopalomenia acuminata.* 1—genital section of coelom; 2—pericardium; 3—coelomoducts; 4—cloaca; 5—spermatheca (after Wirén).

Myzostomidae are not closely related to *Dinophilus*, but their coeloms are very similar in principle. In *Protomyzostomum* (Vol. 1, Fig. 96), which has been studied in detail by D. M. Fedotov (1915), the genital cells originally lie in parenchyma (amorphous peripheral phagocytoblast), within which the coelomic cavity also arises. It expands, and the ovaries gradually become enclosed in its walls. In the adult animal the coelomic cavity lies dorsally to the gut and has numerous ramifications. It opens into the cloaca by an unpaired pore and several pairs of coelomoducts. That is equivalent to opening to the exterior, since the cloaca is an invagination of the external covering. The rudiments of the male efferent ducts are formed from one pair of coelomic outgrowths; separating, they expand and are connected by means of numerous ramifications to the seminal follicles, and open to the exterior by a pair of lateral orifices.

The coelom of oligomerous annelids is therefore a purely genital cavity. The next stage is represented by the class Solenogastres (Fig. 111). Their coelom is differentiated into two sections: anterior (genital) and posterior (pericardial). Genital products develop in the walls of the genital section of the coelom, which functions as a gonad. The posterior section surrounds the hind-gut and the heart, through which runs the hind-gut. A pair of coelomoducts (with a genital function only, as in *Dinophilus*) arise from the posterior section; they do not have an excretory function in Solenogastres. The anterior, genital section of the coelom may be paired or unpaired, and similarly the external openings of the coelomoducts may fuse into one unpaired opening.

In the structure of their coelom Solenogastres stand slightly higher than oligomerous annelids. In both groups the coelom is, above all, a genital cavity. The chief difference is that in Solenogastres its posterior section serves as a pericardium.

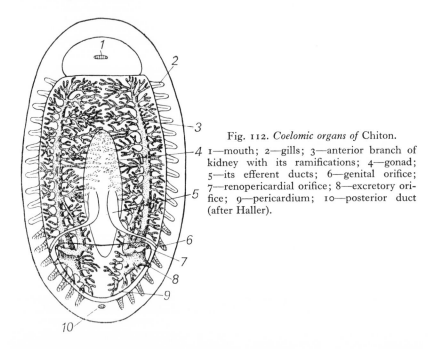

Fig. 112. *Coelomic organs of* Chiton. 1—mouth; 2—gills; 3—anterior branch of kidney with its ramifications; 4—gonad; 5—its efferent ducts; 6—genital orifice; 7—renopericardial orifice; 8—excretory orifice; 9—pericardium; 10—posterior duct (after Haller).

Loricata, which are closely related to Solenogastres, have advanced along the path of separation of the two sections of the coelom, genital and pericardial. The coelom of Loricata (Fig. 112) consists of two independent cavities. The anterior cavity begins as a paired formation, but with further development its two halves fuse into a single gonad, as in *Chaetoderma* among Solenogastres. Only in one genus of Loricata (*Nuttalochiton*) does the gonad remain paired in the adult state. In Loricata each section of the coelom has its own pair of coelomoducts: the coelomoducts of the anterior section function as genital ducts, and those of the posterior, pericardial section as kidneys; they accordingly form numerous ramifications, which serve to increase the surface of excretory epithelium.

Cephalopoda, which are the most highly organised molluscs, have retained some very primitive features in their coelomic structure. In the first place, both *Nautilus* and Decapoda retain the communication between the genital and the pericardial sections of the coelom. In this respect they stand at the same level as Solenogastres. In *Nautilus* both sections are very wide. The genital section of the coelom forms a true body cavity, surrounding the stomach and one of the loops of the gut. It is prolonged into a siphon, extending into a whorl of the shell. The gonad, strictly speaking,

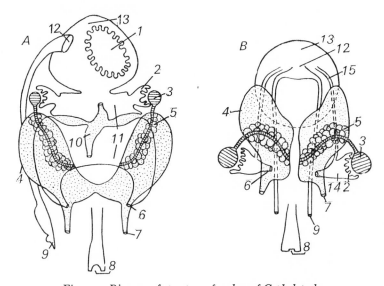

Fig. 113. *Diagram of structure of coelom of Cephalopoda.*
A—*Sepia* (Decapoda), female, rear view. B—*Eledone moschata* (Octopoda), female, rear view. 1—ovary; 2—pericardial glands; 3—branchial hearts; 4—kidneys; 5—venous appendages of kidneys; 6—renopericardial orifice; 7—external excretory orifice; 8—posterior duct; 9—external genital orifice; 10—ventricle of heart; 11—pericardium; 12—opening of oviduct into gonad cavity; 13—genital section of coelom; 14—expansion of coelom around pericardial gland; 15—water-bearing canals, connecting gonad cavity to vestiges of pericardium (from Lang).

occupies only a specialised area of the coelom. The genital section of the coelom opens to the exterior by an asymmetrical coelomoduct, and the pericardial section by a pair of separate coelomoducts. In *Nautilus* the excretory part of the two kidneys is separated from these canals and opens independently to the exterior.

The genital section of the coelom in Dibranchiata Decapoda (Fig. 113, *A*) is widened in almost the same way, and also contains the gonad and the stomach; the pericardial section contains the heart, the branchial hearts, and the pericardial glands. The pericardial section opens by two renopericardial orifices into the kidneys, which open to the exterior. The genital section opens to the exterior by an unpaired genital duct. Both the kidneys and the genital duct are coelomoducts. The external end of the oviduct, however, as well as the shell gland, are formed by the ectoderm.

In contrast to Decapoda, the body cavity of Octopoda (Fig. 113, B) is reduced to a system of narrow, thick-walled canals that connect the gonad cavity with the kidneys and the exterior. The gonad cavity is fairly large; a pair of oviducts or an unpaired seminal duct lead directly from it to the exterior, into the mantle cavity. In addition, a pair of canals run from it to two cavities surrounding the appendages of the branchial hearts (pericardial glands). These cavities are vestiges of the reduced pericardium and communicate with the kidneys by a pair of nephropericardial orifices. The heart lies outside the coelom, right in the parenchyma; among molluscs that location of the heart is found only in Octopoda and Anomiidae (Lamellibranchia). The coelom of Octopoda is therefore reduced and specialised, but it retains the main primitive feature of the coelom of lower invertebrates, a communication between its two sections, genital and renopericardial.

Although in the cephalopod coelom the genital and pericardial sections remain in communication with each other, as in Solenogastres, each section has its own coelomoducts, as in Loricata. In all other molluscs the two sections are completely separated, as in Loricata. In them the gonad cavity no longer plays the role of the general body cavity, as it does in lower Cephalopoda. But the pericardium of most Lamellibranchia and of some Prosobranchia Aspidobranchia surrounds not only the heart but also the hind-gut, which penetrates the ventricle of the heart.

The original number of coelomoducts in molluscs is an extremely important question. We have seen that *Protomyzostomum* has four pairs. The discovery of *Neopilina*, with its six pairs of coelomoducts, makes it very probable that a relatively large number, later subjected to oligomerisation, was primitive for molluscs: Loricata and Cephalopoda Tetrabranchiata have retained two pairs, and other molluscs one pair. In particular, one pair of coelomoducts represents the basic structural type for all Gastropoda and Bivalvia, in spite of all the diversity in the structure of the kidneys and genital ducts observed in these classes.

The most primitive interrelationships between parts of the coelomic apparatus are found in the most primitive members of these two classes: in some Protobranchia among Bivalvia (Fig. 114, C), and in lower Rhipidoglossa (Fig. 114, F) among Gastropoda. In Protobranchia the gonad opens by a pair of canals into the renopericardial ducts, and the genital products are discharged through the kidneys; in Rhipidoglossa the gonad opens through a single asymmetrical canal into the renopericardial duct on the right side, and the genital products are discharged through the right kidney alone. The canals connecting the gonad with the renopericardial duct are clearly nothing else but vestiges of a communication between the genital and pericardial sections of the coelom. Thus the structure of the coelomic apparatus of Protobranchia and Rhipidoglossa is in essence close to that of Solenogastres.

Further development differs in the two classes. In most Gastropoda the right kidney, as such, is reduced, and is transformed into the distal part of the genital duct; its renopericardial connection is broken. The left kidney, which has no connection with the gonad, continues to function as a kidney. Consequently we may consider that the kidney and the genital duct of most Gastropoda represent the left and right coelomoducts of one pair, not of two different pairs. But to argue more strictly, the genital duct and the kidney of Gastropoda are antimerically homologous only in their

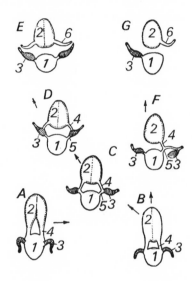

Fig. 114. *Diagram of transformation of the coelomoducts in Lamellibranchia and Gastropoda.*

A—Solenogastres as original prototype. *B*—Prorhipidoglossa (hypothetical). *C*—Protobranchia. *D*—*Pecten*. *E*—Eulamellibranchia. *F*—Rhipidoglossa. *G*—most Gastropoda. 1—pericardium; 2—gonad; 3—kidney; 4—duct connecting gonad with pericardium or coelomoduct; 5—renopericardial duct; 6—secondary genital duct (of compound nature) (after Pelseneer, modified).

distal parts, and their proximal parts are of different nature: the proximal part of the kidney represents the renopericardial communication, which is lacking on the right side, whereas the proximal part of the genital duct represents the vestige of the communication between the gonad and the pericardium, which does not exist on the left side (Fig. 114, *G*).

In contrast to Gastropoda, Bivalvia have a fully-symmetrical urogenital apparatus, all changes in which take place identically on each side. In *Solenomya* (Protobranchia) the gonadal ducts open into a renopericardial canal. Here it is perfectly clear that these are not an independent pair of coelomoducts, but vestiges of the communication between the pericardium and the genital gland. In *Proneomenia* (Solenogastres) the opening of the canal joining the gonad to the pericardium is fairly distant from the funnel of the coelomoduct; in *Solenomya* these two openings are close together. In other Lamellibranchia the gonadal ducts are still farther displaced and open in the kidney itself. In *Lima* (Filibranchia) they open in the proximal section of the kidney; in *Pecten* (Fig. 114, *D*) and *Ostrea* (Filibranchia), in the central part of the kidney; in Nuculidae (Protobranchia), in its distal part; and finally, in Eulamellibranchia the genital glands open quite independently of the kidney, although the genital and excretory orifices lie side by side (Fig. 114, *E*). Thus the genital ducts of Bivalvia are formed in their distal parts by division of the kidney, and in their proximal parts they are, as in Gastropoda, a vestige of the communication between the gonad and the pericardium. The sole genital duct in Scaphopoda opens in the centre of the kidney, as in *Pecten* among Bivalvia. In any case, the possession of two pairs of coelomoducts by Eulamellibranchia is secondary.

Thus the coelom of molluscs is at a somewhat higher level of development than that of oligomerous annelids. It mainly continues to be a genital cavity and has, in addition, become a pericardial cavity. The opinion is often expressed that molluscs have a reduced coelom, that their gonad and pericardium are vestiges of a large and (as some think) a metameric coelomic cavity possessed by their ancestors. There are, however, no traces of reduction of the coelom in mollusc ontogeny, in which molluscs differ from leeches and arthropods, where traces of that process are evident. Comparative anatomy also provides no clear evidence in support of that opinion. The only argument in its favour is the presence of a relatively large coelom in just such archaic forms as *Nautilus* and *Neopilina*. Even in these, however, it may be a secondary adaptation.

Thus there is no conclusive evidence that the ancestors of molluscs possessed a large coelom, and no evidence at all of original metamerism of the mollusc coelom.

POLYMEROUS ANNELIDS

In the prototype the coelom of polymerous annelids is constructed as follows: each segment (except larval segments) contains a pair of coelomic sacs lying at the sides of the gut. Where they met each other on the dorsal and ventral sides their walls form *mesenteries* (Fig. 115, *A*); where they met the cavities lying in front and behind they form transverse septa, or *dissepiments*. Each sac opens to the exterior by a single coelomoduct. The walls of the coelomic sacs consist of coelomic epithelium lining their cavity, and of musculature adjoining it and developing from the same rudiment. The musculature include first the longitudinal muscles of the body wall, as well as the muscles of the dissepiments and the mesenteries, the musculature of the gut, etc.

Such is the prototype. In the great majority of polychaetes, however, the mesoderm of the larval segments also becomes metamerised and epithelised (although rather late), so that only the cephalic and anal lobes remain without a coelom. Although sometimes coelomic canals occur also in the cephalic lobe (e.g. the canals of the palps in *Polygordius*) they are always processes of the coeloms of the first body segment. After coelomisation of the larval segments the whole gut is held between the coelomic sacs, which in the aggregate form the whole body cavity. The coelom of the post-larval segments retains a genital function, since in all post-larval segments, or more often in only some of them, gonads are formed on the walls of the coelom. The coelomoducts, if any, act as gonoducts in these segments. The coelom of polychaetes also has a pericardial function, since the most important vessels lie in the mesenteries, and the peri-intestinal sinus is compressed between the coelom and the wall of the gut. In addition, the coelom has a distributive function (see Chap. VIII); for that reason there are numerous groups of ciliated cells in its epithelium (A. Lubischev, 1912; A. Meyer, 1927, 1929); they constitute the ciliary apparatus of the coelom, which causes circulation of the coelomic fluid (Fig. 115, B). Finally, as N. A. Livanov (1940) points out, a very important function of the coelom is its supporting function; the muscle tension in

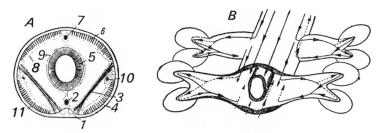

Fig. 115. *Coelom of polychaetes.*

A—diagrammatic cross-section through one of the segments of *Polygordius*: 1—ventral brain in epidermis; 2—ventral blood vessel; 3—oblique muscles; 4—skin; 5—gut; 6—one of the dorsal quadrants of the longitudinal muscles; 7—dorsal vessel; 8—parietal layer of coelomic epithelium (somatopleure); 9—its visceral layer (splanchnopleure); 10—cross-section of nephridium; 11—ventral quadrant of longitudinal musculature (after Reisinger). *B*—two trunk segments of *Tomopteris* (Phyllodocemorpha): diagram of circulation of coelomic fluid under action of ciliated cells of coelomic epithelium (after Meyer).

the body walls causes the fluid in the coelom to give a certain rigidity to the whole body and to its separate parts. In the earthworm (*Lumbricus terrestris*) the hydrostatic pressure in the front body segments is equal to that of a 16-cm column of water, and in the rear segments to an 8-cm column; in an anaesthesised worm it falls to zero (G. E. Newell, 1950). The lower pressure in the rear segments is due to the poorer development of the parietal musculature of these segments (N. Sokol'skaya, 1951).

The diversity of coelom structure in the class Polychaeta has arisen mainly because of reduction of one or both mesenteries, and reduction of some dissepiments and increased development of others (e.g. the so-called diaphragm of Terebellidae). Frequently also longitudinal septa parallel to the frontal plane, or oblique septa, develop in the coelom. In some cases, e.g. in the mud-dwelling Amphinomidae (Nereimorpha), all the dissepiments are almost entirely reduced, so that the coelom loses its metamerism. We see the same condition in the class Echiuroidea, which has completely lost its dissepiments. We must point out that the reduction of dissepiments in Echiuroidea, as in many polychaetes, is an adaptation to burrowing, enabling coelomic fluid to flow from one end of the body to the other ('hydraulic' method of burrowing, L. Zenkevich, 1944).

In burrowing, the front end of the body is first narrowed and elongated by contraction of its annular muscles, and is forced into the soil; then the musculature of the rear end contracts and the fluid is forced into the front end, whose musculature is then relaxed; under the pressure caused by the entire musculature of the rear half of the body, the soil is pushed aside. The same kind of powerful musculo-cutaneous sac and spacious, fluid-filled coelom are possessed by fossorial marine animals of several other groups, such as Priapuloidea, Sipunculoidea, and Synaptidae (Holothurioidea).

Oligochaetes are a very homogeneous group with respect to the coelom, as in their whole organisation. The oligochaete coelom mostly retains the primitive, homonomous-metameric structure characteristic of the annelid prototype. It is true that the dissepiments are usually pierced by apertures (connections), but considerable reduction of all dissepiments is comparatively rare. They are absent in the larval segments of Naididae (D. Lastochkin, 1922) and Lumbricidae, and are much reduced in Aeolosomatidae. Each of the dissepiments of the post-larval segments of Lumbricidae has a single aperture (Fig. 140, D), through which the ventral nerve chain passes, but normally these apertures are closed by sphincters lying within the dissepiments, and there is practically no communication between the coeloms of separate segments (G. Newell, 1950). In oligochaetes the dorsal mesentery is always, and the ventral frequently, reduced. Terrestrial oligochaetes have, besides coelomoducts, and often side by side with them, simple dorsal coelomic pores, which in *Lumbricus* begin on the eleventh body segment. They are regarded as new formations, not homologous with typical coelomoducts; the pores possess sphincters. They occur also in the front part of the body of several aquatic oligochaetes. A very characteristic feature of oligochaetes is restriction of the gonads to a few strictly-defined segments of the body (genital segments), which are constant in each family. In some families, e.g. Lumbricidae, in the segments containing testes the coelom is divided by a longitudinal horizontal septum into two parts; the testes lie in the ventral part and the funnels of the coelomoducts open into it. The mature sperm enters the part that is so partitioned off, and cannot spread into the rest of the cavity of the segment. Coelomoducts occur only in the genital segments.

In leeches the coelom undergoes much more complex changes. The most primitive coelom structure in that group is found in *Acanthobdella peledina*, which was studied in detail by N. A. Livanov (1905) and which occupies a place intermediate between leeches and oligochaetes. Some authors even include it among oligochaetes. The coelom of *Acanthobdella* has different structure in different parts of the body. In the three preclitellar (pre-genital) segments the coelom is well developed, there are no mesenteries, and the dissepiments are perforated on both the dorsal and the ventral sides of the gut. In the clitellar segments the coelom is reduced to two longitudinal canals, dorsal and ventral, connected in each segment by narrow, slit-like rings; the remaining space is filled with coelenchymatous [1] tissue, arising through retroperitoneal [2] movement of cells of the coelomic epithelium. In the rear segments the coelom is again much wider (Fig. 116, *A*). The testes lie outside the coelom and have an envelope resembling the somatopleure,[3] which is united to a solid process of the latter; thus the testes lie, without doubt, in an individualised part of the coelom. The ovaries lie in the walls of the coelom, but the ova do not enter the coelom, since the oviducts are quite separate from it.

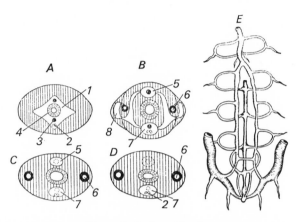

Fig. 116. *Coelom of leeches.*

Diagrammatic cross-sections through: *A—Acanthobdella peledina*; *B—Piscicola* (Ichthyobdellidae); *C—Hirudo* (Gnathobdellea); *D—Herpobdella* (Gnathobdellea) (after Oka). *E*—posterior part of 'lacunar' system of *Aulostoma gulo* (Gnathobdellea) (from Fedotov, after Mikhailovskii). 1—coelom; 2—ventral nerve chain; 3—ventral blood vessel; 4—gut; 5—dorsal lacuna; 6—lateral lacuna; 7—ventral lacuna; 8—lateral vesicles of lacunar system.

In other leeches the coelom is further reduced and converted into a system of canals. In the most typical case there are four longitudinal canals or lacunae: dorsal, ventral, and two lateral. The dorsal and ventral canals already exist in the genital segments of *Acanthobdella*. In the locations of the former somites the longitudinal lacunae are joined together by annular connectives, sometimes of very complex structure. These connectives are the vestiges of coelomic sacs, and the longitudinal lacunae are the former apertures in dissepiments, converted into canals by the thickening of the latter. The dorsal and ventral canals of rhynchobdellid leeches contain blood vessels, and the nerve chain lies in the ventral canal. Besides the longitudinal lacunae and the metamerically-repeated annular

[1] Coelenchyma is amorphous tissue formed from cells of coelomic epithelium that have lost their epithelial connection.

[2] Retroperitoneal (behind the peritoneum) = behind the boundary membrane of the coelomic epithelium, on the side opposite the coelom.

[3] *Somatopleure* = the external wall of the coelomic sac, adjoining the ectoderm, as opposed to the *splanchnopleure*, which adjoins the gut.

connectives, there is a network of peripheral lacunae arising from the former; it is particularly well developed in Glossosiphoniidae among Rhynchobdellea and in Gnathobdellea. That system of peripheral lacunae is so strongly developed that it resembles a true highly-advanced blood vascular system. In Gnathobdellea the process of substitution is completed, the vascular system disappears and is functionally replaced by the system of coelomic lacunae (Fig. 116, C, D). By the way, the term 'lacunae' as applied to the coelomic canals of leeches is very inappropriate, but it is firmly rooted and would be difficult to change.

Because of assumption of the transport function by the coelom of leeches (see Chap. VIII), a need arose for adaptations that would provide for movement of fluid within its canals. Such adaptations have arisen in different groups of leeches. In many Ichthyobdellidae (Rhynchobdellea; Fig. 116, B) contractile vessels with muscular walls have developed, lying along the course of the marginal canals that run parallel to the lateral canals and on the outer side of them. In Gnathobdellea the two main lateral lacunae are contractile.

The first investigators of leeches never doubted that Gnathobdellea had a true blood vascular system. They thought it strange only that the lateral vessels were contractile in Gnathobdellea, whereas in all other annelids the dorsal vessel had that role. The suggestion that the vascular system was reduced in leeches was first made by L. Cuénot (1891); it was developed by A. Oka (1902) and L. Johanson (1896), and partly by N. A. Livanov and V. D. Zelenskii. All these investigations revealed that in leeches the vascular system has been replaced by the system of coelomic canals, one of the most effective instances of substitution in the animal kingdom (D. Fedotov, 1939).

5. REDUCTION OF THE COELOM IN ONYCHOPHORA AND ARTHROPODA

Reduction of the coelom is very characteristic of these two groups. During the embryonic development of Onychophora and most Arthropoda metameric coelomic pouches are formed, similar in location to the coelomic sacs of annelids. In arthropods they are formed not only in the post-larval segments but also, in most cases, in the larval segments, and some-times even in the head (see Vol. 1, Chap. VI). With further development, however, all these coelomic pouches break down into their separate component cells, losing their mutual epithelial connection, i.e. they become coelenchyma. These cells form muscles, supporting and connective tissue, etc. As a result of the breakdown of the walls of the coelom its cavities lose their independent status and coalesce with the surrounding schizo-coelic spaces. Only exceptionally do separate parts of the coelom persist, becoming part of the excretory or genital apparatus or becoming vestigial. Very often no traces of the coelom are left. In some arthropods reduction of the coelom extends also to the earlier stages: the mesodermal bands are not segmented, do not form coelomic pouches, and are converted directly into coelenchyma, as for instance in some parasitic Hymenoptera (O. Ivanova-Kazas, 1959). In arthropods this method of development is doubtless derivative, since they must unquestionably have been derived from annelids with a well-developed coelom.

In most Crustacea the breakdown of the mesodermal bands begins early, and temporary coeloms, if formed, do not last long. They have been described, however, in almost all the best-studied groups: in Anostraca (*Artemia*, N. Nasonov, 1887, and a number of later authors; *Chirocephalus*, H. Cannon, 1927); in Conchostraca (H. Cannon, 1924); in

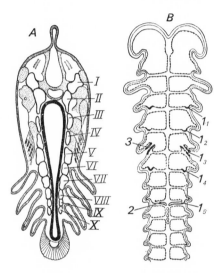

Fig. 117. *Temporary coelom of arthropods.*
A—larva of *Lernaea branchialis* (Copepoda), frontal section: I–X—ten pairs of coelomic sacs (after D. D. Pedashenko). *B*—anterior part of embryo of scorpion (diagram): 1_1–1_5—rudiments of coelomoducts, including gonoducts (2) and coxal glands (3) (from Davydov).

Copepoda (*Cyclops*, F. Urbanovich, 1885; *Lernaea*,* D. Pedashenko, 1899; *Nebalia*, S. Manton, 1934; *Anaspides*, V. Hickman, 1936); in Mysidacea (Yu. Vagner, 1896; S. Manton, 1928); etc. In the embryo of *Gammarus* (P. Weygoldt, 1958) and some others coelomic pouches develop only in the thorax, and in the embryo of *Astacus* (H. Reichenbach, 1886) only in the abdomen. D. D. Pedashenko describes the formation of ten pairs of coelomic pouches in the development of *Lernaea branchialis** (Fig. 117, *A*). They begin there in the maxillular segment, i.e. the first post-larval segment. In this respect *Lernaea*, and also *Cyclops*, resemble some polychaetes (*Prionospio* in the order Spiomorpha) and differ sharply from higher arthropods. Some of the coelomic pouches of *Lernaea* divide into three sections—dorsal, ventral, and lateral. The dorsal sections approach each other, and at the level of the seventh and eighth segments a schizocoelic cavity is formed between them, into which the primary genital cells enter. That cavity represents the gonad cavity, which therefore is a retroperitoneal and not a coelomic cavity, i.e. it is a cavity bounded by the external, basal surface of the peritoneal (coelomic) epithelium; nevertheless, the walls of the gonads are formed by the walls of the coelomic pouches.

* The present name of this copepod is *Lernaeocera branchialis* (Editor's comment).

The shell (or maxillary) glands are formed from the lateral sections of the coelom in the maxillary segment. The antennary glands are formed from the unsegmented mesoderm of the nauplial segments. In spite of this method of development of the antennary glands, both these glands must be regarded as coelomoducts that have assumed an excretory function, like the coelomoducts of molluscs. A terminal vesicle, regarded as a vestige of the coelomic cavity, lies at the proximal end of each coelomoduct. Other vestiges of the coelom besides these vesicles exist in adult Copepoda and in some shrimps and barnacles. In parasitic Copepoda, e.g. *Lernan-thropus*, C. Heider (1879) described a 'vascular system'—paired ventral canals, probably representing a product of fusion of the ventral sections of the coelomic sacs of *Lernaea*. Shrimps (*Palaemon, Crangon, Palae-monetes*) have a small unpaired pouch lying on the dorsal side of the body behind the genital gland, penetrated by the aorta; it is probably a vestige of the dorsal (genital) section of the coelom. In *Laura* (Ascothoracida) the unpaired vestige of the coelom described by N. M. Knipovich (1892) is probably a product of the nephridial sections of the coelomic pouches (D. Pedashenko, 1899).

Crustacea therefore show great variety in their coeloms. In a number of forms, fairly diversified vestiges of the coelom persist even in the adult state. Others retain them only as temporary organs during development. Others still (and these are in the majority), have lost the coelom even during development.

The development of the coelom in Onychophora (Fig. 118, *A–D*) is similar to that in arthropods. Coelomic sacs are formed in all segments of their body, both post-larval and larval. In *Eoperipatus* the foremost pair of well-developed somites belongs to the antennary segment, but R. Evans (1902) describes vestigial pre-antennary somites also in it. The coelomic sacs differentiate into two or three (*Peripatus edwardsii*) sections. When there are two sections, one is dorsal and the other lateral; when there are three, they are dorsal, lateral, and ventral, as in *Lernaea*.

As in *Lernaea* also, the dorsal sections of the coeloms participate in formation of the gonads, although the gonads are formed in an entirely different way. The dorsal sections become constricted off and lie side by side in two metameric rows. Later all the pouches in each row fuse together, thus forming two long tubular gonads. The coelomic pouches of the penultimate body segment form the genital ducts, the ventral sections of the coelom forming the distal parts of the ducts. *Peripatus* therefore differs from Crustacea in that not only are the walls of the genital glands formed of coelomic epithelium but the cavity of the genital gland is formed by fusion of coelomic cavities. In Onychophora, when there are three sections in the coelom, the ventral sections form coelomoducts; when there are two sections, the coelomoducts are formed from the lateral sections. Each coelomoduct with an excretory function opens at its proximal end into a terminal vesicle, which represents a vestige of the coelom. When there are three sections, the lateral sections grow into the appendage, break up into separate cells, and disappear, leaving no trace of their former identity.

Consequently all that remains of the coelom in adult Onychophora is the terminal vesicles of the kidneys and the genital glands. All the other cavities lying between the gut and the body walls in that group, which have fairly complex segmentation, constitute a pseudocoel, the result of merging of the coelom with the schizocoel.

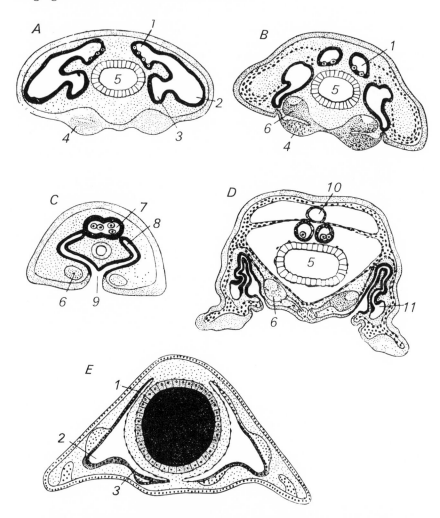

Fig. 118. *Diagrams of development of coelom in Onychophora and Chilopoda.*
A—embryo of *Peripatus*, cross-section; coelomic sacs are divided into three sections—dorsal, ventral, and lateral. B–D—the same; transformation of dorsal sections into gonads, and formation of kidneys and genital ducts from lateral sections. E—cross-section of embryo of *Scolopendra*. 1—dorsal section of coelom; 2—lateral section of coelom; 3—ventral section of coelom; 4—ventral organs; 5—gut; 6—nerve stems; 7—gonads; 8—genital ducts; 9—genital orifice; 10—heart; 11—terminal vesicle of kidney (after Davydov).

The coelom of Onychophora takes no part in formation of the heart.

The development and fate of the coelom in Myriapoda were first studied in detail by R. Heymons (1901) in *Scolopendra*. The embryo of *Scolopendra* has paired coelomic pouches in all body segments and also in the head (antennary and pre-antennary somites). Each of the trunk somites forms three processes, dorsal, lateral, and ventral, as in *Lernaea* and *Peripatus* (Fig. 118, *E*). The dorsal processes grow towards each other and meet above the gut. The heart cavity is formed in the mesentery between them; the musculature of the heart develops from large cells (*cardioblasts*) that become separated in the walls of the dorsal sections. The heart develops in the same way in *Hanseniella* (Symphyla), according to O. W. Tiegs (1940). The ventral sections of the coelom of *Scolopendra* meet on the ventral side of the gut and give rise to the ventral vessel. *Scolopendra* differs from all other arthropods and resembles annelids in that method of formation of the ventral vessel. The rudiments of the genital glands also become pinched off from the dorsal sections of the coelom, from the fifth to the 21st trunk segments; the gonad cavity, as in Onychophora, is a coelomic cavity. The genital ducts are formed from the coelomic cavities of the last two (genital) segments. Excretory coelomoducts are unknown in *Scolopendra;* there are not even the maxillary coelomoducts that are widespread in lower Atelocerata, including Chilopoda Anamorpha.

The coelom persists for a relatively long time in Chilopoda embryos, but its walls finally break down and give rise to musculature, fatty body, etc. In juvenile Anamorpha (e.g. *Lithobius*) the coelomic cavities are well developed in newly-formed segments, which generally have a somewhat embryonic character. In the same way the coeloms of the anal segments persist in Symphyla larvae, providing the material for growth in length of the heart during anamorphosis. The development of the coelom in Diplopoda is generally similar to that just described, but in Julidae, according to F. G. Heathcote (1886), unlike Chilopoda and Onychophora, but like Symphyla, the gonads develop from the ventral sections of the coelom. In Pauropoda the somites are poorly developed, although some of the anterior somites have small cavities. The somites of *Pauropus* give rise to neither gonads nor gonoducts, and because of the absence of a blood vascular system they do not form cardioblasts. The mesoderm of anamorphic segments has no coelomic cavities. The gonads are formed from ventro-medial bands of mesoderm not forming part of the somites. The trunk somites give rise only to parietal musculature (O. Tiegs, 1940, 1947). For the coelomoducts of Myriapoda see Chap. VII.

Among insects the most primitive coeloms are found in the larvae of lower insects, including Hemimetabola, particularly Orthoptera and Blattoidea. The coelomic pouches of Orthoptera give off processes at the bases of the limbs, but in the locust *Stenobothrus* and the cockroach *Phyllodromia* division of the coelomic pouches into three sections, as in *Scolopendra*, has been described (N. Kholodkovskii, 1891). In Holometabola (Coleoptera, Hymenoptera, Lepidoptera) the coelom is much less developed; in Diptera it is very slight, being entirely absent in Muscidae and Sarcophagidae, as it is in parasitic Hymenoptera. Thus we see in insects the same tendency to constantly-increasing reduction of the coelom that we have seen in Crustacea, but in insects the correlation between systematic position and degree of coelom reduction shows up more clearly.

In all winged insects (Pterygota) the coelomic cavities break down early into coelenchymatous elements, and do not take part as such (so far as is known) in the formation of any organs. In Apterygota coelomoducts homologous with the maxillary glands of Crustacea (see Chap. VII) are formed from the mesoderm of the maxillary segments.

Among Chelicerata we find the most complete development of the embryonic coelom in scorpions (Fig. 117, B). The scorpion embryo has coelomic pouches in all body segments. Each of them is divided into the usual three sections. The heart and the pericardium develop from the dorsal section. The cells (which are turned towards each other) of the two dorsal sections of one pair are cardioblasts and give rise to the heart, as in annelids and *Scolopendra;* but a difference from *Scolopendra* is that the dorsal parts of the coelom are constricted off and, in the opinion of some authors, remain to form the pericardial cavity. If that is indeed so, and the pericardium of Arachnoidea represents not a pseudocoel, as in insects, but a true coelomic cavity, we must conclude that the coelom persists to a later stage and in a better form in Arachnoidea than in any other arthropods.

In scorpions the coelomoducts and the genital glands are formed from the ventral sections of the coelom. Some of the coelomoducts serve as excretory organs (coxal glands), others take part in the formation of the genital ducts, others still are purely temporary and disappear without trace. The formation of genital glands from the ventral sections of the coelom distinguishes Chelicerata from all other arthropods except Diplopoda and Symphyla.

Well-developed metameric coelomic pouches are retained in the development of all Chelicerata, including Parasitiformes, in spite of the high level of specialisation of groups in the order. On the whole we may say that Chelicerata stand first among all arthropods in the degree of retention of the embryonic coelom, which, like many other features, shows their highly-individualised position among arthropod classes.

It is curious that the embryos of Pantopoda, according to most authors (J. Meisenheimer, 1902; V. Dogiel, 1911, 1913) and contrary to early statements, totally lack coelomic cavities—one more feature distinguishing Pantopoda from Chelicerata.

We thus see that the coelom has ceased to exist in arthropods as an independent apparatus. In the anatomy of the adult animal it is represented by few vestiges. It has been transformed into a system of purely embryonic organs, whose function is confined to organogenesis. From it are formed the coelenchyma (which is differentiated into musculature, fatty body, blood cells, etc.) and also separate parts of the genital, excretory, and vascular apparatuses.

6. Coelomic Formations in Deuterostomia

As we have seen in Vol. 1, Chap. IX, Deuterostomia larvae have in the prototype three pairs of coelomic sacs. Hemichorda and Pogonophora differ from the prototype in having an unpaired anterior coelom, or protocoel. D. M. Fedotov (1923) and C. N. Dawydoff (1948) consider that that feature is secondary in Hemichorda, particularly in Enteropneusta. The coelom of adult Hemichorda retains, almost completely, the divisions observed in the larvae. In the proboscis region a small vesicle, lying above the cephalic chord and possessing contractile walls, separates from the axocoel. That is the so-called pericardium, probably homologous with the rear part of the right axocoel (protocoel) of echinoderms. Some Pogonophora (order Athecanephria, A. Ivanov, 1955b) have a similar pericardium. In its relation to the vascular apparatus the pericardium of tunicates closely resembles that of Hemichorda, but, unlike it, lies on the ventral side of the body as in Pogonophora (regarding the orientation of the latter see Vol. 1, Chap. IX).

The collar coeloms of adult Enteropneusta are fused together. The trunk coeloms of some Enteropneusta contain lateral mesenteries lying more or less in the frontal plane as well as sagittal mesenteries, so that the whole trunk coelom is divided lengthwise into four sections. In the family Protobalanidae all sections of the coelom retain the lumen and coelomic epithelium. In most other Enteropneusta the coelomic cavities are filled with musculature and coelenchymatous tissue. The whole of the somatopleure in the trunk is consumed in the formation of these tissues, while the splanchnopleure retains its epithelial structure. Coelomic epithelium is practically absent in the collar coelom, but is developed on all the walls of the proboscis coelom. The gonads lie free of the coelom, but their walls develop from the coelomic epithelium.

The coelom structure in Pterobranchia does not differ substantially from that in Enteropneusta, but the coelomic cavities are better developed, corresponding to their less-developed musculature. The lateral mesenteries of the trunk coelom are partly developed in *Cephalodiscus* and are absent in *Rhabdopleura*.

Pogonophora approach Hemichorda in the tripartite division of their coelom (although they differ from Hemichorda in the segmentation of the rear part of the third coelom) and in the method of its formation during ontogeny (Fig. 119) (see Vol. 1, Chap. IX). The development of the coelom in *Amphioxus* has been described, and the coeloms of *Amphioxus* and vertebrates compared with those of Hemichorda, in Vol. 1, Chap. IX. Pogonophora show the same tendency as Enteropneusta towards formation of musculature and coelenchyma from coelomic epithelium.

The gonads of Pogonophora are retroperitoneally located and project into the trunk coelom, into which the genital products are discharged (Vol. 1, Fig. 172). Pogonophora are unisexual. The genital products are expelled through one pair of coelomoducts, which develop as seminal ducts or oviducts. In the structure of their genital apparatus, therefore, Pogonophora resemble vertebrates or annelids more than they resemble other lower Deuterostomia (A. Ivanov, 1958a).

In tunicates the coelomic mesoderm occupies the same situation during development as it does in *Amphioxus*, but it does not form a cavity and

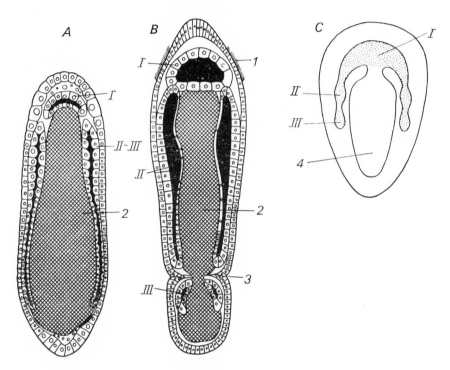

Fig. 119. *Formation of coelom in Pogonophora, and comparison with its formation in* Dolichoglossus (*Enteropneusta*).

A and *B*—*Siboglinum caulleryi* (Pogonophora), two stages of coelom formation, frontal sections (after A. V. Ivanov). *C*—*Dolichoglossus pusillus* (Enteropneusta) (after Korschelt, from Ivanov). I—rudiment of anterior coelom; II–III—common rudiment of central and posterior coeloms; II—rudiments of central coeloms; III—rudiments of posterior coeloms; 1—larval crown of cilia; 2—temporary endoderm of *Siboglinum*; 3—diaphragm formed between collar and trunk; 4—endoderm (rudiment of gut) in *Dolichoglossus*.

gives rise directly to the longitudinal musculature of the tail. The coelomic organs of the adult ascidian are formed only during metamorphosis, in the form of three evaginations on the ventral side of the branchial gut. The unpaired evagination gives rise to the pericardium, and the paired (epicardial tubes) to the perivisceral cavity, which is lined with flat-celled epithelium and lies in the posterior (basal) part of the body.

The coelomic cavities attain a higher degree of complexity in echinoderms. Among the three pairs of coelomic pouches in the dipleurula the second left coelom of the larva, or *hydrocoel*, plays the leading role in ontogeny. In the prototype the second coeloms of the dipleurula are linked with the tentacular apparatus (see Vol. 1, Chap. IX); the hydrocoel of modern echinoderms is also linked with the tentacular apparatus in the form of tube-feet and tentacles, since it gives rise to the ambulacral system. The left hydrocoel curves round the mouth in a horse-shoe shape and forms the circumoral ambulacral ring possessed by all modern echinoderms. Radial ambulacral vessels grow out from the ambulacral ring. Depending on the shape of the body, they either radiate over the oral surface (as in Asteroidea and Ophiuroidea) or, because of reduction of the aboral surface, curve along the meridians of the body (as in Echinoidea and Holo-thurioidea). Transverse canals run from the radial canals to the tube-feet and the ampullae. In some holothurians (Synaptidae, Vol. 1, Fig. 202, A; Molpadiida, Pelagothuriidae) the tube-feet are reduced, except the first ones, which lie round the mouth and are modified into branched circumoral tentacles. In Synaptidae the radial ambulacral canals also are more or less reduced. The oral tube-feet of Crinoidea and Ophiuroidea, as well as some of the tentacles of holothurians, arise directly from the circumoral ring.

Various appendages located in the general body cavity arise from the ambulacral ring in the interradii. Asteroidea, Ophiuroidea, and Holo-thurioidea possess hollow, muscular Polian vesicles, which probably regulate the internal pressure in the ambulacral system; they number from one to four. In Asteroidea Tiedemann's vesicles, dense accumulations of tubules running from the ring canal, apparently glandular in nature, lie beside the Polian vesicles. Cidaroidea (Echinoidea) have a continuous wreath of short tubular appendages along the whole circumoral ring; in other Echinoidea they are concentrated in the interradial Polian vesicles, which thus resemble the Tiedemann's vesicles of Asteroidea in their structure (P. Svetlov, 1924), and in Spatangoidea they are entirely absent.

The interradial appendages of the ring develop in four interradii. The stone canal arises in the fifth interradius, its other end opening into the ampulla (Fig. 120). The ampulla is a product of the left anterior coelom (left axocoel), and the stone canal represents a combination of the hydrocoel and the left axocoel. The ampulla opens to the exterior by a pore corre-sponding to the pore-opening of the left axocoel of the larva. Sometimes that opening remains simple, as in most Ophiuroidea, and sometimes several openings are formed instead of one, resulting in a perforated madreporic plate as in Asteroidea, Echinoidea, Euryalae among Ophi-uroidea, etc.

Besides the ampulla with its pore-canals, part of the axial sinus (left sinus; D. Fedotov, 1924; J. Smith, 1940) develops from the left axocoel. Another part of the axial sinus (right axial sinus) develops from the right axocoel (Vol. 1, Fig. 195). Together they surround the stone canal and the axial organ; in its oral part the latter consists of a dense network of vessels, and in its aboral part of a mass of lymphatic tissue. The oral part of the axial organ is connected with the circumoral ring by a blood vascular system. The axial organ, the stone canal, and the axial sinuses constitute the axial assemblage of organs characteristic of Asteroidea, Ophiuroidea, and Echinoidea (Fig. 120). In all these classes the genital sinus, enclosing

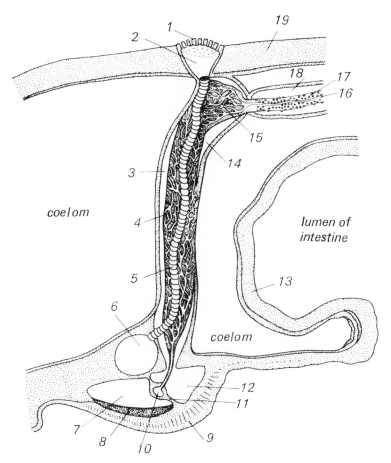

Fig. 120. *Diagram of the axial assemblage of organs in Asteroidea.*
1—pore-canals; 2—ampulla; 3—left axial sinus; 4—oral part of axial organ; 5—stone canal; 6—annular ambulacral canal; 7—external perihaemal ring; 8—annular hyponeural cord; 9—annular ectoneural cord; 10—annular (oral) blood vessel; 11—septum; 12—internal perihaemal ring; 13—wall of gut; 14—right axial sinus; 15—aboral part of axial organ; 16—blood vessel of genital cord; 17—genital cord (genital stolon); 18—genital sinus; 19—body wall (after A. A. Strelkov).

the genital cord, arises from the right axocoel (Fig. 121). The genital cord develops from the genital rudiment and is a tube consisting of epithelially-arranged primary genital cells. A bundle of vessels runs from the aboral section of the axial organ within the genital cord. The axial organ thus probably provides nutrition to the genital cord, among other functions.

The genital sinus separates from the rest of the coelom and closely adjoins the right axial sinus. In Asteroidea it lies mainly in the body wall, forming a complete ring round the aboral side of the animal. It forms a similar ring in Ophiuroidea (Fig. 121, *A*). Along the route the genital sinus and the genital cord give rise to gonads in all the interradii. The genital sinus, expanding, forms the gonadal sacs, and the genital cord forms the gonads themselves (Fig. 121, *B*). The sacs grow to the external wall of the body, where the genital orifices develop, connecting the gonad cavity to the exterior and serving to discharge genital products.

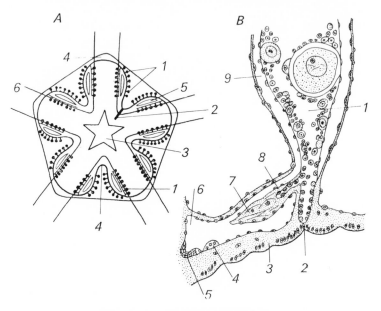

Fig. 121. *Genital sinus of Ophiuroidea.*

A—diagram of topography of aboral annular sinus (genital sinus) of Ophiuroidea (from Lang): 1—gonads; 2—axial sinus; 3—mouth; 4—genital sinus; 5—openings of bursae; 6—arms. *B*—*Ophiura texturata*, cross-section through aboral annular sinus at the point where it bears the ovary: 1—gonad cavity; 2—genital orifice, not yet opened; 3—wall of bursa; 4—nerve ring of aboral sinus (endoneural annular cord); 5—muscle bundle; 6—aboral annular sinus; 7—annular sanguiferous lacuna; 8—genital cord; 9—genital sinus (after L. Cuénot).

The left axial sinus, which communicates at one end with the ampulla, opens at the other end into an annular sinus called the circumoral *perihaemal* ring. From the perihaemal ring a pair of radial perihaemal canals run along each radius (Fig. 69). These canals lie on the oral (actinal) side of the body, directly beneath the radial ectoneural cords. A radial blood vessel is squeezed into the mesentery between the two perihaemal canals of each radius. Embryologically the perihaemal ring is formed in four interradii from the left posterior coelom, separating from it, and one interradius (at the point of origin of the axial sinus) is formed from the left anterior coelom (Fig. 122). The radial perihaemal canals are formed as processes of the perihaemal ring.

From the two posterior coeloms, mainly the left one, there is formed a large body cavity in the strict sense of the term, containing the gut and other internal organs. The gut is suspended on mesenteries, which, however, never divide the coelom into separate sections. A secondary subdivision of the coelom is, however, found in echinoderms. A circumoral lantern sinus separates from the general body cavity in Echinoidea. Ophiuroidea have a perioesophageal sinus, and Echinoidea and Ophiuroidea have a

Fig. 122. *Development of coelomic cavities during metamorphosis of the starfish* Asterina gibbosa (*diagrammatic, after* C. N. Dawydoff); *juvenile with trace of larval preoral lobe (stalk).*

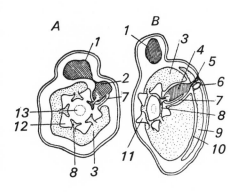

A—view from oral side. *B*—view with half-turn. 1—trace of left anterior coelom of stalk; 3—left posterior coelom; 2 and 4—left anterior coelom (axocoel); 5—pore-canal; 6—right anterior coelom (madreporic vesicle); 7—stone canal; 8—perihaemal annular canal or its rudiments; 9—right posterior coelom; 10—principal mesentery; 11—radial processes of hydrocoel (rudiments of radial ambulacral canals); 12—hydrocoel, here shaped like a pentagonal disc (and not like a horseshoe, as in a typical case); 13—location of future origin of mouth of adult.

perianal sinus. All of these sinuses represent secondarily-separated parts of the general body cavity. The coelomic epithelium of echinoderms is usually covered with cilia, whose activity causes the coelomic fluid to circulate. That fluid is similar in composition to sea-water, but contains proteins and cellular elements.

Some Ophiuroidea, particularly Euryalae, are characterised by a considerable degree of reduction of the coelom, due to the great development of the genital pouches or bursae possessed by all Ophiuroidea (D. Fedotov, 1926, 1927). These bursae are invaginations of the external covering of the disc, opening to the exterior beside the points of origin of the radii; the genital glands open into the cavities of the bursae. In Euryalae these cavities combine into a single 'tertiary body cavity' and grow so much that they almost squeeze out the coelom (Fig. 123). The coelom becomes a system of narrow slit-like spaces lying between the body wall, the gut, and the wall of the bursal cavity. The resulting inter-relationships resemble those between the coelom and the peribranchial cavity in *Amphioxus*. In effect the tertiary body cavity replaces the coelom (D. Fedotov, 1939). The possible physiological significance of the exceptional development of the bursae in Euryalae is discussed in Chap. II.

The above is a brief outline of the system of coelomic cavities in echinoderms. The most characteristic feature of that phylum is the possession of a body cavity, usually spacious, containing all the internal organs, as well as possession of several complex canal systems; the latter arise partly by separation from the above trunk coelom and partly from the left central and the two anterior coeloms. These systems are extremely constant, recurring in their main features in all classes of echinoderms and showing strict regularity in their variations. The cavities of one system never

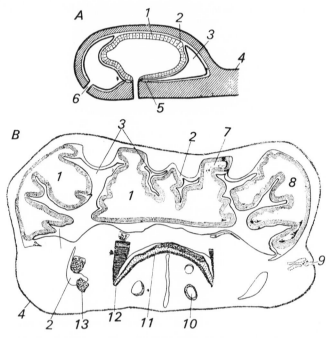

Fig. 123. *Tertiary body cavity in Euryalae* (*Ophiuroidea*), *originating from the genital pouches.*

A—diagram of vertical section through disc of *Gorgonocephalus*. *B*—vertical section through disc of *Euryale aspera*. 1—gut; 2—traces of secondary body cavity; 3—tertiary body cavity, formed of bursae; 4—epidermis; 5—circumoral ring of tertiary cavity; 6—cleft of bursa; 7—radial processes of gut; 8—interradial processes of gut; 9—pores and pore-canals; 10—oral tentacles; 11—ectoneural ring; 12—Tiedemann's vesicles; 13—muscles of first segments of arms (after D. M. Fedotov).

communicate with those of another system, except in the few places listed by us, such as the very ancient union between the stone canal and the ampulla, found in all echinoderms. When new orifices develop, such as the opening of the madreporic orifices in the general body cavity in most holothurians (see Vol. 1, Chap. X), these openings cause the group in which they occur to differ sharply from the rest of the phylum.

The functions of all these systems are somewhat obscure. The best understood is the locomotor and grasping function of the appendages of the ambulacral system (in Asteroidea, Echinoidea, Holothurioidea), but in addition the tube-feet certainly have a tactile function; a respiratory function has also been demonstrated in them. The perihaemal system rather has some trophic function, but it is difficult to be more precise. Rhythmic contractions have been described in the right axial sinus of a number of echinoderms, which therefore fulfils the function of a contractile pericardium and, as we have stated above, is probably homologous with the pericardial sac of Enteropneusta. In the larvae of some echinoderms, such as *Echinus*, the wall of the right axial sinus forms an invagination corresponding to the heart of other lower Deuterostomia (N. Narasimha-murti, 1933).

Crinoids are unique with regard to their coelomic formations. Being representatives of Pelmatozoa, crinoids differ from other modern echinoderms in a number of primitive features. At the same time modern crinoids display a number of far-reaching specialisations, and their coeloms are less primitive than those of most other echinoderms.

We have already seen that primitive Palaeozoic crinoids had a single pore-opening and apparently a single stone canal. Intermediate forms had five madreporic plates and probably five stone canals, like some Asteroidea and Ophiuroidea. Modern crinoids have a tremendous number of pore and stone canals.

The *Antedon* larva has a single ampulla, formed from the left anterior coelom. The single stone canal, which leads to the ambulacral ring, opens into it. The left posterior coelom lies in the oral part of the calyx, and the right posterior coelom in its aboral part. Later all the mesenteries disappear, and all the coelomic cavities—right and left posterior coeloms, and ampulla —coalesce into one continuous body cavity. The latter then fills with a network of cross-bars of connective tissue covered with coelomic epithelium, so that the coelomic cavity does not disappear but only becomes spongy. The right posterior coelom, before coalescing with the left, gives rise to the chambers of the five-chambered organ, which lie radially around the axial organ and extend into the stalk parallel to one another.

After the ampulla coalesces with the rest of the coelom, the pore canal is in direct communication with the general body cavity. The stone canal also now enters that cavity. An unlimited increase in the numbers of both canals thus becomes possible. Along its whole length the ambulacral ring buds off numerous stone canals that open into the coelom, which forms a large number of pore canals opening to the exterior. In the adult *Antedon* we find hundreds of pore canals and hundreds of stone canals (Fig. 124).

Another characteristic of crinoids is the absence of a perihaemal ring; perhaps that is a primitive feature in them, but it may be only a result of the coalescence of all the coelomic cavities; the ring does not appear in the ontogeny of *Antedon*. There are radial perihaemal canals in the arms of crinoids (Fig. 69, *E*). In addition each arm contains four more coelomic canals, all opening into the general coelom of the calyx. One of them contains a radial genital cord, so that the genital products of most crinoids develop in the ramifications of the arms, the pinnules. As in Asteroidea, Echinoidea, and Ophiuroidea, the genital cord is connected with the axial organ, which in crinoids lies along the body axis, within the five-chambered organ.

We may point out one feature of the echinoderm coelom possessed also by Enteropneusta: the development of part of the neural apparatus from the walls of the coelom. Apart from the lower Deuterostomia, we find such development of part of the neural apparatus from the peripheral phagocytoblast only in coelenterates. Both echinoderms and coelenterates

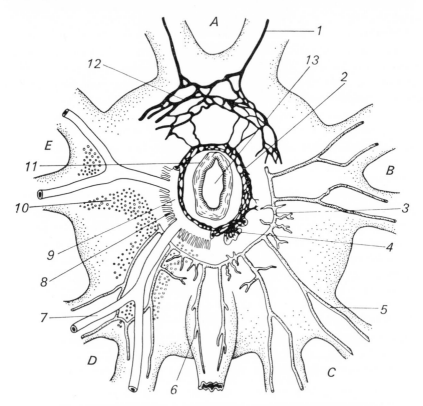

Fig. 124. Antedon bifidum, *internal organs on oral side* (*diagrammatic*).

1—genital lacunae of arms; 2—lacunar ring; 3—annular hyponeural cord; 4—spongy expansions of lacunae; 5—hyponeural cords of arms; 6—nerves of anal papilla; 7—radial ambulacral canals; 8—stone canals; 9—annular ambulacral canal; 10—pore-canals; 11—ectoneural ring; 12—network of lacunae of skin; 13—gullet; letters denote corresponding radii (after L. Cuénot).

are characterized by a high degree of epithelisation of both the central and the peripheral phagocytoblast (where the latter is separated). In both groups that feature is probably very ancient; we possess no data that would prevent us from surmising that Deuterostomia have inherited a high degree of epithelisation of their phagocytoblast, together with the phagocytoblastic section of their neural apparatus, from radial ancestors, although we cannot prove that surmise.

We may suggest that all three main groups of Bilateria (Scolecida, Trochozoa, and Deuterostomia) originated independently from coelenterate ancestors and separated from the phylum Coelenterata at different levels of its evolution. Scolecida separated at the parenchymatous level from coelenterates (or larvae of coelenterates) that were histologically at the planula level and did not yet possess an epithelised gut. It is not surprising that they are completely devoid of the phagocytoblastic section of the neural apparatus, and that their entire neural apparatus as a whole corresponds only to the kinetoblastic section of the neural apparatus of coelenterates.

Trochozoa originated from coelenterates that possessed (we may suppose) epithelised central, and still unepithelised peripheral, phagocytoblast, and only later did they epithelise the latter, with the development of coelomic epithelium. Accordingly some of them apparently have neural elements of endodermal origin (G. Nevmyvaka, 1956) and do not have neural elements of coelomic-epithelial origin.[1] Finally, Deuterostomia originated from some ancient coelenterates that probably did not fall below modern Anthozoa and ctenophores in the degree of epithelisation of their phagocytoblast, and consequently Echinodermata and Enteropneusta have well-developed sections of the neural apparatus, originating from the three main epithelia of their body, cutaneous, intestinal, and coelomic.

We get the impression that at the organisational level of coelenterates all phylogenetically-new epithelia also contain neurosensory cells and consequently are capable of giving rise to parts of the neurosensory apparatus (see Chap. III). On the other hand, at the bilateral level of organisation physiologically-new epithelia (the intestinal epithelium of Scolecida, the coelomic epithelium of Trochozoa) contain no neurosensory cells and are incapable of giving rise to nerve tissue.

In that case, in any group of Bilateria only the epithelia that have been inherited by that group from coelenterates can give rise to nerve tissue. The epithelia that have arisen in that group when it was already at the bilateral stage of its evolution are incapable of doing so.

---

[1] Neural elements of phagocytoblastic origin in annelids and molluscs, however, have been little studied. The fine structure and ontogeny of the different sections of the neural apparatus of echinoderms also are inadequately studied. Therefore any conclusions about the origin of the coelom in different stems of Bilateria based on the characteristics of the neural apparatus are still hypothetical.

Chapter VI

THE MUSCLE SYSTEM AND THE CONTRACTILE-MOTOR APPARATUS

1. Preliminary Remarks

In the popular view, the chief difference between animals and plants is the mobility of the former; and in fact, although immobile animals and mobile plants exist, both are exceptional. A more general characteristic of animals (although here also there are exceptions) is 'animal nutrition', feeding by seizing and swallowing organic food. Movement in space, protective contraction and shrinkage on irritation, and seizure of organic food are the three primary functions whose development and refinement mark the evolution of Protozoa in the direction of an animal type of organisation. At the amoeboid level of organisation all these functions are fulfilled by undifferentiated protoplasm; at the flagellate level permanent contractile structures appear: flagella, mainly used for locomotion, and myonemes, mainly used for protective contraction (although their role is gradually expanding); the swallowing function is retained mainly by undifferentiated protoplasm.

In Metazoa the locomotor function is at first fulfilled by the ciliary apparatus (see Chap. II); in sponges the food-swallowing function is still fulfilled by the amoeboid activity of separate phagocytes and choanocytes. Beginning with coelenterates, however, contractile (muscular) elements appear in the Metazoan body, originally taking part in the seizing-swallowing and protective-contractile functions, but gradually assuming a greater role also in the locomotion of the animal. Finally complete substitution takes place, replacement of ciliary by muscular locomotor organs.

The assemblage of contractile elements of the body constitutes the animal's muscle system. The character of muscular elements in different animal groups (and consequently the character of their muscle systems) varies, so that comparative analysis of muscle systems is required. At the same time the muscular elements in each animal form part of its contractile-motor apparatus. In higher forms the latter also includes skeletal parts and a number of accessory adaptations (vessels, tracheae, etc.). Neural elements (motor, and partly also sensory, endings) form an essential part of every locomotor apparatus, whether primitive or complex, in Enterozoa.

As we have seen in Chap. III, progress in the motor apparatus is the leading factor in the progressive evolution of animals. At the lower stages of evolution of Metazoa the major role in that respect belongs to its seizing and swallowing function; later it passes to the locomotor function, to the hydrokinetic function, and to a multitude of secondary, specialised functions of the motor apparatus, predominantly associated with action on the external environment (constructive and many other activities). In the process of progressive evolution the activities of animals become more complex and diversified, and their motor apparatus correspondingly becomes more complex; but at all levels of evolution animals' activities continue to be the principal factor in further progress. The significance of translational motion in the progressive evolution of animals has been stressed by L. A. Zenkevich (1944).

We shall discuss first the evolution of the muscle system, and then the principal features of the evolution of the contractile-motor apparatus of Metazoa, without attempting to cover the entire range of variation in motor apparatuses in all groups of invertebrates.

2. THE MUSCLE SYSTEM

Contractility, like irritability, is one of the chief properties of undifferentiated protoplasm, such as the protoplasm of amoeboid cells. The contractility of cells is greatly increased when stable contractile formations appear in them, *myonemes* or *myofibrils*. The term 'myonemes' usually denotes contractile fibrils occurring in the body of some Protozoa, and the term 'myofibrils' contractile fibrils found in the contractile cells of Metazoa. There are no essential differences between them. Metaphyta have no contractile structures of that type, and their movements, when there are any, are produced by other adaptations. Among Protozoa, myonemes are possessed by many Infusoria, Gregarinida, Radiolaria Acantharia, and some Flagellata. We can scarcely include in the concept of 'myonemes' the contractile structures of flagella, cilia, and undulating membranes. The myofibrils of Metazoa are usually grouped in bundles; each contractile intracellular formation, whether a separate myofibril or a bundle of myofibrils, may be called a 'contractile fibre'.

Among Metazoa, sponges, which have no nerve system, have accordingly no true contractile elements. Nevertheless some of them are able to make weak and extremely slow movements—some general contraction of the body and the oscula, and closing and opening of pores. The latter results from the contractility of porocytes, in which pore-canals run intracellularly. Myofibrils have not been proved to exist in porocytes, which probably possess only the contractility found in undifferentiated protoplasm. The general contractility of the body is ascribed to the presence in the parenchyma of special fusiform cells that act as myocytes, but their existence has not been conclusively proved.

Among Enterozoa the most primitive muscle system is possessed, not by coelenterates with their epithelised structure, but by some Turbellaria Acoela. In many of the Acoela that are more primitive in this respect (*Pseudoconvoluta, Oligochoerus erythrophthalmus*, Fig. 10) most of the contractile elements of the body are myofibrils or bunches of myofibrils, not linked with separate cells: the cutaneous musculature of *Pseudo-convoluta* is a product of differentiation of its syncytial epidermis, and in *Oligochoerus erythrophthalmus* the epidermis has a cellular structure and the muscle fibres of the skin extend through the territory of many cells. Beyond the limits of Turbellaria Acoela, muscle fibres (longitudinal and annular) embedded in the external epithelium have been described in *Acanthomacrostomum spiculiferum* (Macrostomida) (F. Papi and B. Swedmark, 1959) and in *Carinoma* (Palaeonemertini) (O. Bürger, 1897–1907); the latter, therefore, stands in this respect at the level of the most primitive turbellarians. Frequently the dorso-ventral muscles of the parenchyma extend their dichotomously-branched ends into the proto-plasm of the cutaneous epithelium, reaching the layer of the basal appara-tuses of the cilia, as happens (for example) in *Otocoelis chiridotae* (W. N. Beklemishev, 1915).

A characteristic difference between the myoepithelium of turbellarians and that of coelenterates is that the separate fibres of the cutaneous muscles of turbellarians are not restricted to specific epithelial cells. Moreover, in many Acoela the muscle fibres are entirely embedded in the epithelial layer, instead of forming a special layer of muscle processes as in hydroids. In many Acoela, however (*Aphanostoma pallidum*, etc.) all the fibres of the cutaneous muscles have already moved to the boundary between the epidermis and the parenchyma underlying it.

In higher turbellarians, even within the order Acoela, increasing specialisation of cells has led to formation of true muscle cells among other cell types. Most cells in the external epithelium have lost their con-nection with the fibres of the cutaneous muscles; the muscle fibres are formed from the remainder of the former epithelial cells, which have lost their connection with the surface of the body (Fig. 10); these cells are myocytes or muscle cells. Usually they leave the epidermis altogether and lie subepidermally in the parenchyma. In many turbellarians of different orders, including Rhabdocoela, the individualised cutaneous musculature has at the same time a syncytial character; all the cutaneous muscle fibres then lie in a continuous thin layer of sarcoplasm, adjoining the boundary layer of the cutaneous epithelium on the inner side and containing nuclei (Fig. 125, C).

In most turbellarians parenchymatous muscle cells also are to some extent separated from the rest of the parenchyma. They are often thick fibres, branching dichotomously at the ends and covered with a thin layer of sarcoplasm, with a nucleus lying upon it or attached by a slender neck (Fig. 125, D). In any case, at a certain level of organisation found in many higher turbellarians the chief components of the muscle system are already not independent myofibrils, as in *Pseudoconvoluta* and a number of other Acoela, but separate muscle cells; besides these, however, symplastic muscle formations continue to occur in most turbellarians and even in most flatworms in general, often in precisely those organs that have the strongest musculature (pharynx, male genital apparatus).

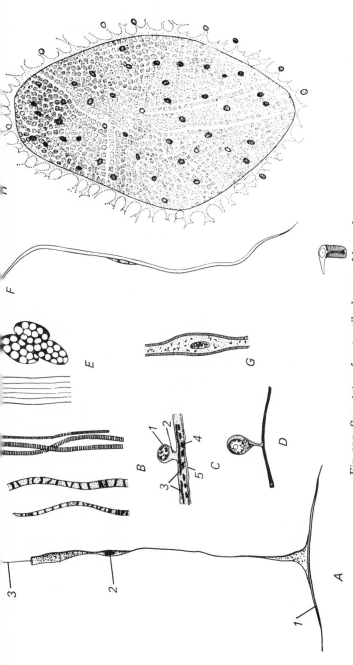

Fig. 125. *Some types of contractile elements of invertebrates.*

A—endodermal epithelial-muscle cell from a longitudinal muscle of a radial septum in *Anemonia sulcata* (Actiniaria): 1—muscle process; 2—cell nucleus; 3—flagellum (after K. C. Schneider). *B*—transverse-striated structure in muscle fibres of *Paraphanostoma* (Acoela), dorso-ventral muscles on left, longitudinal muscles on right (after Westblad). *C*—portion of syncytial cutaneous muscle of *Desmote vorax* (Rhabdocoela Dalyelliida): 1—nucleus; 2—sarcoplasm; 3—longitudinal muscle fibres; 4—annular muscle fibres; 5—basal layer of epithelium (after W. N. Beklemishev). *D*—muscle cell from dorso-ventral muscles of *Desmote vorax* (ends of muscle fibre not shown) (after W. N. Beklemishev). *E*—structure of muscle cell of *Geoplana rufiventris* (Triclada Terricola), profile and cross-section; numerous and irregularly-arranged myofibrils (after von Graff). *F*—muscle cell of earthworm (Oligochaeta Lumbricidae), above—isolated, below—in cross-section (from K. C. Schneider). *G*—part of a muscle cell of *Glossosiphonia* (Hirudinea), longitudinal section (after N. A. Livanov). *H*—polyenergid, transversely-striated muscle fibre of water-beetle *Dytiscus* (Coleoptera), cross-section (from K. C. Schneider).

Some turbellarians also show another type of muscle cell structure, which we must regard as a substantial improvement of their functional structure: each muscle cell of that type contains not one, but a whole bundle of contractile fibres (Fig. 125, E); their arrangement within the cell varies, and often is very regular. With that structure of muscle cells, the separate muscle fibres finally cease to be direct components of the muscle system; that position is taken over by fibre bundles confined to muscle cells.

The structure of muscle cells varies greatly in different invertebrates. The type of structure found in most turbellarians (Fig. 125, D) is widespread among other flatworms, but is found also outside that group.

Nemertines often have muscle cells of Triclada type (Fig. 125, E), i.e. the contractile fibres in each cell are few, and in cross-section show a fairly irregular arrangement. Among higher groups an extremely widespread type is that found among turbellarians in the terrestrial *Dolichoplana feildeni* (Triclada); it has two variants, the *Ascaris* type and the Hirudinea type. In both cases contractile fibres are arranged in a layer along the periphery of a much-elongated muscle cell, and the protoplasm (sarcoplasm) lies along its axis. With the *Ascaris* type the cell is flattened lenticularly and the contractile layer along one of its edges remains discontinuous; the cell nucleus lies outside it, often in a special nucleus-containing process (Fig. 125, F). Muscle cells of this type occur in large nematodes; in many polychaetes: *Polygordius, Saccocirrus* (W. Salensky, 1907), *Scolelepis, Fabricia, Chone, Myxicola, Serpula* (K. Johansson, 1927); in most oligochaetes (V. Izosimov, 1940); rarely in leeches, etc.

In muscle cells of Hirudinea type the contractile layer is continuous and the nucleus lies in the sarcoplasm occupying the axial part of the cell (Fig. 125, G). Muscle cells of this type occur in Priapuloidea, Kinorhyncha, and Acanthocephala (K. Lang, 1953); in some polychaetes: *Spirographis, Protula* (K. Johansson, 1927), Nereidae, Nephthydidae (N. Livanov, 1940); rarely in oligochaetes; and as a rule in leeches, molluscs, and a number of other groups.

For Cnidaria, with their body mainly constructed of two epithelial layers, the primary form of contractile cell is the epithelial-muscle cell (Fig. 125, A), the body of which is part of the epithelium (ectodermal or endodermal) and has on its basal side a muscle process perpendicular to the cell axis and usually containing a single contractile fibre. By means of partial reduction of the epithelial part of the cell and loss of its connection with the epithelium, a purely-muscle cell, lying entirely beneath the epithelium, is formed from the epithelial-muscle cell. Such purely-muscle cells are widespread in several parts of the body of Anthozoa and Scyphozoa.

Ctenophores differ sharply from other coelenterates in lacking epithelial-muscle cells, and in this respect they certainly occupy a higher place. In coelenterate muscle cells the contractile fibres lie along the periphery, and the sarcoplasm and the nucleus lie along the cell axis, so that they belong to a much more highly developed type of contractile elements than do the muscle cells of most turbellarians.

Multinuclear, polyenergid muscle fibres of transversely-striated type, containing numerous myofibril bundles, are characteristic of arthropods. They display considerable convergent resemblance to the striated muscle fibres of vertebrates (Fig. 125, *H*).

In other groups besides arthropods the striated structure of myofibrils occurs sporadically but fairly frequently. We find it in some muscles of Siphonophora, in the annular muscle of the umbrella of Scyphomedusae, in the front-end retractors and the dorso-ventral fibres of several Acoela (E. Westblad, 1948) (Fig. 125, *B*), and in the proboscis retractors of *Rhynchomesostoma* (Turbellaria Rhabdocoela). In many Rotatoria some or all of the body muscles, and in Kinorhyncha the longitudinal (intersegmental) muscles, are transversely striated. Separate muscles composed of striated fibres occur sometimes in polychaetes (*Nephthys, Magelona*). Striated muscles also occur in the musculature of the tongue, pharynx, and other organs in some Gastropoda. The muscles of *Sagitta* (Chaetognatha), the tail muscles of Appendicularia and of larval Ascidiae, and the trunk muscles of Salpae are striated. The only striated fibres in the body of adult Ascidiae are those forming the musculature of the cardiopericardium. In general we may say that the striated structure of muscle fibres has arisen independently in the most diverse groups of Metazoa, and mainly in those muscles that make particularly rapid and energetic movements. Striated muscles are in fact many times more rapid in contraction, and especially in subsequent relaxation, than smooth muscles. Where prolonged contraction of muscles with minimum effort is required, however, smooth muscles are more appropriate; consequently we find them in the adductors of most Lamellibranchia. In Pectinidae the adductor consists of two parts, smooth and striated; the former is used for closing the shell, and the latter for rapid flapping of the shell valves when clearing the mantle cavity of pseudofaeces and foreign particles, and when swimming.

We must point out that, when speaking of the striated structure of invertebrate muscle cells, we are thinking of the transverse striation of myofibrils as such and not of the striation of the cells as a whole. In the relative locations of contractile fibres, nuclei, and sarcoplasm, the transversely-striated muscle cells of different invertebrates usually resemble the smooth muscle fibres of the same forms, and for the most part have nothing in common with the complex, polyenergid muscle fibres of arthropods and vertebrates.

As mentioned above, the epithelial-muscle cells of Cnidaria are formed from both kinetoblast and phagocytoblast. The colonial musculature of Pennatularia develops from the endodermal epithelium of solenia. The purely-muscle cells of Anthozoa are mostly formed from the epithelium of the radial chambers, i.e. from the peripheral phagocytoblast. It is usually considered that the musculature of the mesenchyma and arms of ctenophores also develops from the peripheral phagocytoblast. Outside the subphylum Cnidaria the central phagocytoblast (endoderm) apparently

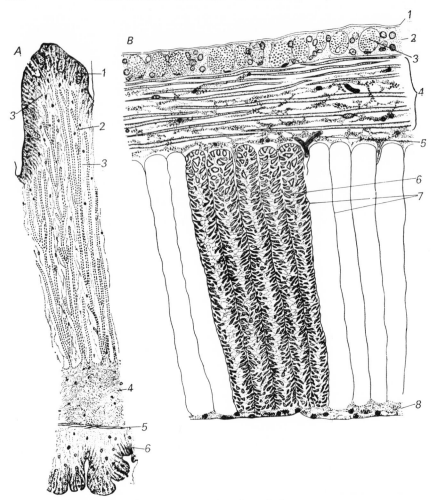

Fig. 126. *Examples of muscle tissues.*

A—part of cross-section through septum of *Anemonia sulcata* (Actiniaria), with longitudinal muscle: 1—endodermal epithelium; 2—fibre of longitudinal muscle (rows of
cross-sections of separate fibres lying along the crests of the boundary layer); 3—basal
parts of endodermal cells, extending like thin fibres towards the boundary layer; 4—
boundary layer between the two layers of endoderm that coat the septum on each side;
5—transverse (radial) muscle fibres, belonging to the external layer of endoderm; 6—
external (turned towards the intercameral space) layer of endoderm (after K. C. Schneider).
B—cross-section through body wall of earthworm *Eisenia* (Oligochaeta Lumbricidae):
1—cuticle; 2—epidermis; 3—mucus glands; 4—annular musculature; 5—capillaries;
6—'capsules' of longitudinal musculature; 7—their septa, formed by boundary membranes;
8—coelomic epithelium (after K. C. Schneider). *C*—pharynx bulbosus of *Provortex
brevitubus* (Rhabdocoela Dalyelliida), medial section; 1—external mouth; 2—dilator
muscles of buccal cavity; 3—grasping edge of pharynx; 4—retractor of front end of body;
5—retractors of pharynx; 6—mouth of gut; 7—gut epithelium; 8—nucleus-containing
parts of cells of epithelium of lumen of pharynx; 9—boundary layer separating the
musculature of the pharynx from the mesenchyma of the body; 10—radial muscles of
pharynx; 11—internal annular muscles of pharynx; 12—external annular muscles of
pharynx; 13—sunken nucleus-containing parts of cells of epithelium of pharyngeal

[*continued opposite*

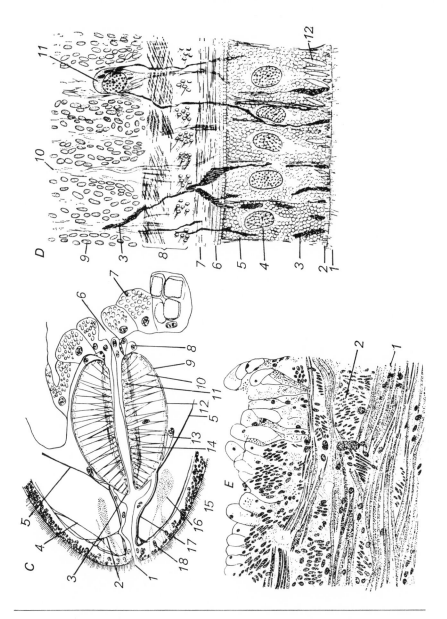

sheath; 14—internal longitudinal muscles of pharynx; 15—small rhabdoids of external epithelium; 16—cutaneous mucus cell with branched duct; 17—pharynx protractors; 18—nuclei of external epithelium (after A. Luther). D—muscle bundles of *Geoplana rufiventris* (Triclada Terricola), cross-section through skin of ventral side of body: 1—cilia; 2—their basal apparatus; 3—ducts of mucus glands, cut across; 4—nucleus of epidermal cell; 5—intercellular substance of epithelium; 6—boundary membrane; 7—annular muscles; 8—layer of diagonal muscles; 9—longitudinal muscle fibres; 10—dorso-ventral muscles; 11—rhabdite cell; 12—rhabdites in skin (after von Graff). E—muscula-ture of foot of *Helix pomatia* (Gastropoda Pulmonata): 1—bundles of muscle fibres; 2—the same, cut across (after K. C. Schneider).

loses the power of giving rise to contractile structures, and outside the phylum Scolecida that power is to a great extent also lost by the kinetoblast (ectoderm).

In polychaetes the endomesodermal rudiment (4d) gives rise, as a rule, to all the body musculature except the trochophoral musculature, which is formed from the so-called ectomesoderm—from the progeny of separate immigrating cells of the second and third quartets. As we have seen, however, these cells are merely peripheral phagocytoblast cells that immigrate from the vegetative pole of the embryo. It is true that in some polychaetes (Spionidae, Serpulidae, some Aphroditidae, etc.) the whole coelomic mesoderm of the post-larval body, including the musculature, develops from the ectoderm (P. Ivanov, 1912; H. Korn, 1959b), but that method of development is without doubt a secondary modification.[1] Certainly the muscles of the sweat glands and of the iris of vertebrates are ectodermal, differentiating at late stages of development. But in the whole animal kingdom, beginning with the higher Cnidaria, most of the body musculature develops from the peripheral phagocytoblast.

Whereas in most Acoela and many other (mostly small) flatworms the muscle system is represented by separate muscle cells, in larger or higher forms these cells begin to group into muscle bundles or layers. Muscle tissue arises in this way and is always a compound tissue, since supporting formations, cells and connective tissue, always lie between the muscle cells that form it. In addition it is always connected to motor nerves, and sometimes to sensory nerves, and in higher forms it is permeated by blood vessels or tracheae.

The tendency towards formation of muscle tissue already appears in higher coelenterates—in many Anthozoa (Fig. 126, *A*) and higher medusae, and also in ctenophores. In many large turbellarians muscle cells are arranged in more or less regular bundles, usually (it is true) fairly loose. Nevertheless such groupings of muscle cells are already muscle tissue. Their presence is particularly characteristic of terrestrial Triclada, the largest and strongest turbellarians. In *Geoplana rufiventris* the longitudinal cutaneous muscles form large, radially-arranged bundles, each containing some tens of muscle cells (fibres); the diagonal bundles each contain 5–10 fibres, and the annular bundles form a layer four or five fibres deep (Fig. 126, *D*). Here we certainly have muscle tissue and organs formed from it, muscles in the form of bundles or layers, which together represent the muscle system of the animal. Combining into units of higher order, the separate muscle cells cease to be direct members of the system, such members being the muscles composed of these cells. We see that here also, according to the general rule, muscle tissue is formed of two kinds of components, muscle cells and supporting formations (cells and variously-

---

[1] P. P. Ivanov (1937) suggests that in these families the part of the egg cytoplasm that later goes into the formation of new somites enters blastomere 2d during cleavage, whereas in other polychaetes it enters blastomere 4d, giving rise to the mesodermal band of the trochophore.

differentiated ground tissues); it is a compound tissue, which it continues to be in all higher animals. Finally we must point out that the muscles of *Geoplana*, and in general of all turbellarians, are still very primitive in the sense that their separation from the surrounding parenchyma from which they have been formed is still incomplete, and the supporting components contained in them are unspecialised. Exceptions are such muscle formations as the musculature of the pharynx bulbosus of Rhabdocoela, which is surrounded on all sides by a boundary layer to which the separate muscle fibres are attached by their ends (Fig. 126, *C*). At the same time, no flat-worm has its entire musculature composed of muscle tissue; together with multicellular muscles composed of muscle tissue, flatworms always also have independent muscle cells lying isolated among other tissues; in this respect the muscle system of large flatworms has a compound, transitional character. The same transitional character is found in the muscle systems of other parenchymatous worms [1] (nemertines, leeches) and molluscs (Fig. 126, *E*). In some molluscs, however, muscle tissue predominates over isolated muscle cells, on account of the large size and considerable muscular power of many of these animals; at the same time molluscan muscles formed of muscle tissue have acquired great individuality, especially those that are attached to the shell or other skeletal parts. In polychaetes and oligochaetes almost no independent muscle cells remain, on account of the predominance in their bodies of polarised tissues and non-cellular boundary formations (see Chap. I); the structure of annelids being of that general nature, the majority of their muscle cells have become part of polarised muscle tissue (Fig. 126, *B*), forming muscle layers, bands, and separate muscles. The musculature of most lower Deuterostomia stands at approximately the same level of tissue formation. In Brachiopoda the whole musculature consists exclusively of muscles composed of muscle tissue (Vol. 1, Fig. 183). The same applies to arthropods; in both these groups the muscles are attached to hard skeletal parts that work as levers, and isolated muscle cells would be ineffective. The direct members of the muscle system of arthropods are well-separated, fully-individualised skeletal muscles formed of muscle tissue (Fig. 145), and also muscle layers and plexuses that form a coating for internal organs.

The muscle system of animals therefore displays three stages of structural complexity: in the first stage it consists of independent myonemes (Protozoa) or myofibrils (some of the lower Acoela, Macrostomida, and Palaeonemertini); in the second stage, of epithelial-muscle or muscle cells (many, usually small, Coelenterata and Scolecida); in the third stage, of muscles composed of muscle elements and supporting elements (Arthropoda, etc.). There are many transitional stages between these three.

---

[1] We use the word 'worms' here not in a taxonomic sense, since there is no such group in the natural system of animals, but in a physiognomic sense to denote a particular facies and level of organisation, a definite life form (see Vol. 1, Introduction).

3. CONTRACTILE-MOTOR APPARATUS. FIRST STAGES OF ITS DEVELOPMENT.
   COELENTERATA

All the components of the muscle system of any animal, located within its body in a definite order, form its contractile-motor apparatus. As well as contractile elements, that apparatus always contains non-contractile formations that fulfil passive mechanical functions, especially the functions of uniting the contractile elements and of resilient opposition, playing the role of antagonists to the musculature contractions. In addition, any energetic contraction of the musculature requires increased solidity at the points where the increased force is applied; therefore the contractile-motor apparatus includes skeletal elements in the wide sense of the term. Subsequently the latter often assume the role of levers, increasing the speed, the power, and the precision of movements, and the role of implements acting on objects in the external environment.

The structure of the contractile-motor apparatus depends to a great extent on the level and type of the animal's organisation, and in turn profoundly affects its entire organisation.

The function of uniting the separate contractile elements is well shown by the boundary membranes of Actiniaria; the contractile parts of the epithelial-muscle cells are attached to these membranes, and thus are combined into functional units of a higher order—muscular areas; these membranes are unusually elastic and have a certain amount of resilience, which enables them to return to their original state after stretching. The forces involved, however, are not great enough for these formations to play the role of antagonists to the musculature. A clear example of a resilient formation acting as antagonist to the musculature is the mesogloea of the umbrella of Scyphomedusae. The entire musculature of the umbrella is located on its subumbrellar side, and by its rapid contraction reduces the subumbrellar cavity, expelling water from it and moving the animal forward jerkily. The slow return of the umbrella to its former shape is caused by the resilience of the mesogloea. The economy of energy obtained by substitution of resilient formations for antagonistic muscles has led to wide distribution of such formations in the animal world.

In medusae resilient resistance is produced by the mesogloea, which is so firm that it gives the animal's body a stable form. For that reason the muscles are attached to it only on one side; they merely bend the mesogloea layer, which later resumes its shape through its own resilience.

Unlike medusae, such animals as Hydrida, Actiniaria, and most flat-worms do not have a constant body form; on the contrary, their body becomes shorter and thicker or longer and thinner, as a whole or partially, depending on the amount of contraction of different groups of muscles. All of these animals possess musculature in the form of a sac formed by their longitudinal and annular muscles and filled with fluid. In polyps that is the fluid of the gastrovascular cavity (practically water), and in flatworms it is the semi-liquid tissues of their body. In both cases, when the mouth is closed the fluid filling the body acts as a support; muscle tension in the sac gives the body rigidity, and different amounts of contraction of the longitudinal and annular muscles produce changes in body shape (Actiniaria). Further improvement of that type of motor apparatus is shown by polychaetes and some other coelomate animals (see below).

A third structural principle of the contractile-motor apparatus arises after the appearance of hard skeletal parts. Skeletal elements include formations of varied origin, structure, and composition, their only common characteristic being increased mechanical stability: boundary membranes, cuticles, solid epidermal secretions, different kinds of internal skeletons such as the skeletal plates of echinoderms, the cartilages of cephalopods, the fibrous and cartilaginous-fibrous endosternites of Chelicerata, etc. Usually all these formations arise in connection with a supporting or protective function, and only secondarily and partially do they begin to be used for muscle attachment, being thus brought into the contractile-motor apparatus. The role of skeletal components in the structure of that apparatus becomes much greater as their number and solidity increase and as the musculature develops into a constantly-growing number of separate, individualised muscles. These two processes are intimately linked together. Among invertebrates they reach the highest level in arthropods, with their complexly-segmented transversely-striated musculature and external skeleton. Parallel improvement takes place in the sensory and motor innervation required to set in motion a complex motor apparatus, and also in the development of adaptations for the nutrition and respiration of energetically-acting muscles.

What was the original structural plan of the contractile-motor apparatus, and what was its original function? We have seen in Chap. III that the question of the first appearance of neural and muscular apparatuses coincides with the question of the origin of the type of organisation characteristic of Enterozoa, and the origin of that type of organisation is clearly linked with the transition of primary Metazoa to nutrition by the swallowing of large prey. Therefore the primary function of the contractile-motor apparatus was not locomotion by the animal itself, but the seizure and swallowing of food: the prehensile and swallowing functions, not the locomotor function, were primary.[1] The original locomotor apparatus of Metazoa was a ciliary, not a muscular, apparatus. Ctenophora, as representatives of primary, free-swimming coelenterates, provide an example of the level of organisation at which the animal travels almost entirely by means of a ciliary locomotor apparatus, but uses its musculature mainly to seize and swallow prey. If we assume, with A. A. Zakhvatkin (1949), that in the adult state primary Metazoa were sessile colonial animals, the connection between the origin of musculature and the prehensile and swallowing function becomes still more evident. Sessile coelenterates, in fact, especially hydroid polyps, represent in their whole

[1] As L. A. Zenkevich (1944) points out, the myonemic apparatus of Protozoa also usually has no locomotor significance.

organisation the fullest embodiment of the prehensile and swallowing function to be found in the animal kingdom.

The muscle system of hydropolyps corresponds fully with their thoroughly-polarised structure, and is formed exclusively of epithelial-muscle cells. The contractile-motor apparatus of hydroids is characterised by the fact that the muscle processes of these cells are longitudinal in the ectoderm and annular in the endoderm; in the terminal region of the hydranth the musculature is more developed on the hypostome, and around the mouth the annular fibres form something like a sphincter. On the other hand, there are practically none in the coenosarc. The body of Hydrida, which has no perisarc, is contractile along its entire length. Freshwater Hydrida represent the simplest variant of a motor apparatus supported by fluid enclosed in a muscular sac. When the mouth is closed the hydra may stretch into a long thin thread or contract into a short thick sac, depending on whether the annular or the longitudinal muscles have more tension. By expelling water it can contract into a small pellet. Metagenetic hydroids have a perisarc closely apposed to the branches of the coenosarc and to the stalks of the hydranths, and consequently these parts are not contractile and lack musculature. The head of a hydranth is usually contractile in all its parts, but sometimes a firm supporting tissue of endodermal origin develops in its basal part, giving the head a certain amount of constancy of shape (e.g. *Tubularia*). The musculature is best developed in the terminal part of the head, in the hypostome, which makes the movements of swallowing.

The hollow tentacles of Hydrida have ectodermal longitudinal and endodermal annular musculature, and are capable of great extension in length; the tentacles of marine hydroids (Leptolida) are solid, and their endoderm has no annular musculature. These tentacles are able to bend, but are incapable of any great extension. The same types of tentacle muscle occur outside the order Hydrida. Actiniaria have hollow and very extensible tentacles, furnished with annular and longitudinal muscles; Bryozoa and Serpulimorpha have tentacles without annular musculature and of more or less constant length. The elongation of hydroid and sea-anemone tentacles is achieved by increase in the pressure of the cavity fluid, produced by increased tension in the body musculature; contraction of the annular musculature of the tentacles prevents them from expanding in diameter, and relaxation of the longitudinal musculature enables them to increase in length.

Besides the functions of seizing, swallowing, and retaining food, the motor apparatus of hydroids has another function, that of shortening, drawing back, and pulling in the tentacles and the head in response to unfavourable stimuli, or, more briefly, that of concealment. All sessile animals are able to contract in one way or another in response to un-favourable stimuli, and probably, together with prehension, that was a primary function of the contractile-motor apparatus. Primitive organs and apparatuses are usually multifunctional.

When Hydrida are not sessile the motor apparatus also has a locomotor function. *Hydra*, as is well known, has several methods of locomotion. *Protohydra* burrows in the sand by pumping fluid from one end of the body to the other (J. Omer-Cooper, 1957), like many worms (see below).

The motor apparatus of Anthozoa (Fig. 127, *A*, *B*) is distinguished by development and partly by differentiation of adaptations designed to fulfil the same basic functions, food-capture and concealment. In addition, solitary forms (Actiniaria), like Hydrida among Hydrozoa, are capable also of progressive movement. The ectodermal musculature there also is longitudinal. In Ceriantharia it is strongly developed and enables the body to contract; accordingly the longitudinal muscles of their dissepiments are relatively weak. In Actiniaria, on the other hand, the ectodermal longitudinal musculature of the stalk is poorly developed, and in the majority it is absent. Functionally it is replaced by the endodermal longitudinal muscles of the newest septa. That substitution became possible in Actiniaria because of the regular location of new septa along the whole circumference of the cross-section of the body. In Ceriantharia all new septa are concentrated on one side of the body, and therefore they cannot replace the strictly-longitudinal musculature of the stalk. On the oral disc the longitudinal muscle fibres are arranged radially and act as dilators of the mouth. The endodermal musculature of the trunk and the pharynx is annular; in many sea-anemones it forms a sphincter beneath the edge of the disc, as a result of which the edge of the body moves over the disc and covers it in a concealment reaction. In the septa (mesenteries) the endodermal musculature forms powerful longitudinal muscles (retractors), the arrangement of which is very important in the morphology of Anthozoa. At both ends of the body the fibres of each bundle diverge fanwise and are attached to the basal and oral discs along their radii (Fig. 127, *B*). On the reverse sides of the septa the muscle fibres run radially: in 'complete' septa (i.e. those attached to the pharynx) the radial fibres act as dilators of the pharynx (Fig. 127, *B*). In the lower parts of the septa they become oblique and are attached by their ends to the basal disc: they are parietobasal muscles. The basal disc of sea-anemones possesses radial and annular musculature of endodermal origin. The musculature of the basal disc itself and also the septal muscles attached to it are used for suctional attachment to the substrate and for crawling.

The septal musculature is very important in digestion, by closing the mesenterial filaments with their digestive epithelium around swallowed prey. The longitudinal septal muscles also play a major role in protective contraction of the animal. Rapid contraction of all the longitudinal muscles of the body and of the radial muscles of the oral disc produces retraction of the disc and of the whole front end, followed by contraction of the disc sphincter, which covers the retracted head with a fold of the body wall. As a result the whole animal may be transformed into a dense flat cake, presenting a minimum of surface to enemies and the elements.

The physiology of sea-anemone movements has been described in a

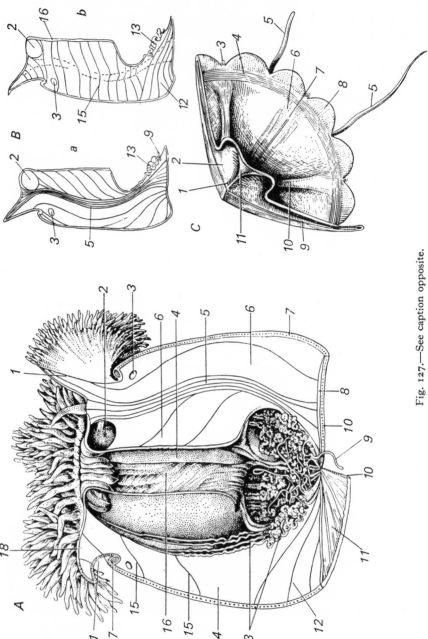

Fig. 127.—See caption opposite.

number of works (E. Batham and C. Pantin, 1950a, 1950b, 1951). On the protective contraction of the animal all the cavity fluid is expelled through the open mouth. The opposite process of filling the gastral cavity with water, without which the sea-anemone could not expand, is performed by the activity of the cilia of the siphonoglyph, which force water into the body. It is also possible to swallow water by peristaltic movements of the pharynx. In any case the expansion of the sea-anemone takes place slowly; its body does not have enough elasticity for automatic expansion. When a sea-anemone (*Metridium senile*) is in a state of rest, the pressure of the cavity fluid is equal to that of a 2–3-mm column of water and is due entirely to the activity of the siphonoglyph cilia. When the parietal musculature contracts the pressure in the gastral cavity increases and compresses the pharynx, which because of its flat shape makes an excellent valve. When the mouth is closed, the tension of the longitudinal and annular muscles of the stalk causes increased pressure within the cavity, which in *Metridium senile* reaches that of 5–6 cm of water. When full of water a sea-anemone can make great changes in its body shape and tentacle length by muscular action. When there is more tension in the annular muscles the pressure of the gastrovascular fluid forces the body to increase in length; more tension in the longitudinal muscles causes increase in body diameter. In the latter case the body is shortened. Thus when the mouth is closed the gastro-vascular fluid of sea-anemones plays the same mechanical role as the coelomic fluid of annelids.

The above facts throw new light on the chief features of Actiniaria architectonics. For instance, the biradiate nature of their symmetry is due to the flattened shape of the pharynx, and the latter is an adaptation to the valvular function of the pharynx, which automatically prevents the empty-ing of the gastral cavity when the pressure within it rises. The wide distribution of biradiate symmetry among Anthozoa therefore has a functional explanation.

Fig. 127. *Diagrams of structure of muscle apparatus of coelenterates.*

A—*Metridium senile* (Actiniaria); the animal is cut in half, but one of the sectors of the basal disc of the removed half is retained: 1—'head' of this sea-anemone; 2—circumoral circular communication running through all complete septa; 3—circular marginal com-munication; 4—siphonoglyph; 5—one of the oral-disc retractors; 6—one of the complete (i.e. grown to the pharynx) septa, not one of the 'directional' ones; 7—annular muscles of the stalk (ectodermal); 8—radial (basilar) muscles of basal disc; 9—acontia; 10—annular muscles of basal disc, arranged in concentric circles; 11—basal disc, on which radial lines denote the former places of attachment of the cut septa; 12—parieto-basilar muscles of septa; 13—mesenteric cords; 14—incomplete (but furnished with a retractor) septum, side view, turned into the intercameral space; 15—radial muscles of septa (pharynx dilators), not all shown; 16—pharynx; 17—sphincter, which closes over the indrawn head when the animal is strongly contracted; 18—oral disc. *B*—*Metridium senile*, complete non-directional septa: *a*—view from the chamber, *b*—view from the intercameral space (after Batham and Pantin): numerals have the same denotation as in *A*. *C*—a Discomedusa, dissected sector of body with the manubrium cut away: 1—gastral filaments; 2—gastral cavity; 3—rhopalia; 4—annular muscles of umbrella; 5—tentacles; 6—deltoid muscles; 7—radial septum; 8—muscles of marginal lobes of umbrella; 9—radial canals; 10—oral ligaments; 11—columella (diagram after Delage and Hérouard).

The physiology of movement in Actiniaria differs sharply from that in higher animals such as arthropods and vertebrates. The musculature of Actiniaria consists mainly of undifferentiated muscle layers that form part of the body wall and its extensions (tentacles and septa). Their contractions determine the shape of the animal. Every movement of the body calls for co-ordinated action by many muscle groups. That is partly true also of Hydrida. The most primitive structure of the neuromuscular apparatus therefore not only does not exclude, but demands, the most complex co-ordination in the work of its parts; and the most differentiated parts of the motor apparatus of Actiniaria, such as the front-end retractors and the disc sphincter, have the simplest movements. Progressive evolution necessarily includes simplification of organs and functions as well as complexity.

Sea-pens (Octocorallia Pennatularia) have a well-developed colonial muscular apparatus. It consists of the external longitudinal and internal annular muscles of the stalk. Both develop in the walls of the perisarc canals (solenia), which there have a regular longitudinal and annular arrangement. In the lower part of the stalk a powerful sphincter, able to compress even the principal canals, develops from the annular muscles. Because of that musculature the stalk is capable of peristaltic movement that enables the colony to dig into the soil with its basal element and even to travel slowly; in this way sea-pens, by collecting water in the gastro-vascular canals through the siphonozooids and then forcing it back, partly use the hydraulic mechanism so widespread in fossorial worms, molluscs, and Enteropneusta.

Unlike ctenophores and the larvae of other coelenterates, which swim by means of cilia or ctenes, medusae (and some siphonophores) have advanced to swimming by means of muscular contractions of the bell, which is the expanded margin of the basal part of the head of a polyp. Rapid contraction of the bell forces water out of its cavity and thrusts the medusa backward. Medusa movement is therefore reactive. This method of movement is as secondary as is the organisation of the medusa itself. In Scyphomedusae the bell contracts by means of annular and radial muscle bands formed by the ectoderm on the subumbrellar side; the elasticity of the bell mesogloea restores its shape; on the exumbrella there is little or no musculature. Possession of elastic mesogloea gives the medusa stability of shape. The locomotor apparatus of medusae therefore differs radically in its organisational principles from that of Actiniaria. The total amount of locomotor musculature in the body of Scyphomedusae is very small: in *Cyanea arctica*, according to I. A. Vetokhin, it constitutes only 1·2% of the body mass (L. Zenkevich, 1944). The prehensile function is fulfilled by the musculature of the marginal tentacles and the arms, which is fairly complex in its nature.

As we have seen, however, many Scyphomedusae (e.g. *Aurelia*) in the adult state feed on small plankton brought by cilia to the orifices that remain after the mouth is overgrown, and the arms and tentacles lose their prehensile function (A. Southward, 1949). We shall not dwell on the peculiarities of the musculature of Hydromedusae and Siphonophora.

The musculature of ctenophores is all embedded in the mesogloea. It consists of longitudinal and annular fibres beneath the skin, the same kind of fibres around the gut, radial fibres running from the gut to the skin, and the powerful musculature of the tentacles. Sometimes there are sphincters around the mouth and the aboral pole. The aboral sphincter, by its contraction, may cover up the aboral sense organ and thus have a protective function, resembling the role of the sphincter of the oral disc in sea-anemones. In the general principle of its organisation (a musculo-cutaneous sac enveloping the connective tissue of the body together with its enclosed internal organs) the muscular apparatus of ctenophores is therefore closer to that of flatworms than to that of Cnidaria. Most ctenophores are predators that swallow their prey; some of them seize it with their tentacles, and others directly with the mouth; the chief role of their musculature is prehension and swallowing, and in addition it has a protective function; in most ctenophores it does not have a locomotor function, but it does acquire that in some specialised groups. *Cestus veneris*, for instance, which has an elongated ribbon-like body of considerable size (over 1 metre) swims by undulating its body, and its musculature (which consists of longitudinal, transverse, and diagonal muscles) is comparatively well developed. In Lobifera the musculature of the lobes is composed of two systems of bundles arranged fanwise and intersecting each other (Vol. 1, Fig. 29, *A*); rapid contraction of these muscles produces flapping of the lobes and retrograde motion of the animal, e.g. when it meets an obstacle (B. Coonfield, 1936), a curious resemblance to Cephalopoda Decapoda. In both cases there are two types of progressive movement: smooth movement, mouth-foremost, by means of fins (Cephalopoda) or ctenes (Ctenophora), and jerky backward movement.

4. CONTRACTILE-MOTOR APPARATUS OF LOWER TURBELLARIANS, AND FIRST STEPS IN ITS DIFFERENTIATION

The most primitive of turbellarians, the lower Acoela, are closer to coelenterate larvae (parenchymulae and planulae) than to adult Cnidaria in the general level of their organisation. In particular, they are characterised by a slight degree of epithelisation of their phagocytoblast, considerable development of amorphous parenchyma, and considerable formative capacity of the latter. Consequently all turbellarians possess, besides epidermal musculature, well-developed parenchymatous, phagocytoblastic musculature, which we have already seen in ctenophores. In its simplest form the epidermal musculature consists of myofibrils, differentiated within the epidermis. They form two or three layers: there are external annular and internal longitudinal fibres; sometimes there is also a layer of external longitudinal muscles (many Anaperidae) or of diagonal muscles. The epidermal musculature of Acoela therefore forms a continuous sac

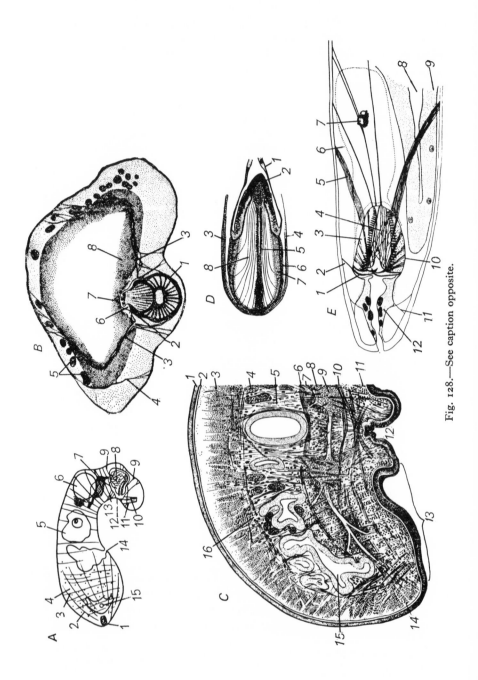

Fig. 128.—See caption opposite.

consisting of two or three layers of muscle fibres and covering the whole body, i.e. it is one variant of the musculo-cutaneous sac. From the mechanical point of view the musculo-cutaneous sac of flatworms is very similar to what we see in Hydrida and Actiniaria. In all these cases it is formed of layers of annular and longitudinal muscles and is filled with fluid; the chief difference is that in polyps the contained fluid is that which fills their gastrovascular canals, whereas in flatworms it is the actual semi-liquid tissues of their body. When the external muscular sac contracts, however, the fluid in flatworms is not capable of being compressed, the pressure within the body rises, and differences in the tension of the separate groups of cutaneous muscles cause changes in body shape. Many turbellarians can stretch their whole body into a thread, contract it into a ball, or protrude the front or rear end of the body. Then all their internal organs undergo either severe temporary deformation or displacement, depending on the consistency of their tissues, but in both cases without suffering harm. Nevertheless the need to undergo such treatment doubtless affects the animal's whole organisation and probably prevents its organs from attaining high functional perfection. On the other hand, a 'mesenchymatous' organisation, in which the whole musculo-cutaneous sac is filled with mesenchyma and the organs contained in it, has several structural advantages: one of these is complete freedom for development of muscle fibres or bundles cutting across the body in any direction and appearing (in the process of evolution) anywhere that the need for them arises.

Fig. 128. *Structure of the contractile apparatus of flatworms.*

A—Pseudoconvoluta flava (Acoela), medial section, showing the scheme of arrangement of the parenchymatous musculature: 1—frontal glands; 2—front-end retractors; 3—dorso-ventral muscle fibres; 4—transverse muscle fibres; 5—medially-located oocyte; 6—copulatory pouch; 7—sperm ducts running to the copulatory organ (8); 9—retractors of the copulatory organ; 10—glandular process of the copulatory apparatus; 11—its protractors; 12—male genital orifice; 13—protractors of copulatory organ; 14—mouth, with incipient pharynx; 15—brain with statocyst (original). *B—Phaenocora polycirra* (Rhabdocoela Typhloplanoidea), cross-section, showing oligomerised type of musculature of rhabdocoelous turbellarians: 1—male copulatory organ; 2—its ventral retractor; 3—its dorsal retractors, penetrating the gut wall; 4—dorso-ventral muscles; 5—branched vitellarium; 6—male genital canal; 7—protractors of copulatory organ; 8—epithelium of gut (after Beklemishev). *C—Artiodactylus speciosus* (Triclada Terricola), cross-section in bothrium region: 1—epidermis; 2—its boundary layer; 3—longitudinal muscles of parenchyma on dorsal side; 4—vitelline follicles; 5—dorso-ventral muscles; 6—transverse muscles; 7—glandular cells, whose ducts open into the crawling grooves; 8—oviduct; 9—stems of the deep intraparenchymal neural plexus; 10—transverse muscles of the ventral side; 11—glands of bothrium; 12—crawling grooves; 13—bothrium; 14—longitudinal muscles; 15—branches of gut; 16—transverse muscles of dorsal side (after von Graff). *D—*proboscis of *Phonorhynchus mamertinus* (Rhabdocoela Polycystidae), sagittal section: 1—proboscis sheath; 2—adhesive epithelium at end of proboscis; 3—dorsal protractor; 4—ventral protractor; 5—ducts of glandular cells, opening at end of proboscis; 6—boundary layer of proboscis; 7—annular musculature of proboscis; 8—longitudinal musculature of proboscis (after J. Meixner). *E—*proboscis of *Gnathorhynchus conocaudatus* (Rhabdocoela Gnathorhynchidae), diagram of structure of front end of body, view from left side: 1—hooks of proboscis; 2—anterior retractor of proboscis; 3—longitudinal muscle crest, acting as support for hook; 4—internal retractors of tip of proboscis; 5—external retractors of proboscis; 6—outline of brain (dotted line); 7—eye; 8—gut cavity; 9—its epithelium; 10—adductor muscles of hooks; 11—dilators of proboscis sheath; 12—epithelium of proboscis sheath (after J. Meixner).

In the simplest case, the parenchymatous muscles of Acoela are re-presented by separate fibres, which either run irregularly in all directions or more or less keep to three main directions: longitudinal, transverse, and dorso-ventral (Fig. 128, *A*). Systems of retractors of the front and rear ends of the body often separate from the longitudinal muscles, and retractors of the pharynx from the dorso-ventral muscles. Even in forms that have separate digestive syncytia, muscle cells or whole bundles of them freely penetrate these syncytia (Fig. 98, *A*); moreover, even outside the order Acoela we find instances of the dorso-ventral muscles penetrating the amoeboid epithelium of the gut (Fig. 128, *B*).

As compared with the highly-polarised structure and regular layers of muscle processes in hydroids, Acoela, with their parenchyma traversed in all directions by diffusely-scattered muscle fibres, certainly represent a more primitive condition, but one containing more varied possibilities for progressive development.

Apart from the function of general contractility of the body, the musculature of lower turbellarians usually serves the mouth and the copulatory apparatus. *Xenoturbella* does not have the latter, and we find in it only those primary functions of the contractile-motor apparatus that we have seen in coelenterates: (*i*) general body contractility and (*ii*) seizing and swallowing prey. In the most primitive turbellarians, which have no epithelial pharynx, the oral orifice is served both by cutane-ous musculature and by a rosette of dorso-ventral fibres attached around it and running to the dorsal side. With the development of a pharynx simplex the swallowing function is much improved and the swallowing musculature becomes more complex. In particular, pharynx retractors appear, attached at one end to the pharynx and at the other to the integu-ment. With the development of the copulatory apparatus (male only in some forms, male and female in others) the musculature of these organs is formed from parenchymatous muscles. When organs of cutaneous armament exist, these also have their own muscles—protractors and retractors. In most Acoela the front end of the body takes part in the capture of prey, as it has an adhesive gland and is capable of considerable extension. It is extended by means of relaxation of its longitudinal and contraction of its annular musculature, and also contraction of the longitudinal and annular musculature of the rest of the body; it is retracted primarily by contraction of its own longitudinal musculature, and also of parenchy-matous retractor muscles (Fig. 128, *A*). In other forms (*Convoluta, Oligochoerus*) the function of prey-capture is fulfilled also by the edges of the body, which curve ventrally and often possess circumoral vesicular glands provided with stylets.

In most Acoela the locomotor function is entirely fulfilled by cilia in the cutaneous epithelium, but in many cases, especially in larger (*Anaperus, Amphiscolops*) or much elongated (*Proporus*) forms, the body musculature also takes some part in locomotion. Elongated forms use serpentine un-dulations of the body in movement. Body contractility and the power of changing shape aid greatly in squeezing into narrow spaces, which is especially important in soil-dwelling forms (e.g. *Convoluta roskoffensis*). In general, with total contractility of the body and movement by means of cilia on the surface of the substratum or within it, the participation of musculature in locomotion arises itself everywhere. It is therefore not surprising that all large turbellarians have, to a greater or lesser degree, made the transition to muscular locomotion; only the cilia of the ventral side are still used in crawling; the activity of the rest of their cilia has only hydrokinetic significance. In fact, only the possibility of using their musculature for locomotion has enabled many turbellarians to attain the large size at which locomotion by means of cilia alone would be impossible.

Four types of muscular locomotion are found in turbellarians: (*i*) crawling by peristaltic movements of the whole body; (*ii*) crawling by means of peristaltic waves in the side of the body apposed to the substrate; that

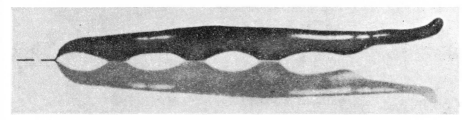

Fig. 129. *Terrestrial planaria of* Rhynchodemus bilineatus (*Turbellaria Triclada*), *moving by means of peristalsis of its creeping sole, view in profile* (after C. F. A. Pantin).

side is then distinguished by marked development of the musculature, especially the dorso-ventral muscles, and forms something of the nature of a creeping sole; sometimes the peristaltic waves in the sole appear in the form of actual transverse ridges ('myopodia'), so that the creeping planaria touches the substrate only with the crests of these ridges, e.g. *Rhynchodemus bilineatus* (Triclada Terricola) (Fig. 129), according to C. F. A. Pantin (1950); as the waves move from the front to the rear along the sole the animal flows with them, as on caterpillar-tractor treads (cf. also J. Pfitzner, 1958); (*iii*) movement by serpentine body flexures, found in elongated forms; we find serpentine creeping associated with relatively weak musculature and considerable body size, best shown in *Rhynchoscolex* (Notandropora); with strong longitudinal musculature it changes into swimming: we see an incipient form of such swimming in *Promonotus* (Alloeocoela Crossocoela; W. Beklemischev, 1927); with weak musculature and small body size, much-elongated forms swim by means of cilia, but with the aid of body flexures; (*iv*) swimming by bending and straightening body appendages. Among turbellarians this method of locomotion is found in many Polyclada with specialised lateral body margins that act as primitive fins. (See J. Olmsted, 1922, regarding neural regulation of locomotion in Polyclada.)

Also among locomotor organs must be classified the attachment organs that develop in many turbellarians. These include the tails of Dal-yelliidae (Rhabdocoela), which have adhesive cells and strong musculature; the tail plates of Macrostomida and Monocelididae (Alloeocoela; true suckers, located on the rear end (*Bdelloura*, Triclada; Temnocephala) or the sides (*Polycotylus*, Triclada) of the body; and so on. Suctorial pads (*Sorocelis*, Triclada) or front-end suckers (*Dendrocoelum*) usually have both an attachment and a locomotor function, and at the same time may be used to seize prey. Caterpillar-type progression by alternate attachment of the front and rear ends of the body of the substrate often occurs, e.g. in many Triclada, but not usually in normal movement, only as a flight reaction to very powerful stimuli.

The musculature of turbellarians is differentiated in various ways according to their methods of progression and prey-capture, but always within the framework of their basic structural plan: a thin continuous layer of cutaneous musculature, plus parenchymatous musculature that permeates the whole body rather loosely. The parenchymatous musculature of Acoela consists of numerous fibres or small bundles running in various directions. The parenchymatous musculature of large turbellarians (Polyclada, Triclada) possesses equally numerous bundles but they are much larger, often deserving the name of muscles composed of muscle tissue. Small turbellarians of higher orders, e.g. most Rhabdocoela, have advanced along the path of oligomerisation, and in many of them the parenchymatous musculature is reduced to a small, constant number of thick fibres with special functions: front-end and rear-end retractors, and muscles moving the pharynx and the copulatory organs. In such cases general body contractility is entirely due to the cutaneous musculature and the retractors of the ends of the body. Marked individualisation, although of unicellular muscles, has been made possible in Rhabdocoela by the strong development and firmness of the boundary layers of the skin, pharynx, etc., which have assumed a skeletal role. In Rhabdocoela the muscles of the pharynx (pharynx bulbosus) often are strictly constant in number and arrangement for particular species. When we recall that in small Rhabdocoela the cutaneous glands (Fig. 11) and the nephrocytes show the same constancy in number and arrangement, we may say that that order has already taken a step towards the constancy in number of cellular elements characteristic of Rotatoria and nematodes.

In higher turbellarians the development of the prehensory and swallowing function is expressed primarily in the creation of diversified and often complex types of pharynx. Strong pharyngeal musculature and an improved pharynx enable turbellarians to capture prey in spite of the weakness of the rest of the body musculature. The grasping tentacles possessed by coelenterates are entirely absent in turbellarians. The cephalic tentacles of Polyclada, perhaps homologous with the aboral tentacles of ctenophores and Narcomedusae, have retained only the sensory function and have totally lost the grasping function. The edges of the body or of its front end may be adapted as accessory organs for the capture of prey.

In primitive cases adaptations of the front of the body for grasping are expressed, as we have seen, in the possession of adhesive glands and retractor muscles. In the suborder Kalyptorhynchia (order Rhabdocoela) there is a complexly-differentiated proboscis at the front end of the body, provided with a muscular sheath into which it can be withdrawn by a retractor muscle running from the tip of the proboscis to the bottom of the sheath; the proboscis is protruded by contraction of the walls of the sheath; external muscles attached to the sheath (retractors and protractors) move the proboscis as a whole. The proboscis may be undivided and have adhesive glands at the tip (Polycystidae, etc.) (Fig. 128, *D*, Fig. 186, *E*), or may be bifid and sometimes have cuticular pincers (Schizorhynchidae, Gnathorhynchidae; Fig. 128, *E*). The proboscis of Kalyptorhynchia is always located at the front end of the body and may be regarded as a modification of that end. In the family Trigonostomidae, however, a similar (although sheathless) proboscis is formed subterminally on the ventral side of the head.

In parasitic flatworms proboscides, similar in structure to those of Kalyptorhynchia but bearing spines or hooks and used as organs of attachment to the host, arise in varying numbers and locations. Many Cyclophyllidea, e.g. *Taenia solium*, have a single apical proboscis on the scolex; all Tetrarhynchidea possess four such proboscides, located on the scolex with radial symmetry (Fig. 130, *C*). The fluke *Rhopalies* (Digenea) has a pair of retractile proboscides, lying on the sides of the oral sucker (K. Skryabin, 1947). These facts do not permit us to regard the grasping and attachment proboscides of flatworms as modifications of the front end of the body; rather we may see in them organs arising through local reconstruction of integument and musculature, not necessarily at the apical end of the body.[1]

The motor apparatus of parasitic flatworms, which is similar in principle to that of turbellarians, is characterised by great development of the attachment organs. Digenea usually have two suckers, oral and ventral; in Monogenea and their larvae the rear end of the body is separated in the form of a cercomere (Vol. 1, Fig. 58, *C*) bearing a complex attachment apparatus of hooks and suckers with corresponding musculature. The cercomere is also retained in a modified form in the development of Cestoda. The scolex of Cestoda also has suckers and other attachment organs.

The motor apparatus of the Digenea larva (cercaria) is of great interest. The rear part of its body is specialised in the form of a narrow 'tail', used as a swimming organ. We therefore have here a new method of locomotion, absent in turbellarians: locomotion by means of a specially-modified part

---

[1] The paired proboscides of Rhopaliidae are homologous (O. Fuhrmann, 1923) with the corner spine-bearing processes of the body of Echinostomidae, whereas the proboscides of Tetrarhynchidea are probably homologous with the accessory suckers of Tetraphyllidea. The homologue of the sucker of *Taeniarhynchus* is the apical proboscis of *Taenia*.

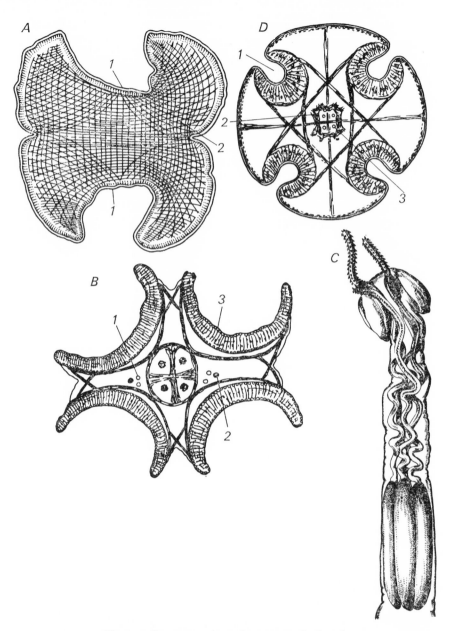

Fig. 130. *Some types of attachment organs in Cestoda.*

A—cross-section through the scolex of Pseudophyllidea, with two suctorial fossae; 1—suctorial fossa (bothrium); 2—muscle fibre. *B*—cross-section through scolex of Tetraphyllidea, with four bothria: 1—sections of nephridia; 2—section of nerve stem; 3—bothridia (from Hyman). *C*—scolex of *Tetrarhynchus* (Tetra-rhynchidea); the sheath of one of the four long proboscides opens at the front edge of each of the four suctorial fossae; the proboscis is furnished with hooks and is withdrawn into the narrow part of the sheath by a long retractor; the proboscis is extruded by contraction of the powerful musculature of the proximal widened part of the sheath. *D*—cross-section through scolex of *Nematobothrium* (Cyclophyllidea), with four suckers (acetabula) and biradiately-arranged musculature: 1—suckers; 2—neural apparatus of scolex; 3—sections of nephridia (after Fuhrmann).

of the body. We find analogous adaptations in a very similar form in tunicates, amphibian tadpoles, etc.

We have already discussed the question of possible homology between the tail of cercariae and the cercomere of Monogenea and Cestoda in Vol. 1, Chap. IV.

The tails of cercariae vary greatly in size, shape, and possession of cuticular appendages. Many species have bifid tails (Fig. 131). The

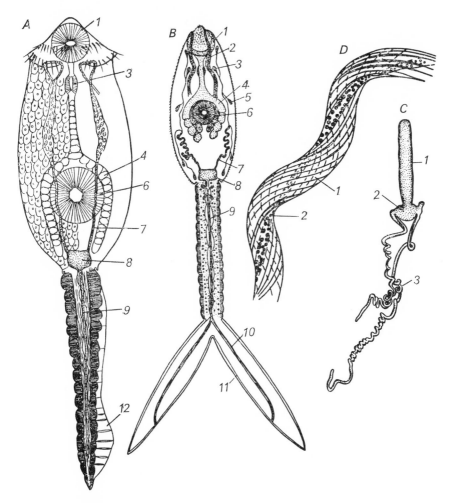

Fig. 131. *Movement by means of caudal appendages in cercariae (larvae of Digenea).* *A*—Cercaria abyssicola of *Valvata piscinalis*. *B*—Furcocercaria of *Planorbis umbilicatus*: 1—oral sucker; 2—front-end glands; 3—pharynx; 4—gut; 5— terminal cells of nephridia; 6—ventral sucker; 7—excretory trunks; 8—urinary bladder; 9—unpaired duct of urinary bladder; 10—paired ducts of urinary bladder; 11—paired lobes of tail; 12—tail fin (after Wesenberg-Lund). *Bucephalus hydriformis*: *C*—general aspect: 1—body of cercaria; 2—tail vesicle (unpaired part of tail); 3—caudal filaments; *D*—fine structure of caudal filaments: 1—muscle fibres; 2—elastic cord (after Vlastov).

internal structure of the tail is equally diversified: one of the extreme types is *Bucephalus hydriformis* (Fig. 131, *C, D*), whose tail consists of an unpaired basal part and two filamentous appendages. Each of the latter contains 14 longitudinal muscle fibres that take a slightly spiral course, as well as an elastic cord of complex structure extending along each filament. Contraction of the muscles produces shortening and corkscrew-like twisting of the caudal filaments; when the muscles relax the filaments elongate and straighten, partly because of the elasticity of the cord and partly, perhaps, because fluid is forced into them from the unpaired section of the tail by contraction of the cutaneous musculature of the latter (B. Vlastov, 1923).

This example shows how far differentiation of the rear end of the body has advanced in cercariae with its conversion into a locomotor organ.

5. Contractile-motor Apparatus of Nemathelminthes, Acanthocephala, and Nemertini

The motor apparatus of nematomorphic groups shows considerable diversity, because of differences in the modes of life and methods of locomotion of the various classes. There are, however, features common to the whole subphylum. Because of the great reduction of parenchyma and (often) the great development of the body cavity the structure of Nemathelminthes has become more or less polarised, which leads to regularity and occasionally to simplification in the cutaneous musculature as compared with that of flatworms. At the same time the parenchymatous musculature has undergone considerable reduction in Nemathelminthes and all that remains of it is separate, specialised muscles—a trend already noted in Rhabdocoela. Finally, the presence of cuticular formations in all groups of Nemathelminthes creates the conditions for development of a motor apparatus consisting of individualised muscles and movably-articulated skeletal parts: Kinorhyncha have advanced farthest in that direction.

All Nemathelminthes and Acanthocephala can be divided into four groups, according to their predominant method of locomotion: (*i*) classes that retain ciliary locomotion (Gastrotricha, Rotatoria); (*ii*) animals moving by serpentine flexures of the body (Nematodes, adult Nematomorpha); (*iii*) animals with a hydraulic method of locomotion (L. A. Zenkevich, 1944), extruding and withdrawing the front end of the body by forcing cavity fluid into it (Kinorhyncha, Priapuloidea); (*iv*) immovably-attached parasites (Acanthocephala).

Gastrotricha, with a very small body, have a much oligomerised and specialised muscular apparatus (Vol. 1, Fig. 63, *A*). Out of the whole cutaneous musculature, they have retained only two or four longitudinal muscles running from the posterior attachment tubules to the front end and causing body flexures used in steering when the animals are swimming. Because of the absence of a musculo-cutaneous sac the body shape is constant, perhaps on account of the cuticle. The front ends of the muscle bands may send branches to the pharynx, enabling them to act also as pharynx-retractors.

The cutaneous musculature of Rotatoria is rather less oligomerised and still more differentiated than that of Gastrotricha. The great differentiation of the musculature of Rotatoria is evidently due to the division of their body, which in the most typical case is divided into head, trunk, and foot (which in turn is segmented), with corresponding segmentation of the cuticle. The cutaneous musculature of Rotatoria contains up to eleven longitudinal muscles and several annular and transverse muscles; in addition there are muscle cells in most organs and also muscles moving the pharynx. In the less differentiated Notommatidae the longitudinal muscles serve as body flexors; in Bdelloidea they take part in telescopic contraction of the body and in caterpillar-like progression; in forms with differentiated foot and head they serve as head-retractors and help in foot movements. In testaceous forms, considerable portions of the longitudinal muscles along the part of the trunk with a hard covering may be reduced. Annular muscles are located in the front part of the body, and sometimes form a sphincter that closes above the withdrawn rotary apparatus when a stimulus produces a concealment reaction in the front end of the body. The rotary apparatus has its own (often quite complex) musculature, which has been little studied. The lateral spines (in *Pedalia*, etc.) are moved by transverse muscles. Within the class Rotatoria both the division of the body into sections and the methods of locomotion (as well as the methods of procuring food) vary greatly, and therefore comparative study of their motor apparatus would be of great interest.

Free-living Nematodes, which are mostly soil-dwellers, have a solid but flexible cuticle, strong but undifferentiated movements, and separate, strictly regular (polarised, see Chap. I) but also poorly-differentiated musculature. In the latter respect they stand well below Gastrotricha and Rotatoria, especially the latter.

The muscular apparatus of nematodes is mostly composed of longitudinal cutaneous muscles; in this class annular cutaneous muscles have entirely disappeared, and all that remains of the parenchymatous muscles of flatworms is the muscles forming part of the digestive or the genital apparatus. In some nematodes the number of muscle cells in the skin is very small and strictly constant (Meromyaria); in other species, mostly large ones, there are many more, and perhaps their number has increased secondarily (Polymyaria). The chief function of the cutaneous muscles is locomotor; because of their lack of annular muscles, nematodes are incapable of peristaltic alterations in body diameter and travel exclusively by characteristic serpentine undulations. In spite of the absence of annular muscles, the trunk musculature of *Ascaris* still works on the principle of action by the musculo-cutaneous sac against the resistance of the cavity fluid (A. Krotov, 1956).

The musculature of the digestive apparatus of nematodes is restricted to the fore-gut and the hind-gut; the mid-gut has no muscular coating.

The musculature of the fore-gut consists of muscles forming part of the pharynx and also muscles moving the pharynx (protractors and retractors). In the hind-gut the muscle fibres form a sphincter. Both the male and the female genital ducts have a coating of muscle cells, which aid the discharge of genital products. The entire assemblage of muscles ensures the movement of the spicules of the male copulatory apparatus.

The musculature of adult Nematomorpha also consists mainly of layers of longitudinal cutaneous muscles.

The hard cuticle of Kinorhyncha, which is divided into separate sclerites, has convergent resemblance to the external skeleton of arthropods. The musculature of Kinorhyncha is accordingly distinguished by an unusually high degree of differentiation, having developed into a number of separate muscles that connect the parts of the exoskeleton with one another and also connect the pharynx with the exoskeleton. Thus the musculature of Kinorhyncha resembles that of limbless arthropods, such as insect larvae, in its degree of differentiation. That convergence indicates the high correlation between the structure of the skeleton and that of the musculature. Like nematodes, Kinorhyncha have no annular musculature; their longitudinal cutaneous muscles, in harmony with the metameric structure of the skeleton, are divided into metameric bundles running from one zonite to another (Vol. 1, Fig. 64, A). Each two adjoining zonites are connected by a pair of dorsal and a pair of ventral bundles. There are also front-end retractors in the form of four bundles running from the head to zonites V–IX. Their antagonistic muscles are the dorso-ventral muscle fibres in each zonite, whose contraction raises the pressure of the body-cavity fluid so that the front end of the body is extruded. The muscular pharynx (pharynx bulbosus) possesses protractors and retractors. Kinorhyncha move by extrusion and retraction of the head, which grips the substrate with spines.

Priapuloidea have a similar means of locomotion and a similar division of the body into a trunk and a retractile 'proboscis' armed with spines. Because of their large size and their lack of a metameric hard covering, however, the cutaneous musculature of Priapuloidea forms a continuous musculo-cutaneous sac with an outer layer of annular and an inner layer of longitudinal muscles. As a result of the absence of parenchyma and the presence of a large body cavity, the musculature forms fully-separated and polarised layers and cords; the absence of a hard skeleton prevents it from developing into individualised muscles. Extrusion of the proboscis is caused by general contraction of the cutaneous musculature, and its withdrawal by the action of numerous strong retractor muscles that run through the body cavity.

We recall that the trunk of larval Priapuloidea bears two shields, dorsal and ventral (Vol. 1, Fig. 64, E, F); it is highly probable that in the larva the increased cavity-fluid pressure required to extrude the proboscis is obtained by the action of the dorso-ventral muscles, as in Kinorhyncha, which they resemble so much externally. The gill outgrowths of

Priapulus have musculature representing a continuation of the cutaneous muscles.

Whereas the proboscis of Priapuloidea, which is used for movement in compact soil, is extruded forward by pressure created by the whole musculature of the body walls, the proboscis of Acanthocephala, which is used solely for attachment to the gut wall of the host, does not require application of so much force. It is located on the front end of the body and is constructed like the proboscis of Taeniidae (Cyclophyllidea), i.e. it has its own muscular sheath and is armed with spines. The proboscis is everted by contraction of the muscular walls of the sheath. The front end of the body, which bears the proboscis, can, together with it, be withdrawn into the body cavity by long retractors running from the bottom of the sheath to the body walls; evagination of the front end is achieved here also by contraction of the musculo-cutaneous sac. In Acanthocephala, as a general rule, that sac consists of an external layer of annular and an internal layer of longitudinal muscles.

We have now seen in a number of flatworms and in Acanthocephala the formation of proboscides, used in free-living turbellarians for seizing prey and in parasitic flatworms and Acanthocephala for attachment to the host; the function of attachment to the host may with some probability be regarded as a modification of the prey-seizing function.

The proboscis of Eukalyptorhynchia (Turbellaria Rhabdocoela) may serve as a prototype from which are derived not only all the above types of proboscis but also the proboscis of nemertines, which is the most highly-developed glandular grasping proboscis with a sheath found in Scolecida.

The proboscis is the most important organ of nemertines and is the most characteristic feature of their organisation (Fig. 132). When withdrawn it is a long tubular invagination of the head covering, directly below the frontal organ. Like the proboscis of turbellarians and Trigonostomidae, therefore, the proboscis of nemertines is a subapical, ventral organ, and consequently cannot be regarded as a modification of the front end of the body. The external opening of the withdrawn proboscis is called the *rhynchostome*. The withdrawn proboscis lies in the cavity of its sheath (*rhynchocoel*). The proboscis sheath also has a tubular form. Its walls are muscular and are lined with epithelium. The sheath usually begins behind the brain, so that the foremost part of the proboscis (*rhynchodaeum*) has no sheath and cannot be evereted; the rear end of the proboscis is attached to the rear end of the sheath by the retractor muscle. Eversion and ejection of the proboscis are performed by contraction of the muscular walls of the sheath, and withdrawal by the retractor. The extruded proboscis is covered with ectodermal epithelium, followed by a layer of annular and longitudinal muscles; then comes the sheath epithelium of the proboscis, and, finally, its cavity; the number of muscle layers varies in different orders. Usually the proboscis is not everted for its entire length; in Metanemertini that is impossible, since in the posterior third of the proboscis there is a

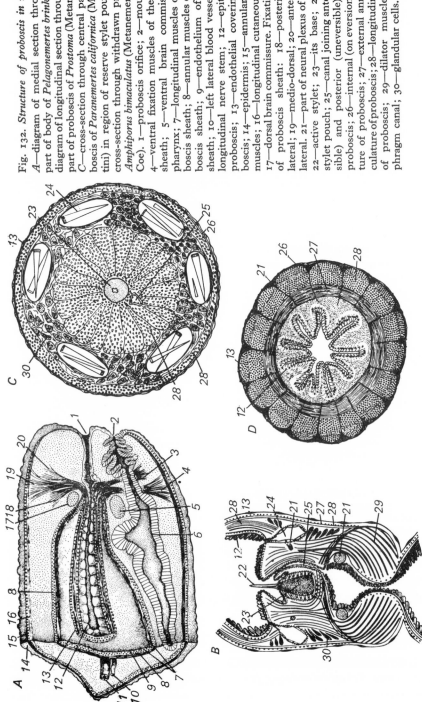

Fig. 132. *Structure of proboscis in nemertines.*
A—diagram of medial section through front part of body of *Pelagonemertes brinkmani*. B—diagram of longitudinal section through central part of proboscis of *Prostoma* (Metanemertini). C—cross-section through central part of proboscis of *Paranemertes californica* (Metanemertini) in region of reserve stylet pouches. D—cross-section through withdrawn proboscis of *Amphiporus bimaculatus* (Metanemertini) (after Coe). 1—proboscis orifice; 2—mouth; 3 and 4—ventral fixation muscles of the proboscis sheath; 5—ventral brain commissure; 6—pharynx; 7—longitudinal muscles of the proboscis sheath; 8—annular muscles of the proboscis sheath; 9—endothelium of proboscis sheath; 10—left lateral blood vessel; 11—left longitudinal nerve stem; 12—epithelium of proboscis; 13—endothelial covering of proboscis; 14—epidermis; 15—annular cutaneous muscles; 16—longitudinal cutaneous muscles; 17—dorsal brain commissure. Fixation muscles of proboscis sheath: 18—posterior dorsolateral; 19—medio-dorsal; 20—anterior dorsolateral. 21—part of neural plexus of proboscis; 22—active stylet; 23—its base; 24—reserve stylet pouch; 25—canal joining anterior (eversible) and posterior (uneversible) parts of proboscis; 26—internal (on eversion) musculature of proboscis; 27—external annular musculature of proboscis; 28—longitudinal muscles of proboscis; 29—dilator muscles of diaphragm canal; 30—glandular cells.

diaphragm bearing a stylet with which the animal pierces its prey, and the proboscis is everted only far enough for the diaphragm with the stylet to be at the end of the everted part. The epithelium of the posterior, uneverted part of the proboscis secretes a poisonous substance, which is discharged into the wound made by the stylet. Heteronemertini merely wind their proboscis around their prey (e.g. an annelid), and the secretion of the proboscis epithelium poisons and paralyses it.

The proboscis of nemertines is therefore similar in general structural plan to that of Eukalyptorhynchia; it differs from it, first, in its great absolute and relative size and corresponding differentiation in details, and secondly in possession of a large rhynchocoel, lined with epithelium, within the sheath.

We may note that the proboscides of similar general structural plan that we have seen in various flatworms, Acanthocephala, and nemertines must have arisen quite independently in all these groups. The ability to form such a proboscis is one of the characteristics of Scolecida and does not occur outside that phylum. The everted 'proboscides' of polychaetes, leeches, and snails are modifications of the muscular pharynx and may be equated rather to the pharynx plicatus of turbellarians (Fig. 101) than to the proboscides of Scolecida.

Nemertines have three methods of locomotion: (*i*) swimming by means of cilia, found in young individuals and a few minute forms; (*ii*) crawling, which is typical of the majority; and (*iii*) swimming by means of muscular movements, found in some benthic and all pelagic species. Crawling, as in turbellarians, is performed by using both cilia and body musculature, and is accompanied by profuse smearing of the substrate with mucus. Most nemertines have extremely elongated bodies, which greatly increases the speed of peristaltic crawling. The organisation of nemertines represents one of the highest levels in the adaptation of animals to progressive movement without the aid of special appendages or of sections of the body used as locomotor organs. Speed of movement without special locomotor organs can be attained only by extraordinary development of the musculature. In fact, as L. A. Zenkevich (1944) has demonstrated, in the proportion of muscle tissue to body volume nemertines occupy one of the highest places among all invertebrates (average proportion not less than 50%, and in *Carinina linearis* up to 60%), being surpassed in that respect only by leeches. Zenkevich divides all animals, according to the amount of motor musculature in the body expressed as a percentage of total body volume, into five groups: *amyarian* (without musculature, e.g. sponges); *oligomyarian* (less than 10%, e.g. medusae); *mesomyarian* (10 to 30%); *polymyarian* (30 to 50%); and *hypermyarian* (over 50%). Nemertines stand at the threshold of hypermyarianism.

Besides having a locomotor function, the trunk musculature of many nemertines takes a major part also in food-swallowing: while Metanemertini, which have external digestion, suck their prey gradually, Heteronemertini swallow it whole: *Lineus* can swallow a polychaete larger

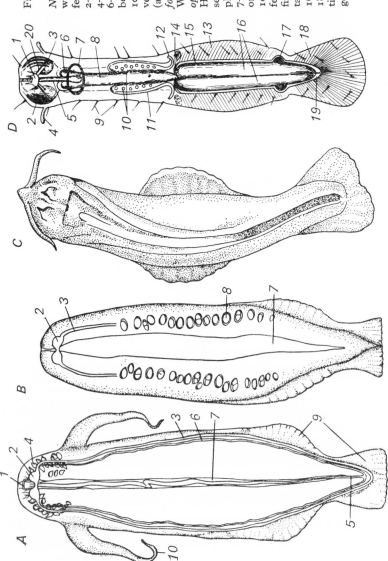

Fig. 133. *Bathypelagic nemertines and animals of similar facies.*
Nectonemertes: A—mature male with well-developed tentacles; *B*—female: 1—annular vessel of head; 2—brain; 3—lateral nerve stems; 4—testes; 5—dorsal blood vessels; 6—lateral blood vessel; 7—proboscis sheath; 8—ovaries; 9—fins; 10—tentacle; proboscis and blood vessels are not shown in female (after Coe). *C*—*Amiskwia sagittiformis* (Lower Cambrian, after Walcott); *D*—*Spadella cephaloptera* (Chaetognatha) (from L. H. Hyman): 1—brain; 2—grasping setae; 3—eyes; 4—tentacles; 5—pharynx; 6—outgrowths of gut; 7—dorsal ciliated ring (sense organ); 8—gut; 9—tactile hairs; 10—ovaries; 11—oviduct; 12—female genital orifice; 13—lateral fin; 14—septum between trunk and tail section; 15—seminal duct; 16—testes; 17—seminal vesicles; 18—caudal fin; 19—posterior section of coelom; 20—circumpharyngeal connectives.

than itself, literally coating its prey (M. Gontcharoff, 1948; C. Beklemishev, 1955). Such swallowing of large prey is achieved by peristaltic movements of the whole body, and therefore the musculature of Heteronemertini is, to a considerable extent, unique.

All nemertines have cutaneous musculature consisting of an external layer of annular and an internal layer of longitudinal muscles. Some Palaeonemertini (*Tubulanus* spp., *Carinina* spp.) and some Heteronemertini also have a layer of internal annular muscles immediately surrounding the gut and the rhynchocoel. On the other hand, all Heteronemertini have a thick layer of external longitudinal muscles lying between the skin and the annular muscles (we have seen similar, but very weak, muscles in Anaperidae among Acoela). Many nemertines also have two systems of mutually-penetrating diagonal fibres, specially developed in those forms that on strong contraction assume a corkscrew-like shape (*Cephalothrix* among Palaeonemertini, *Lineus socialis*). All the annular and longitudinal muscles of nemertines form thick layers of muscle tissue, much thicker than those in turbellarians. Nemertines are the best illustration of the rule that great development of musculature in an animal leads to segregation of muscle cells and appearance of muscle tissue. All nemertines also have dorso-ventral fibres, more strongly developed in flat forms, where they are mainly located in the spaces between the metameric outgrowths of the gut, sometimes forming almost continuous septa between them, e.g. in Heteronemertini (*Cerebratulus lacteus*). The more flattened the nemertine and the greater its power of active swimming (which increases in parallel with the flattening of the body), the more developed is the dorso-ventral musculature. It reaches its highest level in the fins of bathypelagic Metanemertini, such as *Nectonemertes* (Fig. 133).

Finally, there are longitudinal fibres penetrating the internal mesenchyma and very strongly developed in several Palaeonemertini. Altogether, half or more than half of the cross-section of a nemertine is occupied by muscle tissue.

In accordance with the parenchymatous structure and the absence of skeletal parts, nemertines have almost no separate muscles; the only large individualised muscle in their body is the proboscis retractor, the segregation of which is increased by reduction of the connective tissue in the proboscis sheath.

The pharynx of nemertines is a pharynx simplex, and therefore its musculature is not separated from the surrounding parenchyma, that is, it is much less individualised than the pharynx musculature of Rhabdocoela and of most Nemathelminthes. The poor differentiation of the nemertine pharynx is probably due to the fact that it fulfils only a swallowing function, while the prey-seizing function is fulfilled by the proboscis; whereas in other Scolecida the pharynx often fulfils both functions and thereby receives a strong stimulus to progressive development.

6. CONTRACTILE-MOTOR APPARATUS OF MOLLUSCA

Molluscs, like flatworms and nemertines, have well-developed mesenchyma, which naturally leads to persistence of at least some muscle cells in a diffused state. At the same time the large size of many molluscs and the considerable amount of musculature in separate parts of their bodies produce a tendency to segregation of muscle cells and formation of muscle tissue; and the development of hard parts, especially a shell, is a stimulus to formation of separate, individualised muscles attached to these hard parts. Therefore most molluscs retain, besides well-separated muscles, diffuse muscle fibres in the mesenchyma, just as we see in them a diffuse neural plexus together with a separate central section of the neural apparatus, and a general network of lacunae together with arterial and venous systems.

The most characteristic feature of molluscs is possession of a foot and a shell, with profound effects on the structure of their musculature. Among lower molluscs the only exception is Solenogastres, which have no shell and either no foot (order Chaetodermatoidea) or only a vestige of it (order Neomenioidea). Consequently Solenogastres have a more or less evenly developed musculo-cutaneous sac. It is very probable, however, that they lost the foot secondarily, and that the musculo-cutaneous sac is equally secondary in them.

The foot is the primary and principal locomotor organ in molluscs. We apparently find a primitive type of foot in Loricata and in most Gastropoda, where it is a flat muscular creeping sole developed along the blastopore side of the body. Traces of the sole are retained in the foot of the primitive bivalves Protobranchia, whereas in most Lamellibranchia it has a cuneiform or digitiform shape and is used less for creeping and jumping than for digging. Some of the varied modifications of the gastropod foot and its unusual structure in Cephalopoda will be discussed briefly below.

Primitive molluscs are slow-moving, grazing animals which developed heavy protective armour. Most members of the phylum have retained that type of specialisation, but a number of separate groups have taken the path of increased activity, accompanied by lightening or total loss of the shell, development of distant receptors, refinement of the central section of the neural apparatus, and general raising of the level of organisation. Cephalopoda have advanced farthest along that path. Evolution of the locomotor apparatus has taken different directions in separate groups of molluscs, according to their general trend in development.

Loricata have a shell composed of eight movably-articulated plates metamerically arranged along the animal's dorsal side. The shell plates are moved by a number of well-individualised muscles—longitudinal, oblique, and transverse (Fig. 134). The fibres of the last-named are, in effect, dorso-ventral, joining the edges of plates that lie one above the other (A. Ivanov, 1946). The flat foot of Loricata serves as an organ of

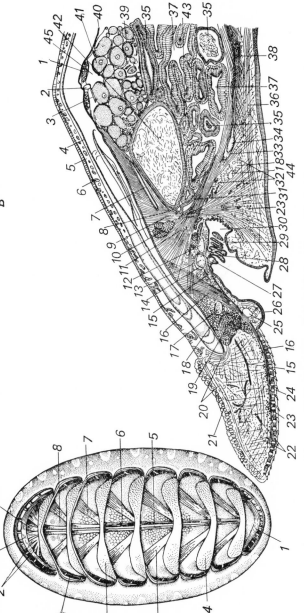

Fig. 134. *Motor apparatus* of Tonicella marmorea (*Loricata*).

A—musculature of dorsal side (plates of shell removed): 1—anterior and posterior ends of longitudinal lateral muscles; 2—oblique muscles; 3—part of longitudinal lateral muscle; 4—transverse muscles; 5—aorta; 6—straight muscles; 7—genital gland visible through integument; 8—place of attachment of dorsal retractors of radula; *B*—cross-section of female through centre of body, diagrammatic: 1—straight muscle; 2—aorta; 3—genital artery; 4—articulamentum; 5—tegmentum; 6—aesthetes; 7—lamina suturalis; 8—internal branch of kidney; 9—internal branch of latero-pedal muscles; 10—third central bundle of latero-pedal muscles; 11—external bundle of lateropedal muscles; 12—branchial artery; 13—pleuro-visceral nerve stem; 14—neuro-lateral sinus; 15—pallial nerve; 16—internal pallial muscle; 17—branchial vein; 18—storage cells; 19—longitudinal lateral muscle; 20—muscle fibres of mantle, running in different directions; 21—integument of marginal zone with "packets"; 22—blood vessels of mantle; 23—bundles of longitudinal muscle fibres; 24—papillary epithelium of mantle; 25—blood vessel; 26—lateral line; 27—efferent branchial vessel; 28—branchial nerve; 29—gill leaflets; 30—pallial groove; 31—external branch of kidney; 32—lacunae; 33—neuro-pedal sinus; 34—pedal nerve stem; 35—gut; 36—transverse muscle fibres of foot; 37—lobules of liver; 38—pedal commissure; 39—follicles of ovary; 40—dorsal integumentary epithelium; 41—oocyte; 42—pyloric section of stomach; 43—visceral artery; 44—lateral sinus; 45—gonocoel (after A. V. Ivanov).

locomotion and of suction against the substrate. It is permeated by muscle fibres running in all directions—longitudinal, transverse, oblique, and dorso-ventral. The last are attached by their dorsal ends to the plates of the shell and, bending round the medially-located internal organs on both sides, spread fanwise over the foot. The dorsal retractors of the vestigial foot of Neomenioidea (Solenogastres) appear to be homologues of these muscles.

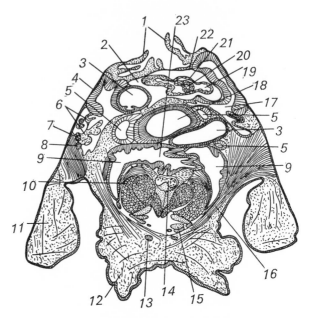

Fig. 135. *Motor apparatus of the primitive gastropod with cap-shaped shell* Puncturella noachina (*Prosobranchia Rhipidoglossa Fissurellidae*). *Cross-section at level of pharynx and heart* (after A. V. Ivanov).

1—lobe of mantle; 2—pericardium; 3—gut; 4—hypobranchial gland; 5—right kidney; 6—ctenidium; 7—branchial ganglion; 8—subintestinal ganglion; 9—oesophageal sacs; 10—radular cartilage; 11—muscle fibres of mantle edge; 12—foot; 13—pedal nerve stem; 14—radular sheath with radula; 15—pharyngeal musculature; 16—columellar muscle; 17—mantle cavity; 18—stomach; 19—left auricle of heart; 20—rectum; 21—wall of ventricle; 22—wall of pericardium; 23—oesophagus.

The edges of the mantle, which surround the body of Loricata on all sides, are also permeated in all directions by muscle fibres; consequently when the animal is applying suction to the substrate the edges of the mantle can be closely apposed to the smallest irregularity in the latter.

The buccal musculature of Loricata has attained extreme complexity, consisting of 37 paired and six unpaired muscles, as a result of the great complexity of the oral apparatus. Here we encounter for the first time a specialised phytophage, capable of using tough plant food unavailable to lower animals and possessing the adaptations required for that purpose (radula, etc.)

Tryblidiida, like Loricata (Vol. 1, Chap. VIII), have paired metameric retractors of the head and foot, connecting these to the shell in two symmetrical rows; Nuculidae still retain four pairs of these muscles, and other Lamellibranchia two pairs. The larvae of lower Gastropoda possess homologues of the paired foot-retractors. The columellar muscle of adult Gastropoda, which is used to retract the foot, is asymmetrical, as

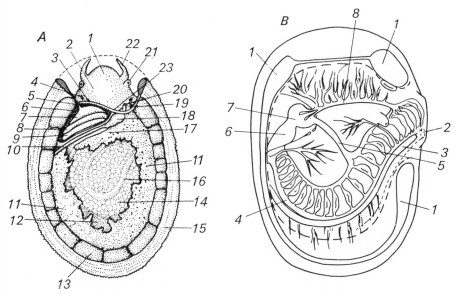

Fig. 136. *Forms of columellar muscle in Gastropoda with conical (cap-shaped) shells.*
A—*Patella pontica* (Prosobranchia Docoglossa), dorsal view; shell, part of mantle, and epithelium of visceral hump removed, pericardium opened; columellar muscle symmetrical, horse-shoe-shaped (after A. V. Ivanov): 1—snout; 2— occipital region of head; 3—pallial vessels; 4—osphradial tubercle; 5—pallial vein; 6—pericardium; 7—auricle; 8—ventricle; 9—bulbus arteriosus; 10— rectum; 11—right kidney; 12—stomach; 13—bundles of columellar muscle; 14— liver; 15—mantle; 16—gut; 17—anterior lobe of right kidney; 18—left kidney; 19—papilla of right kidney; 20—anal orifice; 21—eye; 22—tentacle; 23—papilla of left kidney. B—*Siphonaria* (Pulmonata Basommatophora), dorsal view, shell removed (after Hubendick): 1—columellar muscle; 2—anal orifice; 3—osphradium; 4—afferent branchial vein; 5—efferent branchiorenal vein; 6—ventricle of heart; 7—auricle; 8—anterior pulmonary vascular plexus. Outline of mantle cavity shown by broken line.

it is formed from one of the paired retractors of the larva. In the comparatively-primitive modern Docoglossa, however, its place of attachment to the shell forms a broad, fully-symmetrical horse-shoe, open anteriorly (Fig. 136, *A*). The Pulmonata with a similar cap-shaped shell (*Siphonaria*, etc.) have a markedly-dissymmetrical columellar muscle (Fig. 136, *B*).

The creeping movement of most Gastropoda convergently resembles that of large Triclada: basically it uses waves of muscle contraction running along the sole, with plentiful lubrication of the substrate with mucus, and in small forms with some participation by the cilia of the ciliated epithelium that always covers the sole. Fissurellidae and Docoglossa practically never emerge at all from the shell, and the range of their possible movements is very limited.

In most Prosobranchia, which have a spiral shell, the columellar muscle is attached by its upper end to the columella, and the lower end partly expands fanwise in the foot and is partly attached to the operculum. Besides helping in creeping it plays the role of a retractor, withdrawing the foot and the head into the shell in time of danger and covering its opening with the operculum. In Pulmonata separate retractors also run from the columellar muscle to the cephalic tentacles and the pharynx (Fig. 139, *A*). The antagonist of that system of retractors is the cutaneous musculature, especially the annular muscles, whose contraction causes protrusion of the head and trunk. Consequently the locomotor apparatus of higher Gastropoda enables them to make much more varied and (often) more rapid movements, and, above all, to protrude farther from the shell.

When the shell is reduced the columellar muscle may disappear (e.g. in shell-less slugs among Pulmonata and in some Opisthobranchia), or only the ventral parts of its fibres may persist in the form of actual dorsoventral fibres in the foot.

Among Pectinibranchia extreme development of the foot musculature, and often division of the foot by transverse grooves into two or three parts (propodium, mesopodium, and metapodium) or by a longitudinal groove into two symmetrical halves, leads to the appearance of new forms of creeping, using a walking movement (*Pomatias* among Taenioglossa), or to pulling-up on the protruded propodium, to a peculiar jumping-walking movement (Strombidae; C. Yonge, 1932, 1937; Fig. 137, *A*), and finally to the swimming of the purely-planktonic Heteropoda (Taenioglossa); in the most specialised Heteropoda (*Carinaria*, Fig. 137, *B*; *Pterotrachea*) the propodium is converted into a separate fin, flattened in the sagittal plane, and the metapodium forms the posterior part of the fusiform body of the animal; the columellar muscle acts as a swimming muscle. In *Firoloida* the metapodium is reduced, and the whole foot becomes merely a fin.

In the subclass Opisthobranchia the transition to swimming is one of the chief trends in evolution. Many benthic forms among both Tectibranchia and Nudibranchia are able to swim; some purely-planktonic groups have arisen from both of them.

Tectibranchia swim by means of paired lamellar expansions of the edges of the sole of the foot, the so-called parapodia (not at all homologous with the parapodia of polychaetes). It is strange to see how the large heavy *Aplysia* (Fig. 137, *C*), crawling slowly along the floor of an aquarium, suddenly leaves the floor and begins to glide easily through the water by means of undulatory movements of its fins. *Acera bullata*,[1] on transition

[1] Usually referred to Bullomorpha, but J. E. Morton and N. A. Holme (1955) believe that it belongs to Aplysiomorpha.

to swimming, bends its parapodia to the dorsal side, folds them funnel-wise, and then uses the umbrella so formed almost as a medusa does, making up to 45 strokes per minute and moving head-foremost (J. Morton and N. Holme, 1955). In *Notarchus* (Fig. 137, *D*) the parapodial lobes

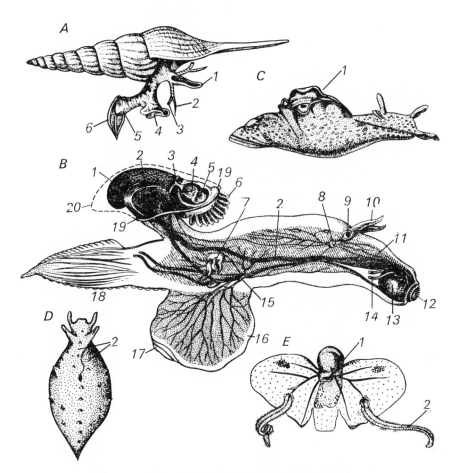

Fig. 137. *Different types of foot structure in Gastropoda.*

A—Rostellaria rectirostris (Taenioglossa Strombidae), protruding from shell, view from right side; foot adapted for walking (limping) movement: 1—proboscis; 2—cephalic tentacles; 3—eye; 4—anterior part of foot; 5—posterior part of foot; 6—operculum. *B—Carinaria mediterranea*, male (Prosobranchia Taenioglossa Heteropoda); planktonic form, foot converted into an unpaired fin: 1—testis; 2—gut; 3—aorta; 4—ventricle of heart; 5—auricle; 6—gill; 7—copulatory organ; 8—cerebral ganglion; 9—eye; 10—tentacle; 11—stomach; 12—mouth; 13—pharynx; 14—mucus glands; 15—pedal ganglion; 16—propodium; 17—mesopodium; 18—metapodium; 19—edge of mantle; 20—shell. *C—Aplysia punctata* (Opisthobranchia Tectibranchia), with lateral appendages of foot (parapodia—1) used for swimming. *D—Notarchus punctatus* (Opisthobranchia Tectibranchia), dorsal view, parapodia (2) bent dorsally and partly grown together. *E—Desmopterus papilio* (Opisthobranchia Pteropoda), parapodia converted into paired wing-like fins, remainder of foot reduced: 1—mouth; 2—tentacle-like appendages of fins (from Pelseneer and Hofmann).

grow together above the back, forming a parapodial cavity open to the front; contraction of the parapodial musculature forces the water out and the animal swims with its rear end foremost. In Pteropoda the central (crawling) part of the foot is vestigial and the parapodium forms a pair of wing-like appendages used for swimming. The columellar muscle of Pteropoda Thecosomata is attached to the shell at one end, and at the other end bifurcates and enters the two lobes of the parapodium, forming most of its musculature. Pteropoda are not a natural group: the predatory Gymnosomata are related to Bullomorpha, and the microphagous Thecosomata to Aplysiomorpha. Nudibranchia have no parapodia. If their benthic forms do swim, it is by rapid lateral flexures of the whole body (*Tethys leporina, Dendronotus arborescens*). The same method is used by the purely-planktonic *Phyllirhoë*, which has a laterally-flattened fish-like body, with the foot entirely eliminated.

Possession of complexly-differentiated musculature of the oral organs is very characteristic of Gastropoda, with also (in higher Gastropoda) complex musculature of the copulatory apparatus.

The motor apparatus of Bivalvia has two main sections, one associated with the foot and the other with the mantle and the shell. As already mentioned, their foot has four (Nuculidae) or two pairs of retractors, attached to the shell and homologous with the foot-retractors of Tryblidiida and Loricata. There are also many muscle fibres in the foot, running in different directions. In the digging foot of *Ensis* (Fig. 138) they form a true musculo-cutaneous sac, resting on the central blood sinus. Before the foot is protruded the heart pumps blood into its sinus, and when the foot is withdrawn the blood is forced out of the sinus into the renal lacunae and the volume of the foot is greatly decreased (A. Graham, 1931). Such changes in the volume of organs by means of blood transfer are widespread among molluscs.

In most Gastropoda the edge of the mantle is simple and secretes the external layer of the shell; in *Turritella* a row of tentacles runs along it, preventing mud from entering the mantle cavity (A. Graham, 1938). In Bivalvia and in *Neopilina* the edge of the mantle is divided by two grooves running along it into three ridges: the outermost, adjoining the shell, is secretory—its inner surface secretes the periostracum, and its outer surface secretes the prismatic layer of the shell; the central ridge is sensory in Bivalvia, and all the sense organs of the edge of the mantle are formed from it; the innermost ridge (velum) is muscular. Often the muscles permeating it are attached to the shell and are retractors of the velum. The velum is used to close the mantle cavity and to regulate the flow of water by all forms in which the edges of the mantle do not grow together.

In a typical case the two halves of the shell are connected by two strong adductor muscles, one anterior and one posterior, which close the shell. The shell opens passively by means of the elasticity of the ligament, which is compressed between the two halves of the hinge when the shell is closed. There is a widespread tendency within the class towards reduction of the anterior adductor and development of unequal-muscled (anisomyarian) and one-muscled (monomyarian) forms. The posterior adductor, when it alone remains, usually migrates to the centre of the shell, as in *Pecten* or *Ostrea*. The processes of reduction of the anterior adductor, which have developed independently in different groups of Bivalvia, have

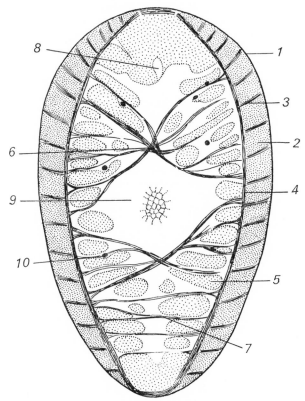

Fig. 138. *Cross-section of foot of* Ensis siliqua (*Bivalvia*).
1—external epithelium and subepithelial layer of annular muscles; 2—external longitudinal muscles; 3—oblique muscles penetrating them; 4—layer of dorso-ventral muscles; 5—internal longitudinal muscles; 6—oblique muscles; 7—transverse muscles; 8—pedal artery; 9—venous sinus of foot; 10—nerves running from pedal ganglion (after A. Graham).

been brilliantly analysed by C. M. Yonge (1954). In primitive cases the two adductors lie at the ends of the ligament, at the point where the lines of attachment of the pallial muscles of the right and left sides cross each other. The adductors arose there through strengthening of the pallial muscles and fusion of the fibres of the right and left sides, which came into contact through the folding of the mantle along the hinge-line (G. Owen, E. Trueman, and C. Yonge, 1953; G. Owen, 1958).

We have seen in Chap. II that in very many Bivalvia the edges of the mantle fuse together over a varying distance, in an extreme case leaving only three orifices leading into the mantle cavity: inhalant and exhalant siphons, and an opening for the foot. In some cases the pallial muscles on the two sides fuse together, as in the formation of adductors, and thus several secondary muscles arise, e.g. the cruciform muscle of Tellinacea, which lies in front of the inhalant siphon (Fig. 8, *C*) (C. Yonge, 1949). The retractor of the siphons is also formed from the pallial muscles (C. Yonge, 1958).

Other organs of Bivalvia also have muscles, e.g. the gills; the latter are highly developed when the movement of the gills has a hydrokinetic effect, e.g. in Nuculanidae (Protobranchia, C. Yonge, 1941) and Septibranchia.

In Cephalopoda the foot is derived from several pairs of rudiments; in Dibranchiata four or five pairs move forward and form arms arranged around the mouth, while one pair remains in the rear. The two halves of the latter grow together and form the infundibulum. In *Nautilus* (Tetrabranchiata), however, the infundibulum remains in the form of two separate halves. Morphologically the arms of Cephalopoda are homologous with marginal outgrowths of the sole of the foot, but A. Naef (1921–23), following A. Willey, regards the infundibulum as the homologue of the epipodial folds of lower Gastropoda. That interpretation, however, is not the only one. The shell served Tetrabranchiata, and still serves their most recent representative *Nautilus*, as a dwelling. In Decapoda it is overgrown by the mantle and converted into a skeletal plate lying on the physiologically-dorsal side of the animal. In some Decapoda it has become vestigial, and in Octopoda it has completely disappeared. Besides the shell, or in place of it, an internal skeleton has developed in Cephalopoda, formed of cartilaginous tissue resembling that of vertebrates. The internal skeleton contains the following parts: (*i*) cephalic cartilage, surrounding the brain and the statocysts, and resembling the cartilaginous skull of vertebrates; (*ii*) orbital cartilages, protecting the eyes; (*iii*) occipital cartilage, lying at the morphologically-anterior edge of the mantle; (*iv*) cartilages lying at the base of the arms; (*v*) the cartilages of the fins; (*vi*) the cartilages behind the infundibulum, etc. All these skeletal parts serve as places of attachment for muscles (Fig. 139, *C*).

Tetrabranchiata possess, first of all, retractors attached at one end to the internal wall of the dwelling-chamber of the shell, and at the other to the lateral parts of the cephalic cartilage. These muscles are homologous with the columellar muscles of Gastropoda. In Decapoda a whole row of well-separated retractors runs from the lower surface of the shell (morphologically its inner surface) to the cephalic cartilage and the infundibulum. There are also retractors of the infundibulum, running to it from the cephalic cartilage, and elevators of it, running to it from the cervical cartilage. The mantle consists of a thick layer of transverse muscles lying between two thinner layers of longitudinal muscles. The arms contain very complex and powerful musculature. The arms bear numerous suckers,

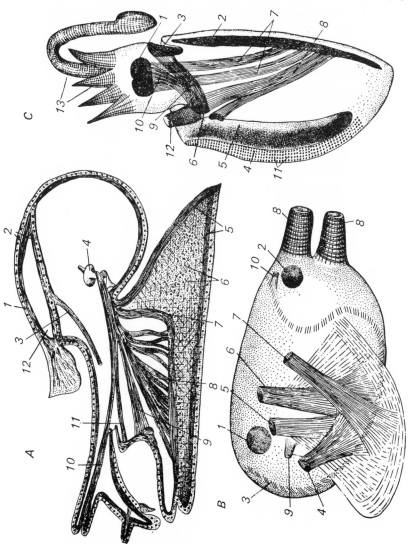

Fig. 139. *Diagrams of structure of contractile-motor apparatus of molluscs.* A—Gastropoda, edible snail (*Helix pomatia*) in sagittal section: 1—longitudinal cutaneous muscles; 2—annular cutaneous muscles; 3—penis retractor; 4—protuberance on columella to which the columellar muscle is attached; 5—longitudinal muscles of foot; 6—oblique muscles of foot; 7 and 8—foot retractors; 9 and 11—pharynx retractors; 10—retractor of posterior tentacles; 12—lung cavity. B—Lamellibranchia: 1 and 2—anterior and posterior adductor muscles; 3—muscles of mantle edge; 4—foot protractor; 5 and 7—foot retractors; 6—foot elevator muscle; 8—annular and longitudinal muscles of siphon; 9—mouth; 10—posterior duct. C—Cephalopoda Dibranchiata, sagittal section, shell and cartilages in black: 1—cephalic cartilaginous capsule; 2—shell; 3—occipital cartilage; 4—mantle; 5—mantle cavity; 6—infundibulum cartilage; 7—lateral and medial head retractors; 8—infundibulum retractor; 9—infundibulum protractor; 10—cervical muscle; 11—row of transverse muscle fibres of mantle, cut across; 12—infundibulum; 13—arms with longitudinal and annular muscle fibres (from V. A. Dogiel).

and in some groups also tentacles. Both have their own musculature. All Decapoda and some Octopoda have lateral fins on the visceral hump.

The locomotion of Cephalopoda is much diversified. Many of them can swim head-foremost by undulatory movements of the fins. All Cephalopoda can swim rapidly with the tip of the visceral hump in front, forcibly ejecting water from the mantle cavity through the infundibulum (jet propulsion!). The complex musculature that governs the movements of the infundibulum enables it to regulate the direction of the stream and consequently the course of the swimming animal, so that many cephalopods can move even head-foremost by jet propulsion. The arms serve not only as grasping instruments but also, in benthic forms, as organs of locomotion.

When the animal is resting, its mantle still makes continuous rhythmic pulsations that change the water in the mantle cavity around the gills (external respiratory adaptations: see Chap. II); thus the pallial musculature of Cephalopoda has a hydrokinetic as well as a locomotor function.

The mouth-parts of Cephalopoda, with their strong jaws and radula, also have well-differentiated musculature.

Cephalopoda far exceed all other molluscs in the degree of differentiation of their musculature, in the proportion of well-individualised muscles in it, in the development of an internal skeleton, and in the varied methods and speed of their locomotion. These facts are in full accordance with the general high level of organisation in this class, with the excellent development of their neural apparatus, with their possession of capillary blood-circulation, etc.

7. Contractile-motor Apparatus of Annelids and Onychophora

In the general principle of organisation of their motor apparatus polychaetes and oligochaetes resemble Actiniaria, although they stand incomparably higher in their organisational level. As with Actiniaria, the body of polychaetes is a muscular sac consisting of longitudinal and annular muscles and exerting pressure on the fluid in the cavity. In Actiniaria, however, that fluid lies in the general gastrovascular cavity, which is divided only into a central (digestive) and a peripheral (distributive) section. In polychaetes the gut and the coelom are completely separated from each other, and the mechanical function of creation of body turgor under the action of cutaneous muscle tension falls entirely on the coelomic fluid. From this point of view the polychaete coelom has a great advantage over the gastrovascular cavity of Actiniaria, since it is a closed cavity, whereas the gastrovascular cavity of Actiniaria opens through the mouth, which, as we have seen, seriously affects its functioning as a supporting apparatus. Actiniaria find it necessary to empty themselves from time to time and to refill with water, whereas the annelid coelom operates with a constant quantity of fluid, so that the animal's total body volume remains constant; there is only shortening and thickening, or lengthening and thinning, of the whole body or parts of it, depending on whether the tension in the longitudinal or in the annular muscles predominates. When the pressure of the coelomic fluid is increased the gut should not be affected: an adaptation protecting the gut from compression is formation of numerous metameric dissepiments, in which many forms have radial muscles capable of supporting the gut when the annelid is elongated. It is curious that the homology of the radial chambers of Actiniaria and the coelomic chambers of annelids, of which we spoke in Chap. V, is displayed in the similar roles they play in the activity of the motor apparatus. Apart from the common principle of the hollow muscular sac, however, annelids and Actiniaria are entirely different in all the rest of their motor apparatus, and we shall not make further comparisons between them.

Polychaetes are among the animals with the most completely polarised structure; because of their large coelom, their supporting and connective tissues are reduced to bounding formations: sheaths and membranes lying between the epithelial and muscle layers and almost without cellular elements. Such a structure makes possible the segregation of muscle cells and the formation of muscle tissue. In fact, the greater part of the musculature in the polychaete body is in the form of perfectly regular layers of annular and longitudinal muscles in the body wall and almost equally regular muscle bundles in the dissepiments (Fig. 140, *A*). There are also layers of muscle cells in the walls of the gut and partly in those of the blood vessels; fairly complex musculature in the oral apparatus; and well-individualised muscles moving the setae and other parts of the parapodia and also the cephalic appendages.

The annular muscles of the body lie in one or more layers beneath the external epithelium; in many Errantia they are relatively slightly developed and remain only in certain parts of the body. The longitudinal muscles are much more strongly developed; they form four longitudinal bands, two dorsal and two ventral. In small forms, such as *Polygordius*, ribbon-like muscle cells of *Ascaris* type stand edgewise on the body wall and lie in a single layer. In some cases, e.g. in Serpulimorpha, the boundary layer separating the annular from the longitudinal muscles communicates with the latter by a large number of thin longitudinal membranes, to which the muscle cells lying between them are attached by their narrow ends; in a cross-section the latter appear to be arranged pinnately. Thus a supporting structure is created, regulating and fixing the arrangement of muscle fibres and making possible the existence of thick muscle bands. Blood vessels also penetrate along the boundary formations into the depth of the musculature.

From the coelomic epithelium, on the contrary, neither longitudinal body muscles nor the muscles of the gut are sharply defined, on account of the coelomic-epithelial, retroperitoneal formation of these muscles in ontogeny. There are no boundary layers between them and the coelomic epithelium (N. Livanov, 1914, 1940). In some cases there are coelomic-epithelial muscle cells; in other cases we find coelenchymatous cells, amoebocytes, in the longitudinal musculature between muscle cells;

thus the muscle tissue of polychaetes, like almost every muscle tissue, is a compound tissue.

Most of the muscle cells in the longitudinal muscles extend through more than one segment and are not interrupted at the sites of the dissepiments; the bands of longitudinal muscle are merely penetrated there by the dorso-ventral muscle bundles of the dissepiments. In *Eunice*, however,

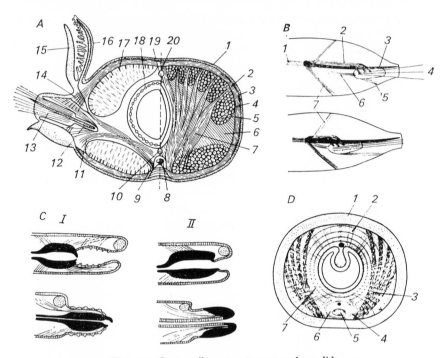

Fig. 140. *Contractile-motor apparatus of annelids.*

A—diagram of cross-section through a segment of *Eunice*, on left through the centre of the segment, on right through a dissepiment: 1—epidermis; 2—layer of annular muscles; 3—boundary membrane of epidermis; 4—boundary membrane between annular and longitudinal musculature; 5—longitudinal muscles of body wall; 6—dissepiment; 7—its dorso-ventral muscle bundles; 8—ventral nerve chain; 9—its boundary membrane; 10—ventral blood vessel; 11—oblique muscles; 12—aciculum; 13—setae of neuropodium; 14—vestige of notopodium; 15—dorsal cirrus; 16—gill; 17—boundary membrane in longitudinal muscle band; 18—intestinal vascular plexus; 19—dorsal vascular plexus; 20—dorsal mesentery (after N. A. Livanov). *B*—diagram of arrangement of parapodial muscles in *Ichthyotomus sanguinarius* (Eunicemorpha), bundle of setae extended in upper drawing, withdrawn in lower: 1—retractors of supporting seta; 2—setae extensors; 3—aciculum; 4—setae; 5—protractor, and 6—retractor of setae; 7—protractor of aciculum (after Eisig). *C*—diagrams of the pharyngeal musculature of errant polychaetes: I—pharynx is everted by simultaneous action of musculo-cutaneous sac and protractors; II—pharynx is everted solely by action of musculo-cutaneous sac (increased pressure of coelomic fluid). In both cases the pharynx is drawn back by contraction of the retractor muscles (after Lang). *D*—diagram of arrangement of musculature in dissepiment of *Lumbricus terrestris*: 1—body wall; 2—radial muscles of dissepiment; 3—oblique muscles; 4—sphincter of ventral orifice; 5—ventral nerve chain; 6—ventral opening in dissepiment; 7—annular musculature (after Newell).

the dissepiments divide the longitudinal muscles (N. Livanov, 1914) and the latter therefore become metameric, as in arthropods, *Amphioxus*, and vertebrates.

Abundant appendages—sensory, locomotor, respiratory—are characteristic of polychaetes. These include first the cephalic appendages described above and the parapodia. Parapodia, arranged metamerically in pairs, are as characteristic of polychaetes as the foot and the shell are of molluscs. Although some polychaetes (*Polygordius*) have no parapodia, their absence there is secondary, like the absence of the foot and the shell in some molluscs (e.g. in *Rhodope*, order Nudibranchia).

A typical parapodium consists of the following parts: (*i*) two lobes, ventral, or *neuropodium*, and dorsal, or *notopodium*, used as organs of locomotion and armed with setae;[1] (*ii*) two tactile tentacles, dorsal and ventral; (*iii*) a single gill, lying dorsally to the dorsal tentacle; and (*iv*) a lateral sense organ.

Outside the class Polychaeta the homologues of parapodia are probably the limbs of other Articulata and the epipodia of Prosobranchia Rhipidoglossa. Epipodia contain proper epipodial folds running along the sides of the foot, and also tentacles comparable to parapodial tentacles and sense organs similar to the lateral organs of polychaetes. It is less clear whether we can regard the fold of an epipodium itself as the homologue of a row of polychaete notopodia. In any case, because of the development of the foot the epipodium of lower Gastropoda entirely lacks a locomotor function and has retained only the sensory and ciliary-hydrokinetic functions (J. Risbec, 1955).

While accepting the homology of the mollusc epipodium with part of the parapodial assemblage of organs in polychaetes, we are far from being inclined to derive polychaetes from molluscs, or vice versa; it is quite possible, however, that these organs already existed in one form or another in the ancestral group from which all Trochozoa originated. It is still uncertain whether, following Sedgwick (see Vol. 1, Chap. VI), we can see in the parapodial tentacles of polychaetes (and the epipodial tentacles of Rhipidoglossa) homologues of the circumoral tentacles of coelenterates.

A typical parapodium with a full complement of components occurs in a few polychaetes (e.g. in the families Glyceridae, Nephthydidae, Spionidae); usually some part is lacking (e.g. the gills in Nereidae, the notopodium in Eunicidae, etc.) and consequently others are hypertrophied ('balancement des organes', Cuvier). In many families of sessile polychaetes complete dissociation of parapodia has taken place, and the parts of them remaining (e.g. the notopodium and the neuropodium in Terebellidae) are

---

[1] In annelids each seta is secreted by a single chaetogenous cell, lying deep in a chaetiferous pouch formed by the epidermis. In both these features the setae of annelids resemble those of the mantle edge of Brachiopoda and differ from those of Pogonophora (which are secreted by the chaetogenous syncytium; A. Ivanov, 1957) and from the compound spines of *Enantia* (Polyclada) (L. von Graff, 1889) and of Acoela (W. Beklemischev, 1929), each of which consists of many combined rods or tubules, each secreted by a separate chaetogenous cell.

widely separated. Among the spines of each of the branches of the para-podium of errant polychaetes one (the aciculum) is larger than the others, and serves as a supporting axis for the corresponding branch.

The possession of rows of parapodia by errant polychaetes has produced a break in their annular musculature along lateral lines, and division of their longitudinal musculature into two bands, dorsal and ventral. More-over, muscles moving the parapodia have developed in each segment: these muscles are composed of assemblages of well-individualised bundles. Some of these bundles are clearly derived from the annular musculature, e.g. the flexors and extensors of the parapodia. Usually there are also protractors and retractors of the setae (Fig. 140, B). Contraction of these muscles in different combinations enables the parapodia and, in particular, the setae of polychaetes (especially errant polychaetes) to perform quite complex movements. Rigidity of the parapodia is achieved by the hydro-static pressure of the coelomic fluid, which in turn is caused by tension in the body-wall muscles.

H. Eisig (1887) classifies the methods of locomotion used by polychaetes as follows:

1. *Running* (Fig. 141). The two parapodia of a single body segment are moved simultaneously in opposite directions: one moves forward as the other moves backward. As a parapodium moves forward the setae are withdrawn, and as it moves backward they are protruded and spread out fanwise. That rowing movement of the parapodia is accompanied by lateral movements of the body: the animal makes serpentine undulations to right and left, the parapodia on the convex side of each bend being moved forward and those on the concave side backward. The animal thus creeps or runs with a flowing movement on the surface of the substrate. Often only the neuropodia take part in that movement, the notopodia being used for defence, as in Amphinomidae. Most errant polychaetes use this method of locomotion, which explains the predominance of longitudinal musculature in them. The running method is most highly developed in short predatory forms: some Aphroditidae approach arthropods in their agility. In most polychaetes swimming in the water is a variant of this method of locomotion.

2. *Swimming* by means of rapid undulatory flexures of the body, as in nematodes. This occurs in several small forms.

3. *Caterpillar-like crawling*. Rarely occurs.

4. *Peristaltic crawling*, performed by alternate shortening and lengthen-ing of separate parts of the body with simultaneous gripping of the substrate by setae, mostly of the neuropodia (Fig. 141, B).

Movement within the soil is achieved mostly by the last method, with flow of the coelomic fluid from one end of the body to the other (in forms with reduced dissepiments). It is more efficient mechanically for a digging animal of cylindrical form to have as large a diameter as possible, but a long slender worm can crawl more rapidly on the surface (G. Chapman, 1950). Peristaltic crawling is found in Drilomorpha (Capitellidae, Areni-colidae, Maldanidae) and also in Terebellidae, Sabellidae, and some

Eunicidae. Phyllodocidae, Glyceridae (Phyllodocemorpha), and Ariciidae (Spiomorpha) are capable of both running ('rowing') and peristaltic crawling. The latter method requires more or less equal development of both the longitudinal and the annular musculature of the body wall.

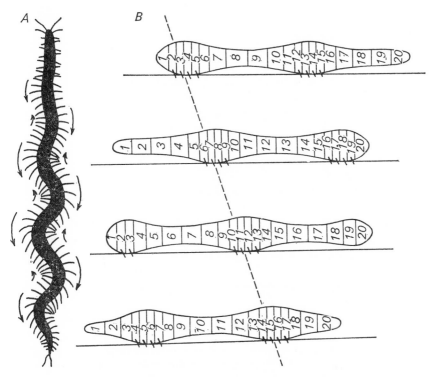

Fig. 141. *Principal methods of locomotion in annelids.*

A—undulatory-rowing locomotion of *Nereis diversicolor*: while the front end of the body is extended straight, waves of lateral curves pass from front to rear along the rest of the trunk; on the convex parts of the body the parapodia make strong backward strokes (large arrows), and on the concave parts they move slowly forward (small arrows). At any given moment the directions of the two parapodia of any segment are always directly opposite each other (after Buddenbrock). *B*—peristaltic crawling of earthworm—four successive positions of the animal. Only those setae pushing against the substrate at a given moment are shown (after J. A. Ramsay).

The appearance of parapodia armed with setae marked a great advance in the locomotor apparatus, greatly increasing its efficiency. That evidently explains the small amount of musculature in the bodies of polychaetes and oligochaetes as compared with nemertines and molluscs, and the considerable increase in the relative amount of musculature in leeches, which have lost their setae. All 'Gephyrea' [1] (Priapuloidea,

[1] We use this word not as the name of any group in the natural system, but as a designation of the life-form of a fossorial marine animal with a musculo-cutaneous sac and an undivided coelom.

Echiuroidea, Sipunculoidea) also have relatively small musculature, which is explained by the high efficiency of their 'hydraulic' method of locomotion (L. Zenkevich, 1944).

All Serpulidae, and also some Sabellidae (e.g. *Spirographis*) and some members of other families of sessile polychaetes (e.g. *Owenia fusiformis* in the family Oweniidae) usually do not leave their tubes, and perish if they are removed from them. They are all true sessile animals, actually less mobile than Actiniaria, although they move within their tubes. Most Sedentaria and many Errantia display a series of transitional stages in the development of a sessile mode of life. We cannot describe that whole series in detail, but we may point out that, simultaneously with loss of mobility and transition from the mode of life of a predator to that of a filter-feeder or a deposit-feeder, the sensory apparatus is first reconstructed (degradation or disappearance of cephalic eyes, often development of trunk ocelli, etc.); later the neural apparatus is altered (degeneration of the brain), and finally the motor apparatus: general weakening of the musculature, simplification of the musculature of the parapodia and the pharynx, disappearance of the pharyngeal sac (regarding the evolution of other adaptations to a sessile mode of life found in polychaetes, see Vol. 1, Chap. VII).

Oligochaetes have no parapodia, and their dorsal and ventral bunches of setae are widely separated. Their peculiarities of muscle structure do not go beyond the limits of their variation in polychaetes. In earthworms (family Lumbricidae) the muscular sac occupies 40 to 50% of the body volume (polymyaria). Their annular musculature is particularly well developed in the front eight segments, which play a major role in penetrating the soil: there it constitutes up to 50% of the total thickness of musculature. Towards the rear end the absolute thickness of the whole parietal musculature and the relative thickness of the annular layer decrease, but even there the latter does not fall below 25% (N. Sokol'skaya, 1951), in consequence of their peristaltic type of crawling.

Terrestrial oligochaetes, Lumbricidae in particular, have more than once served as the vehicle for study of the mechanism of peristaltic crawling in annelids. We have seen that the annular and longitudinal muscles of their bodies are mutually antagonistic. That antagonism is possible only because of their possession of a closed body cavity and the unalterable volume of their cavity fluid. In fact, an earthworm's nephridia and gonoducts never release fluid from the cavity to the exterior under any pressure, since they act as valves. The dorsal pores of the coelom are closed only by sphincters, but with the pressures existing in the worm's body they also do not allow fluid to pass. In *Lumbricus terrestris* the dissepiments begin after the fifth (last larval) segment; they are all pierced by apertures through which the ventral nerve chain passes. These apertures, however, are always closed by their sphincters, and fluid does not pass from segment to segment even when the worm is moving (as is well seen by X-raying a worm after injecting bismuth salts into the cavity of one or more segments). Thus the hydraulic method of locomotion (in the form in which it occurs in fossorial polychaetes with reduced dissepiments) is not found in earthworms. Burrowing, like peristaltic crawling, is performed by them by successive changes in the shape of separate segments; the fluid plays only a supporting role. A necessary condition for peristaltic crawling by animals with a large body cavity is the possession of dissepiments, since only that makes possible alternating, mutually-independent changes in the shape of successive segments. Nematodes and other animals with the life-form of 'Gephyrea' are incapable of peristaltic crawling.

The annular, oblique, and radial muscles of dissepiments (Fig. 140, D) maintain the septa in a state of tension while the body diameter changes. The radial muscles, in addition, protect the gut from being squeezed by the pressure of the coelomic fluid; the dorso-ventral muscles of the mesenteries flatten the body (G. Newell, 1950).

Among other oligochaetes, the chief methods of locomotion in the family Naididae are swimming and crawling; consequently Naididae have weaker musculature than earthworms; like many polychaetes, they have relatively weak annular musculature. The family Tubificidae, which burrow in mud with the front end of the body and make oscillatory (respiratory) movements with the rear end, have strong musculature at the front and weak at the rear end, the annular musculature at the rear end being particularly weak (N. Sokol'skaya, 1951).

An ecologo-morphological analysis of the organisation of leeches, including their locomotor apparatus, has been made by N. A. Livanov (1945). Most leeches are blood-suckers, periodically attacking larger animals; as a consequence, the development of an attachment organ in the form of a sucker located on the ventral side of the seven posterior somites (four in *Acanthobdella*) is very characteristic of them. Most leeches also have an anterior sucker. Attacks on large animals require quick movements, and leeches can travel swiftly by alternately attaching the front and rear suckers to the substrate, or swim even more swiftly by undulatory flexures in the dorso-ventral plane.[1] A leech attached by its rear end to its victim can bend and extend in all directions. As a result of these methods of locomotion leeches have lost their setae, and their coelom has lost its supporting function and become reduced to a system of canals within the coelenchymatous tissue. The body musculature consists of an external layer of annular muscles, two systems of spiral muscles, and a large mass of longitudinal musculature. In addition there are well-developed dorso-ventral muscles, musculature of the suckers and the pharynx, specialised cutaneous muscles producing eversion of cutaneous papillae, etc. On the whole leeches surpass all other invertebrates in the amount of muscle elements in the body (L. Zenkevich, 1944).

In the general character of their histological structure leeches differ

---

[1] The medicinal leech (*Hirudo medicinalis*) is stimulated into action by water movements, starts swimming about, and attaches itself by suction to a man or animal that has entered the water (G. Shchegolev, 1951).

radically from other annelids in their great development of amorphous coelenchyma. Nevertheless the large amount of musculature leads in them, as in nemertines, to the formation of continuous layers of muscle. That applies particularly to the powerfully-developed longitudinal muscles. The dorso-ventral muscles are represented by separate, loosely-arranged bundles running between the mesenchyma and the longitudinal musculature and dividing the latter into separate sections. Because of the absence of parapodia and setae, there are almost no individualised muscles in the body of leeches.

Besides the locomotor function, the motor apparatus also has a water-moving (hydrokinetic) function.

That function is fulfilled by three different methods: (*i*) by peristaltic changes in body diameter; (*ii*) by undulatory flexures of the body; (*iii*) by rowing movements of the body appendages. Examples of the first method of creating water currents are provided by *Arenicola* and *Sabella* (see Vol. I, Chap. VII); probably many other sessile tube-dwelling polychaetes, with well-developed annular musculature, behave similarly. Creation of currents by means of undulatory flexures of the body takes place when the animal is attached either by its front end (tube-dwelling oligochaetes) or by all the parapodia of the body (*Nereis*), or when it is attached by a posterior sucker (several leeches). Body appendages have a water-moving function in (for instance) *Chaetopterus*.

Tube-dwellers (*Tubifex*, etc.) that burrow in mud with their front end make oscillatory movements with their freely-protruding rear end, creating water currents that bring oxygen (of which there is very little in the water layers immediately adjacent to the soil) from above; thus the hydrokinetic activity of these animals, like that of leeches, has a purely respiratory purpose, and dispenses with current-directing adaptations (see Chap. II). *Nereis virens*, which lives in U-shaped burrows, is immovably fixed to the burrow walls by all its parapodia and uses the parapodial musculature to make oscillatory movements of the body. Dorso-ventral body flexures are thus produced, whereas in crawling *Nereis* uses lateral body flexures (A. Lindroth, 1938). A strong water current is created in the tube and brings the animal respiratory oxygen, food, odours, and also food particles (proved for *Nereis diversicolor* by M. Harley, 1950); the latter are filtered out by a mucus funnel, which the animal constructs at the tube entrance from the secretion of the parapodial glands, and then the animal eats the funnel together with the attached particles. A similar method of food-procurement by oscillation of the animal in a U-shaped tube and use of a mucus filter is known in *Urechis caupo* (Echiuroidea; A. Redfield and M. Florkin, 1931) and some chironomid larvae (Insecta Diptera), e.g. *Chironomus plumosus*, and has therefore arisen convergently in widely-differing groups of Articulata. In polychaetes of the family Chaetopteridae (Spiomorpha) the tube also serves as a current-directing adaptation, while the notopodia of the three thoracic segments have grown together into unpaired 'palettes' whose movement creates a current; here also, as stated in Vol. 1, Chap. VII, food is captured by periodically-secreted mucus nets (G. Wells and R. Dales, 1951).

Onychophora, like annelids and unlike the great majority of arthropods, have a continuous musculo-cutaneous sac. Their epidermis, which is

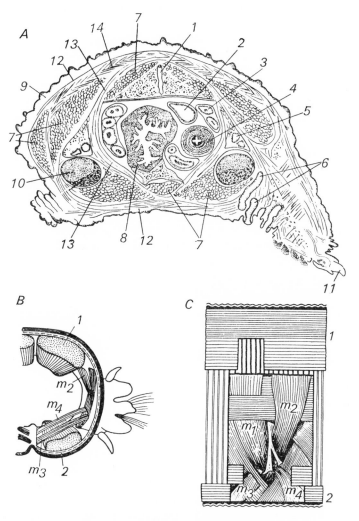

Fig. 142. *Musculature of Onychophora.*

A—male *Peripatus edwardsii*, cross-section through fourth segment from end of body: 1—heart; 2—anal glands; 3—efferent seminal ducts; 4—ejaculatory duct; 5—kidney; 6—crural gland; 7—longitudinal mucles; 8—gut; 9—connective-tissue layer of skin; 10—nerve stems; 11—claws of limbs; 12—annular muscles; 13—transverse muscles; 14—diagonal muscles (from Kükenthal). *B* and *C*—comparison of musculature of limbs of *Peripatoides* and parapodia of polychaetes: *B*—diagrammatic cross-section through segment of polychaete; *C*—musculature of limb of *Peripatoides*, view from within, muscles of body wall partly removed: 1—dorsal longitudinal muscles; 2—ventral longitudinal muscles; m_1—tergal limb promotor; m_2—tergal limb remotor; m_3—sternal limb promotor; m_4—sternal limb remotor (from B. N. Shvanvich).

coated with a thin chitin-containing cuticle, is underlaid by a fibrous cutis, which increases the mechanical stability of the integument. Next comes the cutaneous musculature, consisting of three layers: (*i*) a layer of annular fibres, continuous, but not very thick; (*ii*) two criss-cross layers of diagonal fibres; (*iii*) a thick layer of longitudinal muscles. The longitudinal musculature is developed in separate bands, as follows: one pair of dorsal, two pairs of lateral, and one pair of ventral muscles, and one unpaired ventro-medial muscle band. In addition there is the so-called transverse musculature, encircling the body cavity (pseudocoel) and separating the bands of longitudinal muscle with its bundles (Fig. 142). The well-developed musculo-cutaneous sac of Onychophora, which maintains the pressure of the cavity fluid, evidently ensures the rigidity of the body, as in annelids, and provides a solid support for the limbs. At the same time the body musculature probably takes part in ejection of the adhesive secretion of the salivary glands, by means of which Onychophora capture their prey.

The limb musculature is formed from the diagonal and transverse muscles of the body; the longitudinal musculature of the body takes no part in its formation. The muscles of a limb mainly run parallel to its axis; there are also fairly numerous fibres running transversely to that axis. Claws are formed by the cuticle and brought into action by special groups of muscles consisting of adductors, remotors,[1] and levators.

The pharynx of Onychophora has a thick muscular wall, and the gullet is much thinner; the mid-gut also has musculature composed of longitudinal and annular muscles.

In the degree of segregation of muscle elements and individualisation of muscle bundles Onychophora stand perhaps closest to leeches, and in general below polychaetes, as a result of reduction of the coelom and possession of coelenchyma.

When in motion *Peripatopsis capensis* keeps its body straight and does not use body flexures, in contrast to the method of crawling used by errant polychaetes and Scolopendromorpha (Fig. 144, *D*). In Onychophora the propulsive force is created entirely by the extrinsic muscles of the limbs and not by the longitudinal muscles of the body. In walking, Onychophora have several methods of moving the limbs (several kinds of gait), and change in rate of movement is achieved mainly by change of gait (S. Manton, 1950).

In general the structure of the musculature of Onychophora, like their whole organisation, confirms the uniqueness of their position among the subphyla of the phylum Articulata.

8. CONTRACTILE-MOTOR APPARATUS OF ARTHROPODA

An outstanding feature of arthropod structure is the cutaneous skeleton, formed by the cuticle of the external epithelium and as a rule divided into

---

[1] Remotor is defined here as the muscle that causes backward movement of the limb, its antagonist being referred to as promotor (Ed. comment).

more or less rigid sclerites joined by flexible membranes. Consequently the basic principle of organisation of the motor apparatus of arthropods differs substantially from that of annelids. The musculo-cutaneous sac and the closed coelomic pouches that serve as its support have completely disappeared, and the body has lost the power of extension and contraction. All movements are restricted to flexures of the body and its appendages, made by muscles attached to parts of the external skeleton either directly or by means of its solid projections (apodemes) or flexible apophyses, which also are cuticular outgrowths and act as tendons. In Crustacea, however (e.g. *Praunus*, Mysidacea), there are true tendons, also serving for attachment of separate muscles to the inner side of skeletal parts (A. Mayrat, 1955). Chelicerata and some Crustacea have a true internal skeleton, constructed of fibrous and fibrous-cartilaginous tissue of meso-dermal origin and also serving for muscle attachment. It is particularly well developed in Xiphosura and scorpions (Fig. 146). In them its most important part is the so-called endosternite—a plate in the prosoma with a number of projections, one pair (the rearmost) of which encircles the ventral nerve chain like a ring. The endosternite of Araneina is shown in Fig. 64. The similarity in the structure of their internal skeleton is a further proof of the mutual relationship of Xiphosura and terrestrial Chelicerata (E. Ray Lankester, W. Benham, and E. Beck, 1885; W. Schimkewitsch, 1894; S. Pereyaslawtzewa, 1901). Among Crustacea two unpaired endo-sternites are found, for instance, in the head of *Calanus* (Copepoda) (E. Lowe, 1936); they serve for attachment of some of the muscles of the antennae and the oral appendages. It is possible that the internal skeleton of Crustacea and Chelicerata is inherited from common Precambrian ancestors.

When a group of muscles, by contracting, changes the relative positions of two body segments, or of two limb segments, it invariably has its antagonists, whose contraction changes the relative positions of the same parts in the opposite direction. On the other hand, elastic antagonists take no part, as a rule, in the work of the motor apparatus. If the cavity fluid is used to create body rigidity, it is not on the annelid principle (pressure upon it by a continuous musculo-cutaneous sac) but on the Kinorhyncha principle (pressure due to mutual approach of plates of the cutaneous armour, produced by contraction of the individualised muscle bundles connecting them).

We have seen that in animals so different in all other respects as Loricata and Kinorhyncha the appearance of a segmented skeleton leads to breaking up of the musculature into separate individualised muscles, attached to different skeletal parts and moving them relatively to each other. The same takes place in arthropods, which far surpass all other invertebrates in the complexity and refinement of their motor apparatus. The refinement of the motor apparatus in arthropods is due largely to their possession of segmented limbs articulated to the body, built on the compound-lever principle, and also to the possession of wings by higher insects. In spite

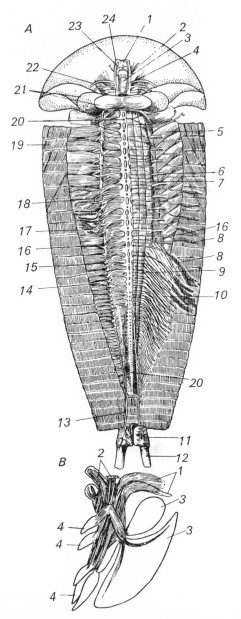

Fig. 143. *Musculature of* Triops cancriformis (*Phyllopoda Notostraca*).
A—the animal opened along the dorsal side and flattened out; in the
left half of the trunk, part of the musculature has been removed and
the nerves cut, in order to show the external muscles; in the right
half the course of the peripheral nerves has been partly uncovered:
1—brain; 2—circumpharyngeal connectives; 3—gullet; 4—ganglia
of antennae II (tritocerebrum); 5—diagonal muscles; 6—internal transverse muscles; 7—nerves to ventral bands of longitudinal muscles;

[*continued opposite*

of the reduction of the coelom, arthropods differ markedly from leeches and Onychophora in their relatively slight development of amorphous and cellular connective tissue, which in leeches functionally replaces coelomic fluid as a support for the musculo-cutaneous sac. The poor development of connective tissue in arthropods intensifies the separation and individualisation of their muscles.

Among modern arthropods, two groups differ from annelids less than the others in the structure of their musculature, and throw some light on the methods whereby the typical motor apparatus of arthropods arose: these are Phyllopoda among Crustacea and Chilopoda among Myriapoda.

Triops cancriformis (Phyllopoda Notostraca) (E. Zaddach, 1841) (Fig. 143) has a thin integument, almost without sclerotisation in the front part of the trunk. Accordingly its metamerisation is primary, i.e. each of the longitudinal muscle fibres lies within a single segment, attached to the annular intersegmental folds of the integument. The very fact of the division of longitudinal muscles into metameric sections represents considerable progress beyond the level of annelids and Onychophora: among annelids such division takes place only in some polychaetes. In the thoracic segments the longitudinal musculature of *Triops*, like that of polychaetes, is divided into four bands by the heart, the ventral chain, and the rows of limbs; in the abdomen—i.e. in the posterior, limbless segments— the longitudinal muscles form a uniform continuous layer. We may say that here there is a true musculo-cutaneous sac. Admittedly *Triops* has no independent annular musculature, but in that respect it resembles many errant polychaetes, which have lost their annular musculature on giving up peristaltic crawling.

In correspondence with the metameric characteristics of Notostraca their dorsal and ventral longitudinal muscles behave rather differently: the fibres of the dorsal bands run, as expected, from the front to the rear boundary of each segment. In the first eleven thoracic segments the fibres of the ventral bands do the same, but in the succeeding polypod segments they run from one pair of limbs to another. The ventral bands do not continue into the limbless segments, and the entire continuous muscular covering of these segments is formed by widening of the dorsal bands of longitudinal muscle.

Besides longitudinal muscles, the thoracic section of *Triops* possesses the following systems: (*i*) posterior oblique muscles, forming a continuous

8—nerves to dorsal muscles; 9—ventral bands of longitudinal muscles; 10—posterior oblique muscles; 11—anal lobe; 12—beginning of caudal filaments; 13—nerves to anal lobe; 14—external dorsal muscles of limbs; 15—ligament; 16—genital duct (cut across); 17—nerves to dorsal muscles of limbs; 18—external ventral muscles of limbs; 19— anterior (dorsal) oblique muscles; 20—ventral nerve chain; 21— muscles of mandibles; 22—nerves of antennae II; 23—nerves of antennae I; 24—muscles of upper lip. *B*—musculature of limb: 1— dorsal pair of external muscles of limb; 2—ventral pair of external muscles of limb; 3—exites; 4—endites (after E. Zaddach).

layer connecting the ventral and dorsal bands of longitudinal muscles; they are probably formed from the longitudinal musculature; (*ii*) anterior, or dorsal, oblique muscles, lying on the dorsal side of the sixth and seventh trunk segments; they lie deeper than the longitudinal muscles and cover the dorsal bands of the latter internally; their front ends are attached to the head; (*iii*) diagonal muscles, lying along the sides of the trunk in the spaces between the dorsal and ventral bands of longitudinal muscle; behind the eleventh segment they are covered, on the body-cavity side, by the posterior oblique muscles, which they cross almost at right angles; the diagonal muscles are in turn covered by the muscles moving the limbs; (*iv*) muscles moving the limbs, which lie closest to the skin: each pair of limbs has two adductor muscles running from the bases of the limbs to the centre line of the body, and two abductor muscles beginning near the lateral edges of the dorsal longitudinal bands. The muscles moving the limbs are therefore very similar to the corresponding muscles of polychaetes and Onychophora (Fig. 142, *B*, *C*), and like them are doubtless products of the annular musculature. In the aggregate the limb muscles form an almost continuous layer of transverse musculature on the ventral and lateral sides of the body; (*v*) finally, on each of the first eleven thoracic segments there is a pair of dorso-ventral muscles. The head musculature is also very simple, but consists of separate specialised muscles, the largest of which are the mandibular muscles.

In the trunk of Notostraca, therefore, there are well-developed longitudinal musculature and products of the annular and diagonal musculature of annelids; the dorso-ventral muscles, perhaps, are the remains of the musculature of the dissepiments. A primitive feature is the development of the musculature in the form of almost-continuous layers, with fibres lying in the same directions and with the same relative locations as in the corresponding musculature of annelids and Onychophora.

The musculature of the limbs is extremely primitive, not even having rudimentary segmentation, which also does not exist in the cutaneous cover of the limbs; the structure of Phyllopoda limbs shows no traces of secondary origin of their leaf-shaped limbs from the segmented limbs of other Arthropoda. E. W. MacBride (1914) and L. A. Borradaile (1926) contend that the leaf-shaped limbs of Phyllopoda are primitive. The opposite view is taken by, among others, H. L. Sanders (1957), who regards as prototypes the segmented limbs of Cephalocarida. A progressive feature of Phyllopoda is the breakdown of the whole trunk musculature into separate bundles, separate muscles, attached by their ends to the body coverings. That feature was the starting-point for further evolution of musculature in Crustacea.

Other classes of Arthropoda probably followed similar paths to a certain extent.

The smaller Anostraca took the path of oligomerisation of the number of muscle bundles: instead of four broad flat bundles of longitudinal muscles underlying the whole skin, they have in each segment individualised muscle bundles, two dorsal and two ventral, and the whole musculature assumes an aspect more customary for arthropods.

Somewhat different primitive features have persisted in the motor apparatus of Chilopoda.[1] Sclerotisation of the integument has advanced further in them than in *Triops*, but Chilopoda still represent one of the early stages in that process. The integument of the head, the rear end of the body, and the limbs have already undergone far-reaching sclerotisation, but in each trunk segment there are still many small sclerites, which in the further evolution of the group fuse together into larger shields. That process takes place first in the dorsal part of the segment, which is most exposed to unfavourable environmental influences, and where tergal shields are generally formed: two (anterior and posterior) in Geophilomorpha, or three (one medial and two lateral) in some Scolopendromorpha, or one in the rest of Chilopoda. The ventral side comes next: it is apposed to the substrate and usually has one or more small intersternal sclerites as well as a large sternal shield. The pleural region falls furthest behind in skeletal development: in most Chilopoda it has a fairly large number of separate sclerites (more than ten in *Scolopendra*) separated by wide intervals of unsclerotised cuticle (Fig. 144, *A*). The original function of the sclerites was protective, as is seen from the existence of separate sclerites that do not yet serve as points for muscle attachment. But very early—sometimes, perhaps, at the first appearance of sclerites—the muscles of the trunk and the limbs began to be attached to them, and the sclerites assumed the role of skeletal parts as well as that of armour-plates. That connection arising between sclerites and musculature has led to corresponding changes in both. As a result of the attachment of muscles the sclerites increased in size, and later whole groups of them fused into large shields; at the same time the primitive continuous musculature divided into separate muscle bundles as a result of its attachment to skeletal parts.

The most notable primitive feature in the trunk musculature of Chilopoda Epimorpha is retention in the pleural region of a continuous and almost intact layer of annular muscles (Fig. 144, *C*); beneath the tergites and sternites the annular muscles are naturally reduced. The regularity of arrangement of the annular muscles is also somewhat disrupted at the places of attachment of the limbs because of the development of some of the limb muscles from them, as we have seen in polychaetes. Beneath the annular muscles lies a layer of longitudinal muscles, grouped in bundles; they are best developed in the tergal and sternal regions, but also exist in the pleural region; some of the bundles are short and run between adjacent segments, others are long and are attached through several segments; the existence of these long muscles produces anisotergy in Chilopoda (see Vol. 1, Chap. VII), that being a feature of the specialisation in their structure. Deepest-lying is the 'transverse' musculature, consisting

---

[1] The structure and character of the motor apparatus of Chilopoda have been analysed in detail by E. G. Becker (1926, 1949, 1950), by W. Bücherl (1940), and especially by S. M. Manton (1950–61).

of dorso-ventral and pleuro-sternal cords and being, perhaps, to some extent homologous with the transverse musculature of Onychophora (Fig. 144, B).

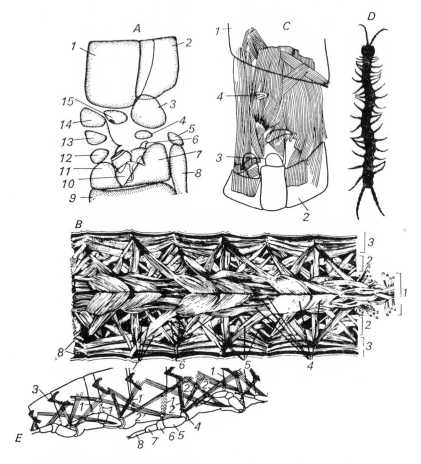

Fig. 144. *Motor apparatus of the myriapods Chilopoda and Symphyla.*
A—external skeleton of a trunk segment of Geophilomorpha (Chilopoda), view from right side: 1—posterior tergite; 2—anterior tergite; 3, 4, 5, 12, 13, and 14—various sclerites of the pleural region (pleurites); 6 and 10—sclerites of coxa of limb; 7—precoxal pleurite; 8—intersternite; 9—sternite; 11—metacoxal pleurite; 15—pleurite bearing spiracle (after E. G. Becker). B—trunk musculature of *Scolopendra* (Chilopoda), animal opened along dorsal side: 1—tergal region; 2—pleural region; 3—sternal region; 4—dorsal longitudinal muscles. Muscles of ventral side of body: 5—dorso-ventral; 6—ventro-pleural; 7—dorso-pleural; 8—longitudinal (after Bücherl). C—pleural musculature of stigmatiferous segment of juvenile *Scolopendra* (right half, from exterior): 1—tergite; 2—sternite; 3—place of attachment of limb; 4—spiracle (after E. G. Becker). D—crawling *Scolopendra* (after E. G. Becker). E—*Symphylella vulgaris*, pleural musculature of central part of body, view from right side: 1—still-independent pleural (dorso-ventral) muscles; 2—limb muscles; 3—paranotal projections of tergites; 4-8—limb segments: 4—coxa; 5—trochanter; 6—femur; 7—tibia; 8—tarsus (after E. G. Becker).

In this way, while a specially-primitive feature of the musculature of *Triops* is the almost-continuous and uniform layer of longitudinal muscles, a specially-primitive feature of the musculature of Chilopoda Epimorpha is (besides that) possession of slightly-differentiated annular musculature in the pleural region of the trunk. Another primitive feature of Chilopoda is the large amount of both skeletal parts and muscle bundles contained in the motor apparatus of each trunk segment, with relatively homonomous metamerism of the trunk. The trunk of Notostraca, with its almost-unsclerotised integument, is still more primitive in that respect.

The trunk musculature of Chilopoda Epimorpha takes part, together with the limb musculature, in the locomotion of the animal. The crawling of these myriapods (Fig. 144, *D*) is accompanied by serpentine flexures of the trunk; the limbs move in harmony with these flexures, as do the parapodia of errant polychaetes in a similar method of locomotion. The limbs of *Scolopendra* are very imperfect as walking legs: the external skeleton of their basal segment (coxa) consists of several sclerites not fused into a continuous cuticular ring, and the trochanter has the same structure; the articulation of these segments is equally imperfect. Crawling on dry land by means of imperfect limbs calls for great expenditure of muscular effort and leads to extreme development of the trunk musculature (polymyaria). Scolopendromorpha have not yet emerged from the critical stage of the locomotor apparatus that arose with migration to dry land, of which L. A. Zenkevich (1944) speaks with regard to the first terrestrial invertebrates. Thus Scolopendromorpha are, as it were, the reptiles among Atelocerata.

We have noted above (Vol. 1, Chap. VII) the rapid progress made by the locomotor apparatus within the class Chilopoda, concluding with the appearance of the highly-developed locomotor organs of Scutigeromorpha. The latter have advanced from crawling to rapid running, and their trunk musculature has lost its locomotor application and become much reduced. The same trend in development appears in Symphyla (Fig. 144, *E*), in which the muscles remaining in the pleural region of the trunk are mainly those that move the limbs (E. Becker, 1950).

In higher insects, because of concentration of the locomotor function in the thorax, there has been considerable simplification of the skeleton and the musculature in the abdomen, while those in the thorax have become much more complex. Because of decrease in the number of limbs and general shortening of the trunk, the methods of locomotion have changed: the role of the limbs and their musculature in crawling and walking has increased, while the role of the trunk musculature has diminished. With the development of flight the dorsal longitudinal and dorso-ventral muscles of the thorax of most winged insects have assumed the function of principal flight muscles (indirect wing muscles; Fig. 145, *B*, *C*), and the pleural musculature of the thorax has assumed the functions of flight control, altering the angle of the wings, folding and straightening the wings, etc.

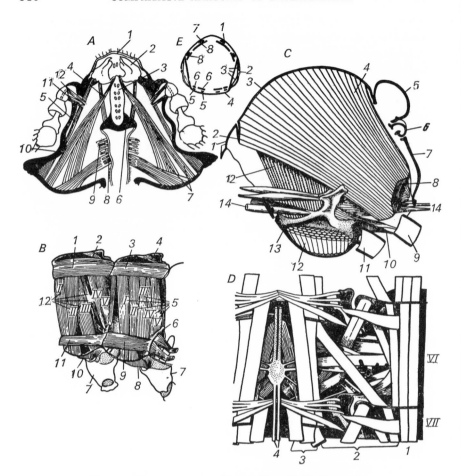

Fig. 145. *Motor apparatus of insects* (from B. N. Shvanvich).

A—diagram of frontal section of head of the hair-eating *Gyropus ovalis* (Anoplura Mallophaga): 1—upper lip (labrum); 2—mandibles; 3—pharynx; abductor muscles of mandibles: 4—short, 5—long; 6—sclerite of pharynx; 7—adductor muscles of mandibles; 8—sphincter of gullet; 9—gullet dilators; 10—antennal fossa of head capsule; 11—adductor muscle of antenna; 12—abductor muscle of antenna. *B*—diagram of musculature of mesothorax and metathorax of the alder fly *Sialis* (Neuroptera), an example of primitive structure of locomotor musculature in Holometabola: first (1) and second (2) longitudinal muscles of mesothorax; 3 and 4—the same of the metathorax; 5—dorso-ventral muscles (indirect wing muscles) of the metathorax; 6—furcal processes; 7—coxae; 8—sternal crest of metathorax; 9—ventral longitudinal muscles of metathorax; 10—sternal crest of mesothorax; 11—ventral longitudinal muscles of mesothorax; 12—dorso-ventral muscles of mesothorax. The Roman numerals within the drawing denote the numbering of the dorso-ventral muscles. *C*—thoracic musculature of the bee *Apis mellifera* (Hymenoptera), an example of highly-developed locomotor musculature in Holometabola; simultaneously with integration of the thorax and inclusion of one abdominal segment in it, differentiation and oligomerisation of the flight musculature has taken place; as a result of loss of independence by the rear wings, the flight musculature of the metathorax has disappeared: 1—tergite of pronotum; 2—anterior phragma; 3—scutum (anterior part of tergite of mesothorax); 4—first longitudinal muscle of mesothorax;

[*continued opposite*

The wing is attached to both the tergite and the sternite; therefore if the tergite moves dorsally, leaving the upper edge of the pleurite, the end of the wing descends. The tergum is arched, with the longitudinal muscles spanning the arch and making the tergum still more humped when they contract, raising the tergal point of attachment of the wing and lowering the wing. The dorso-ventral muscles flatten the arch of the tergum, lower the tergal point of attachment of the wing, and so raise the wing. The arm of the lever is very short, and therefore the wing can move very rapidly. At the same time the short lever arm requires a very small amount of linear shortening of the muscles, which greatly facilitates rapid alternation of contraction and relaxation. While the indirect wing muscles alternately raise and lower the wing, the direct wing muscles cause it to move forward and backward and rotate it around its longitudinal axis. There are usually four pairs of direct muscles, attached at one end to different parts of the wing base and at the other to different sclerites of the sternal and pleural regions. In Odonata, unlike all other insects, the direct wing muscles play a major role in wing movement.

An important part in the motor apparatus of the head and thorax of insects in played by the internal skeleton, formed by projections of the cuticle that serve for muscle attachment and to increase the strength of the skeleton as a whole (Fig. 145, B, C).

The musculature and the internal skeleton of the head of insects are extremely complex and diversified, particularly because of the complexity and diversity of the oral apparatus in different groups of insects: Fig. 145, A represents a comparatively simple case.

We may assume that in the prototype of the arthropods the musculature consists of four muscle systems: (i) a continuous layer of annular, longitudinal, and perhaps spiral body-wall muscles; (ii) a system of transverse and dorso-ventral muscles, homologous with the dissepiment muscles in annelids; (iii) limb muscles; and (iv) muscles of the internal organs.

The evolution of the locomotor apparatus consists mainly in: (i) breakdown of the continuous layers of cutaneous musculature into separate bundles, a trend intensified particularly as a result of the appearance of a number of sclerites in the cutaneous skeleton and of the development of an internal skeleton from projections of the sclerites into the body (Fig.

5—scutellum (central part of tergite of mesothorax); 6—postnotum (posterior part of tergite of mesothorax); 7—tergite of abdominal segment I, included in thorax; 8—phragma; 9—coxae of third legs; 10—sternite of metathorax; 11—coxae of second legs; 12—dorso-ventral muscle of mesothorax; 13—sternite of mesothorax; 14—ventral longitudinal muscle. D—musculature of abdominal segment of *Heterojapyx gallardi* (Campodeoidea), an example of primary complexity of that musculature, which resembles the musculature of Chilopoda; the animal is opened along the central line of the back: 1—dorsal longitudinal muscles; 2—musculature of pleural region; 3—ventral longitudinal muscles; 4—ventral nerve chain; VI and VII—tergites cut across. E—diagram of cross-section through an abdominal segment in higher insects; the musculature is much simplified as a result of loss of the abdominal locomotor function: 1—tergite; 2—stigma; 3—dorso-ventral muscles; 4—sternite; 5 and 6—ventral longitudinal muscles, external and internal; 7 and 8—dorsal longitudinal muscles, external and internal.

146); (*ii*) increase in the role of the limb musculature and in the complexity of its structure; (*iii*) heteronomous reconstruction of the entire musculature and skeleton as a result of increasing specialisation of the segments and tagmatisation of the body. All of these processes developed independently in the three main stems of arthropods—Crustacea, Atelocerata, and Chelicerata. On the whole, the locomotor apparatus of the higher arthropods is the most complex and perfect among all invertebrates, in correspondence

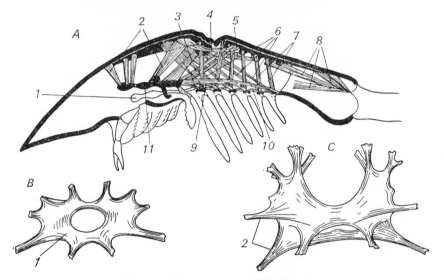

Fig. 146. *Motor apparatus of Chelicerata.*

A—trunk musculature and internal skeleton of Xiphosura (*Limulus polyphemus*): 1—central section of neural apparatus; 2—muscles running from endosternite to presomatic shield; 3—flexors of mesosoma; 4—intertergal muscle; 5—muscles running from endapophyses of prosoma to those of mesosoma; 6—extensors of mesosoma; 7—dorso-ventral muscles serving the mesosomatic limbs; 8—various groups of muscles moving the metasomatic spine; 9—ventral longitudinal muscles; 10—intersternal muscles; 11—endosternite (from Grassé). Endosternite of scorpion (*Buthus eupeus*): *B*—rear view, *C*—view from above: 1—subneural cap; 2—lamellar processes (after A. V. Ivanov).

with the extreme diversity of their activities. Besides engaging in search for and consumption of food, migration, and sometimes very complex copulatory behaviour, many arthropods display care for their young and building activities; the highest level of complexity in their behaviour is found in social Hymenoptera and termites. Because of that diversity of activities, the locomotor apparatus of higher arthropods is characterised by differentiation of function and consequent differentiation of the apparatus itself into individualised sections: organs for locomotion, grasping, mastication, swallowing, etc. Often the locomotor organs also break down into independent complexes, used in different methods of locomotion: crawling and flying in insects, walking and swimming by abdominal flexures in Astacura (Crustacea Decapoda), etc. To a certain extent similar differentiation of the motor apparatus, with retention of its internal unity, is found in other invertebrates, but arthropods attain higher levels in this respect, surpassing even Cephalopoda.

The motor apparatus of most Crustacea and also of some aquatic insects has a hydrokinetic as well as a locomotor function. Water currents are created by the activity of the limbs of Crustacea and serve to change the water around the gills or to bring food particles to the mouth or the capturing apparatus, or for both purposes together.

In primitive cases (Anostraca, Notostraca, and Cephalocarida) all the locomotor appendages bear both branchial epipodites and food-capturing endites, but these crustaceans are not filter-feeders in the true sense of the term (H. Cannon, 1928a; H. Sanders, 1957). In Conchostraca and Clado-cera the locomotor function has passed to the antennae II and the work of the thoracic limbs is reduced to creation of water currents, with the bivalvular shell fulfilling a current-directing role. The respiratory role is played by the epipodites of the limbs, while the crests of hairs on their endites act as filters, catching food particles (E. Naumann, 1921; O. Storch, 1924). Very many Copepoda Calanoidea also are filter-feeders, but in them the afferent water currents are created by the antennae II, as well as by the mandibular palps and the maxillae I; the filter is formed by the maxillae II; the thoracic limbs are used for rapid swimming, but the action of the filtering apparatus also produces slow forward motion (H. Cannon, 1928b).

Primitive Malacostraca (*Gnathophausia*, *Hemimysis* among Mysidacea, *Nyctiphanes* among Euphausiacea, *Anaspides* and *Paranaspides* among Syncarida) also are filter-feeders. Water currents are created by the maxillae II, and the hairs of their proximal endites form the filter. The maxillipedes (first thoracic limbs) bear combs that remove particles from the filter and pass them on to the maxillae I, which transfer them to the mouth. Gill appendages are developed on the thoracic limbs, and the water currents to the gills are created by the exopodites of the same limbs (H. Cannon and S. Manton, 1927b, 1929).

Higher members of all the superorders of Malacostraca have given up filter-feeding, and with it the hydrokinetic role of the maxillae II, but they have kept the water currents to the gills, usually creating them by the limbs on which the gills are seated. These are generally thoracic limbs, but in Isopoda and Stomatopoda they are abdominal limbs, which in these orders also take part in the work of current-creation.

In Crustacea the role of a current-directing adaptation is played by the bivalvular shell or carapace, which is formed by folds of skin that descend from the back over the sides of the body. That is a very ancient formation possessed by most of the primitive orders. We must conclude that its current-directing function is primary, and the protective function secondary. In fossorial forms such as Stomatopoda (R. Serène, 1951) the burrow also may fulfil a current-directing role.[1]

---

[1] In Stomatopoda the hydrokinetic function is fulfilled by the epipodites of thoracic limbs 2, 3, 4, and 5, and of abdominal limbs 1, 2, 3, 4, and 5 (R. Serène, 1951).

The hydrokinetic adaptations of insects also may be used to change the water around the gills (e.g. movement of gill leaflets of Ephemeroptera larvae, accumulation of water in hind-gut of Odonata larvae, etc.), and to bring food particles to the filtering apparatus (movement of brushes of upper lip of Culicidae larvae), or both purposes together (oscillation of some Chironomidae in their tubes).

9. CONTRACTILE-MOTOR APPARATUS OF LOWER DEUTEROSTOMIA

In spite of all the primitive features of lower Deuterostomia their motor apparatus shows much specialisation, corresponding to the mode of life of each group. Nevertheless it is easy to demonstrate that the motor apparatus of Hemichorda has the most primitive structure and approaches the prototype from which the motor apparatuses of other groups of lower Deuterostomia are derived.

The motor apparatus of Enteropneusta is adapted to their fossorial mode of life. All Enteropneusta live in U-shaped burrows, within which they can move rapidly back and forth. Their chief digging organs are the proboscis and the collar. In digging the proboscis is extended, enters the soil, and expands, and the body is drawn after it; the collar in turn becomes longer and thinner and pushes forward behind the proboscis, and then forcibly shortens, expands, and widens the burrow. The trunk does not participate actively in digging; but in a completed burrow or on the surface the animal travels mainly by serpentine and peristaltic trunk movements.

The proboscis (Fig. 147, *A*) possesses cutaneous musculature—annular and longitudinal—and also dorso-ventral fibres in a septum running from the stomochord to the ventral surface of the body. When the proboscis pores are closed and the cutaneous musculature is contracted, the proboscis becomes very rigid. Its coelom then plays the same mechanical role as the coelom of annelids does. The action of the musculature of the collar and the trunk is based on the same principle.

The collar (Fig. 147, *B*) contains, besides powerful pharyngeal musculature, a layer of longitudinal and some annular muscles in its lateral walls, radial fibres running from the body wall to the pharynx (in the anterior and posterior parts of the collar), and two systems of fibres lying in the front wall of the collar and diverging fanwise from the neck of the proboscis. In addition paired pharyngeal processes, running forward parallel to the notochord, contain strong longitudinal muscles, flexors and extensors of the proboscis.

While the fluid contained in the proboscis coelom, according to the experiments of W. Bateson (1886), is secreted by body tissues (by processes of the heart into the proboscis coelom—glomerulus) and the proboscis pore serves only to convey excess fluid to the exterior, the coelom of the collar is in communication with the surrounding water, which enters and leaves it through the collar pores, increasing or decreasing the volume of the collar—how clearly we see the homology between the collar coelom and the hydrocoel of echinoderms!

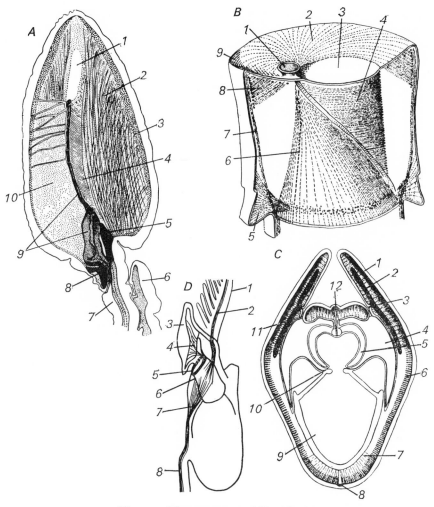

Fig. 147. *Motor apparatus of Hemichorda.*

A—medial section of proboscis of *Schisocardium brasiliense* (Enteropneusta):
1—proboscis cavity; 2—longitudinal muscles of proboscis; 3—annular muscles
of proboscis; 4—dorso-ventral muscles; 5—cardiopericardium; 6—collar; 7—
dorsal wall of pharynx; 8—skeleton of proboscis; 9—notochord; 10—ventral
mesentery of proboscis. *B*—diagram of musculature of collar of Enteropneusta:
1—base of proboscis; 2—parietal muscles of front wall of collar; 3—mouth;
4—annular muscles of pharynx; 5—posterior radial muscles; 6—longitudinal
pharyngeal muscles; 7—longitudinal parietal muscles; 8—anterior radial muscles;
9—annular parietal muscles. *C*—cross-section of trunk of *Ptychodera clavigera*
(Enteropneusta) in branchial region: 1—dorsal folds; 2—gonads; 3—genital
orifice; 4—gill pouch; 5—tongues of branchial arches; 6—longitudinal muscles of
body wall; 7—radial muscles, penetrating coelom and attached to gut; 8—ventral
nerve cord; 9—gullet; 10—internal and 11—external orifices of branchial sacs;
12—dorsal nerve cord (from Delage and Hérouard). *D*—diagram of arrangement
of musculature in *Rhabdopleura normanni* (Graptolithoidea), lateral view; 1—right
arm; 2—longitudinal muscle of arm; 3—cephalic shield; 4—its muscles (retractors); 5—pharyngeal musculature; 6—gullet; 7—ventral nerve cords (paired);
8—stalk muscles (paired) (from A. Shchepot'ev).

The trunk possesses musculature consisting of a weak external annular layer and a somewhat stronger internal longitudinal layer. On the ventral side numerous radial muscles run from the body wall to the gut, passing directly through the lumen of the coelom. There are similar fibres in the dorso-ventral folds developed in the branchial section of the trunk of *Ptychodera* (Fig. 147, *C*). Their role is probably the same as that of the radial muscles in the dissepiments of annelids, protecting the gut from being flattened by the pressure of the coelomic fluid created by the contraction of the musculo-cutaneous sac.

In accordance with the marked difference in the mode of life of the two classes of Hemichorda, the motor apparatus of Pterobranchia has little resemblance to that of Enteropneusta. At first glance the resemblance is limited to the coelomic-epithelial origin of the musculature.

Rhabdopleura (Fig. 147, *D*) has one pair of strong muscle bundles in the stalk, which reach the trunk and run along its ventral side, encircle the gullet, and pass into the collar. Three muscle systems originate from the dorsal side of the gullet: (*i*) retractors of the lophophore arms; (*ii*) pharyngeal muscles; (*iii*) the muscles of the shield, which are the strongest: they form two bundles running fanwise to the ventral wall of the shield (the sole). There are also muscle fibres in the wall of the pericardium, in the mesenteries, and in the walls of the collar (A. Shchepot'ev, 1907). All that musculature is used mainly for four groups of functions: (*i*) withdrawal of the animal into its tube, performed by contraction of the stalk; (*ii*) withdrawal of the lophophore arms and their re-extension, the latter achieved by raising the hydrostatic pressure in the collar cavity by contraction of the walls of the latter; (*iii*) crawling by means of the shield musculature; (*iv*) vegetative functions served by the musculature of the mouth, the gut, and the pericardium.

The musculature of *Cephalodiscus* is very similar in structure but more strongly developed. The stalk contains annular fibres that elongate it, as well as strong longitudinal muscles. In *Rhabdopleura* the stalk is apparently elongated by its elasticity, since annular muscles have not been described in that genus.

Eversion of the arms of Pterobranchia by the forcing of fluid into them from the collar coelom, which communicates by pores with the external environment, is again a feature that reminds one of the hydrocoel of echinoderms. A primitive feature possessed in common with Enteropneusta is the possession of musculature in all three sections of the body, which we no longer see in Chordata.

The locomotor apparatus of *Amphioxus* (Acrania) consists of longitudinal musculature, skeletal parts, and fins. The longitudinal musculature consists of two large masses of longitudinal muscle fibres formed from the epithelium of the myotomes (i.e. metameric dorsal parts of the trunk coelom) and lying along the sides of the animal's axial organs—the notochord and the neural tube. The longitudinal muscles are divided by boundary layers (myocommata) into a number of sections (myomeres).

The various species of Acrania have from 50 to 85 pairs. It is characteristic that the myomeres of the right and left sides lie alternately, and not opposite to each other.[1]

Since the myocommata are not simple transverse septa but have an infundibular form, the myomeres also have the form of semi-cones inserted into one another, their tips pointing forward. The skeleton of *Amphioxus* is formed of a resilient notochord surrounded by a mass of gelatinous-fibrous matrix, almost devoid of cellular elements; the same matrix underlies the skin and the epithelium of the body cavity, and penetrates and envelops the musculature in the form of a system of myocommata. The fins of *Amphioxus* consist of an unpaired fold beginning on the medial line in front of the mouth, curving round the rostrum, running along the back and round the tail, and continuing along the ventral side as far as the orifice of the peribranchial cavity; from that point paired folds (metapleures) run forward along the sides of the ventral surface.

Because of the above-described structure of its motor apparatus *Amphioxus* can swim rapidly, flexing its body like a fish to right and left; it can also burrow into the sand: that is possible because of the rigidity of the front end of the body, which is due to the fact that the notochord extends to the extreme tip of it.

Besides locomotor musculature, *Amphioxus* has a fairly strong layer of transverse muscle fibres on the ventral side of the body between the metapleures: these muscles are used to compress the peribranchial cavity. Finally, it has well-developed musculature in the walls of the buccal cavity and sphincters around the internal oral orifice (in the wall of the velum) and the anus. The entire longitudinal musculature of *Amphioxus* corresponds only to the longitudinal musculature of the trunk of Enteropneusta. Nothing remains of the proboscis musculature in the organisation of *Amphioxus*, and all that is left of the collar musculature is the musculature of the buccal cavity. These facts are in accordance with the complete reduction of the first two pairs of coelomic sacs and their products in *Amphioxus*. We have already discussed the metamerism of the longitudinal musculature of the trunk of *Amphioxus* in Vol. 1, Chap. IX.

In Appendicularia (see Vol. 1, Fig. 177) and larval Ascidiae (see Vol. 1, Fig. 176), as a result of the marked tagmatisation of the body, the musculature (like the axial organs—notochord, neural cord or neural tube) persists only in the tail. An exception is found in the large appendicularian *Megalocerca*, in which annular fibres have been described in the trunk walls. The caudal musculature of Appendicularia is completely homologous with the longitudinal musculature of the tail of *Amphioxus*, but because

---

[1] This arrangement is an example of alternating symmetry based on the presence of a plane of sliding reflection (see Vol. 1, Chap. VI), but in the organisation of *Amphioxus* only the longitudinal musculature and the points of origin of the spinal nerves are subject to that type of symmetry: in other words, the sliding-reflection symmetry of *Amphioxus*, unlike that of *Tatria* (Vol. 1, Fig. 93, *G*), is very incomplete.

of the animals' small size and their tendency towards constancy in the number of cellular elements it is much simplified. Each of the two symmetrical muscle bands is formed of ten metamerically-arranged muscle cells. Thus the caudal musculature of Appendicularia, with all its simplicity, is metameric. We may point out that the methods whereby myomerism has arisen in the development of *Amphioxus* and in that of Appendicularia differ widely, which is explained by the determinate cleavage and the small number of cells in the body of Appendicularia; nevertheless it is possible that the myomerism of the tail of Appendicularia is a modification of the metamerism of Acrania, just as the cleavage and organ-formation in Tunicata are modifications of the type of these processes seen in the development of Acrania.

We have seen that the trunk of Appendicularia is almost or quite devoid of musculature; all the musculature is concentrated in the tail. In contrast, fairly strong musculature develops in the trunk of Salpae and adult Ascidiae, corresponding to the mode of life of each of these groups.

The body walls of Ascidiae contain an external layer of longitudinal and an internal layer of annular muscles, and also sphincters in both siphons. The arrangement of the musculature varies in different genera (Fig. 148, C, D), sometimes taking the form of a musculo-cutaneous sac. In that case it is clearly a new formation, arising within the group concerned. Here once again we see how easily such a muscle arrangement can arise (as V. L. Vagin, 1947, points out) and how unimportant it is as a characteristic of large systematic groups.

Salps have a peculiar method of jet propulsion. Closing the mouth by means of a sphincter, they eject water forcibly through the cloacal orifice, that being possible because of a system of muscles encircling the body like a hoop and contracting in regular succession from front to rear. Then the muscles relax, the mouth opens, and the elasticity of the tunic causes the body to resume its former shape and volume, water entering the mouth and refilling the pharyngeal and cloacal cavities. The form and arrangement of the muscles differ in salps (Desmomyaria) and in Doliolida (Cyclomyaria) (Fig. 148, A, B). The locomotor musculature of salps and Doliolida is homologous with the musculature of the siphons and perhaps with that of the inter-siphon areas of Ascidiae (J. Godeaux, 1957–58).

The motor apparatus of echinoderms shows great diversity and, for all its low efficiency, sometimes remarkable complexity. Its original form is the much-reduced motor apparatus of Cystidea (e.g. *Aristocystis;* Vol. 1, Fig. 185, A), sessile animals with an immovable shell and without any movable appendages, except perhaps a circumoral wreath of ambulacral tentacles. As we have seen in Vol. 1, Chap. X, most echinoderms have reacquired some measure of mobility. In this they have taken two different paths: first, conversion of the immovable shell into a movable articulated skeleton, the plates of which serve for attachment of the muscles moving them; and secondly by improving the ambulacral apparatus. Often both of these principles are applied simultaneously.

The first principle, i.e. conversion of the shell plates into movably-articulated skeletal parts, is particularly important when the body of an echinoderm is extended into long arms (in Crinoidea) or rays (in Asteroidea and Ophiuroidea), or forms a segmented (and often also branched) stalk

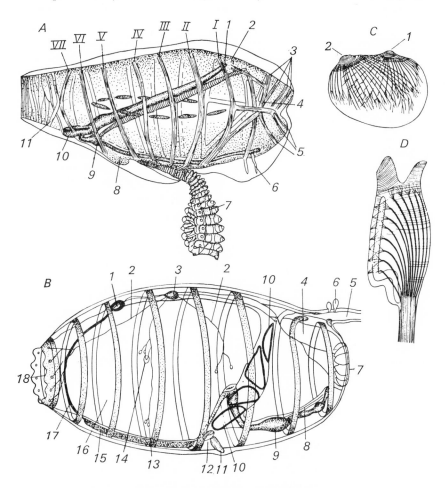

Fig. 148. *Motor apparatus of Tunicata.*

A—Cyclosalpa pinnata (Thaliacea Desmomyaria), oozooid, view from right side: I–VII—trunk muscles; 1—posterior duct; 2—cerebral ganglion; 3—muscles of upper lip; 4—anterior arcuate muscle; 5—muscles of lower lip; 6—processes of epidermis in tunic; 7—stolon with chain of blastozooids; 8—pericardium; 9—beginning of gullet; 10—blind outgrowths of stomach; 11—muscles of cloaca (after Metcalf). *B—Doliolum rarum* (Cyclomyaria), oozooid, view from left side; 1—ciliated fossa (neural gland); 2—peripheral nerves; 3—cerebral ganglion; 4—cloaca; 5—dorsal appendage, on which buds (6) are seated; 7—cloacal orifice; 8—pyloric gland; 9—gut; 10—gill-slits; 11—stolon; 12—heart; 13—sensory nerve-endings; 14—statocyst; 15—endostyle; 16—pharynx; 17—ciliated arch; 18—oral orifice (after Neumann). *C—Pyura antilarum* (Ascidiae), showing musculature of siphons and of body wall: 1—oral and 2—cloacal siphons (after van Name). *D—Clavellina sigillaria* (Ascidiae), musculature of body (after Michaelson).

(in many Pelmatozoa of all classes). Movement by means of the stalk, converted into a caudal appendage, is especially characteristic of Carpoidea.

Crinoidea move exclusively by means of their arms and cirri, and Ophiuroidea by means of their rays. In Ophiuroidea of the order Euryalae the rays are branched, the articulations between their 'vertebrae' are highly developed, and the rays are converted into a very efficient grasping (capturing) apparatus; in Euryalae and the rest of Ophiuroidea the rays are at the same time comparatively efficient locomotor organs.

The appendages of the shell, which possess a skeleton, musculature, and excellent articulations, are very mobile: they are spines and pedicellariae in Asteroidea and Echinoidea, and spines in Ophiuroidea (see Chap. II). The spines of Echinoidea take an active part in the crawling of these animals. In soft Echinoidea of the family Echinothuridae (referred by T. Mortensen, 1935, to the order Lepidocentroidea, but usually placed in the order Diadematoidea) the plates of the shell are movably articulated with one another, covering the body, which is sacciform, as in other Echinoidea; they travel by means of tube-feet, but the articulation of the shell gives the body great plasticity. Probably at one time Holothurioidea followed the same path, their shell being reduced to a circumoral skeletal ring consisting of radial and inter-radial plates, mostly flexibly joined together by connective tissue. That ring supports the distal parts of the radial ambulacral canals and the ampullae of the oral tentacles. Five longitudinal radial muscles are also attached to it and are attached at their other ends to the rear end of the body, around the anus. These are the main retractors of the body. Apart from the ring, only small sclerites scattered through the skin remain. The muscles of the shell are combined into a general musculo-cutaneous sac, functionally identical with that of 'Gephyrea'. Most holothurians still crawl by means of tube-feet, but Synaptida have lost these (except the circumoral tentacles) and have almost lost the radial ambulacral canals (vestiges remain: see D. Lastochkin, 1914, 1915). Consequently Synaptida have changed almost entirely to movement within the soil by the hydraulic method, and only small and young forms of them can crawl 'on their heads', using their circumoral tentacles.

In Asteroidea, where the shell is reduced on the abactinal (aboral) side of the body, longitudinal muscles develop in the skin and take part in the bending of the rays.

The ambulacral system possessed by all echinoderms is a system of canals formed from the left hydrocoel, which sends branches into the externally-projecting muscular tentacles and tube-feet. The tentacles are evidently primary, arising directly from the ambulacral ring (in the typical form, in holothurians); intensification of the function of that apparatus has been achieved by the outgrowth from the ring of radial canals bearing a large number of tube-feet. The tentacles and tube-feet often have grasping, tactile, respiratory, and in some classes (Asteroidea, Echinoidea, Holothurioidea, and probably Ophiocystia) also locomotor functions.

The action of the ambulacral system is based on the same principle of contraction of a muscular sac around contained fluid that we have already seen in Actiniaria, Annelides, etc. The musculature of a separate tube-foot in a starfish (e.g. *Asterias*) is very similar in its arrangement to that of the stem and basal disc of a sea-anemone: the mechanism of extension, contraction, attachment to the substrate, etc., is the same in both. In echinoderms, however, we encounter a different application of that principle: the muscular sac there envelops not the whole body but only the system of coelomic canals that forms part of the motor apparatus. As a mechanical system we cannot compare a sea-anemone to a whole starfish but only to a single tube-foot, and partly to the entire ambulacral system as a functional whole—a branched muscular sac filled with fluid. Such separation of a locomotor section of the coelom, whose fluid is compressed not by cutaneous musculature but only by the musculature of its own walls, ensures great economy of energy. Moreover, with that structure all the rest of the body is released from incessant massage by the musculo-cutaneous sac, which also is a definite advantage.

In the ancestors of echinoderms the ambulacral system evidently served originally only to operate the tentacles, which had a grasping or hydro-kinetic (not a locomotor) role, as we see even now in modern Grapto-lithoidea (Pterobranchia). By the time that the sacciform ancestors of Eleutherozoa had made the transition to an active mode of life the ambulacral apparatus was probably their only remaining motor apparatus; it was natural to use it for locomotion, as a result of which it underwent considerable development. At first that was a satisfactory solution to the problem of transition from a sessile to an active mode of life. There were not, however, possibilities for much improvement in the ambulacral apparatus: it does not provide for speed of movement or manoeuverability. In many echinoderms evolution followed the path of replacement of it by other locomotor adaptations (the arms of Ophiuroidea, the musculo-cutaneous sac of Synaptida; partly the spines of Echinoidea), but these adaptations are still not very efficient. Consequently no echinoderm has yet overcome the effects of its ancestors' attached mode of life and advanced along the highway of morphological progress.

10. Concluding Remarks

To summarise briefly, we see that the motor apparatus of lower Metazoa had three basic independent functions: (*i*) concealment; (*ii*) grasping and swallowing; (*iii*) spatial locomotion and movement of water. Later a number of more specialised functions, which we shall not dwell on here, were added.

The function of concealment is fulfilled by the following methods.

1. General contraction of the longitudinal musculature of the body, producing either reduction of the body surface area (which often is a definite protection from enemies in unfavourable conditions) or with-drawal of the body or of its vulnerable parts into some kind of protective

structure. That is the simplest form of protective reaction, possessed by polyps, turbellarians, and other lower animals.

2. Closing of sphincters over vitally-important parts of the body. In this way Actiniaria protect the oral disc (with simultaneous contraction of the longitudinal musculature); some ctenophores, the mouth and aboral organ; holothurians, the tentacular apparatus; ascidians, the external body orifices, and so on.

3. Withdrawal of the front part of the body by means of special retractors. This method of protection is a refinement of (1). Bryozoa withdraw polypids into cystids in this way; the zooids of *Rhabdopleura* retreat into their thecae by contraction of the stalk retractor; testaceous Rotatoria withdraw the head end into the shell; molluscs withdraw into their shell by means of the columellar muscle; Priapuloidea, Sipunculoidea, and Holothurioidea withdraw the front end of the body, and the last two groups withdraw the tentacles also; and so on. Front-end retractors and protective contraction of them occur also in mobile, shell-less forms, such as some turbellarians.

4. Retreat into a tube. Many animals living in tubes, within which they may move about, disappear into the tubes on the slightest warning: here the concealment function is fulfilled by locomotor organs, sometimes much specialised (Serpulidae and other polychaetes, *Phoronis*, *Lingula* among Brachiopoda, etc.).

5. Closing of a bivalvular shell by means of adductor muscles. This is observed in Lamellibranchia, Brachiopoda, and some Crustacea (Ostracoda, Conchostraca, etc.).

6. Rolling up on the ventral side, in forms with jointed dorsal armour: Loricata, trilobites, some woodlice (among Oniscoidea), *Glomeris* (Diplopoda), etc.

7. Erection of spines or bristles, e.g. in Amphinomidae (Nereimorpha).

8. Autotomy of a part of the body seized by an enemy, performed by powerful muscle contraction, and a number of other, still more specialised methods of protection possessed mainly by the higher invertebrates, Articulata and Mollusca.

Grasping was originally exclusively applied to food; as developed later, it can take the form of active defence (grasping an enemy) or play a part in copulation (grasping the sexual partner), in construction, or in other activities directed towards the environment.

The function of grasping (mostly grasping prey) is fulfilled in various ways:

1. Grasping and swallowing by the mouth alone, without accessory adaptations, take place in tentacle-less hydroids (*Protohydra*), in tentacle-less ctenophores (e.g. in Beroidea), in primitive turbellarians without a complex pharynx (e.g. *Xenoturbella*), in Enteropneusta, and in other forms.

2. Grasping by specialised oral and pharyngeal apparatuses. These are characteristic of most turbellarians and Nemathelminthes. We must particularly mention the complex pharyngeal equipment of Rotatoria.

Errant polychaetes and many Gastropoda (mainly Prosobranchia) have a complex pharynx, capable of eversion and often bearing some kind of armament, used for seizing food; Asteroidea have an eversible gut; the most complex masticatory apparatus is the Aristotle's lantern of Echinoidea. The pharynx of Cephalopoda, like that of many Gastropoda, is armed with strong jaws as well as a radula.

3. The front end of the body is often converted into a grasping organ even if the oral orifice is not located on it. It is then often furnished with adhesive glands and musculature enabling it to be considerably extended and rapidly withdrawn (many turbellarians); sometimes glandular pads or true suckers develop on it, as in many Triclada. In Echiuroidea the well-developed pre-oral lobe often serves as an organ for collecting detritus from the soil surface; the large *Bonellia* scavenges the sea bottom with its 'proboscis' for a full metre around its burrow, which it does not leave.

4. Intricately-differentiated grasping proboscides develop, as we have seen, on the front end of the body of many turbellarians; in nemertines these adaptations attain great size and complexity.

5. Grasping tentacles are extremely characteristic of coelenterates. The tentacles of Bilateria usually have a hydrokinetic function and are not used for grasping (Bryozoa, *Phoronis*, Pterobranchia, Sabellidae, and many others). Among organs far removed from, but, perhaps, still homologous with the tentacles of ctenophores, the ambulacral appendages of echinoderms have partly retained a grasping function, particularly the oral tentacles of many holothurians, and also the tube-feet of Asteroidea and Crinoidea.

6. The limbs of arthropods. As we have seen, grasping is one of the primary functions of the primitive arthropod limb, where it is fulfilled by the basal endites; it may be fulfilled secondarily by the distal parts of the limb. In different arthropods the grasping function of the limbs is put to the most varied uses and often produces complex adaptations.

7. The foot of molluscs and its various products are to some extent also used for grasping; in this respect the grasping function reaches its highest development in Cephalopoda.

8. The pedicellariae of Asteroidea and Echinoidea represent one of the most peculiar types of grasping organs, since they may be scattered over the whole body and are used to seize live animals and inanimate particles arriving on its surface. The avicularia of Bryozoa Cheilostomata are similar pincers scattered over the whole body, but morphologically they are zooids of a colony, not organs of a metazoan individual.

9. The arms of Ophiuroidea and particularly the dichotomously branched 'rays' of the predatory ophiuroids belonging to the order Euryalae (*Gorgonocephalus*, etc.) constitute a very peculiar organ of prehension. The rays of Ophiuroidea are, in fact, merely projections of the edge of the animals' body extending in five radial directions; but after prolonged evolution they have been converted into highly-perfected grasping organs in Euryalae.

Finally, the locomotor and hydrokinetic functions of the motor apparatus may be fulfilled in many different ways. I have based a classification of methods of locomotion by Metazoa on a scheme proposed by L. A. Zenkevich (1944), with some modifications. The following are the principal methods of locomotion:

1. Locomotion by means of flagella and cilia: found in many larvae, and among adults in ctenophores, Rotatoria, Gastrotricha, small turbellarians and nemertines, and certain annelids and molluscs. Cilia may also play an accessory role in peristaltic crawling.

2. Locomotion by means of alteration of body shape.

(a) Crawling by means of peristaltic changes in body shape or by peristaltic waves passing along the ventral surface of the body (the 'sole'). Crawling on the sole or basal disc occurs in Actiniaria, terrestrial Planariae, Loricata, Gastropoda, and Pterobranchia. Crawling by means of peristalsis of the whole body is found in nemertines, annelids, and some other groups. In nemertines it occurs in a pure form; in oligochaetes and some errant polychaetes it is improved by use of setae that take hold of the substrate and push the animal, this representing a combination of two methods of locomotion, by peristalsis and by means of appendages.

(b) A number of polychaetes, Echiuroidea, Sipunculoidea, Priapuloidea, and other fossorial forms move by rhythmic flow of coelomic fluid from one end of the body to the other (the hydraulic method of burrowing). Other forms of this hydraulic method of travelling in the soil are found in the solitary hydroid *Protohydra* (J. Omer-Cooper, 1957), some fossorial molluscs, Enteropneusta, etc.

(c) Swimming or crawling by means of undulatory flexures of the body. This is found in Cestidea (Ctenophora), many turbellarians and nemertines, all nematodes, leeches, Chaetognatha, Acrania, and many other groups. In polychaetes it is complicated by participation of the parapodia.

(d) Swimming by means of undulatory flexures of the tail: cercariae, Appendicularia, larval Ascidiae, amphibian tadpoles, etc.

(e) Caterpillar-like progression by alternate attachment of the front and rear ends of the body to the substrate. Occurs somewhat rarely in different groups: Hydrida, Bdelloidea (Rotatoria), leeches, some polychaetes, and insect larvae. The stepping and pressing method of locomotion of some Gastropoda (*Aporrhais, Struthiolaria, Pteroceras* among Taenioglossa, etc.) has a certain resemblance to this method.

(f) Reactive (jet) propulsion, achieved by ejecting water from some cavity of the body by abrupt muscle contraction. Occurs in medusae, Cephalopoda, some Opisthobranchia (*Notarchus*), Salpae, Odonata larvae, etc.

3. Locomotion by means of body appendages.

(a) Swimming by means of undulatory movements of longitudinal fins (without body flexures): Heteropoda (Taenioglossa), many Opisthobranchia, Cephalopoda.

(b) Swimming or flight by means of rowing strokes of paired appendages, not limbs. This occurs in very diverse forms: Ctenophora Lobifera; *Pecten*, *Lima*, and other Lamellibranchia, which swim by flapping the shell; winged insects; etc.

(*c*) Rowing or walking by means of paired limb-type appendages. Such appendages include the parapodia of polychaetes; the limbs of Tardigrada, Onychophora, and Arthropoda; the movable paired spines of some Rotatoria (*Hexarthra*, etc.); the tube-feet of Holothurioidea, Asteroidea, and Echinoidea; and the walking appendages of *Monobryozoon ambulans*. In primitive cases locomotion by means of limbs may be combined with crawling or swimming by means of serpentine body flexures (polychaetes, Myriapoda, etc.). In higher arthropods swimming, walking, and running by means of limbs become independent and reach high levels of diversification and perfection. The most perfect of all methods of locomotion by invertebrates, however, is insect flight.

We have not included all methods of locomotion in the above list, since an intelligible description of a number of peculiar and uncommon methods found in molluscs, echinoderms, and other groups would require too much space.

As stated above, the contractile-motor apparatus may also have a hydrokinetic function. In most cases the hydrokinetic action is a modification of the locomotor action. In order to create water currents, various animals employ those methods of locomotion that use water as a support; an immovably-attached or slow-moving animal sets water in motion by thrusting it away. That result is attained by undulatory movements of the whole body (*Nereis*, Tubificidae) or of the tail (Appendicularia); by peristaltic changes in body diameter within a narrow tube (*Arenicola* and several other polychaetes; *Labidoplax* among Synaptida); by using organs of reactive propulsion (respiratory movements of Cephalopoda, Odonata larvae, etc.); and by swimming movements of locomotor appendages: the thoracic parapodia ('palettes') of *Chaetopterus*, the limbs of Crustacea, etc.

Sometimes water currents are created by the use of appendages not functioning as locomotor organs but working on the same principle, e.g. the tracheal gills of Ephemeroptera or the labral lobes of mosquito larvae (*Anopheles*, etc.). Some hydrokinetic organs have no connection with locomotor organs, either in their origin or in their serial homologues. Such organs are, for instance, the water-pumping movable gills of Septibranchia and Nuculanidae (= Ledidae) among Protobranchia (see Chap. II). There are even opposite cases, where organs and movements that were originally hydrokinetic are used for locomotion: the swimming of *Pecten* or *Lima* by means of flapping of the shell valves originated in the flapping of these valves for a purely hydrokinetic purpose, the clearing of foreign particles from the mantle cavity (C. Yonge, 1936). We may point out that in this case there is replacement only of the secondary function of shell-closing, since the primary function, protection, remains unaltered.

In a number of cases the hydrokinetic function, when fulfilled by parts of the contractile-motor apparatus, is intensified by the presence of current-directing adaptations. These may be either parts of the animal's body (shells, folds of the shell of Crustacea), or structures (the burrows of *Nereis*, the tubes of *Chaetopterus* or Chironomidae, etc.), or something intermediate between a part of the body and a structure (the dwellings of Appendicularia, see Chap. II). The use of peristaltic changes in body diameter to create water currents is generally possible only when current-directing adaptations are in the form of tubes open at both ends.

Some Ophiuroidea (*Ophiothrix*, *Amphipholis*) produce respiratory movements with the aboral wall of the disc, making water flow into the bursae and out again. This type of respiratory movement is not a modification of the locomotor movements of the animal.

On land, the analogue of hydrokinetic movement is the respiratory movements of several insects, Arachnoidea, and terrestrial Gastropoda—also not modifications of locomotor movements. In birds and higher insects, however (e.g. *Anopheles*, according to O. N. Vinogradskaya, 1960), contraction of the locomotor (particularly the flight) muscles simultaneously compresses the air-sacs lying between them (in birds) or the thick stems of the thoracic tracheae (in Diptera). Thus the amount of work done by the flight muscles and the intensity of aeration of the respiratory apparatus are automatically made to correspond. The role of flight musculature in the aeration of insect tracheae was described long ago by H. Rathke (1842).

Chapter VII

ORIGIN AND DEVELOPMENT OF THE
EXCRETORY APPARATUS

1. PRELIMINARY REMARKS

Excretion is an aggregate of processes directed towards regulation of the composition of the internal environment of an organism by removing from it, or rendering harmless, injurious, superfluous, or unnecessary substances. It includes: (*i*) excretion in the strict sense of the word, i.e. removing or rendering harmless the waste products of the organism's own metabolism; (*ii*) regulation of osmotic pressure in the internal environment; (*iii*) regulation of the ionic composition of the body fluids; (*iv*) removal of unnecessary or harmful particles.

By excretion in the strict sense of the word we usually mean removal from the organism of solid and liquid products of metabolism; these do not include CO_2, which is regarded as an essential part of the respiratory process; but in the case of aquatic animals it is clear that that is an academic distinction, since they excrete CO_2 in a dissolved form, like other metabolic products (katabolites).

In the most primitive cases the terminal product of nitrogen exchange is ammonia. In aquatic invertebrates, even in such groups as annelids, Sipunculoidea, higher Crustacea, and Cephalopoda, more than 50% of the nitrogen is excreted in the form of ammonia, it being removed from the body by diffusion almost as easily as CO_2.

Besides ammonia, less poisonous end-products are also formed: urea, uric acid, guanine, etc. The power to synthesise these substances is particularly important in terrestrial forms (snails, woodlice, insects, arachnids) in view of the slower circulation of water through their bodies (H. Delaunay, 1934; E. Dresel and V. Moyle, 1950). Therefore the excretory function includes preliminary treatment of waste, and sometimes removal from the organism of several kinds of excreta. In higher animals these partial functions are usually distributed among different organs.

The need for osmotic regulation does not arise in most marine biotopes, with their relatively-stable salinity. Accordingly the internal environment of the lower phyla of marine invertebrates is not separated osmotically from the surrounding water, and in experimental conditions the concentration of their cavity fluid fluctuates with the concentration of the surrounding water, sometimes being somewhat hypertonic.

Protozoa that live in water with low or variable salinity (littoral areas, estuaries, continental waters) evacuate water absorbed by their bodies from the surrounding hypotonic medium by means of contractile vacuoles; they have the power of osmoregulation of the intracellular environment.

Some lower Metazoa without the power of osmoregulation of their body's internal environment also can live in brackish or fresh water, entirely through osmoregulation of their component cells. Choanocytes and some other cells of freshwater sponges, like freshwater Protozoa, have contractile vacuoles (S. Kent, 1880; M. Jepps, 1947). Special organs regulating the osmotic pressure of the internal environment of multicellular animals first appear in flatworms; the power of active maintenance of hypertonia of the internal environment in brackish or fresh water exists to some extent in higher turbellarians (Rhabdocoela, Triclada) and is well developed in some annelids, gastropods, and crustaceans. Substantial improvement of the power of osmoregulation, however, is found only in arthropods, chiefly in higher crustaceans and insects; fully homoiosmotic animals (i.e. those with relatively constant osmotic pressure in their internal environment, no matter what the external salinity may be) include *Artemia salina*, Mysidacea, euryhaline shrimps, and the larvae of a number of mosquitoes (A. Krogh, 1939; G. Belyaev, 1950, 1951).

Thus the power of osmoregulation of the internal environment depends on possession of osmoregulatory organs, and develops as a result of life in water of variable or low salinity. Osmoregulatory organs were originally quite independent of excretory organs, but in higher animals they are usually closely associated with them, forming a single apparatus, so that we shall discuss them both in a single chapter.

All animals can regulate the ionic composition of their internal environment; lower marine animals (e.g. *Aurelia*, ctenophores) maintain a specific ionic relation (different from that of sea-water) in their tissues. The highest degree of such regulation is found in cephalopods, higher crustaceans (J. Robertson, 1957) and Culicidae larvae (V. Wigglesworth, 1950). In these groups its mechanism is associated with entrance of ions into the blood through the gills and with their differential excretion by the kidneys.

The function of removal of foreign particles from the organism is closely associated with excretion, since grains of insoluble waste are often so removed together with micro-organisms that penetrate the internal environment and with inanimate particles of external origin. All of these particles are phagocytised by cells retaining the power to do so, and often they are ejected from the body through the skin, the gut, or the kidneys.

In primitive Metazoa excretion, in the strict sense, and regulation of osmotic pressure and ionic composition are performed not by special organs but by all or most of the cells in the organism. In many lower Metazoa and large Protozoa a major role in the removal of CO_2 and excreta is played by intracellular symbiotic algae (see Chap. II). The excretory function of coelenterates may be largely concentrated in the endoderm, which has assumed the tasks of removal of metabolic products from other tissues and their discharge to the exterior. The epithelium of the gut partly retains that role even in higher animals (see Chap. V). At the same time the peripheral phagocytoblast plays a substantial excretory role in all animals that possess it. There the excretory function is assumed, on the one hand, by separate cells and groups of cells of unpolarised phagocytoblast, and on the other hand by the coelomic epithelium and the walls of the coelomoducts. The kinetoblast takes part in excretion mainly through the formation of true ectodermal nephridia from it.

The more highly developed types of nephridia and excretory coelomoducts, together with excretory appendages of the gut (Malpighian tubules), are the most perfect excretory organs, with nephrocytes arranged in the form of regular epithelium and with efferent ducts: they have the general name of *emunctoria*.

While the excretory organs of lower Metazoa remove the products of excretion from tissue cells and therefore must have a diffuse or branched structure, the situation changes with the development of a distributive apparatus. A single internal environment is created, into which all body cells discharge their excreta, and the task of the excretory organs becomes removal of the excreta from the internal environment and also regulation of its salt balance. When an efficient distributive apparatus exists, the excretory apparatus often acquires a centralised structure, and in accordance with the general increased intensity of life processes its efficiency rises considerably.

2. Excretory Organs Without Efferent Ducts

A. O. Kovalevskii (1889–97) was the first to make extensive use of the vital-staining method to determine the excretory function of various organs in many groups of animals. Later the method was widely adopted and, with others, is still used.

It has been found, mainly with the vital-staining method, that in both coelenterates and lower turbellarians the excretory function has been largely assumed by the phagocytoblast. Placing the calcareous sponge *Grantia* in a suspension of carmine, A. T. Masterman (1894) observed amoebocytes loaded with carmine grains creeping out of the parenchyma. There are cells that concentrate carmine also in the mesogloea of *Alcyonium*. Most coelenterates have no anatomically-distinct excretory organs. The few exceptions have already been mentioned in discussion of the excretory function of the gut (Chap. V).

An excretory function has been ascribed to the intestinal pores of coelenterates, but that suggestion has insufficient physiological foundation. Morphologically these pores are probably, as we have seen, homologues of the coelomoducts that fulfil an excretory function in many animals.

In Turbellaria Acoela excretion is performed both by the digestive syncytium (central phagocytoblast) and by several cells of the peripheral parenchyma (nephrocytes), which accumulate excreta (E. Westblad, 1923). The excreta accumulated in these cells are converted into pigment and into crystals of uric acid, and remain there throughout life, helping to create the animal's colour and the pattern of its coloration. The excreta are rendered harmless, not by removal from the body, but by conversion into an insoluble form. They constitute the so-called 'storage kidneys'. Such storage kidneys are found in many animals, but as the sole excretory adaptation they usually occur only in short-lived animals. They occur in many worms, arthropods, echinoderms, etc. In insects the role of storage kidneys is played mainly by the fatty body, which also is a product of the peripheral phagocytoblast.

Among other orders of turbellarians even such large and complex forms as Triclada do not yet have anatomically-distinct excretory organs. Their gut has an excretory function (E. Westblad, 1923); it retains that function also in Rhabdocoela. The latter have in addition well-developed nephrocytes (parenchymatous cells with an excretory function) of two kinds: some form diffuse 'storage kidneys', and others fill up with grains of excreta, creep out through the skin or the gut, and perish with their contents. These cells accumulate waste matter taken from other cells and transport it out of the organism. Generally speaking, the first excretory adaptation to be expressed morphologically in Scolecida takes the form of separate excretory cells scattered through the peripheral phagocytoblast. Such cells are called *atrocytes*, *excretophores*, etc., but in fact they are scattered nephrocytes.

With further development of the excretory apparatus the nephrocytes mostly concentrate in definite parts of the body. In animals possessing unepithelised peripheral phagocytoblast, the coelom being absent or lost, loose aggregates of nephrocytes are formed (arthropods); when there is a coelom, separate parts of its epithelium or of the walls of the coelomoducts often assume an excretory character (annelids, molluscs). We shall discuss the excretory role of coelomoducts later; at present we shall dwell on nephrocytes not contained in coelomoducts or nephridia.

Among annelids the formation of supplementary nephrocytes from the coelomic epithelium is observed in some polychaetes and in oligochaetes. In polychaetes these cells coat the vessels, mainly the ventral vessel; in oligochaetes they coat the gut and some vessels. They are called *chloragen cells*. In leeches the nephrocytes form the *botryoidal tissue*, whose cells group around the branched coelomic canals.

In molluscs nephrocytes develop from the pericardial epithelium; the latter often forms outgrowths that increase the surface of excretory epithelium (the pericardial glands of gastropods, the Keber's organ of lamellibranchs, the appendages of the branchial hearts of cephalopods, Fig. 166). When these nephrocytes disintegrate, the contained excreta usually enter the coelom and are conveyed to the exterior by the nephridia or the coelomoducts. In lamellibranchs foreign particles (such as India ink) carried in the blood stream are phagocytised by wandering amoebocytes, which then creep into the mantle cavity or the pericardium and are carried away to the exterior. Among arthropods, a fairly substantial role in excretion is played in Chelicerata and Crustacea by nephrocytes scattered in the body cavity, which are able to concentrate ammoniacal carmine. The so-called pericardial glands, which lie beside the heart, have the same function in insects and myriapods. In addition, urates accumulate in the fatty body.

In echinoderms, which generally have almost no separate excretory organs, a major role in excretion is played by wandering amoebocytes, which load up with waste products and then leave the organism, creeping out through the surface of the skin, the gut, the respiratory tree (in holothurians), the bursae (in Ophiuroidea), etc. At the same time excreta

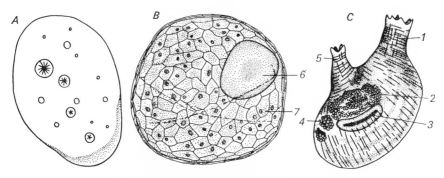

Fig. 149. *Storage kidneys of Ascidiae.*

A—blood cell of *Ascidia mentula. B*—one of the follicles of a storage kidney of *Ascidiella aspersa* (after Azema, from Prenant). *C*—*Molgula*, entire animal, freed from tunic (after Lacaze-Duthiers, from Prenant). 1—oral siphon; 2—genital gland; 3—kidney; 4—liver; 5—cloacal siphon; 6—concretion; 7—epithelium of renal follicle.

accumulate in many parts of the body in an insoluble form. Therefore echinoderms stand at the same level as coelenterates and sponges with regard to excretion, as V. A. Dogiel (1938) correctly remarks, and that character of their excretion corresponds fully to the general low level of their organisation. The level of the excretory organisation of tunicates is no higher. Many Ascidiae (e.g. *Ciona*) have a large number of nephrocytes that accumulate grains of excreta and resemble diffuse storage kidneys. Other Ascidiae possess perfected storage kidneys in the form of numerous small sacs or even one large sac (family Molgulidae), whose walls are composed of excretory epithelium and whose cavity contains concretions (Fig. 149) consisting chiefly of urates. The renal sac of Molgulidae always contains symbiotic fungi, which apparently feed on the accumulated excreta (H. Lacaze-Duthiers, 1874; A. Giard, 1888; P. Büchner, 1930, 1953). The storage kidney of Molgulidae is one of the most peculiar excretory organs in the animal kingdom.

3. PROTONEPHRIDIA

Protonephridia first appear in turbellarians: originally they were organs of osmoregulation, but they very soon became connected with excretory cells and involved in the excretory function (E. Westblad, 1923). Coelenterates have no nephridia, and homologues of them are unknown in that group; neither do nephridia occur in Acoela or Xenoturbellida. They rarely occur in Polyclada, being poorly developed and little studied there. They are found generally in all other orders of Turbellaria.

The protonephridia of most Triclada are anatomically the most primitive. They consist of eight slender longitudinal canals lying in the parenchyma.

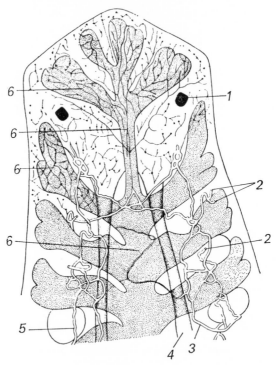

Fig. 150. Procerodes lobata (*Triclada Maricola*), *protonephridial canals of front end of body.*

1—eye; 2—external excretory orifice; 3—dorso-medial longitudinal canal of excretory apparatus; 4—ventral nerve stems; 5—dorso-lateral longitudinal canal of excretory apparatus; 6—gut (after Lang).

Four lie on the dorsal and four on the ventral side (Fig. 151, *A*). They are all joined by transverse commissures and give out numerous branches penetrating the whole body (Fig. 150). Numerous short efferent ducts run from the principal canals towards the skin, opening to the exterior by small pores. The protonephridia of Triclada have no excretory function. More advanced and smaller forms display architectonical simplification due to oligomerisation: the numbers of longitudinal canals, of transverse commissures, and of external orifices are all reduced (Fig. 151). This has already taken place in the dwarf triclad *Pentacoelum caspicum* (W. Beklemishev, 1954). Alloeocoela have four, not eight, canals, only two transverse commissures, and three or four pairs of external orifices. Rhabdocoela retain only two canals, with two (sometimes only one) external orifices; they have no transverse commissures.

Only one longitudinal canal remains in Notandropora (family Catenulidae). The canals are intricately branched and form a network of capillaries.

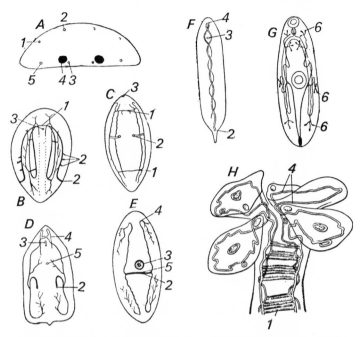

Fig. 151. *Diagrams of anatomical structure of protonephridial apparatus of flatworms. A—Euplanaria polychroa* (Triclada Paludicola), cross-section: 1, 2, 3, and 5—main excretory canals (dorso-lateral, dorso-medial, ventro-lateral, and ventro-medial); 4—ventral nerve stems (after Mikoletskii). *B—Baicalarctia gulo* (Alloeocoela) (after G. M. Fridman). *C—Prorhynchus stagnalis* (Alloeocoela) (from G. M. Fridman). *D—Phaenocora stagnalis* (Rhabdocoela) (after Fuhrmann). *E—Rhynchomesostoma rostratum* (Rhabdocoela) (after Luther). *F—Stenostomum tenuicauda* (Notandropora) (from G. M. Fridman). *G*—Digenea (after Fuhrmann). *H— Orygmatobothrium versatile* (Cestodes Tetraphyllidea), protonephridial canals of scolex (from Skryabin). 1—commissures between trunks of the two sides of the body; 2—excretory orifices; 3—mouth; 4—loops of main trunks in brain region; 5—genital orifice; 6—terminal cells (in Digenea—limited in number and constant in location).

In the architectonics of their protonephridia trematodes stand at the same level as Rhabdocoela, that is, they have one or two canals without transverse commissures, with one or two external orifices. Cestodes have a large number of orifices and transverse commissures between the trunks, so that many of them resemble Triclada in the organisational level of their protonephridia. In Typhloplanidae (Rhabdocoela) the external orifice of the protonephridia is united with the oral (Mesostomatini, Typhloplaninae) or the genital orifice (*Rhynchomesostoma*).

Histologically (Fig. 152) the simplest structure of nephridia has been described by E. Reisinger (1923) in *Protomonotresis centrophora* (Alloeo-coela Holocoela). The canal walls facing the lumen are evenly and con-tinuously covered with cilia. The canals therefore represent tubular involutions of the external ciliated epithelium of the animal. *Baicalarctia* (Holocoela) has the same canal structure, but the cilia are arranged in three longitudinal rows (G. Fridman, 1933). We find an extreme stage in reduction and differentiation of the ciliary apparatus of the proto-nephridia in some Rhabdocoela (Fig. 152, *B, C*): the canal walls have the same structure, but there are no cilia as such. Only here and there on

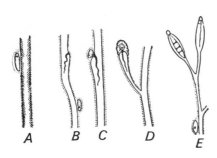

Fig. 152. *Water-moving adaptations in nephridia of turbellarians.*

A—excretory canal of *Protomonotresis* (Alloeocoela), evenly covered with cilia. *B*—flame cell in canal lumen, not con-nected with individual nucleus, in *Typhlo-planella halleziana* (Rhabdocoela). *C*—the same, but always attached to a particular nucleus, in *Protoplanella simplex* (Rhabdo-coela). *D*—unicellular terminal apparatus in *Phaenocora unipunctata* (Rhabdocoela). *E*—syncytial terminal apparatuses, not connected with particular nuclei, in *Mesostoma productum* (Rhabdocoela) (after Reisinger).

the thickened walls are there long bunches of cilia fused together, which lie in the canal lumen and drive water along it by their undulations. If we assume that such a bundle of cilia sinks into an evagination of the canal wall we obtain the typical terminal apparatus possessed by most proto-nephridia. We obtain a small ampulla, at the bottom of which lies a flame cell, i.e. a bunch of fused cilia projecting into the canal and constantly moving (Fig. 152, *D, E*). Digenea are characterised by a limited number and regular arrangement of the terminal apparatuses of the nephridia (cf. Fig. 131 for cercariae).

Such is the high degree of differentiation of protonephridia in flatworms. What is the function of this apparatus? Usually an excretory function is ascribed to it. But the original and principal function of protonephridia was not excretion but regulation of osmotic pressure in the animal's body (E. Westblad, 1923), and above all removal of the water that is being continually absorbed from the surrounding hypotonic solution (which fresh water would be for all animals). Therefore protonephridia are very poorly developed in marine as compared with freshwater turbellarians. In Triclada the protonephridia still have no connection with the excretory function. In some Rhabdocoela atrocytes lie along the main canals of the protonephridia, and discharge the excreta accumulated in them into the current of fluid that constantly flows along the canals. Thus excretory elements come into association with the protonephridia. Finally, in some Rhabdocoela (*Gyratrix, Acrorhynchus*) and in *Stenostomum* the epithelium in certain parts of the canals themselves acquires an independent excretory function. Only then do the protonephridia become excretory organs. In this way they have a dual function in higher turbellarians, not only osmoregulatory, but also excretory. The terminal apparatuses have no excretory function.

The question of the original, most primitive form of protonephridia is of interest. Protonephridia are always formed from ectoderm (e.g. in Polyclada; A. Lang, 1884). Their structure in *Protomonotresis* indicates that originally they were invaginations of external ciliated epithelium. Therefore we should naturally think that a multitude of separate tubules opening to the exterior by independent orifices, like the tracheal system of Onychophora, was the original type of protonephridia; but no nephridial system of that type is known to us in flatworms. Only annelids have many pairs of metamerically-arranged, independent, slightly-branching nephridia. The nephridia of Triclada show the next stage: originally-independent bunches of protonephridia have fused into a single system, which, however, retains the multitude of external orifices. All other types of nephridial apparatus in flatworms represent further stages of integration and architectonical simplification of that original type. One's eye is struck by a certain parallelism to the integration and architectonical simplification of the tracheal system of Atelocerata. That parallelism, obviously, is due merely to the fact that in both cases we have a system of fine tubules representing invaginations of the integument, whose function is to ensure the inflow or outflow of substances between the external environment and the body tissues.

Protonephridia are fairly widely distributed outside the subphylum of Platyhelminthes. Among the closest relatives of flatworms they occur in both orders of Gastrotricha (H. Wilke, 1954). Then we find them in Rotatoria (Fig. 153, *A*), Kinorhyncha, some Acanthocephala (Fig. 153, *B*), and Priapuloidea (Fig. 153, *C*). Nemertines also have protonephridia. Protonephridia are therefore found in all groups of Scolecida except nematodes and Nematomorpha, which are devoid of cilia.

We should point out that in nematodes the osmoregulatory function has been assumed by unicellular skin glands, the so-called cervical glands; in addition they have partly an excretory function, which is also possessed by the gut, as in turbellarians. In *Ascaris* the cervical glands extend in lateral lines along almost the entire body, and open to the exterior by a common duct on the ventral side near the front end of the body.

The paired protonephridia of Gastrotricha Chaetonotoidea and Kinorhyncha, because of the minute size of these animals, each possess only one terminal ampulla. While the protonephridia of flatworms form a complex system of ramifications penetrating the whole body, those of Rotatoria and nemertines (see Vol. 1, Fig. 68) are short stout canals with a few short branches covered with a large number of closely-packed terminal apertures. In large Rotatoria (*Asplanchna*, Fig. 153, *A*) the number of these ampullae sometimes reaches several tens, and in small forms there are from two to

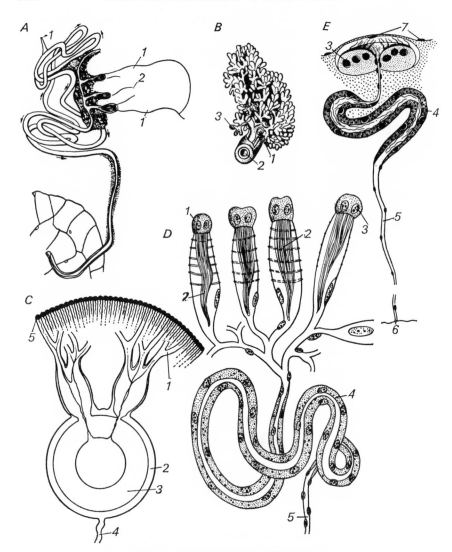

Fig. 153. *Protonephridia of Scolecida.*

A—Asplanchna (Rotatoria): 1—suspensory cord; 2—flagella projecting into body cavity. *B—Hamaniella microcephala* (Acanthocephala): 1—nuclei of walls of nephridia; 2—nephridial canal; 3—separate terminal apparatuses with flame cells. *C—Halicryptus spinulosus* (Priapuloidea), diagrammatic cross-section through urogenital[1] canal and protonephridium, showing only a few of the branched nephridia that are actually connected to all of the terminal apparatuses: 1—protonephridial ducts; 2—coelomic epithelium; 3—epithelium of urogenital canal; 4—mesentery; 5—terminal apparatuses in the form of tubular cells, each with a nucleus and a flagellum (solenocytes) (combined from two drawings by Lüling). *D*—one of the branches of a nephridium of *Geonemertes agricola* (Metanemertini). *E*—one of the nephridia of *Cephalothrix major* (Palaeonemertini) (diagrammatic, after Coe). 1—terminal cell; 2—flame cell; 3—nucleus of terminal expansion; 4—winding (excretory) part of duct; 5—efferent section of duct; 6—nephropore; 7—wall of blood vessel.

eight on each protonephridium. The difference between the protonephridia of flatworms and those of Rotatoria and nemertines is due to the fact that those of flatworms are embedded in parenchyma, within which diffusion of water and excreta is difficult, and therefore the drainage system must penetrate to all extremities of the body. In Rotatoria, on the contrary, the nephridia lie within a large schizocoel, which envelops all the organs, and the fluid filling it serves as a direct link between the tissues and the protonephridia. In just the same way the protonephridia of nemertines are embedded among the vessels of the blood circulatory apparatus (Fig. 161, A), which itself drains the whole body, making concentration of the nephridia possible. Thus the concentrated arrangement of the proto- nephridia of Rotatoria and many nemertines, unlike the protonephridia of flatworms, which extend through all or almost all the body, is due to the greater perfection of the distributive apparatus of the two higher groups. In most Rotatoria the excretory orifice is united with the genital and anal orifices; in Acanthocephala and Priapuloidea the protonephridia open into the genital ducts. We have already seen similar union of the excretory and genital orifices in *Rhynchomesostoma* among Turbellaria Rhabdocoela.

A pair of typical protonephridia is frequently found in the trochophore larvae of annelids and molluscs (Fig. 154, A, D). In molluscs protoneph- ridia occur only in the larval stage, adult molluscs never having them. The nephridia of annelid trochophores also are mostly reduced during meta- morphosis, but in further development a pair of nephridia is formed in each segment. In the prototype, therefore, annelids have a very characteristic arrangement of the nephridial system: a large number of metameric pairs of nephridia embedded in the coelomic sacs, one nephridium in each sac. In the simplest form each of these nephridia is constructed on approxi- mately the same pattern as the nephridia of Rotatoria and nemertines, i.e. it is a bunch of canals projecting into the body cavity and covered with terminal apparatuses. The terminal apparatuses undergo further alteration in adult polychaetes and become tubular cells ending in a nucleated expansion and containing, instead of flame cells, a single long flagellum (Fig. 154, B). Because of their tubular shape these cells are called *solenocytes*.

The protonephridia of simplest structure are those of the family Phyllodocidae, e.g. in the genus *Phyllodoce* (Fig. 154, B). Among other adult polychaetes, typical closed protonephridia furnished with solenocytes occur in Nephthydidae and Glyceridae. In these families their structure has several complex features (A. Lyubishchev, 1924; E. Goodrich, 1945), which we shall not dwell upon. The more extensively modified nephridia are those that E. Goodrich (1945) suggests calling *metanephridia*. Meta- nephridia are found in adult Nereidae, Hesionidae, and *Polygordius*. The *Polygordius* larva has two pairs of typical protonephridia with terminal apparatuses. The first pair disappears together with all the larva's temporary organs. The second pair loses its terminal apparatuses, but its distal parts become part of the first pair of nephridia of the adult. As the body segments are formed a pair of nephridia develops in each of them, each nephridium consisting of a small winding canal lying in the coelom and opening at one end to the exterior and at the other end into the cavity of the preceding segment by a small *nephrostome*. Only one pair of the definitive nephridia lacks nephrostomes.

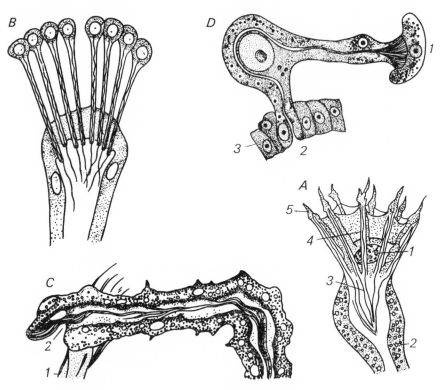

Fig. 154. *Nephridia of trochophore larvae and polychaetes.*

A—trochophore of *Polygordius*: 1—nucleus of terminal cell; 2—wall of nephridial canal; 3—flagella; 4—ducts in processes (5) of terminal cell (after Goodrich). *B*—*Phyllodoce paretti*, end of a branch of a protonephridium with solenocytes (after Goodrich). *C*—*Trypanosyllis*, proximal end of opened nephridium: 1—dissepiment; 2—nephrostome (after Goodrich). *D*—embryo of *Limnea* (Gastropoda): 1—terminal cell; 2—external orifice; 3—ectoderm (after Meisenheimer).

Because of their possession of nephrostomes, the nephridia of *Polygordius* and other forms similar in that respect (Fig. 154, *C*) resemble coelomoducts, with which they were confused before Goodrich's investigations. In polychaetes the anatomical differences between the two appear mainly in the fact that a nephrostome is a narrow orifice, whereas coelomoducts usually open into the coelom by a wide ciliated funnel. The embryological differences are more distinct: nephridia are formed from ectoderm (the external skin) or from ectomesodermal nephroblasts, and develop from the periphery towards the coelom. Coelomoducts are formed from the coelomic epithelium and develop towards the external skin. The ectodermal origin of nephridia has been demonstrated, in particular, for *Polygordius*. In oligochaetes they are formed from ectodermal teloblasts, as has been demonstrated for *Criodrilus*. B. Sukatschoff (1900) has demonstrated the same for *Nephelis* among leeches.

Before Goodrich's works appeared, the excretory organs of invertebrates were divided into protonephridia, possessing terminal bulbs, and metanephridia, possessing funnels opening into the coelom (B. Hatschek, 1888). That division is now entirely unacceptable, since it has been shown that the canals with funnels opening into the coelom are of different types: some of them are nephridia, others coelomoducts. Although Goodrich revealed these differences more than half a century ago (1897–1901) and his data have been confirmed by many authors (L. Fage, 1906; A. Lyubishchev, 1924), the old confusion of terms still appears in popular literature and even in textbooks. The terms 'protonephridia' and 'metanephridia' should now be used for the two types of true (ectodermal) nephridia; closed nephridia of the *Phyllodoce* type being called *protonephridia*, and nephridia of the *Polygordius* type, possessing nephrostomes, being called *metanephridia*.

True ectodermal nephridia occur in all polychaetes in one form or another, but mostly they enter into some kind of communication with the coelomoducts, and are entirely independent of them only in the family Capitellidae, members of which have a single pair of metanephridia and a single pair of coelomoducts in each segment. Other families have independent nephridia only when the coelomoducts are reduced (Nereidae, Glyceridae, etc.), or in sexually-immature individuals with still-undeveloped coelomoducts, or in certain body segments that lack coelomoducts, e.g. in the anterior sterile segments of *Poecilochaetus* (family Disomidae, order Spiomorpha). Oligochaetes usually have a single pair of nephridia, with nephrostomes (Fig. 155, *A*), in each body segment except the larval segments.[1] But in some large terrestrial forms the original paired rudiments of nephridia in the embryo give rise to a large number of independent *micronephridia* (families Perichaetidae, Acanthodrilidae, Cryptodrilidae, etc.). In some species polymerisation of funnels occurs while they retain a common nephridial duct, e.g. *Thamnodrilus crassus* (Glossoscolecidae); but the common plexus of nephridial tubules in the body walls, described by F. E. Beddard (1895) in Megascolecidae under the name of *plectonephridia*, actually does not exist in oligochaetes (K. Bahl, 1947). In *Allolobophora antipae* (Lumbricidae) some of the nephridia in each segment fuse into a longitudinal canal, and the two canals enter the hind-gut; a curious case of convergence with the protonephric ducts of vertebrates. We may point out, however, that the excretory apparatus of vertebrates, which consists of coelomoducts, is not homologous with that of oligochaetes, which consists of nephridia.

---

[1] In the opinion of P. G. Svetlov (personal communication), the nephridia of oligochaetes are actually mixonephria; that would explain their absence in the larval segments.

The ecological significance of the above adaptation in oligochaetes is also quite different: it lies in economy of water, which is partly absorbed by the gut, so that the excreta are discharged in a dehydrated form (K. Bahl, 1938). A similar feature is widespread in tropical terrestrial worms of the family Megascolecidae. Instead of opening to the exterior their nephridia lead into the gut (in some cases only in the posterior body segments, in others in most segments, and in *Travoscolides* in all segments). In the most primitive forms the few pairs of nephridia opening into the gut do so directly; in *Megascolex campester* the nephridia open into a pair of ureters lying along the sides of the dorsal blood vessel: they are connected in each segment by a transverse duct, and open into the gut by a single pair of orifices (Fig. 155, C). Among the species examined, the maximum degree of integration of the nephridial apparatus appears in *Travoscolides:* it has an unpaired ureter that descends into the typhlosole and receives the metameric nephridia, opening into the gut by unpaired metameric pores (K. Bahl, 1946).

The *nephromixia* (for this term see the next section of this chapter) of the terebellid *Lanice conchilega* (E. Meyer, 1901) are connected in the same way by longitudinal ducts: the three prediaphragmal nephridia on each side are connected with each other, as also are the four postdiaphragmal nephridia; but they all open to the exterior, not into the gut; naturally a polychaete living in the sea does not need to economise water.

In *Octochaetus*, *Acanthodrilus*, and some other oligochaetes several pairs of nephridia open into the pharynx (*peptonephridia;* F. Beddard, 1895). *Octochaetus* also has bundles of micronephridia directly entering the hind-gut, which has led to erroneous attempts to identify the Malpighian tubules of arthropods with modified nephridia.

In the parasitic oligochaete *Branchiobdella* the number of nephridia falls to two pairs, and in some Naididae nephridia are entirely absent. A number of oligochaetes of various families show reduction of the nephrostome and formation of closed nephridia—a feature widespread in leeches.

The nephridia of leeches (Fig. 155, B) usually have blind proximal ends. At the blind end of a nephridium there is a ciliated funnel opening into the coelom and leading into a small capsule; the capsule is attached directly to the blind end of the nephridium. The cavity of the capsule contains a large number of phagocytes. The homology of the funnel is not quite clear: some assign a coelomic origin to it, and regard it as a homologue of the ciliated organs of the coelom of polychaetes (see below); others regard it as a homologue of the nephrostome. N. A. Livanov (1940) and E. S. Goodrich (1945) vigorously support the latter view.

Very many Ichthyobdellidae (Rhynchobdellea) display the same kind of multiplication of nephridia that we have seen among oligochaetes in Perichaetidae and others, but in Ichthyobdellidae all these nephridia remain connected with each other, forming a dense network of ramifications in every segment. There also they are called plectonephridia. Some authors regard plectonephridia as the primary type of nephridial apparatus in annelids, equating it to the branched nephridia of flatworms. From the purely morphological point of view, however, an aggregate of separate small nephridia is more primitive than an integrated canal system (see

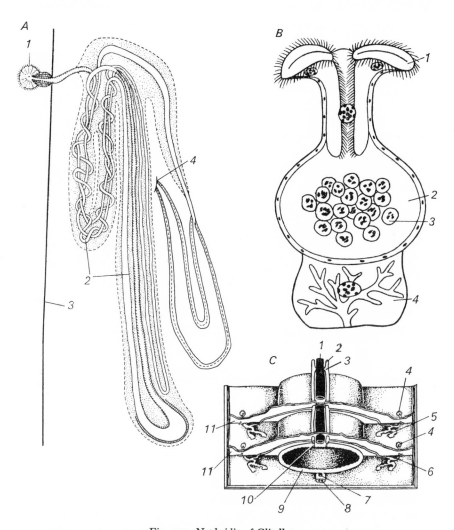

Fig. 155. *Nephridia of Clitellata.*

A—diagram of nephridium of earthworm: 1—nephrostome; 2—winding canal; 3—dissepiment; 4—excretory pore (after Meisenheimer). *B*—terminal apparatus of nephridium of *Clepsine*: 1—funnel; 2—capsule; 3—leucocytes; 4—terminal cell of nephridium with blindly-ending branched duct (after Meisenheimer). *C*—diagram of structure of excretory organs of *Megascolex campester* (Oligochaeta) behind the 20th body segment: 1—dorsal blood vessel; 2—lateral supraintestinal excretory duct; 3—transverse connection between the two excretory ducts; 4—funnel of nephridium; 5—excretory sac of nephridium; 6—nephridial duct leading to dissepiment; 7—ventral blood vessel; 8—ventral nerve chain; 9—gut; 10—canals leading from longitudinal excretory trunk into gut; 11—dissepiment. Numerous small meronephridia, opening to the exterior independently of the enteronephridial apparatus, are not shown (after K. N. Bahl).

above, re nephridia of flatworms), and from the historical point of view it is easier to imagine that the plectonephridia of leeches developed secondarily from the simple nephridia of most annelids, because they occur in separate, highly-specialised groups and are totally absent in primitive classes, in polychaetes in particular.

True nephridia are very rare outside the annelid subphylum. They occur in Kamptozoa and in *Phoronis* (Fig. 156, *A*) (E. Goodrich, 1903).

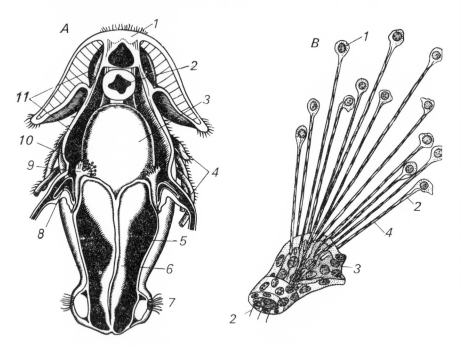

Fig. 156. *Protonephridia found outside of Scolecida, Annelides, and Mollusca.*
A—diagrammatic frontal section through actinotrocha (*Phoronis* larva): 1—apical plate; 2—gullet; 3—stomach; 4—tentacles of metatroch (larval tentacles); 5—trunk coelom; 6—gut; 7—telotroch; 8—anterior section of coelom; 9—nephridium; 10—solenocytes; 11—schizocoel (after Goodrich). *B*—part of nephridium of *Amphioxus lanceolatus* (Acrania), semi-diagrammatic: 1—nuclei of solenocytes; 2—flagella; 3—canal of nephridium; 4—solenocyte tube (after Goodrich).

Neither adult molluscs nor arthropods ever have true nephridia, although E. S. Goodrich (1946) regards the excretory glands of Crustacea as nephridia. No Deuterostomia, except *Amphioxus*, have true nephridia.

The presence of typical solenocytes in the nephridia of *Amphioxus* (Fig. 156, *B*) was demonstrated by E. S. Goodrich (1902), as well as the absence in them of the nephrostome that had been described by Boveri (1892), who discovered these organs. R. Legros (1898) suggested that the nephridia of *Amphioxus* develop from the coelom walls, and in that case (in spite of the presence of solenocytes) they would be coelomoducts. But E. S. Goodrich (1934) showed that in fact the nephridia of *Amphioxus* develop from unicellular rudiments lying in pairs in each segment between the coelomic sacs and the gut. The original nephroblasts, in Goodrich's opinion, are formed in the growth zone of the embryo from ectodermal and ectomesodermal material, and one pair of their progeny exists in each segment. In the development of the nephridia the solenocytes are invaginated into the coelom, and the nephropore opens into the branchial gut.

The presence of true nephridia in *Amphioxus* and their absence in all other Deuterostomia is an amazing fact. The sharp dividing line between typical Protostomia and Deuterostomia is beyond question. As we have seen, the two groups have developed independently from a coelenterate prototype. Not only that, but the different groups of Protostomia (Scolecida, Trochozoa) have also developed independently from that prototype. No modern coelenterate possesses protonephridia. In modern flatworms we see all stages of development of protonephridia, from simple tubes lined with ciliated synepithelium to organs possessing characteristic terminal apparatuses. At the same time it is clear that flatworms did not give rise to any group outside Scolecida. All these facts support the independent origin of protonephridia in the different stems of Bilateria.

4. COELOMODUCTS AND THEIR ROLE AS EXCRETORY ORGANS

As we have seen, the peripheral phagocytoblast of many animals contains excretory elements. It is therefore not surprising that coelomoducts, which are formed from the peripheral phagocytoblast, often assume an

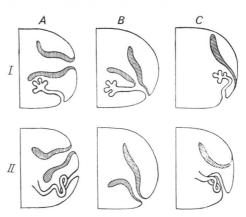

Fig. 157. *Diagram of relationships between nephridia and coelomoducts in polychaetes.*

IA—hypothetical stage with independent genital funnel and protonephridium; IB—Phyllodocidae; IC—*Nephthys*; IIA—*Dasybranchus caducus* (Drilomorpha Capitellidae), with separate genital funnel and nephridium; IIB—most Polychaeta (nephromixium); IIC—Nereidae; the walls of the nephridia are drawn in black, and the coelomoducts are shaded (after Goodrich).

excretory function. The coelomoducts may then replace entirely the true protonephridia, as happens in molluscs, or enter into various relationships with them, as in polychaetes. These inter-relationships were described first by E. S. Goodrich, and later by L. Fage and a number of other authors.

The coelomoducts of polychaetes did not originally have an excretory function, being merely genital funnels. In some cases, e.g. in *Dasybranchus* (Capitellidae, Fig. 157, *A*) or in the genital segments of oligochaetes, the same body segments contain both nephridia and coelomoducts that are quite independent of each other. The former serve as excretory organs and the latter as genital ducts.

In some polychaetes the coelomoducts are vestigial and do not have the function of genital ducts. They are funnels opening into the coelomic cavity and ending blindly, the so-called *cilio-phagocytic organs* (term of L. Cuénot, 1902).

· In *Nereis* they are located independently of the nephridia on the dorsal side of the coelom, whereas in *Glycera*, for instance, these funnels lie in the immediate vicinity of the nephridia. Together they form a compound organ functioning as follows: the funnel collects dead cells, crystals of excreta, and other waste matter from the coelom by ciliary action. These are all phagocytised, and it is surmised that the products of their chemical decomposition are washed out through the adjacent nephridia. The first link between the two organs is thus established. Similar relationships perhaps exist in leeches, as we have seen above. Some polychaetes, such as *Polygordius*, have no coelomoducts. In many Phyllodocidae, Goniadidae, and other lower polychaetes, in which the coelomoduct funnel opens into the distal half of the nephridium, the distal part of the nephridium is used for the discharge of genital products. In *Alciopa* (Phyllodocemorpha) (Fig. 158, *B*) a similar communication between the genital funnel and the nephridium is established only when the animal attains sexual maturity.

Finally, in most polychaetes the genital funnel opens into the nephridium not at its side but at its proximal end, growing directly to the nephrostome (Fig. 158, *C*). Here we see, as it were, a single organ, but one actually formed of two parts of different origins and different morphological signifi-cance—the genital funnel and the nephridium.

If the genital funnel and the nephridium fuse incompletely and are well distinguishable, the product of their fusion is called a *nephromixium;* if the two component parts are fully integrated we speak of a *mixonephrium*, i.e. of a nephridium of mixed origin (E. Goodrich, 1945). Physiologically the funnel and the canal serve for discharge of genital products, but the canal also has, as before, an excretory function. In Syllidae, Spionidae, Ariciidae, and some others the funnels of the coelomoducts grow to the metanephridia only when the animal attains sexual maturity, as a result of which distinct nephromixia are formed. Eunicidae, Aphroditidae, Amphinomidae, and most families of sessile polychaetes have constant nephromixia or mixonephria. In a number of cases heteronomy and division of functions between these develop, so that some mixonephria retain only the genital function and the others only the excretory function. *Sternaspis* (Drilomorpha), for instance, has only two pairs of mixonephria: the anterior pair is excretory and the posterior pair genital.

The combinations of nephridia, coelomoducts, and nephromixia in polychaetes are thus extremely diversified. But in any case it is character-istic of polychaetes, as of all annelids, that their coelomoducts do not have an excretory function, but are combined in one way or another with the nephridia; the excretory function remains with the latter.

Among molluscs, Solenogastres resemble polychaetes in that their coelomoducts have no excretory function and are merely genital ducts. The coelomoducts of all other molluscs have become true excretory

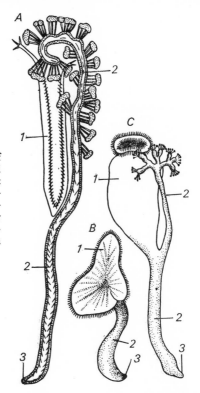

Fig. 158. *Nephromixia of polychaetes.*

A—Vanadis formosa, protonephridium of sexually-immature female, with genital funnel lying beside it and yet independent of the nephridium. *B—Irma latifrons.* 1— genital funnel; 2—nephridial canal; 3— efferent orifice of nephridium (after Goodrich). *C—Alciopa cantrainii*, nephridium of sexually-mature female with adnate genital funnel.

organs, kidneys. As a result they have greatly expanded, ramified, and become closely associated with the blood-circulatory apparatus. In Cephalopoda the so-called venous appendages, expansions of the large veins (Fig. 166), project into the kidneys.

The morphology of the coelomoducts of molluscs is discussed in Vol. 1, Chap. VIII, and Vol. 2, Chap. V.

Onychophora have a very complete coelomoduct apparatus (see Chap. V). Most of them (Fig. 159, *B*) have an excretory function. They open to the exterior near the bases of the legs, on the medial side of the latter. The coelomoducts of the mandibular segment do not develop. Those of the oral-papillae segment (Fig. 159, *C*) are converted into salivary glands that stretch along the whole body; they open by a single common duct into the oral cavity. The coelomoducts of the penultimate segment of the body are converted into gonoducts, and those of the last segment into anal glands.

· The use of coelomoducts as salivary glands by Onychophora presents a certain analogy to the similar use of nephridia by several oligochaetes.

Arthropoda, like Onychophora, have no coelom in the adult state, but partly retain coelomoducts. Excretory coelomoducts usually persist in Crustacea and Atelocerata in separate segments of the head section, and in Chelicerata in separate segments of the prosoma. Among all the classes of Arthropoda they persist most constantly in Crustacea. The latter, as a rule, have excretory glands in two segments, those of antennae II and of

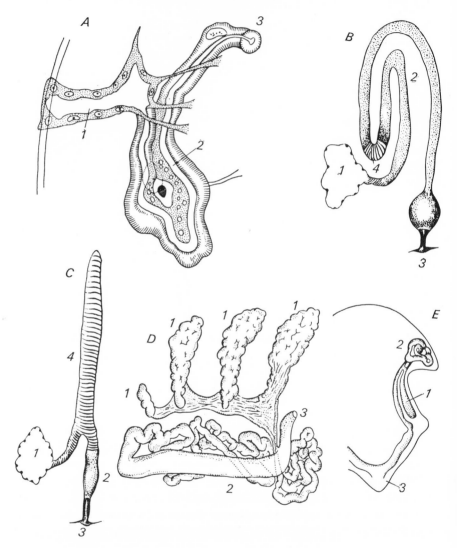

Fig. 159. *Excretory coelomoducts of Arthropoda and Onychophora.*
A—antennary gland of *Cypris strigata* (Ostracoda). *B*—diagram of typical coelomoduct of Onychophora. *C*—coelomoduct of segment of oral papillae of Onychophora, converted into a salivary gland. *D*—coxal gland of *Limulus polyphemus* with four metameric coelomic sacs, but with a single duct. *E*—labial gland of *Machilis*.
1—coelomic sac; 2—coelomoduct canal; 3—external excretory orifice; 4—secretory part of canal in Onychophora (from Dogiel).

maxillae II. In adults, however, both pairs persist only in marine Ostracoda, Leptostraca, and lower Mysidacea of the family Lophogastridae. Other Crustacea retain only one pair: the maxillary in all Entomostraca and in Isopoda, Cumacea, and most Mysidacea among Malacostraca, and the antennary in other Malacostraca. In Copepoda the antennary gland functions at all larval stages, and the maxillary in adults. Freshwater Ostracoda at first have both pairs of glands, but the antennary degenerates at the end of larval life. In adult Cirripedia Thoracica the antennary glands are converted into the so-called cement glands, which are used to attach the animal to the substrate. The excretory glands of Crustacea consist of two sections: a terminal sac representing, as we have already seen, the remains of the coelom, and an efferent canal (Fig. 159, A). The latter opens into the sac by a funnel and is mostly formed by an outgrowth of the coelom wall (D. Pedashenko, 1899; Yu. Vagner, 1896; P. Butschinsky, 1894).

In Decapoda the coelomoduct canal breaks down into a labyrinth and a urinary bladder; in the freshwater Astacus the labyrinth in turn breaks down into a labyrinth proper and a joining canal. The latter serves as a place of reabsorption of salts; while the contents of the coelomic sac and the labyrinth are isotonic with the blood, the contents of the urinary bladder in Astacus are hypertonic in relation to the blood (in Carcinus they are isotonic).

In all Crustacea only the terminal efferent part of the canal is formed from ectoderm. In some higher Crustacea, however (Homarus, Paguristes), the greater part of the canal is formed from ectoderm, and in Ostracoda the whole canal is purely ectodermal. That variation in the method of development of the canal of the excretory gland is probably a case of metorisis. By this term V. M. Shimkevich denotes a transfer of boundaries, replacement of one of the rudiments of a complex organ by another. Evidently the development of the canal from mesoderm, as observed in the most primitive orders (Copepoda, Mysidacea), is primary, and consequently the canals themselves are coelomoducts. Contradictory embryological data, however, at present allow another interpretation to be made: for instance, Goodrich regards all the excretory glands of Crustacea as nephridia.

The coelomoducts of Crustacea have an osmoregulatory as well as an excretory function.

The excretory coelomoducts of Chelicerata open at the bases of the legs and are called coxal glands. Anatomically they are fairly similar to the kidneys of Onychophora, and likewise consist of a coelomic sac and a winding canal. In some cases, e.g. in Xiphosura (Fig. 159, D) or dipneumonous arachnids (Agelena), the excretory part of each coxal gland is formed by the fusion of mesodermal rudiments belonging to from two to four segments, and opens to the exterior by a single duct. The distal parts of the ducts are of ectodermal origin. In Scorpiones the rudiments of the coelomoducts are also formed in four segments, but only one pair of them gives rise to coxal glands. The multiplicity of the metameric rudiments of the coelomoducts results in the latter being developed in different body segments in different Chelicerata; the ducts usually open in the first or third leg segments. Generally an entire order is characterised by a definite location for the coxal glands, but sometimes the location varies within an order (Amblypygi, Araneina, Parasitiformes, Acariformes). Coxal glands corresponding to the antennary glands of Crustacea occur very rarely in the cheliceral segment (in some Acariformes), and rather more often in the second leg segment, which corresponds to the maxillae-II segment (Pseudoscorpionoidea, various members of Acariformes and Parasitiformes). Thus the coelomoducts have been reduced in different metameres in Chelicerata and in Crustacea. Table 2 presents a comparison of the metameric locations of the coelomoducts occurring in all classes of Arthropoda.

Among Atelocerata coelomoducts persist as excretory organs in Myriapoda and Apterygota; their existence in higher insects has not been proved, and in most cases they are certainly absent there. They are absent also in Chilopoda Epimorpha. The coelomoducts of Atelocerata usually belong to the maxillae-II segment and are in the form of the so-called tubular cephalic glands; they have thin-walled terminal sacs, which represent the remains of the coelom (Fig. 159, *E*). The existence of these glands has been demonstrated in Chilopoda Anamorpha, Symphyla (K. Fahlander, 1938–40), Diplopoda (L. Brunz, 1903; V. Isaev, 1911), Pauropoda, Collembola (G. Quiel, 1915), Thysanura (L. Brunz, 1908), and Campodeoidea (J. Philiptschenko, 1908). In Julidae the 'labyrinths' of the two maxillary glands form loops that extent along the gut almost to the rear end of the body. In other Myriapoda they lie entirely in the head. A very primitive condition is found in Pauropoda, which retain the glands in the antennae-II segment (premandibular) as well as the maxillary glands; but while their maxillary glands are excretory, the antennary have changed their function and become salivary. *Scutigerella* (Symphyla) has glands in the maxillae-I segment as well as those of the maxillae-II segment. *Hanseniella* (Symphyla) has glands belonging to the maxillae-I and the premandibular segments. The labial glands of *Lithobius* each have two nephrostomes, since they include homologues of the coelomoducts of the maxillae-I segment, which exist in *Hanseniella*. The absence of maxillary glands in Chilopoda Epimorpha is explained by K. Fahlander (1938) as being due to reduction in the size of their head, as a result of the fossorial mode of life of most members of that group. An excretory function has been demonstrated for the maxillary glands of *Scutigera* (N. Palm, 1954), Pauropoda, a number of Chilopoda and Diplopoda, Collembola, and Thysanura; the gland sacs excrete ammoniacal carmine and the canal excretes indigocarmine, as do the coxal glands of Chelicerata and the kidneys of Crustacea.

Among higher insects, the mesodermal origin of the labial glands has been demonstrated for *Pontania* (Tenthredinidae, Hymenoptera) (O. Pflugfelder, 1934), Psocoidea (Weber, 1938), and several other forms.

These data for *Pontania*, however, are not confirmed by O. M. Ivanova-Kazas (1959).

Thus most Atelocerata retain excretory coelomoducts in the same body segments as Crustacea do. In this respect, as in many others, these two groups are closer to each other than either of them is to Chelicerata.

TABLE 2. Metameric location of coelomoducts in Arthropoda

Groups	Serial nos. of postantennular segments					
	1	2	3	4	5	6
	Appendages					
	Antennae II, chelicerae	Mandibles, pedipalps	Maxillae I, legs I	Maxillae II, legs II	Thoracic legs I, legs III	Thoracic legs II, legs IV
Xiphosura	—	(+)	(+)	(+)	+	—
Scorpiones	—	(+)	(+)	(+)	+	—
Telyphones	—	—	—	+	(+)	—
Amblypygi Charontinidae	—	—	+	—	+	—
Amblypygi Tarantulidae	—	—	+	—	—	—
Araneina Mesothelae and Tetrapneumones	—	—	+	—	+	—
Dipneumones (Dysderidae and some others)	—	—	+	—	+	—
Dipneumones (majority)	—	—	+	(+)	(+)	(+)
Opiliones	—	—	(+)	—	+	—
Pseudoscorpiones	—	—	—	—	+	—
Ricinulei	—	—	+	—	—	—
Solifugae	—	—	+	—	—	—
Schizopeltidia	—	—	+	—	—	—
Palpigradi	—	—	+	—	—	—
Argasidae (Parasitiformes)	—	—	+	—	—	—
Leptostraca; Lophogastridae (Mysidacea), marine Ostracoda	+	—	—	+	—	—
Entomostraca, Stomatopoda, some Peracarida	(+)	—	—	+	—	—
Malacostraca (majority)	+	—	—	(+)	—	—
Thermosbaena	—	—	—	—	—	—
Pauropoda	+	—	—	+	—	—
Diplopoda	—	—	—	+	—	—
Hanseniella (Symphyla)	+	—	+	—	—	—
Scutigerella (Symphyla)	—	—	+	+	—	—
Chilopoda Anamorpha	—	—	(+)	+	—	—
Apterygota	—	—	—	+	—	—

If we regard the coelomoducts of Protostomia as distant homologues of the intestinal pores of coelenterates (see Vol. 1, Chap. IX), they are also homologues of the coelomoducts of Deuterostomia. The latter may be represented either by coelomic pores (the proboscis pores and collar pores of Hemichorda, the pore-canal of echinoderms, the Hatschek's pit of *Amphioxus*) or by genital ducts (the third coelom of Pogonophora), or by emunctoria (the prosomatic coelomoducts of Pogonophora). Synaptidae (Holothurioidea) have ciliated funnels formed by the coelomic epithelium (Fig. 160, *A*). These funnels end blindly and are located on stalks along the mesenteries of the gut. They are called *urnules*. In both structure and

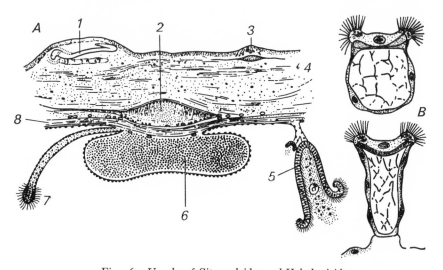

Fig. 160. *Urnules of Sipunculoidea and Holothurioidea.*
A—cross-section through body wall of *Synapta galliennei*. 1—calcareous corpuscle of skin; 2—radial ectoneural cord; 3—glandular-sensory papilla; 4—cutis; 5—urnule; 6—longitudinal radial muscle; 7—ciliated claviform process; 8—annular muscles of body wall (after L. Cuénot). *B*—diagrammatic section through urnule of *Sipunculus nudus*, attached (lower) amd floating (upper) (after W. D. Selensky).

function (collection of particles suspended in the coelomic fluid) the urnules of Synaptidae greatly resemble the ciliary and ciliary-phagocytic organs of annelids, of many polychaetes and leeches, and particularly the urnules of Sipunculoidea (Fig. 160, *B*). The latter are partly located on the coelom walls, like the urnules of Synaptidae, and partly break away and float freely in the cavity fluid.

Usually all such ciliary-phagocytic organs are regarded as homologues of the genital funnels. V. D. Zelenskii (1915), on the contrary, regards them as an independent type of coelomic-epithelium product. That is quite probable with respect to the urnules of holothurians, especially since that group also possesses ciliated organs of another type, far removed from coelomoducts in their structure—claviform ciliated tentacles (Fig. 160, *A*).

Another example of parallelism between Protostomia and Deuterostomia is provided by the coelomoducts of vertebrates. It is well known that all vertebrates possess typical coelomoducts, like those of annelids and molluscs, with both genital and excretory functions: these are the renal canals of vertebrates. In the vertebrate prototype there is a pair of these canals in each body segment. They are formed from the coelomic epithelium and open into the coelom by a ciliated funnel. The other ends of the renal canals on both sides open into a common efferent duct.

TABLE 3. Coelomoducts of Deuterostomia

	Belonging to	Form	Excretory function	Genital function
Echinodermata	Coe_1	Short straight pore-canals	—	—
Enteropneusta	Coe_1 and Coe_2	Short straight pore-canals	—	—
Acrania	Coe_1	Hatschek's pit	—	—
Pogonophora	Coe_1	Looped canals	+	—
	Coe_3	Looped canals	—	+
Vertebrata	Coe_1	Short straight pore-canals	—	—
Vertebrata	Products of Coe_3	Looped canals	+	+

The renal canals of vertebrates, however, differ from the coelomoducts of annelids in having an excretory function, which the latter never have. They differ from nephromixia in being formed entirely from coelomic epithelium, without any participation by ectodermal rudiments. Their resemblance to annelids in that respect is therefore not so deep as it was thought to be before Goodrich's analysis of excretory organs.

Amphioxus has no coelomoducts except Hatschek's pit, but it has typical nephridia with solenocytes, which no longer exist in any other Deuterostomia. Some vertebrates retain the pores of the premandibular somites (see Vol. 1, Chap. IX), the left one of which is homologous with the opening of Hatschek's pit. At the same time vertebrates totally lack true ectodermal nephridia, but they have typical coelomoducts, which are absent in *Amphioxus* and the great majority of other Deuterostomia.

Among lower Deuterostomia, Pogonophora are nearest to vertebrates in the nature of their coelomoducts. They alone have long looped canals, with excretory and genital functions, instead of short pore-canals. But in vertebrates both functions are fulfilled by the renal canals, which are products of the third coeloms, whereas in Pogonophora the coelomoducts of the third coeloms are only genital ducts and those of the first coeloms are only emunctoria. The fact that the coelomoducts of the first coeloms have assumed (even if in only one group) the form of excretory organs argues in favour of their serial homology with the excretory coelomoducts of vertebrates.

Thus we see that both Protostomia and Deuterostomia possess phylogenetic possibilities that permit the independent appearance of extremely similar organs in each of them: among Deuterostomia we find the nephridia of *Amphioxus*, the coelomoducts of vertebrates, and the urnules of Synaptidae. It would be much more difficult to assume that all such formations were possessed by common ancestors, since a number of facts (some of them already discussed above) oblige us to ascribe to these common ancestors a very low level of organisation, at which the possession of nephridia or coelomoducts is scarcely conceivable.

Chapter VIII ORIGIN AND TYPES OF CIRCULATORY APPARATUSES

1. PRIMITIVE FORMS OF CIRCULATORY APPARATUS

Every living organism is a whole composed of parts; the life process always presupposes interaction (including chemical interaction) between spatially-separated parts. Therefore the movement of substances within the body is one of the basic functions of every living creature. In Protozoa and small Metazoa the chemical interaction of parts is usually achieved without the use of anatomically-separated circulatory adaptations. The large size attained by some multicellular animals and vascular plants has become possible only through the development of circulatory apparatuses in both these groups. The appearance and development of a circulatory apparatus (together with a central neural apparatus) is one of the chief methods whereby the bodies of Metazoa become integrated and their level of organisation is raised.

Animals without a circulatory apparatus necessarily stand at a low organisational level. Lack of centralisation of all the chief functions is characteristic of them. Above all, their digestive function is decentralised, it being fulfilled in primitive forms by scattered wandering phagocytes, as in sponges and several Acoela; when there is a digestive apparatus it also is decentralised and distributed throughout the body. The excretory function consists in tissue excretion alone, or else there are excretory organs that remove waste matter directly from the tissues. There are no specific respiratory organs, and separate incretory organs are not known and probably do not exist. Humoral interaction of cells certainly takes place, but only from cell to cell. The neural apparatus of lower Metazoa is in the form of a diffuse plexus, and its effects on the integration of the organism are also very imperfect.

In hydroid polyps the decentralisation of the digestive apparatus is due to the fact that, apart from the gut, almost the whole body consists of a single layer of ectodermal epithelium closely apposed to the gut, which furnishes nutriment directly to the tissues and apparently removes excreta from them. Respiration is performed by the whole body surface, and no special circulatory apparatus is required. In hydroid colonies and other coelenterates, however, the coenosarc cavity has to some extent assumed the role of a colonial circulatory apparatus. That is particularly evident in polymorphic colonies, in which some individuals have a digestive, some an excretory, some (in hydroids and Siphonophora) a food-seizing or a sexual function, and so on; exchange among all these individuals is maintained by the coenosarc canals. Thus the organisation of coelenterate colonies is often also physiologically at a higher level than that of the individual zooids.

The peripheral sections of the gastrovascular apparatus of Anthozoa, medusae, and large turbellarians (Figs 94–96) no doubt constitute a primitive circulatory apparatus. As we have seen in Chap. V, the chief function of the gastrovascular canals of medusae is certainly that of distribution. The movement of the flagella of their epithelium creates currents that distribute digested food through the body. But here the distribution is done very roughly: the canals of the gastrovascular system are large and only slightly branched, and do not carry the contained fluid to every cell. There are large portions of the body between the canals whose cells exist in an uncentralised internal environment. The radial chambers of Anthozoa evidently have a distributive function; there is also definite circulation of fluid in the gastrovascular apparatus of that group, caused by the activity of the flagella of endodermal cells.

In flatworms the gut is still less centralised than in higher coelenterates. In large forms, such as Polyclada or Triclada, the gut ramifies and penetrates all parts of the body. Each part receives nutriment from the nearest branch of the gut, and discharges its excreta into it. If there are nephridia they also are decentralised, ramifying throughout the body. There is no special circulatory apparatus. That produces several structural peculiarities in flatworms. The branches of the gut are necessarily much coarser than blood-carrying capillaries might be, and consequently they cannot penetrate into separate massive organs. Therefore neither massive muscular organs nor massive complexes of glands can exist in flatworms, since the central parts of such massive organs would be placed in unfavourable nutritional conditions. Large complexes of unicellular glands in flatworms are always loosely organised, and their cells are, as it were, scattered in all directions.

The absence of a special circulatory apparatus is therefore only partly compensated for by the ramification of the gut, and it restricts the level of differentiation of organs that can be attained by groups of animals that lack such an apparatus.

The first signs of an anatomically-distinct circulatory apparatus appear in Turbellaria Rhabdocoela in the form of the schizocoelic cavities possessed by many of them. These cavities make easy movement of fluid possible, especially under the action of muscular contractions of the body. Nutrients diffusing from the gut are easily passed through these cavities by simple mixing, without any regular circulation. The simplified form of the gut in Rhabdocoela, which in some species (e.g. *Mesostoma ehrenbergii*) takes the form of a relatively-small narrow tube (Fig. 34, *C*), is partly due to their possession of a schizocoel.

The circulatory function of the schizocoel is very evident in Rotatoria, in which it is highly developed. The fluid contained in it washes all organs in the body and serves as an intermediate internal environment connecting them with each other. For that reason the gut of Rotatoria is compact and centralised, and the nephridia also are concentrated in relatively-small bundles (Fig. 153, *A*).

Thus the possession of a schizocoel, which serves as a primitive circulatory apparatus, enables Rotatoria and other Nematomorpha to centralise both the gut and the excretory apparatus. In coelomate animals, such as annelids, the same distributive function is fulfilled by the coelom. The coelom of annelids is primarily an excretory and circulatory organ. That is shown by the demonstrated presence of urates in the coelomic fluid of polychaetes, and also by the location of the nephridia, which are embedded in the coelom and collect excreta from it. It is true that the walls of the nephridia of large polychaetes and oligochaetes contain numerous blood vessels, from which they also obtain excreta. But in both large and small forms in primitive families the nephridia have no blood vessels and evidently receive excreta only from the cavity fluid. The cavity fluid of polychaetes is rich in various cellular elements, including phagocytes, which play a major role in cleansing the coelom from foreign particles, parasites, grains of excreta, etc., and also including *trophocytes*, which accumulate reserves of nutrients in their protoplasm. In Sabellidae, Terebellidae, and a number of other polychaetes the whole coelomic fluid has the nature of a dense suspension of *oleocytes*, large trophocytes full of globules of fat and granules. Dissolved nutrients, especially proteins, are scarce in the coelomic fluid of polychaetes and oligochaetes, and, as usual in such cases (E. Liebmann, 1946), the coelom participates in the circulation of nutrients mainly by means of its cellular elements.

In addition, the coelom of many polychaetes is important in gaseous exchange. The blood of Aphroditidae and Glyceridae does not contain haemoglobin, and their blood vessels are only slightly developed, but at the same time there is well-organised circulation of coelomic fluid (by means of a system of ciliated coelomic-epithelial cells, as in the gastrovascular apparatus of coelenterates); Glyceridae have erythrocytes in their coelomic fluid and possess coelomic gills—thin-walled outgrowths of the body wall, which ensure gas-exchange between the coelom and the external environment. In that family, therefore, there is no doubt about the respiratory and circulatory functions of the coelom. We see the same in Capitellidae, in which, moreover, the blood vascular apparatus is totally eliminated: as a result, evidently, the coelom of several Capitellidae projects into the longitudinal musculature, dividing it into separate bundles. Nereidae have a well-developed blood vascular apparatus, but nevertheless their longitudinal musculature is almost devoid of blood vessels, and the coelom projects at the sides between the longitudinal and the annular muscles; evidently oxygen is provided to the musculature of Nereidae by the coelomic fluid.

In general there is a certain antagonism in annelids between the coelom and the blood vascular apparatus and when the latter is slightly developed the coelomic fluid takes part in gaseous exchange. When there is no blood vascular apparatus system all distributive functions are assumed by the coelom. We may point out that in all invertebrates from annelids to echinoderms cells coloured with respiratory pigment usually occur only in cavities of coelomic origin and are found only exceptionally in blood vessels, e.g. in *Phoronis*, in the polychaete *Magelona* (A. Lindroth, 1938), in some species of *Glycimeris* (= *Pectunculus*) (H. Griesbach, 1889), and in some Molpadiida.

The clearest example of antagonism between the blood vascular apparatus and derivatives of the coelom is the replacement of the blood vascular apparatus by a system of coelomic canals (lacunae) in Gnathobdellida, which we have mentioned in Chap. V. The complexly-dissected lacunar system of leeches, which is provided with special pumping organs, is the most perfect adaptation of the coelom to the distributive function.

2. THE BLOOD VASCULAR APPARATUS

GENERAL PRINCIPLES

The blood vascular apparatus has arisen from an aggregate of schizocoelic cavities, and is the most perfect adaptation of the latter to fulfilment of the distributive function. This view was first expressed by O. and R. Hertwig, and was proved by (in particular) M. Fernandez (1904); N. A. Livanov also contributed much to correct understanding of the circulatory apparatus of invertebrates. According to Livanov (1914), blood vessels are fissures in the mesenchyma, around which its connective tissue becomes organised and which take the form of regular canals. Blood is a compound tissue, with basic material liquefied. In the pilidia of nemertines the archihaemal cavity is formed, according to O. Bürger, by liquefaction of the gelatinous basic material of their supporting and connective tissue; its walls must clearly be formed by the unliquefied part of the same material. The mesenchymal cells that enter the lumina of the vessels form the cellular elements of the blood; they may also give rise to the lining of the vessels (the *vasothelium*) and to the valves. Often the vessels have no cellular lining at all. Mesenchymal muscles may form the musculature of the vessels. If the vessels encircle the coelom, the walls of the latter may give rise to coelomic-epithelial musculature of the vessels, which is characteristic of the circulatory system of annelids and also of the hearts of molluscs, chordates, etc.

There are two chief types of blood circulatory apparatus in invertebrates, depending on the general character of the histological structure of a group. One type is found in animals whose supporting tissues are predominantly cellular in nature, and the other in animals with predominantly non-cellular, gelatinous supporting tissues. The first type includes echinoderms, molluscs, brachiopods, and arthropods; the second, nemertines, annelids, Phoronoidea, and all lower chordates. Some groups of coelomate animals (e.g. Priapuloidea, Sipunculoidea, Bryozoa, and Chaetognatha) have no blood circulatory apparatus.

NEMERTINES

The most primitive group of animals possessing a blood circulatory apparatus is the subphylum Nemertini. On that account E. Haeckel believed at one time that they were the ancestors of all animals with blood circulation, which of course is incorrect. Nemertines are an independent group of lower worms, which in spite of a number of primitive features are highly specialised, and they have given rise to no other group of modern animals.

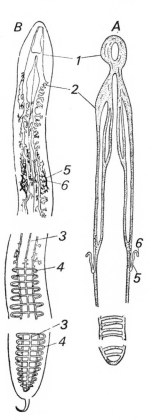

Fig. 161. *Blood vascular apparatus of nemertines.*
Diagram of circulatory apparatus of: *A—Carinoma* (Palaeonemertini) (after W. R. Coe); *B—Cerebratulus lacteus* (Heteronemertini) (O. Bürger). 1—cephalic vascular loop; 2—lateral vessels; 3—dorsal vessel; 4—transverse commissures; 5—protonephridia; 6—their external orifices.

The spaces between organs in nemertines are filled with gelatinous connective tissue, or mesenchyma; it consists of abundant intermediate matter, containing fairly numerous cells and fibres. Some nemertines (*Cerebratulus* among Heteronemertini) have irregular schizocoel cavities in the jelly, filled with fluid. At the same time all nemertines have a true vascular apparatus. In Palaeonemertini it consists of two lateral longitudinal vessels joined in front and behind by anastomoses (Fig. 161, *A*). They form expansions and vascular plexuses in the head, providing for the nutrition of the pre-oral part of the body, into which the gut does not extend. The walls of the vessels are formed of a dense layer of jelly, lined here and there with endothelially-arranged cells. The musculature of the vessels consists merely of separate annular fibres. There is no regu..r blood circulation, the blood being moved backwards and forwards only as a result of contractions of the body musculature.

In other nemertines both the architectonics of the circulatory apparatus and the structure of the walls are more complex. The walls of the vessels possess annular and longitudinal muscles and have become contractile. A third longitudinal vessel (dorsal) has appeared, with metameric transverse vessels connecting it with the lateral vessels (Fig. 161). With this structure, true circulation of the blood has been established: the blood runs from rear to front in the dorsal vessel and from front to rear in the lateral vessels. In some nemertines, however, a periodic reversal of the direction of blood flow has been observed (W. Coe, 1943). The existence of such reversals is characteristic of primitive forms of circulation and occurs sporadically in various groups of animals (for example, fairly often in insects). It occurs in all Tunicata, and to a smaller extent in *Amphioxus* and Myxini.

Nemertines have nothing resembling the peri-intestinal blood sinus of annelids.

ANNELIDS

Typical annelids are characterised by a high degree of tissue polarisation: their bodies are constructed of epithelia and muscular and neural tissues. All the elementary organs formed by these tissues are attached to each other by comparatively thin bounding membranes. The dense basic material of these membranes is almost devoid of cellular elements. Bounding formations occur everywhere that two elementary organs come into contact, e.g. two layers of coelomic epithelium (J. Spengel, 1893). All the bounding membranes in the body are joined together and form a single system of bounding formations (N. Livanov, 1914). Blood vessels constitute, as we have seen, a system of canals developing within the mesenchyma. In annelids the latter is compressed between the walls of the coelom and other organs, and converted into a system of bounding formations. Along with it the vascular system, which in annelids lies within the bounding formations and permeates them, is also compressed.

The walls of annelid blood vessels are formed simply of the substance of the bounding layers. The musculature of the vessels is formed of coelomic-epithelial cells that migrate in a retroperitoneal direction. The valves of the vessels and the 'corpora cardiaca' (which also serve as valves) are formed of the same coelomic epithelium, which grows into the vessels. The chief function of the vascular apparatus of polychaetes is trophic, as is seen from the close association of that apparatus with the gut; their blood is rich in dissolved proteins and relatively poor in cellular elements. The protein and glucose content of the blood of terrestrial oligochaetes is comparable with that of human blood (K. Bahl, 1947). The blood of annelids may also take part in gaseous exchange: it often contains in solution blood elements such as haemoglobin or chlorocruorin, but almost never contains coloured cell elements. It does contain colourless phagocytes. As A. A. Zavarzin (1934) has shown for earthworms, the small phagocytes in the blood are cambial elements, from which there develop during the animal's life all other types of amoebocytes and also all cellular elements (fibroblasts, etc.) occupying the supporting tissue (bounding formations). That supports the view of E. Mechnikov that the trophic function of the peripheral phagocytoblast is primary and the supporting function is secondary.

The original function of the respiratory pigments of polychaetes was not the transfer of oxygen from the respiratory epithelia to the tissues, but the accumulation of oxygen reserves against periods of temporary cessation of respiratory movements. In the more primitive families of polychaetes, therefore, haemoglobin is contained not in the blood but in the neural tissue, which is most sensitive to oxygen shortage (M. Romieu, 1922, with regard to Phyllodocidae and Aphroditidae). Haemoglobin occurs also in the neural tissues of nemertines (M. Prenant, 1924). In higher families of polychaetes, such as Nereidae, haemoglobin already exists in the blood, but its main function continues to be the creation of temporary reserves of oxygen (A. Lindroth, 1938). The existence of respiratory pigments in the blood easily leads to extension of their functions towards participation in the distribution of oxygen, which later has become their principal function.

Besides primitive structure of the vessel walls, the vascular apparatus of typical annelids also displays complex architectonical differentiation (Fig. 162). The most constant component of their vascular apparatus is the *intestinal sinus*—a narrow slit-like space enclosed in the bounding layer of the gut between the intestinal epithelium and the splanchnopleure. The two walls of the bounding layer enclosing the sinus are joined together by trabeculae of matrix passing through the sinus cavity (V. Zelenskii, 1915). Large annelids often have, instead of a continuous intestinal sinus, a peri-intestinal plexus of vessels covering the gut like a cocoon. The intestinal sinus or peri-intestinal plexus is developed only in the post-larval segments, not extending into the larval segments. Instead of it, a dorsal vessel runs forward from the dorsal part of the intestinal sinus, running along the larval segments and lying in the dorsal mesentery. A corpus cardiacum often lies at its base in sessile polychaetes, in the form of a cylindrical outgrowth of the vessel wall, lying in the lumen and pointing forward. When the vessel walls contract peristaltic waves pass from rear to front, and the contracted parts of the vessel press against the corpus cardiacum and so prevent reverse flow of blood to the rear. Thus the corpus cardiacum functions as a valve lying where the dorsal vessel arises from the sinus.

Besides the intestinal sinus, most annelids have an independent ventral or subintestinal vessel in the ventral mesentery. In many forms a dorsal vessel also separates from the intestinal sinus along its whole length. In these forms the dorsal and ventral vessels fuse with the sinus only in the immediate vicinity of the rear end of the body. In addition, the vascular plexus or the longitudinal vessel usually lies along the ventral nerve chain. In the intestinal sinus and the dorsal vessel the blood flows from rear to

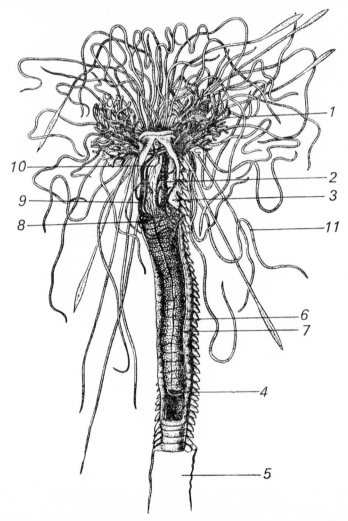

Fig. 162. *Blood vascular apparatus of* Polymnia nebulosa (*Polychaeta Terebello-morpha*); *front part of body opened.*

1—gills; 2—ventral vessel; 3—nephromixia; 4—ventral vessel of gut; 5—dwelling-tube; 6—dorsal vessel on gut; 7—gut, cut off at rear; 8—vascular ring round gullet; 9—dorsal vessel; 10—pharynx; 11—grasping tentacles (after Cuvier).

front; in the ventral vessel and the perineural plexus, from front to rear. These longitudinal routes are connected with each other by metameric vascular arches, originally lying in dissepiments, i.e. intersegmentally. Each metamere or angiosomite (S. Timofeev, 1923) contains two principal pairs of arches, visceral and parietal (Fig. 163). The visceral arches carry

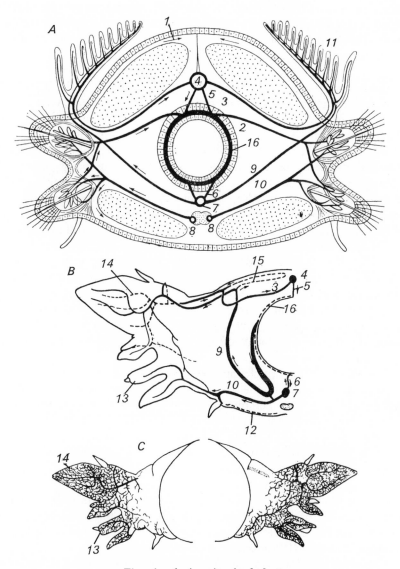

Fig. 163. *Angiosomite of polychaetes.*

A—diagrammatic cross-section through a gill-bearing trunk segment of an errant polychaete; the blood vessels are shown in black; small vessels in the body wall and in the musculature are not shown (after Dogiel). *Nereis virens: B*—angiosomite; *C*—capillary network in skin of parapodia (after Lindroth). 1—parietal vessels; 2—parapodio-intestinal; 3—parapodio-dorsal; 4—dorsal; 5—dorso-intestinal; 6—intestino-ventral; 7—ventral; 8—neural; 9—ventro-parapodial; 10—neuro-parapodial vessels, or their homologues; 11—gills; network of cutaneous capillaries: 12—ventral, 13—neuropodial, 14—notopodial, 15—dorsal; 16—intestinal sinus (or intestinal plexus).

blood from the dorsal vessel to the walls of the gut, and carry blood saturated with nutrients from the gut to the ventral vessel. The parietal arches carry blood from the ventral and perineural vessels to the body walls and the gills, where the blood vessels break down into capillaries. In the skin (Fig. 163, C) and the gills the blood becomes saturated with oxygen. Thence it returns partly to the dorsal vessel and partly directly to the intestinal sinus. The capillaries, which run within the body walls, provide nutrition and gaseous exchange for the musculature, etc.

Obviously, with blood circulation of this type in the body segments, the dorsal vessel contains arterial blood, which it conveys to the supra-pharyngeal ganglion and the cephalic sense organs. In all polychaetes the vascular apparatus of the anterior segments of the body shows sub-stantial differences from the rest, because of the vessels running to the pharyngeal apparatus. Besides, the basic plan of the angiosomite undergoes innumerable modifications, firstly in different families of polychaetes and oligochaetes, and secondly when there is heteronomy in different parts of the body or different segments of the same animal.

The chief path of development of the vascular apparatus in annelids leads to development of heteronomy of angiomerism, producing increasing integration and centralisation of the entire vascular apparatus. However, as a result of the extreme diversity of the heteronomous differentiation of body segments, which is characteristic of polychaetes, the directions taken by the development of the vascular apparatus also are very diversified.

Annelids are therefore characterised both by a fairly low level of histo-logical differentiation of the vascular apparatus and by complete closing and far-reaching architectonical differentiation of that apparatus. There is true blood circulation, but the pumping function is not well centralised. Annelids do not have a single heart that forces blood to move through the vessels. Their blood is moved by the contractility of large sections of the longitudinal vessels and often by that of the anterior arches. The rhythms of contraction of separate vessels are very imperfectly synchronised, and in general the work of the whole apparatus is very little integrated. The arterial pressure of *Nereis* at rest is only 1·0 to 1·5 mm of mercury above the pressure of the coelomic fluid; when the worm moves, the difference increases to 6 mm (J. Ramsay).

The reduction of the vascular apparatus in many forms and its functional replacement by the coelom have been discussed above.

Comparing the vascular apparatus of annelids and nemertines, we find considerable resemblance in histological structure and sharp differences in the general plan of arrangement of the vessels. Nemertines have nothing resembling the intestinal sinus, which is the principal and the most constant part of the apparatus in polychaetes. Besides the intestinal sinus, unpaired vessels (dorsal, ventral, perineural) are well developed in annelids, whereas nemertines have paired lateral vessels. The paired longitudinal vessels of polychaetes (if any) are only secondary in the most specialised types of vascular apparatus (e.g. in Terebellidae), and even there they occur only along a short part of the body length. The radical difference in the structural plans of the vascular apparatuses of nemertines and annelids corresponds fully with the view of the independent appearance of these apparatuses in the two groups, which in turn arose from the premise (proved in Vol. 1, Chaps IV and V) that Scolecida and Trochozoa are not closely related.

MOLLUSCS

Molluscs differ abruptly from annelids in that their supporting and connective tissue is mostly cellular in nature. That mesenchymatous tissue contains a large number of fissures and lacunae of schizocoelic origin, filled with blood. Phylogenetically the vascular system of molluscs has developed from these cavities. Its development consisted in gradual *vascularisation* of the lacunae (i.e. in their gradual conversion into blood vessels); the originally-irregular cavities gradually took the form of sinuses or canals, along which the blood could easily move in a definite direction: later the sinuses decreased in diameter and their walls, which were formed of intermediate substance of connective tissue, became organised and acquired regular contours—the sinuses had become converted into blood vessels. Endothelium has also been described in some molluscan blood vessels.

The stimulus to the above development of the lacunar system was the commencement of fluid circulation in it. We find the simplest form of vascular apparatus among molluscs in Solenogastres (Fig. 164, *B*). It contains no vessels. There are two longitudinal sinuses, ventral and dorsal. The posterior part of the dorsal sinus is invaginated into the dorsal side of the coelom and possesses musculature derived from the coelom wall. It is therefore contractile, and forces blood forward along the dorsal sinus. This contractile part of the dorsal sinus is called the ventricle of the heart. Solenogastres have no auricle of the heart; blood is pumped into the ventricle from the efferent branchial sinuses, which still have no proper walls. The heart, as in all molluscs, is covered externally with peritoneal epithelium. On the side turned towards its cavity the wall of the heart is formed directly by its musculature. The same heart-wall structure is found also in other molluscs. The structure of the rest of the vascular apparatus is very imperfect, and it is able to produce only very slight blood circulation.

Among Loricata, *Nuttalochiton hyadesi* also has no definite blood vessels except a heart. In all other Loricata the dorsal sinus has become a true blood vessel with well-formed walls, and is called the dorsal aorta (Fig. 164, *A*). The aorta runs straight forward and discharges into the cephalic sinus, a system of schizocoelic cavities surrounding the brain. On the way the aorta gives off a number of branches, to the genital gland, etc. The cephalic sinus is separated from the rest of the body by a diaphragm, a tubular invagination of which forms the visceral artery, which supplies blood to the gut and the liver (in other molluscs these viscera receive blood directly from the aorta). The rest of the blood circulation is purely lacunar. The blood collects from small into large lacunae, mainly into the three longitudinal sinuses of the foot; thence it flows into the afferent branchial sinuses, which run along the inner margin of the mantle cavity.

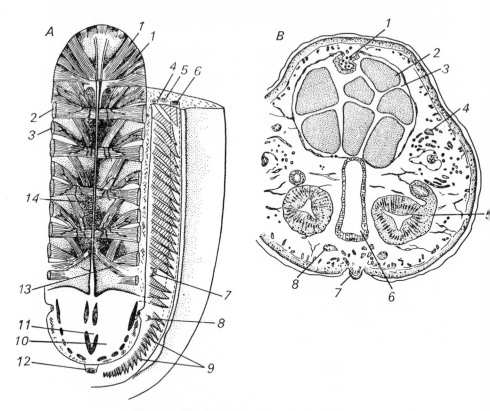

Fig. 164. *Blood vascular apparatus of Amphineura.*

A—Acanthopleura echinata (Loricata): 1, 2, 3—muscles of dorsal side (note their metameric arrangement); 4—afferent branchial vessel; 5—pleuro-visceral nerve stem; 6—efferent branchial vessel; 7—genital orifice; 8—urinary orifice; 9—gills; 10—auricle; 11—ventricle of the heart; 12—posterior duct; 13—oviducts; 14—aorta, sending metameric arteries to the muscles and opening at its front end into the cephalic sinus. *B*—diagrammatic cross-section through *Stylomenia* (Soleno-gastres): 1—heart; 2—genital products; 3—coelom; 4—pleuro-visceral nerve stems; 5—gonoducts; 6—hind-gut; 7—vestigial foot; 8—pedal nerve stems (from Zenkevich and Dogiel).

Passing through the gills, the oxygenated blood collects in the efferent branchial sinuses and through them reaches the auricle. The muscular auricles pump the blood into the ventricle.

The combination of thick-walled ventricle with thin-walled auricle is widespread in the animal kingdom. The physiological significance of this adaptation is: to create blood flow in the arterial system, a ventricle with fairly thick muscular walls is required; considerable pressure, which cannot

exist in the venous system, is required to stretch it during diastole. The
auricle has musculature strong enough to stretch the relaxed walls of
the ventricle and thin enough to be capable of stretching under the action
of venous pressure.

Thus we see in Loricata the beginning of canalisation of the systems
of sanguiferous lacunae, which takes two directions: first, organisation of

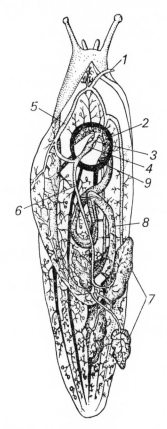

Fig. 165. *Blood vascular apparatus of* Limax
(*Pulmonata Stylommatophora*).

1—genital artery (cut); 2—respiratory orifice;
3—auricle; 4—ventricle of the heart; 5—cephalic
aorta; 6—visceral aorta; 7—liver; 8—gut; 9—
annular venous vessel of branchial cavity. Venous
vessels in black (from Lang).

an arterial system carrying blood from the heart; second, organisation of
a venous system carrying blood first to the gills and then to the heart.
The parts of both systems nearest to the heart are organised first, and
those farther away later.

Among Gastropoda we still find forms with relatively-short and poorly-
branched arterial and venous trunks; the cephalic sinus is most fully
retained in the most primitive Aspidobranchia (*Fissurella, Haliotis*).
At the same time higher Gastropoda have much-branched arterial and
venous systems (Fig. 165).

In most Gastropoda venous blood collects in a system of perivisceral
sinuses surrounding the gut, the liver, and the genital gland. Some of the
blood flows thence directly into the afferent branchial sinus or vessel; some

goes there after passing through the kidney (renal portal system); and some goes directly into the efferent branchial vessel and through it into the auricle. Thus in most Gastropoda the blood in the heart is predominantly, but not entirely, arterial. In some Opisthobranchia, however, (e.g. *Gastropteron*), all the venous blood passes through both the kidney and the gills before entering the heart. Therein they resemble Cephalopoda. The location of the afferent and efferent branchial vessels depends entirely on the structure of the branchial apparatus, and varies greatly in the different groups of Gastropoda.

In Pulmonata and other higher Gastropoda the blood movement produced by withdrawal of the body into the shell is much more forcible than any that could be produced by contraction of their hearts (J. Ramsay, 1952), but on the other hand the transfer of blood from one part of the body of these animals to another as a result of heart action is an important component element in their locomotor activity (see Chap. VI).

The blood of Gastropoda sometimes contains dissolved respiratory pigments: haemoglobin in *Planorbis*, haemocyanin in *Helix*. The blood of *Loligo* (Cephalopoda) also contains haemocyanin.

Among Cephalopoda capillaries appear in *Nautilus*, but only in the skin, and only there does the blood pass from arteries into veins, bypassing the lacunae. Dibranchiata, especially Decapoda, have a capillary system not only in the skin but also in the musculature. In the head region, however, they have a large sinus that collects venous blood from the head and the foot. That sinus gives rise to a well-formed cephalic vein that ascends into the visceral hump and there divides into two venae cavae that enter the branchial hearts, contractile sacs lying at the base of the gills (Fig. 166). The venae cavae receive a number of veins from the viscera. Passing near the kidneys, the venae cavae and other venous trunks enter them by botryose expansions (the venous appendages of the kidneys); they are used to cleanse the blood of excreta. The branchial hearts force the blood through the vessels of the gills, after which it enters the vessels through the branchial vein and flows from there into the auricles and then into the ventricle. All blood entering the ventricle passes first through the kidneys and the gills. Octopoda differ from Decapoda only in details. Thus Cephalopoda have very complete and perfect blood circulation. Like fish, they have only one blood circuit, but whereas in fish the heart contains venous blood, in Cephalopoda it contains arterial blood. The difficulties caused by the need to force blood through two systems of capillaries, body capillaries and gill capillaries, have been overcome in Cephalopoda by their possession of supplementary branchial hearts.

The perfected blood-circulation system and, in particular, the fine ramification of blood vessels in Cephalopoda, reaching partially the level of closed blood circulation, is linked with the general high level of organisation of the group and is one of the features that permit some of them to reach gigantic size. Very large animals can exist only if they possess a capillary system, since only then is the nutrition and respiration of massive organs possible. In general, the maximum body size attainable by any group depends directly on the degree of perfection of their circulatory apparatus. Vertebrates, which possess a highly-developed circulatory apparatus including capillary blood circulation, produce the largest animals on earth. Cephalopoda, and in particular Dibranchiata, possessing partial capillary circulation and a highly-developed blood-circulation system, take second place with such giants as *Architeuthis*. As we shall see later, this rule also obtains in various groups of arthropods.

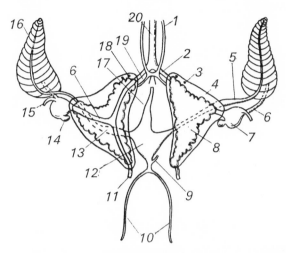

Fig. 166. *Heart and principal blood vessels of* Spirula (*Cephalopoda Dibranchiata*).

1—visceral nerves; 2—junction of visceral nerves; 3—renopericardial orifice; 4—kidney; 5—efferent branchial vessel; 6—afferent branchial vessel; 7—branchial heart; 8—venous appendages of kidney; 9—genital artery; 10—pallial artery; 11—left ventral vein; 12—ventricle of heart; 13—left auricle; 14—appendage of branchial heart; 15—pallial vein; 16—gill; 17—anterior aorta; 18—kidney orifice; 19—branchial nerve; 20—vena cava (after Pelseneer).

The centre of organisation of the circulatory system of molluscs is the activity of the heart. All molluscs have a heart, except a few of the most reduced forms (parasitic Pectinibranchia, Entoconchidae, the dwarf degraded nudibranch *Rhodope*). The heart lies in the pericardium (coelom) in all molluscs except Octopoda and *Anomia* (Lamellibranchia Fili-branchia). The pericardium provides freedom of heart movement and at the same time plays a hydraulic role: during systole of the ventricle the volume of fluid within the pericardium decreases, so that negative pressure is created; this produces a flow of blood from the veins into the auricle, which also lies within the pericardium (J. Ramsay, 1952). The heart usually lies dorsally to the hind-gut, and rarely ventrally to it; in Rhipidoglossa and most Lamellibranchia the ventricle of the heart is penetrated by the hind-gut, i.e. it has a peri-intestinal location. The latter type of structure is probably primary. We may assume that the ventricle of molluscs is homologous with the intestinal sinus of annelids, as described in *Dinophilus* (P. de Beauchamp, 1910), where, in the absence of post-larval segments, it is located in the larval body, like the mollusc heart. Having decreased in size and assumed the role of a powerful pumping organ, the intestinal sinus has become the heart ventricle. The latter has become free of the gut in most molluscs, usually moving dorsally, rarely ventrally. It is characteristic that forms retaining the peri-intestinal position of the ventricle include Rhipidoglossa (and *Neopilina*), which are so primitive in many other respects.

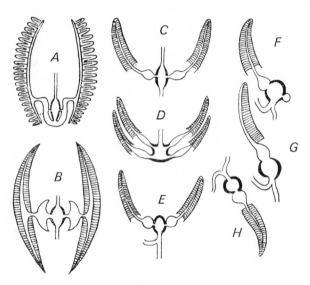

Fig. 167. *Diagram of relationships between heart, gills, and aorta in molluscs.*

A—Loricata; *B*—Lamellibranchia; *C*—Cephalopoda Dibranchiata; *D*—Cephalopoda Tetrabranchiata. Prosobranchia: *E*—Rhipidoglossa (among Diotocardia Zygobranchia); *F*—Azygobranchia; *G*—Prosobranchia (among Monotocardia); *H*—Opisthobranchia Tectibranchia (after Hescheler).

As a rule two auricles open symmetrically into the ventricle. In *Neopilina* and *Nautilus* (Fig. 167, *D*) there are four, and in a number of Gastropoda only one. The auricles are muscular sections of the efferent branchial vessels. In most molluscs a single anterior aorta (aorta cephalica) arises from the heart: these include Amphineura (Fig. 167, *A*), Gastropoda, *Neopilina*, *Nautilus*, and a few relatively-primitive Lamellibranchia. In Gastropoda the aorta gives off a large branch, the arteria visceralis. In Cephalopoda Dibranchiata (Fig. 167, *C*) and most Lamellibranchia (Fig. 167, *B*) two aortae, anterior and posterior, arise from the heart. In some Cephalopoda several other arteries, which usually arise from the aorta, arise directly from the ventricle.

So we have traced the development of the circulatory apparatus within the phylum Mollusca from an irregular system of schizocoelic cavities, with an incipient heart just beginning to create circulation, up to the highly-developed vascular apparatus of Cephalopoda. No other invertebrate phylum displays among its modern representatives such a range of progressive changes in the vascular apparatus.

Comparing the architectonics of the vascular apparatuses of molluscs and of annelids, we observe, in spite of all the differences, some very important features of resemblance: in both cases a longitudinal dorso-medial tract is original and primary, blood flowing through it from rear to front, and its origin being closely associated with the intestinal sinus. The architectonical differences in the vascular apparatuses of the two groups are due to the possession by annelids of a well-developed metameric coelom, between the parts of which all the vessels are compressed and which determines their location, whereas the vessels of molluscs develop freely by means of organisation of lacunae within the voluminous mesenchyma that fills the spaces between the organs.

The similarity in the principal architectonical features of their vascular apparatuses confirms the generally-held view of the close relationship between the ancestral forms of annelids and molluscs, and the differences provide a very clear picture of the far-reaching divergence between the two groups.

ARTHROPODS AND ONYCHOPHORA

We see an entirely different course of development of the vascular apparatus in arthropods, and also in Onychophora, which resemble them in this respect. Arthropods are derived from an annelid-like prototype with well-developed closed blood circulation; they have all lost the closed blood circulation, and many of them show different degrees of simplification and sometimes reduction of the vascular apparatus. The fate of the vascular apparatus in arthropods is closely associated with that of the coelom. When the coelom walls disintegrated they gave rise to musculature and also to loose coelenchymatous tissue that filled the interstices between the organs, containing little intermediate substance but with many spaces full of haemolymph. In this respect the histological structure of arthropods resembles that of molluscs more than that of annelids. For that reason they also lack the closed vascular system of annelids, which runs within the system of bounding layers. Arthropods have retained only the most important parts of the vascular apparatus of annelids: the dorsal vessel, and sometimes the ventral vessel and the lateral arches; they have totally lost the capillaries and the small blood vessels. All the large vessels have become arteries, along which the blood flows from the dorsal vessel or from the heart, which has developed from it. From the last ramifications of these vessels the blood flows into lacunar spaces, by which it returns to the heart. Venous trunks are formed in only a few arthropods, doubtless being of secondary origin and representing the same kind of canalisation of the lacunar spaces as occurs in the venous trunks of molluscs.

The view that a tendency to gradual reduction of the whole vascular apparatus is characteristic of all arthropods is correct only with regard to insects and dwarf forms of other classes.

Like molluscs, arthropods show correlation between body size and the degree of development of the vascular apparatus: the smaller the animal,

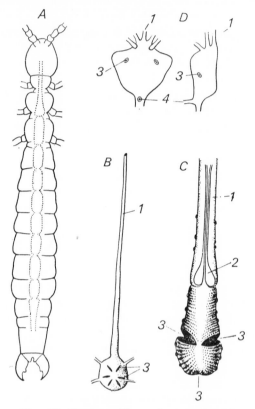

Fig. 168. *Structure of heart of some arthropods.*
A—Japyx (Campodeoidea) (after Grassé). *B—Haemato-pinus* (Anoplura) (from Weber). *C—Eucalanus* (Copepoda): 1—aorta; 2—valve of aorta; 3—ostia (after Claus). *D—Astacus* (Decapoda), dorsal and lateral views: from the front end of the heart arise the aorta and two pairs of lateral arteries, from the rear end the abdominal artery (posterior aorta), and from the ventral side the arteria descendens (4) (after Huxley).

the less branched is its arterial system. Small arthropods retain only the heart out of the whole vascular apparatus, e.g. in Cladocera, in Gamasides among Parasitiformes, in Calanidae among Copepoda (Fig. 168, *C*); or even lose all traces of the apparatus, as in most Copepoda, Ostracoda, and Acariformes. The poor development or total reduction of the vascular apparatus in small arthropods, as in small molluscs, is because in such forms all organs lie very close together and the distributive function can be fulfilled entirely by the lacunar cavities and by the movement of fluid in these produced by body movements. We may point out that in small annelids the vascular system always remains closed, and only the amount of ramification of the vessels is reduced as a result of their small size.

In its method of development the dorsal blood vessel of arthropods is homologous, as we have seen, with that of annelids. In the most primitive forms it extends along the whole, or almost the whole, length of the trunk. Such forms are Phyllopoda and Stomatopoda among Crustacea, scorpions among Chelicerata, Chilopoda and Diplopoda among Myriapoda, and Blattoidea among insects. The dorsal vessel similarly extends along the whole trunk in Onychophora. The dorsal vessel of Onychophora and Anostraca is contractile along its whole length, and possesses metamerically-arranged paired *ostia*. *Branchipus* has the full number of ostia, 18 pairs.

Ostia are afferent orifices with valves, through which the blood is drawn into a vessel. The contractile dorsal vessel, which is provided with ostia, is called the heart in arthropods. In many forms valves lie metamerically along its course, permitting blood flow in only one direction, from rear to front. In a number of other forms the dorsal vessel is differentiated: its front end loses its musculature, ostia, and valves, and is thus converted into an anterior aorta. Sometimes the posterior aorta is formed in the same way, as in some molluscs. Heteronomous structure of the dorsal vessel thus develops. Its contractile part, or heart, lies in different parts of the body (different tagmata), in different classes and orders. Sometimes the heart is much shortened, mostly in Crustacea, but also in some Chelicerata (Pseudoscorpiones, Opiliones, Acariformes) and insects (e.g. Anoplura; Fig. 168, *B*). In Crustacea shortening of the heart takes place not only with reduction of the vascular apparatus (small Entomostraca) but also with a high organisational level of that apparatus. For instance, Decapoda (Fig. 168, *D*) have a shortened, sacciform heart with three pairs of ostia, and a very highly developed vascular apparatus. In them the shortening of the heart is an extreme example of differentiation of the dorsal vessel and centralisation of the pumping function; that is, it is without doubt a progressive feature.

The shortening and differentiation of the dorsal vessel represent the chief progressive tendency in the evolution of the vascular apparatus in arthropods. That process shows a certain analogy to the shortening of their ventral nerve chain, and is likewise due to the development of heteronomy in the entire metamerism of the body.

A few arthropods have a ventral vessel, e.g. Myriapoda, Scorpiones, Xiphosura, and a number of higher Crustacea (Figs 169, 170). In Xiphosura the ventral chain is entirely embedded in the ventral vessel (ventral aorta, Fig. 172). In scorpions the ventral vessel lies between the nerve chain and the gut, forming the supraneural artery; in Malacostraca it is usually a subneural vessel. In Chelicerata and some Myriapoda the ventral vessel is joined to the front end of the heart or to the anterior aorta by a single pair of arches. In most Chilopoda it is joined to the dorsal vessel by two pairs of arches: the anterior arch arises from the first chamber of the heart (in the maxillipede segment), and the posterior arch from the last chamber of the heart. In *Scolopendra* the ventral vessel develops in the ventral mesentery, like the ventral and perineural vessels of annelids. It is hard to establish its precise homology with either of these vessels. The ventral vessel of arthropods forms part of the arterial section of the vascular apparatus, and sends arteries to various organs (to the limbs in Myriapoda).

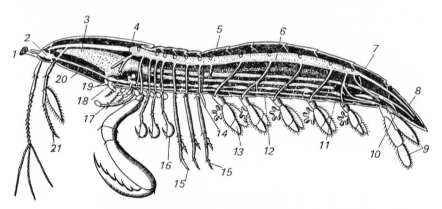

Fig. 169. *Diagram of structure of* Squilla (*Stomatopoda*), *male.*

1—eye; 2—brain; 3—masticatory stomach; 4—carapace; 5—first segment of abdomen; 6—heart, with anterior and posterior aortae and 15 pairs of arteries arising from it, and also with ostia; 7—last abdominal segment; 8—telson; 9—uropods; 10—anus; 11—ventral nerve chain; 12—subneural artery; 13—first pair of pleopods; 14—genital papilla; 15—legs; 16—last pair of maxillipedes; 17—first pair of maxillipedes; 18—mandibles; 19—mouth; 20—circumpharyngeal connectives; 21—antennae II (after R. Siewing from A. Kaestner).

In Stomatopoda and Isopoda the ventral vessel (arteria sternalis) unites with the dorsal branches of a number of metameric arteries (Fig. 170, *A*); in Mysidacea and Decapoda (Fig. 170, *B*) that connection is made by means of a single unpaired artery (arteria descendens), which, however, is homologous with half of one of the pair of lateral arteries. The possession of several metameric pairs of arches by higher Crustacea gives a somewhat annelid-like character to the system of their ventral vessel.

The most primitive type of a well-developed peripheral arterial system is found in forms with numerous homonomous metamerically-arising arteries. Chilopoda, with metameric pairs of arteries arising both from the heart and from the ventral vessel along the full length of both, approach most closely to that type (Fig. 174). Branches of the cephalic aorta also run metamerically to all head segments, apparently including the trito-cerebral segment. Lateral arteries also arise more or less homonomously in scorpions and Stomatopoda (Fig. 169), although the degree of homonomy is lower in both of these than in Myriapoda. Most higher Crustacea are characterised by heteronomous rearrangement of the arterial section of the vascular apparatus (due to centralisation of the pumping function), as well as by shortening of the heart.

Heteronomous rearrangement of the arterial section is found also to some extent in all Arachnoidea. The very intricately branched arterial section in Xiphosura (Fig. 172) is much less homonomous than that in scorpions.

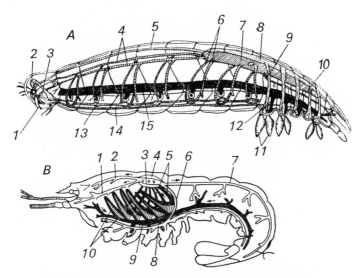

Fig. 170. *Diagrams of vascular apparatus of Malacostraca.*

A—Isopoda: 1—mouth; 2—brain; 3—circumpharyngeal connectives; 4—lateral arteries of anterior thoracic segments; 5—aorta; 6—lateral arteries of sixth thoracic segment; 7—heart; 8—pericardial sinus; 9—branchiopericardial duct; 10—abdominal arteries; 11—pleopods; 12—ducts conveying blood from ventral sinus to branchial appendages of pleopods; 13—ventral nerve chain; 14—subneural artery; 15—ventral sinus. *B*—*Homarus* (Decapoda): 1—lateral cephalic artery; 2—aorta; 3—heart; 4—pericardium; 5—branchial veins; 6—arteria descendens; 7—dorsal abdominal artery; 8—arteria subneuralis; 9—ventral sinus; 10—gills. Lateral visceral artery is not shown (from Lang).

Insects, like lower Crustacea, show considerable reduction of the arterial section as a whole. Insects rarely possess arteries: a primitive metameric arrangement of arteries, comparable to that in higher Crustacea or Myriapoda, has been described in cockroaches, which have two pairs of thoracic and four pairs of abdominal arteries (Fig. 171); their relatives, Mantoidea, have four pairs of abdominal arteries; in some other insects the abdominal arteries are replaced by several pairs of *efferent ostia* provided with valves. The aorta usually divides (in the head only) into two or more short branches. Rarely there are arteries running rearward from the heart (e.g. in Ephemeroptera).

The venous section of the vascular apparatus is entirely lacking in simpler cases. In insects, for instance, the heart lies in the abdomen, where the digestive section of the gut and the Malpighian tubules are also located. The heart draws blood, filled with nutrients and purified from excreta, through the ostia, and then forces it along the aorta into the head, where the brain and the masticatory musculature are located. Acted upon by the difference in pressure so created, the blood flows from the head to the thorax, where it washes the locomotor musculature, and returns from there to the abdomen. There are no strictly-differentiated paths for its return flow, but the connective tissue forms a number of membranes, the

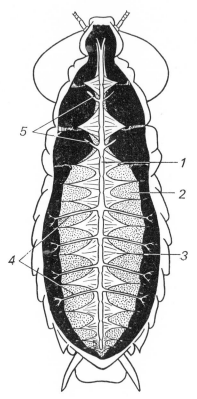

Fig. 171. *Heart and arterial system of* Blaberus (*Insecta, Blattoidea*), *ventral aspect.*

1—heart; 2—dorsal diaphragm; 3—alary muscles; 4—paired abdominal arteries; 5—thoracic arteries (after W. N. Nutting).

so-called diaphragms, which to some extent direct the flow of haemolymph along more definite routes. The ventral, or supraneural, diaphragm is the best developed, and runs along the ventral side of the abdomen; it usually possesses muscle fibres and then takes part in the pumping function, driving blood from front to rear.

The heart of insects, like the dorsal vessel of annelids, generally drives blood from rear to front. Since the time of Malpighi and Réaumur, however, it has been known that there are periodic reversals of the direction of blood flow in larval insects (J. Gerould, 1929). L. V. Yaguzhinskaya

(1954) has described a similar phenomenon in adult mosquitoes (*Anopheles maculipennis*), in which the heart periodically changes the direction of its peristalsis, sending blood for a certain period from the abdomen to the head, and then for a somewhat shorter period drawing blood from the head through the aorta and discharging it into the abdomen.

With a more highly developed vascular apparatus, as in Decapoda (Fig. 170, *B*), blood flows from the arteries into small peripheral sinuses and then gradually collects in a large ventral sinus. From there it flows

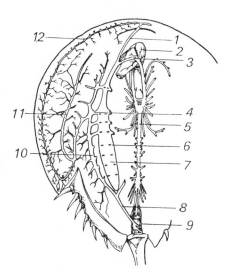

Fig. 172. *Arterial section of vascular apparatus of Atlantic king crab* (Xiphosura polyphemus). *The heart has been moved to the left of the medial location.*

1—frontal artery; 2—pharynx; 3—anterior aortic arches; 4—left lateral artery of second pair; 5—abdominal artery; 6—ostia; 7—heart; 8—dorsal artery of mesosoma; 9—rectum; 10—collateral artery; 11—hepatic artery; 12—marginal artery of prosoma (from Grassé).

to the gills, and from them along the branchiopericardial canals to the pericardium. The pericardium of arthropods, unlike that of molluscs or chordates, is a blood-containing cavity—the pericardial sinus, more or less separated from the rest of the sinus system. The heart opens into the pericardium by ostia. When the venous system is fairly well developed the pericardium loses its direct connection with the sinuses of the body, and the branchiopericardial or pulmonopericardial vessels open into it. Besides higher Crustacea, Xiphosura and lung-breathing Arachnoidea (Scorpiones, Telyphones, Amblypygi, Araneina) also possess such vessels. In these Arachnoidea blood from the tissues collects in the large ventral sinus, which washes the lung-books. There the blood is oxygenated, and from there it flows through the pulmonary veins into the pericardium and thence through the ostia into the heart (Fig. 173).

Among Atelocerata only Scutigeromorpha have rudiments of a venous section of the vascular apparatus, in the form of lacunae that carry blood to the pericardial sinus, in which the tracheal lungs are embedded.

The poor development of the vascular apparatus in insects is doubtless due to the development of tracheal respiration in them. The vascular apparatus of large Malacostraca and some Chelicerata not only ensures supply of nutrients to tissues and removal of excreta from them, but also supplies them with oxygen and removes CO_2 from them. The blood of higher Crustacea (Decapoda, Stomatopoda) contains dissolved respiratory pigment (haemocyanin). To some extent, we find the same relationship also in other groups. Among Chelicerata, for instance, the groups with tracheae (Opiliones, Pseudoscorpionoidea, Ricinulei, Solifugae) have a simplified vascular apparatus: it is most simplified in Solifugae, which have a particularly well-developed tracheal system. They retain only the heart and the anterior and posterior aortae, in spite of the relatively large size of Solifugae. In the same way Onychophora, which have a primitive

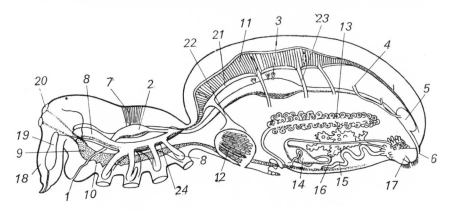

Fig. 173. *Diagram of structure of* Aranea diadematus (*Araneina*).

1—mouth; 2—pharynx; 3—hepatic ducts; 4—Malpighian tubules; 5—expansion of hind-gut; 6—posterior duct; 7—pharynx dilators; 8—blind appendages of gut; 9—suprapharyngeal ganglion; 10—subpharyngeal nerve mass; 11—heart with three ostia; 12—lung; 13—ovaries; 14–16—silk glands; 17—colliculi; 18—chelicerae; 19—poison glands; 20—eye; 21—pericardium; 22—pulmonary veins, opening into pericardium; 23—metameric arteries; 24—aorta (from Warburton).

but extensive tracheal system, have no blood vessels except the heart. In Scutigeromorpha, with their centralised respiratory apparatus (tracheal lungs), the blood apparently takes a large part in oxygen distribution, and the vascular apparatus (Fig. 174) is better developed than in other Myriapoda: the vessels are thicker and more branched, the heart muscula-ture is stronger, and the heart is shortened (it extends rearward only to the segment of the 13th legs) and has no internal valves, i.e. it is becoming more compact.

In insects the vascular apparatus still takes part in the trophic and excretory functions, and in gaseous exchange for small organs only; the tracheal system is used for gaseous exchange in the larger and more massive organs.[1] The absence of respiratory organs produces absence of

[1] This is especially evident if one compares small and large closely-related forms. Thus the ventral-chain ganglia in Tipulidae are permeated by tracheae, whereas the same ganglia in smaller bloodsucking mosquitoes (Culicidae) have no tracheae (T. S. Detinova).

a ramified system of arteries. There are no respiratory pigments in insect blood, if one overlooks a few specialised cases such as *Chironomus plumosus* larvae.

These larvae live in lake mud, in oxygen-deficit conditions, and have haemoglobin dissolved in their haemolymph, where it serves as a carrier of oxygen through the body and as an accumulator of reserve oxygen for

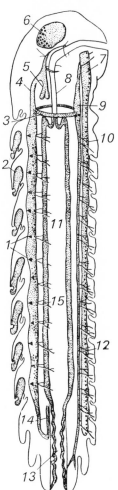

Fig. 174. *Vascular apparatus of* Scutigera (*Chilopoda*). *Diagram of sagittal section of animal.*

1—ostia of heart; 2—fan-shaped tracheae ('tracheal lungs', possessed by Scutigeromorpha); 3—arches of aorta; 4—cephalic aorta; 5—pumping apparatus; 6—cerebrum; 7—ventral nerve chain; 8—oesophagus; 9—ventral vessel; 10 and 12—its ramifications; 11—mid-gut; 13—hind-gut; 14—artery of hind-gut; 15—lateral arteries of heart. Note the number of tergites, only half the number of sternites (from Dogiel).

periods of intermission between respiratory movements (B. Walshe, 1950), precisely as in the polychaete *Nereis virens* (A. Lindroth, 1938); like *Nereis*, the larva of *Chironomus plumosus* lives in a U-shaped tube and creates a water current in it by oscillating movements. Such a mode of life leads to close convergence in the use of respiratory pigments.

Indirect evidence of participation by the vascular apparatus in distribution of oxygen through the bodies of insects is seen in their possession of special adaptations for saturating the haemolymph with oxygen.

One example is the posterior chamber of the heart in Culicidae larvae, which is entirely covered with small branched tracheae, and another example is the 'blood gills' of many Diptera larvae.

The development of a tracheal system, and the cessation of participation by the vascular apparatus in the gaseous exchange of large organs and its consequent simplification, are particularly typical of insects; they are among the principal organisational characteristics of that class. The tracheal system provides tissue respiration, while the simplified vascular apparatus ensures nutrition of, and removal of waste products from, the tissues of the small insect body. Nutrition of massive organs with that type of blood circulation would be scarcely possible, that being the cause of the modest body size of even the largest insects. The largest body sizes among arthropods are attained precisely by the groups that have the most completely ramified circulatory apparatus: Xiphosura and Eurypteroidea among aquatic Chelicerata, scorpions and myriapods among terrestrial arthropods, and Decapoda (e.g. *Macrocheira* in Anomura) among Crustacea. Some Eurypteroidea were three metres long. The limb-spread of *Macrocheira* is up to three metres. Among scorpions, *Pandinus imperator* attains a length of 20 cm. Some Myriapoda also have respectable dimensions: *Scolopendra gigantea* 26·5 cm, *Scaphiostreptes* (Diplopoda) 28 cm, *Acantherpestes* (Carboniferous) 50 cm.

Among insects the largest Coleoptera reach 15 cm, and the longest Phasmatodea and Odonata 13 cm, but the great majority of insects are much smaller, being measured in millimetres. We now see that that is not mere accident: it corresponds to a basic tendency in insect development, and is due particularly to the imperfection of their circulatory apparatus.

Among arthropods with a well-developed vascular apparatus, however, the very great difference in maximum sizes of terrestrial and aquatic forms is striking: the largest Eurypteroidea were up to three metres long, and the largest scorpion is 20 cm long; in other words, taking into account the more massive body structure of Eurypteroidea, the body volume of the largest of them was 4,000 times as great as that of the largest scorpion. A factor limiting the body size of all terrestrial arthropods is the softness of their body covering during the moult: a large body with a soft external skeleton would be much deformed, and would deform its exoskeleton by its weight (C. Kennedy, 1927; W. Hoskins and R. Craig, 1935).

DEUTEROSTOMIA

Among Deuterostomia we find the same two basic types of structure of the vascular apparatus that we have seen in higher Protostomia: open in animals with cellular supporting and connective tissue, and closed when the matrix predominates in the supporting and connective tissue. The first type includes echinoderms, and the second Hemichorda, Tunicata, and Acrania. Vertebrates actually belong to the first type, but like Cephalopoda they have developed a capillary system and closed blood circulation.

The circulatory apparatus of echinoderms is complex, but little integrated morphologically and not very efficient physiologically. It consists, first, of lacunae and vessels of schizocoelic origin, and secondly of coelomic canals and cavities (see Chap. V).

The bodies of echinoderms contain well-developed connective tissue with abundant cells, containing a large number of small slit-like spaces or lacunae filled with fluid. The network of lacunae, which permeates the entire echinoderm body, is combined with a system of more or less definite lacunar canals, representing a differentiated part of the vascular apparatus.

The vascular apparatus of Asteroidea consists of a circumoral lacunar ring from which five radial lacunar vessels arise. The lacunae of the axial organ arise from the circumoral ring, and an aboral lacunar ring is connected with the aboral end of that organ. The aboral ring is enclosed in the genital rachis, and vessels (as well as ramifications of the genital rachis) run from it to the gonads. From the aboral end of the axial organ connecting ducts run to the system of lacunae lying in the gut walls, which are the most important lacunae physiologically. The chief function of the vascular apparatus of echinoderms is the transport of nutrients. Their haemolymph is rich in proteins. All the lacunae (except those of the gut) are closely apposed to the canals of the sinus system. The radial vessels lie in the mesenteries between the paired radial sinuses (see Fig. 69): the walls of the latter compensate for the vessels' lack of walls of their own, and give them a certain amount of differentiation from the surrounding tissue lacunae. In the same way the axial lacunar plexus, which forms the axial organ, is surrounded by sinuses. Since the aboral sinus is unpaired, the blood vessel enters it and is, as it were, suspended from the mesentery. In view of the close connection between the vascular system and the sinus system, the latter is often called the *perihaemal system*. It should be noted that most of the lacunar canals of Asteroidea show very little regularity and differentiation, that there is no true circulation of fluid through the vascular system, and that the whole circulatory apparatus is at a very low level of development.

The vascular apparatus of other echinoderm classes shows a number of differences from that of Asteroidea, these being mainly due to differences in the structure of the perihaemal system (see Chap. V), on which we shall not dwell here. We shall merely point out that the vascular apparatus reaches its highest level of development in Echinoidea and especially in Holothurioidea. A complex system of well-formed blood vessels (among which two longitudinal vessels, provisionally called dorsal and ventral, are dominant) is developed in the gut walls of holothurians (Fig. 175). Transverse connections often form between the vessels of different loops of the gut, freely passing through the body cavity (and surrounded, of course, by coelomic epithelium). In some groups of holothurians the transverse connections of the dorsal vessel form a very complex network, called the rete mirabile: it is particularly well developed in Holothuriidae (Aspidochirota) and in families related to them.

The vascular apparatus of all echinoderms therefore stands at a very

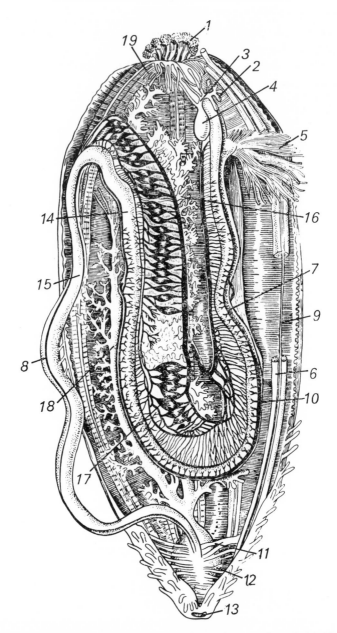

Fig. 175. *Organisation of* Holothuria nubulosa. *Blood vessels marked in black.*
1—oral tentacles; 2—stone canals; 3—ambulacral ring; 4—Polian vesicle; 5—
gonad; 6—longitudinal muscle; 7—descending loop of gut; 8—ventral vessel of
gut; 9—radial ambulacral canal; 10—dorsal vessel of gut; 11—rectum-dilator
muscles; 12—rectum; 13—anus; 14—ascending loop of gut; 15—second descending loop of gut; 16—right lung; 17—rete mirabile; 18—left lung; 19—ampullae
of tentacles (after Lang).

low level. The absence of musculature and valves gives it an especially primitive character. There is no true circulation of fluid in the lacunar system of echinoderms. Some movement of haemolymph can take place only with contraction of the whole body or of some of the large muscles. It is true that a number of echinoderms show signs of rhythmic contractions of the right section of the axial sinus (see Chap. V), which should to a certain extent produce movement of haemolymph in the axial lacunae, but these contractions cannot produce regular circulation of fluid.

The poor development of the vascular apparatus in echinoderms is partly offset by the participation of various cavities of coelomic origin in the distributive function. The epithelium of the general body cavity usually has cilia that set the coelomic fluid in motion. Erythrocytes are known in several Dendrochirota and in all Molpadiida. They contain the pigment erythrocruorin and are found in the coelomic fluid, in the ambulacral system, and (in some Molpadiida) even in the blood vessels of the gut (L. Cuénot, 1948). The fact that cavities of coelomic origin participate in the distributive function partly explains why echinoderms are able to attain such a comparatively large size.

While the open vascular system of echinoderms resembles the vascular system of molluscs to a certain extent, the vascular system of Hemichorda equally resembles that of annelids. Like the latter, Hemichorda have a well-developed system of bounding layers and great reduction of cellular supporting and connective tissue. As a result they have a closed vascular system consisting of regular canals running within the bounding formations. Their vascular apparatus is constructed as follows (Fig. 176). A dorsal vessel, through which (as in annelids) blood flows from rear to front, lies in the dorsal mesentery. In front it enters the cardiac lacuna, lying between the notochord and the pericardium, which lies dorsally to the notochord. The pericardium is a sac formed of coelomic epithelium and musculature; as we have seen, it is the homologue of part of the right anterior coelom of echinoderms. The so-called glomerulus lies in front of the cardiac lacuna: it is a dense network of numerous small vessels, which some regard as equivalent to the axial organ of echinoderms. From it arises a pair of vessels, which form the circumpharyngeal ring by uniting on the ventral side of the collar. The ventral vessel arises from their meeting-point and runs along the ventral mesentery; blood flows from front to rear through the ventral vessel.

Blood flow is caused by the musculature of the pericardium. When it contracts its ventral wall straightens out, and blood from the cardiac lacuna is forced forward through the glomerulus and the annular vessels into the ventral vessel. The blood returns from the ventral vessel, first into the lateral longitudinal vessels through a large number of loops passing through all the bounding formations in the body, and then into the dorsal vessel through the commissural ducts.

At first glance the above-described apparatus presents considerable resemblance to the vascular apparatus of annelids. With regard to its histological structure, that resemblance is due to the similar nature of the supporting tissue; as for its architectonics, the resemblance is due to the vermiform body with its paired coelomic sacs. But something quite foreign to annelid structure appears in the closed pericardial sac, which serves as a pumping organ, a formation characteristic of most Deuterostomia. In general, the resemblance to annelids is no more than very rough convergence.

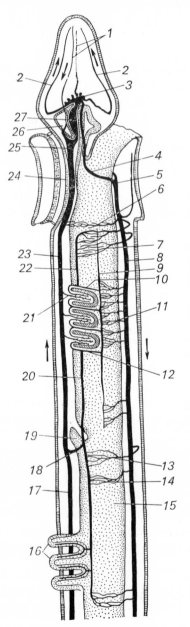

Fig. 176. *Diagram of structure of vascular apparatus in Enteropneusta.*

1—veins of proboscis; 2—arteries of proboscis; 3—glomerulus; 4—circumpharyngeal arches; 5—beginning of ventral longitudinal vessel; 6—capillaries of diaphragm; 7 and 13—capillaries of body wall; 8—ventral longitudinal vessel; 9—marginal branchial vessel; 10 and 12—branchial arteries; 11—capillaries of pharynx; 14—capillaries of gut; 15—gut; 16—hepatic appendages; 17 and 23—dorsal vessel; 18—commissural vessel; 19—postbranchial process of gut; 20 and 22—lateral vessels; 21—branchial veins; 24—dorsal neural tube; 25—front end of dorsal vessel; 26—cardiopericardium; 27—cardiac lacuna (after van der Horst, from Grassé).

The vascular system of Pogonophora (Vol. 1, Fig. 172) is also closed, and consists of a dorsal vessel through which blood flows from front to rear and a ventral vessel through which blood flows from rear to front, as in vertebrates. At the base of the tentacles the ventral vessel possesses a muscular heart, contraction of which forces blood into the afferent vessels of the tentacles. In some Pogonophora (*Siboglinum*, *Oligobrachia*) a sacciform pericardium, very similar to that of Hemichorda, lies dorsally beside the heart. The blood enters the dorsal vessel from the efferent

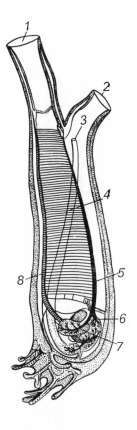

Fig. 177. *Vascular apparatus of* Ciona intestinalis (*Ascidiae*).

1—mouth; 2—cloaca; 3—anus; 4—hind-gut; 5—dorsal longitudinal vessel, sending branches to oesophagus (6) and gonads (7); 8—hypobranchial vessel (from Dogiel).

vessels of the tentacles. Three blind branches arise from its front end, running forward into the cephalic lobe and supplying blood to the brain. In females the longitudinal vessels of the paired ovaries arise from the dorsal vessel behind the muscular diaphragm. The dorsal and ventral vessels communicate with each other in the posterior half of the trunk section by numerous ramifying transverse branches (A. Ivanov, 1955b).

The pericardial sac of Tunicata (Fig. 177) differs from that of Enteropneusta and Pogonophora in that it first forms a groove around the cardiac lacuna and then welds together encircling the latter into a two-layered tube, thus forming a cardiopericardium. Its external or somatic layer is purely epithelial; the internal layer is also epithelial, but it contains longitudinal and annular muscle fibres, and while it is the visceral layer of the pericardium it also forms the muscular wall of the heart. Unlike Enteropneusta, Ascidiae have the heart on the ventral side (Vol. 1, Fig. 178). The hypobranchial vessel runs forward from it along the ventral side of the

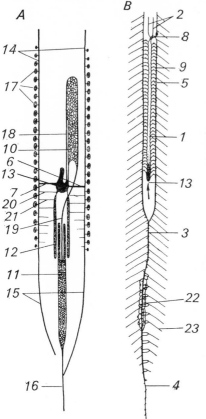

Fig. 178. *Vascular apparatus of* Amphioxus

A—venous section; B—arterial section; diagrammatic, in projection on frontal plane; gill arches are shown foreshortened and much simplified. 1—artery of endostyle (ventral aorta); 2—labial arteries; 3—dorsal aorta; 4—caudal artery; 5—branchial artery; 6—right Cuvierian duct; 7—left Cuvierian duct; 8—glomus; 9—radices of aorta; 10—plexus of hepatic vessels; 11—plexus of intestinal vessels; 12—parietal plexus of vessels; 13—venous sinus; 14—anterior cardial veins; 15—posterior cardial veins; 16—caudal; 17—genital veins; 18—hepatic vein; 19—subintestinal vein; 20 and 21—transverse veins; 22—intestinal artery; 23—septal arteries (from P. Drach).

pharynx and is the homologue of the ventral aorta of Chordata; the visceral vessel runs rearward. The hypobranchial vessel is connected with the dorsal vessel through the vessels of the gill arches. The dorsal vessel also runs to the viscera. Numerous branches of the vessels ramify within both the viscera and the body wall, and also penetrate into the tunic. All Tunicata are subject to periodic changes in the direction of blood flow. Their hearts alternately contract from rear to front and from front to rear; the whole blood circulation changes correspondingly: in the first case the blood flows from rear to front in the ventral vessels and from front to rear in the dorsal vessel, as in *Amphioxus* and vertebrates; in the second case it flows from front to rear in the ventral vessels and from rear to front in the dorsal vessel, as in most invertebrates. In this respect, therefore, Tunicata have a type of blood circulation transitional between those of invertebrates and vertebrates, and we may say that they make the transition in the most crudely eclectic manner. From time to time, however, separate waves of antiperistalsis have been observed also in the contractile vessels of *Amphioxus* and Myxini (E. Skramlik, 1938).

Histologically the vascular apparatus of Ascidiae resembles that of Enteropneusta. The vessels run within bounding formations, have regular walls, and mostly lack endothelium, which occurs only in the largest vessels. Regarding the penetration of the tunics by blood vessels, see Chap. II.

Acrania also have a closed vascular apparatus, the vessels of which run within bounding formations and lack endothelium; their blood does not contain erythrocytes. Histologically, therefore, their circulatory apparatus resembles that of Hemichorda and Tunicata, differing from both of these in the absence of a cardiopericardium.

In its architectonics (Fig. 178) the vascular apparatus of *Amphioxus* is also very close to those of Ascidiae and Enteropneusta; it resembles them in having hypobranchial and hyperbranchial vessels connected by gill-arches. But it is still closer architectonically to the vascular apparatus of vertebrates, even to the possession and location of a hepatic portal system.

The pumping function is not centralised in *Amphioxus;* many vessels are contractile, namely: (*i*) the venous sinus (corresponding more or less to that of vertebrates); (*ii*) the endostyle artery (homologue of the ventral aorta); (*iii*) small enlargements of branchial vessels (of the aortic arches); (*iv*) the subintestinal vein; (*v*) the hepatic vein; and several others. The blood circulation of *Amphioxus* is extremely primitive, the peristaltic movements of the various contractile vessels are poorly synchronised, and the physiological integration of the circulation is very imperfect (E. Skramlik, 1938).

The presence of a continuous endothelial lining in the vascular apparatus of vertebrates, and of erythrocytes in their blood, are two features [1] that have induced and still induce several authors to doubt its homology with vascular apparatuses of schizocoelic origin and (following E. Haeckel) to ascribe a coelomic origin to it. The development of the blood vessels of vertebrates, however, argues against that hypothesis, and the architectonical resemblance between the vascular apparatuses of Acrania and of Vertebrata is so great that one cannot doubt their homology, in spite of their tremendous difference in histological structure; and one is obliged to reject the hypothesis of the coelomic origin of the blood vessels of vertebrates.

---

[1] We may point out that both these features are possessed by the vascular apparatus of Phoronoidea, which, however, in the retroperitoneal location of the vessels, in the presence of a plexus in the stomach walls, and above all in the method of its development in the actinotrocha, is a true vascular apparatus but certainly not of coelomic origin.

Chapter IX
REPRODUCTIVE ORGANS AND THE GENITAL APPARATUS

1. PRELIMINARY REMARKS

Separate genital cells already appear in *Volvox* and in the most primitive Metaphyta.

It was not a mere accident that specialised sex elements, or, broadly speaking, *propagatory* elements (used for reproduction), appeared early in phylogeny, with the development of multicellular organisms. It was an essential condition for further progress. In fact, reproduction even in Protozoa inevitably leads to separation of cells or somatellae, and in loose, poorly-individualised colonies of unicellular animals (e.g., in *Gonium* or *Pandorina* in the order Phytomonadina) to breakdown of the colonies. Both processes terminate the individual existence of the animal or the colony. The need for periodical breakdown of the entire colony prevents specialisation of its elements and far-reaching integration of them into a single organism, and makes the progressive development of a multicellular whole impossible. The appearance of specialised propagatory cells opened the path to progressive development, and only those protozoan colonies that took that path could give rise to multicellular animals. In other words, phylogenetically-early differentiation of propagatory elements was an essential condition for the transformation of a colony of Protozoa into a multicellular organism.

With regard to the time of differentiation of genital cells in the process of ontogeny, a large number of embryological and experimental works have proved that there are two types of organisms among Metazoa: those with early and those with late separation of genital elements, late separation being the more primitive. It is found in animals with indeterminate cleavage. Sponges are the best example. Cells with wide potentialities exist in the bodies of sponges as archaeocytes. On regeneration archaeocytes may, by differentiation, give rise to various other tissue elements. When gemmules or sorites are formed a certain number of archaeocytes, collecting into a cluster, give rise asexually to a new sponge. Finally, archaeocytes give rise to genital elements. In sponges, therefore, genital elements separate during adult life from tissue cells that retain wide potentialities.

Early separation of genital cells is found mostly in animals with determinate cleavage, but also occurs in some groups for which determinate cleavage has not been proved (Cephalopoda, many arthropods). In some cases (e.g., in nematodes) the appearance of genital cells has been traced to the very beginning of cleavage; their definitive separation from somatic cells in nematodes takes place at the fourth division, at the 16-blastomere stage (T. Boveri, 1899). It is obvious that nematodes, with their mosaic cleavage and their constant cellular composition in the imaginal stage, display a secondary and extremely specialised type of development.[1]

There are, of course, all possible intermediate forms between the extremes described above.

By the genital apparatus we mean the whole assemblage of adaptations serving the genital cells and removing them from the organism. The latter is necessary not only to maintain the species but also to preserve the individual, at least when the individual has a relatively long life. The functioning of the genital cells, indeed, is not always co-ordinated with the interests of the organism and is sometimes drastically antagonistic to them. A number of examples from comparative anatomy show that that is not a mere figure of speech. In *Polygordius*, for instance, the genital products fill the body cavity, and the coelomoducts that are used by many annelids to discharge them are lacking. In the Malayan *P. epitocus* the oogonia (which are located in the last five or six segments), on reaching a certain size, begin to phagocytise the gut of the mother. They consume the gut, then the nephridia and the musculature, everything, until nothing is left of the last segments of the maternal organism but a thin cuticle. That cuticle breaks open and liberates the ova (K. Davydov, 1905).

In the above instance the antagonism between the genital products and the maternal organism takes such a violent course because of the absence of special adaptations for discharging them to the exterior. The adaptation there consists in the sacrifice of the individual—if not as a whole, at least in large part—to the species. Most animals have taken another path and created adaptations that provide for removal of genital products from the organism.

In the more primitive forms all genital products, ova and spermatozoa, are ejected into the water, and their meeting with each other is a matter of chance. The probability of their meeting is therefore directly related to the numbers of the two kinds of gametes, being proportional to their numerical abundance.[2] Therefore all animals reproducing in this way must discharge very many ova and spermatozoa into the water, so that the ova must necessarily be very small. The small size of the ova imposes definite limitations on the methods of development, causing prolongation of larval life and heavy pre-imaginal mortality. Therefore any adaptations that increase the probability of encounter between ova and spermatozoa are beneficial for the species. Such adaptations are very numerous.

The simplest case occurs in many sessile or slightly-mobile marine benthic forms, such as many Echinodermata and Lamellibranchia, some polychaetes (e.g. *Podarke*), and many Enteropneusta (C. Burdon-Jones,

[1] Even earlier separation of the genital rudiment, at the 2-blastomere stage, has been described by V. S. Eplat'evskii (1910) for *Sagitta* (Chaetognatha).—Ed. (L. A. Z.)

[2] Numerical abundance is the average number of individuals per unit of space (W. Beklemishev, 1931), in this case the number of ova or spermatozoa per unit of volume of water.

1951). When a population reaches sexual maturity only one individual needs to discharge its genital products into the water for its neighbours of the other sex immediately to do the same; and an explosive wave of spawning sweeps through the whole population. Simultaneous discharge of an enormous quantity of genital products is thus ensured, as a result of which the chances of a meeting of gametes and the percentage of ova fertilised are greatly increased.

Many filter-feeders (all sponges, many lamellibranchs) are characterised by external-internal fertilisation (M. S. Gilyarov's (1958) term); they discharge only spermatozoa into the water, retaining ova in the body or in the mantle cavity; in filtering, each individual inevitably picks up spermatozoa of other individuals of its own species, which fertilise mature ova within its body. Some non-filter-feeding Anthozoa have internal fertilisation of this type as a result of water currents brought into the gastral cavity by the siphonoglyphs for respiration and nutrition.

In more mobile forms individuals of different sexes come together in groups or pairs during the reproductive period, and only then do they discharge their genital products into the water. Rising to the sea surface in masses and spawning there, the swarms of epitokous polychaetes (Nereidae, Syllidae, Eunicidae) provide an example of group mating; sometimes it becomes paired mating, e.g. in *Sphaerosyllis* (Nereimorpha) the swimming male discharges a cloud of spermatozoa on approaching the female, which discharges her ova on entering the cloud (G. Thorson, 1946). In other polychaetes a similar procedure takes place on the sea bottom: the male *Nicolea zostericola* approaches the female's tube and discharges spermatozoa into her tentacles; in response the female discharges ova and mucus, which coagulates around the fertilised ova in the form of capsules. In many polychaetes (e.g. in benthic generations of Nereidae) spawning takes place in the close quarters of a tube into which both the male and the female enter. In many nemertines and some polychaetes (e.g. *Scolecolepis fuliginosa* among Spionidae) two or more spawning individuals surround themselves with a common mass of mucus, into which they discharge their genital products; both the fertilisation and the development of the ova take place within the mucus.

In *Sagartia troglodytes* (Actiniaria) the male and the female come together and discharge their genital products only in each other's immediate vicinity (K. Nyholm, 1943). Similar behaviour is observed in many primitive Gastropoda, such as *Helcion pellucidus* (Docoglossa) and *Gibbula tumida* (Rhipidoglossa, family Trochidae), and also several holothurians (see Fig. 179). In such cases of close approach of individuals spawning in the water we speak of pseudocopulation.

We shall see a number of other methods whereby ova meet spermatozoa when we discuss separate groups. The most perfect of these is fertilisation through introduction of spermatozoa into the female's body by the male. This function is fulfilled by the copulatory apparatus, which arises in many diverse forms in different branches of the animal kingdom.

With internal or external-internal fertilisation, a new possibility arises: the fertilised ovum may be immediately expelled from the mother's body and continue to develop in the external environment, or it may remain in the mother's body and there undergo its entire development. Vivipary then occurs. Vivipary may produce various adaptations (sometimes very complex) for the respiration and nutrition of the developing embryo.

Vivipary has dual significance: in the first place, it is one of the most efficient ways of protecting the developing embryo from various dangers, and consequently reducing 'infant' mortality; in the second place, it provides constancy of environmental conditions for the embryo. Animals whose ova develop in sea-water need almost no adaptations for stabilising

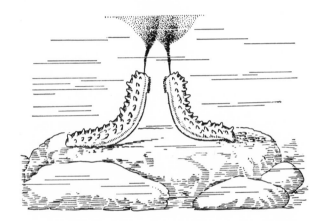

Fig. 179. *A spawning pair of* Stychopus japonicus. *Spawning is of pseudocopulation type. Drawn from nature, by a diver* (after A. F. Dmitriev).

the conditions of embryonic development, in view of the stability of oceanic conditions. With transition to development on the bottom or out of sea-water the embryo becomes exposed to fluctuating conditions and a number of hazards, and various stabilising adaptations are required to ensure a normal course of ontogeny. These include: a large amount of yolk in the ovum itself; envelopes and capsules surrounding the ovum; nutritive secretions, as in the egg-capsules of earthworms; food-ova to be eaten by the developing embryo (e.g. in many gastropods and polychaetes); yolk cells (in flatworms); and, finally, vivipary. All these adaptations reduce mortality in the early stages of development; but since they all require expenditure of the substance of the maternal organism, the development of these adaptations is regularly accompanied by a decrease in fecundity. The formation of embryonic membranes (amnion and serosa) is very characteristic of ontogeny in higher members of terrestrial groups: these occur in a remarkably similar form in terrestrial vertebrates (hence called Amniota), in winged insects (Pterygota), and in scorpions, while being absent in all aquatic animals.

Thus the genital apparatus includes: (*i*) genital glands, or gonads, which are the location for multiplication and development of genital products; (*ii*) genital ducts; (*iii*) copulatory organs; (*iv*) adaptations for creation of envelopes, capsules, etc., for ova; (*v*) adaptations for bearing offspring in viviparous forms.

A number of organs may have some connection with the reproductive function, enabling the sexes to meet (locomotor, tactile, vocal, olfactory, or luminous organs, etc.), aiding in care for the young, etc., but they do not belong to the genital apparatus and are not discussed in this chapter. The males of many animals (especially arthropods) have special clasping organs that serve to hold the female during copulation; these have arisen independently in various (mostly small) groups, and we also refrain from discussing them here.

2. First Steps in Development of Genital Apparatus: Sponges and Coelenterates

Sponges have no genital apparatus. Genital cells arise from archaeocytes and may be located in any part of the body (Fig. 180). A single archaeocyte gives rise to either one oogonium or a bundle of spermatozoids. The latter are often surrounded by a special nutritive cell. In some sponges a few cells form something of the nature of a follicle around each oogonium, which feeds by phagocytising these cells. The spermatozoids are formed mostly in the external layers of the parenchyma, and the oogonia near the flagellate chambers. Sponges are either hermaphroditic or unisexual. Mature spermatozoids are expelled through the efferent canals to the exterior, and then, partly through the afferent canals and the flagellate chambers, enter the parenchyma of mature females, where they fertilise mature ova. The fertilised ova develop there, and ultimately the larvae emerge to the exterior.

The genital cells of sponges are therefore widely distributed in the body, and the only special adaptations affecting them are the food-cells. The chief role in discharging genital products and in ensuring the meeting of gametes is played, firstly by the genital products' own mobility, and secondly by the irrigation apparatus, which provides almost all the communications between the sponge and the external world.

Coelenterates (if we speak of individuals, not colonies) are only one

Fig. 180. *Diffuse distribution of genital cells in lower Metazoa.*

A—oocytes in the body of the sponge *Sycandra*: 1—choanocytes forming the epithelium of the flagellate chambers; 2—spicules; the canals of the irrigation system, large oocytes, and small parenchymatous cells are visible in the parenchyma (from P. P. Ivanov). *B*—longitudinal section through the hydrocaulon of *Eudendrium racemosum* (Leptolida, Athecata): 1 and 2—wandering oogonia in ectoderm and endoderm; 3—young oogonium; 4—interstitial cell in endoderm (after H. Mergner). *C*—*Oxyposthia praedator* (Turbellaria Acoela), location of oogonia and oocytes: 1—tactile sensillae; 2—frontal glands; 3—statocyst; 4—mouth; 5—oogonia; 6—oocytes; 7—male gonopore (after A. V. Ivanov).

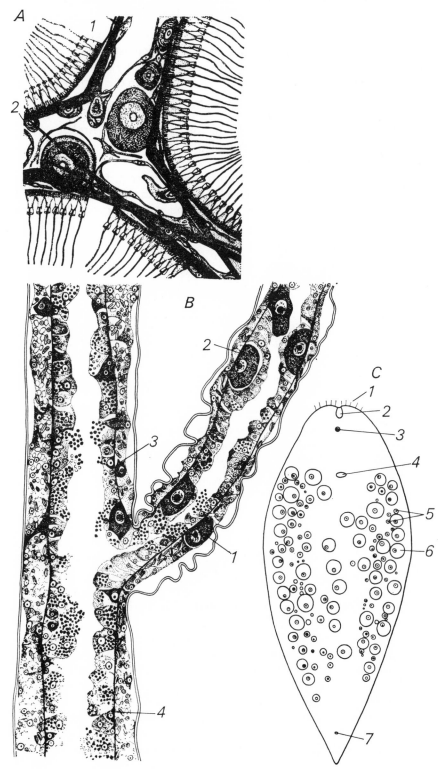

Fig. 180.—See caption opposite.

step higher than sponges. We find cells resembling archaeocytes in hydroids, wandering in the interlayer between ectoderm and endoderm. These cells settle in definite parts of the body and, by multiplying, form small ovaries or testes. In hydroids these gonads lie in the ectoderm. When the genital products mature the ectoderm covering them splits and they emerge. In hydromedusae the gonads are always located near the gastro-vascular apparatus, either in the walls of the proboscis (Anthomedusae, Narcomedusae) or on the subumbrella, along the radial canals (Lepto-medusae, Trachymedusae).

In Scyphozoa the gonads develop in the endoderm, but the ectoderm on the subumbrellar side forms a pocket beside each of them (the subgenital pit), which apparently brings sea-water, and with it oxygen, to the immediate vicinity of the gonads. Genital products are released, by splitting

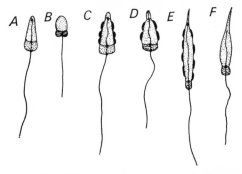

Fig. 181. *Primitive types of spermatozoids in Metazoa.*
A—Grantia compressa (Spongia, Calcarea). *B—Xenoturbella bocki* (Turbellaria, Xenoturbellida). *C—Tubularia mesem-bryanthemum* (Hydrozoa). *D—Paracentrotus lividus* (Echin-oidea). *E—Styela partita* (Ascidiae). *F—Anguilla vulgaris* (Pisces) (after O. Tuzet).

of the endoderm, into the gut cavity, and thence through the mouth to the exterior. In Anthozoa the gonads are also formed in the endoderm, within septa in the radial canals, and the genital products (and sometimes com-pletely-formed larvae) reach the exterior through the mouth.

The spermatozoids of Coelenterata are very similar in structure to those of sponges (Fig. 181), which argues against the derivation of either from the various groups of Flagellata, and supports the view that sponges are an early side-branch of the main stem of Metazoa. Many Bilateria have highly-specialised forms of spermatozoids, but some Protostomia, such as one of the most primitive turbellarians, *Xenoturbella bocki* (E. Westblad, 1949), and also many lower Deuterostomia (Echinodermata, Ascidiae) are close to coelenterates in the structure of their spermatozoids.

We have seen above that Cnidaria, like sponges, have almost no genital apparatus in the strict sense of the term. That refers, however, only to individuals. Colonies, at least hydroid colonies, usually have colonial genital organs in the form of gonophores, blastostyles bearing these gonophores, etc. Occasionally the colonial genital apparatus attains considerable complexity; it is enough to mention the corbulae of Plumulariidae (Thecaphora) and similar formations (see Vol. 1, Chap. III). We have already mentioned the higher organological differentiation of colonies, as compared with individuals, in connection with other apparatuses.

Ctenophores, unlike most other coelenterates, are hermaphrodites. The genital products develop in the endoderm of the meridional canals. It is usually believed that both ova and spermatozoa are discharged through the mouth. A. Totton (1954b) declares, however, that in most ctenophores the ova leave the meridional canals through a split in the ectoderm covering them, and not through the mouth. In Platyctenida the testes have independent seminal ducts opening directly to the exterior; T. Komai (1922) describes fine canals in *Coeloplana bocki*, opening to the exterior by eight rows of orifices and ending towards the ovaries in vesicles that sometimes contain spermatozoa. Apparently these are spermathecae containing spermatozoa of another individual. For the first time, therefore, we find copulatory organs, although still very imperfect, in crawling ctenophores.

3. Origin and Development of Genital Apparatus in Scolecida

A more complex genital apparatus first appears in flatworms. We shall discuss its development in more detail in that group, and shall point out only its main outlines in other phyla, outside Scolecida. A single line of development of the genital apparatus does not, indeed, run through the whole animal kingdom. In each of the groups of Bilateria we find separate examples of very primitive stages in its development; in almost every group it gradually becomes more complex, and adaptations for nutrition, parturition, and copulation arise. In spite of the great diversity of these adaptations, which appear independently in the various groups, the problems involved are repeatedly solved by similar methods, and that in animals far apart in the system. It is noteworthy that flatworms, the first to develop a complex genital apparatus, display almost the greatest diversity in its structure.

The most primitive stage in the development of the genital apparatus of flatworms is found in the remarkable turbellarian *Xenoturbella bocki*. *Xenoturbella* is wholly at the coelenterate level in the structure of its genital apparatus (Fig. 182). Like the great majority of flatworms it is hermaphroditic. Both oogonia and groups of spermatogonia are diffusely scattered in the peripheral parenchyma, as in sponges. There are no efferent ducts or copulatory organs, either male or female. Mature genital products actively enter the gut and reach the exterior through the mouth. Apparently the animal affixes packets of spermatozoids by means of its mouth to the skin of another individual; they penetrate the body of the latter and fertilise mature ova there (E. Westblad, 1949). With that type of sexual behaviour *Xenoturbella* already displays copulation and internal fertilisation, although without any anatomical adaptation for that purpose. It differs from coelenterates only in behaviour: instead of ejecting spermatozoids from its mouth into the water, *Xenoturbella* apparently affixes them with its mouth to the skin of the other individual. If that is confirmed, we may say that a change in behaviour provided the impetus that started the long and complex evolution of the genital apparatus in flatworms. That would be an excellent example of the pre-eminence of behaviour, and consequently the pre-eminence of the neural apparatus, in the evolution of animals.

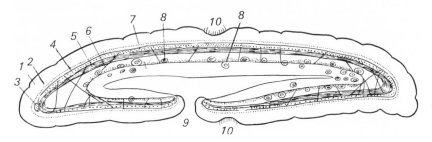

Fig. 182. Xenoturbella bocki (*Turbellaria, Xenoturbellida*), *a turbellarian without genital ducts or copulatory organs, the first stage in the development of a genital apparatus in flatworms. Medial section through adult.*

1—epidermis; 2—neural plexus; 3—statocyst; 4—annular muscle fibres; 5—longitudinal muscle fibres; 6—parenchymatous muscles; 7—epithelium of gut; 8—ova; 9—mouth; 10—ciliated groove (sense organ) (after E. Westblad).

The first steps in development of the genital apparatus are shown by other lower turbellarians, some Acoela. There also in the most primitive cases the genital cells are diffusely scattered in the parenchyma; the spermatogonia lie in small groups, mostly in the dorsal part of the parenchyma, and the oogonia lie in the ventral part; many forms, e.g. *Haploposthia rubropunctata* (Acoela), have only separate groups of oogonia, but a pair of compact ovaries develops in most forms. There are no female efferent ducts in any members of either of the above orders. Maturing ova enter the digestive parenchyma or the gut, and then are ejected either through the mouth or through a split in the external epithelium. In some species the split occurs in a definite place, e.g. in *Aphanostoma sanguineum* in the mid-ventral line of the posterior half of the body (W. Beklemishev, 1915). The chief difference from *Xenoturbella* consists in appearance of a special male copulatory organ, which is found in all Acoela. They have no seminal ducts, however, and the spermatozoids that mature in the parenchyma migrate actively through it to the copulatory organ. In doing so they sometimes form continuous streams which may simulate true seminal ducts.

Male copulatory organs arise in lower turbellarians by several different methods, some of which have been fully traced. These are especially cases where copulatory organs have arisen through change in function of the organs of cutaneous armament (Fig. 183). Organs of cutaneous armament are found in two orders of lower turbellarians, Acoela and Polyclada. They are of various types, two of which, pyriform organs and glandular spines, can give rise to copulatory organs. Pyriform organs, originally organs for attacking prey, are vesicular multicellular glands formed by the outer skin (Fig. 183, *A*); their walls are muscular, and their external duct is often furnished with a stylet formed by modification of the basal membrane (membrana basilaris) of the duct. By means of protractor muscles the stylet can protrude and impale prey, into which the secretion of the gland is injected. Among Polyclada, *Apidioplana* has a large number of pyriform organs scattered over the whole ventral side of the body (Fig. 183, *B*), with no connection with the copulatory apparatus. Members of the family Polyposthiidae also have numerous pyriform organs on the ventral side of the body, but seminal ducts open into some of them and they are converted into copulatory organs; the others remain independent. *Anonymus* (Fig. 183, *C*) has from 10 to 15 pyriform organs on each side, all converted into male copulatory organs, but they are still used as organs for attacking prey. In *Pseudoceros* only one pair of such copulatory organs remains, and in many other Polyclada, such as *Stylochus*, only one organ (A. Lang, 1884). A similar but shorter developmental series appears in Acoela. *Oligochoerus bakuensis* has five or six pairs of pyriform organs along the sides of the body from the mouth to the male gonopore, and one unpaired one behind that orifice. Many species of *Convoluta* and *Oligochoerus* have retained two or three pairs beside the mouth and the male gonopore; *Convoluta adaica* retains only one pair on the margins of the gonopore. Thus a functional connection has been proved to exist in Acoela between some of the pyriform organs and the male gonopore, but they are not used as copulatory organs; it is usually believed that they have the function of stimulating the sexual partner.

Male copulatory organs of the *Stylochus* type occur also in all other orders of turbellarians, mainly in the most primitive members of each group. Sometimes pyriform organs that are not used for expulsion of spermatozoa occur along with them (Fig. 183, *E*). In some forms both are used for attacking prey (e.g. *Gyratrix*, Fig. 186, *E*). Outside of Turbellaria, homologous copulatory organs occur in Monogenea and perhaps in Gastrotricha Macrodasyoidea.

Some Turbellaria Acoela (family Anaperidae) and some Polyclada (family Enantiidae) have organs of cutaneous armament of another type— glandular spines, also subject to change in function and conversion from organs of attack and defence into copulatory organs. A glandular spine is formed from a bundle of thin rods or tubes, each of which is secreted by a separate cell, like the setae of annelids (W. Beklemischev, 1929) (Fig. 183). In contrast to the latter, the rods of turbellarians do not occur singly but always form part of compound spines. The chaetogenous cells that secrete them belong to the external epithelium, and in *Enantia spinifera* (Polyclada) form part of it, so that the spines secreted by them are located on the skin

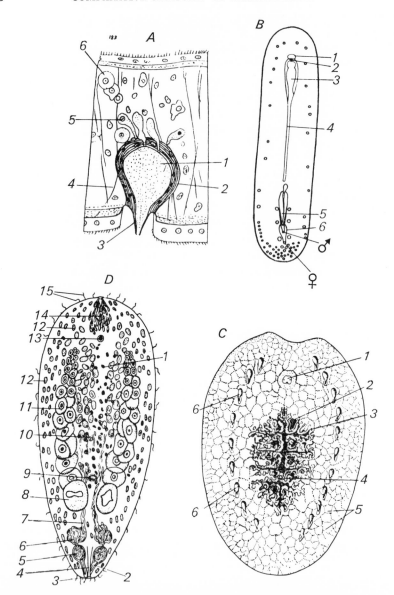

Fig. 183. *Cutaneous armament organs of Turbellaria and copulatory organs arising from them.*

A—pyriform organ of *Apidioplana mira* (Polyclada), part of cross-section of animal: 1—secretion; 2—muscular wall of pyriform organ; 3—stylet of pyriform organ; 4—protractors; 5—glands of pyriform organ; 6—juvenile ovary (after S. Bock). *B*—location of pyriform organs in *Apidioplana mira*: 1—brain; 2—mouth; 3—pharynx; 4—main gut; 5—penis; 6—copulatory pouch; ♂—male and ♀—female gonopores; pyriform organs are shown as small circles (after Bock). *C*—*Anonymus virilis* (Polycada): two rows of pyriform organs, converted into male copulatory organs but continuing to function also as weapons: 1—brain; 2—main gut; 3—pharynx; 4—sucker; 5—net-like anastomosing branches of gut; 6—pyriform organs (after

[continued opposite

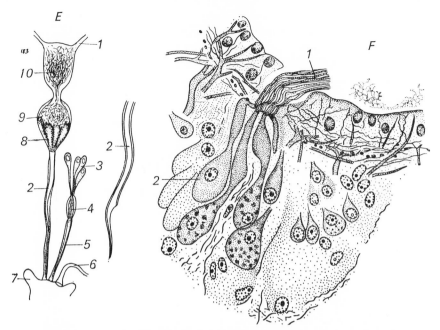

A. Lang). *D—Childia groenlandica* (Acoela), one pair of pyriform organs, converted into male copulatory organs: 1—drops of fat in digestive parenchyma; 2—penis stylets; 3—female gonopore; 4—female atrium; 5—seminal vesicle; 6—'false' seminal vesicles; 7—sperm streams in parenchyma, widening into 'false' seminal vesicles; 8—mature ova; 9—mouth; 10—testes; 11—ovaries; 12—pigmented cells of parenchyma; 13—statocyst; 14—frontal glands; 15—tactile flagella (after v. Graff). *E—Phonorhynchoides flagellata* (Rhabdocoela), two pyriform organs, only one of which is converted into a male copulatory organ, but both open into the genital atrium: 1—seminal ducts; 2—penis stylet (its tip, highly magnified, is shown alongside); 3—glands of pyriform organ; 4—reservoir of accessory pyriform organ ('poison organ'); 5—its stylet; 6—tube of copulatory pouch; 7—genital atrium; 8—glandular epithelium of sac of copulatory organ; 9—sac of copulatory organ; 10—'false' seminal vesicle (after W. N. Beklemishev). *F—Achoerus caspius*, section through wall of genital atrium and glandular spine: 1—glandular spine; 2—its glands (after W. N. Beklemishev).

surface (L. v. Graff, 1889). In Acoela the chaetogenous cells are sunk beneath the epithelium and the spine bases lie below the latter, whereas the ends of the spines may protrude freely to the exterior. In the Caspian *Achoerus ferox* one of the glandular spines acts as a copulatory organ: a large accumulation of spermatozoa (a seminal vesicle, but lacking proper walls) is attached to its base, and the spine itself is penetrated transversely by a sperm-ejecting duct and is much larger than the other spines, which do not take part in sperm ejection (W. Beklemishev, 1937). Copulatorʎ organs of the same type are found in the Atlantic *Paranaperus* and *Paraphanostoma; Paraphanostoma cycloposthium* uses an entire group of spines as penes (E. Westblad, 1942).

Turbellarians possess other forms of copulatory organs besides these two that arise from cutaneous armament organs. The simplest of these are the copulatory organs of *Nemertoderma* (Fig. 184, *A*) and some other Acoela, which are simple tubular skin formations opening into the parenchyma. Spermatozoa form considerable accumulations at their free ends. It is possible that they are everted during copulation (E. Westblad, 1937). In the genus *Polychoerus* (Acoela) a sac is formed around each tube from the parenchymatous musculature (Fig. 184, *B*); contraction of the sac walls

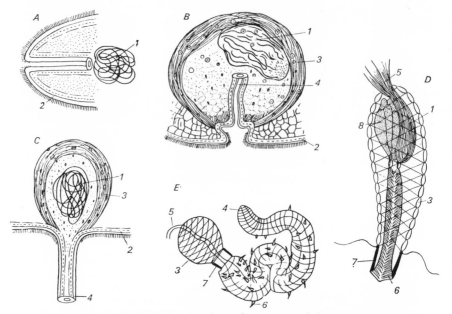

Fig. 184. *Evolution of male copulatory organ of cirrus type.*
A—diagram of copulatory organ of *Nemertoderma bathycola* (Turbellaria Acoela). Cirrus of *Polychoerus caudatus* (Acoela): *B*—retracted, *C*—everted (diagrammatic). Cirrus of *Opisthomum arsenii* (Rhabdocoela): *D*—retracted, *E*—everted (slightly diagrammatic). 1—seminal vesicle; 2—external epithelium; 3—muscular wall of sac of copulatory organ, formed in *Polychoerus* of irregularly-interwoven fibres, and in *Opisthomum* of two layers of spiral muscular bands; 4—cirrus; 5—seminal ducts; 6—spines of cirrus; 7—cuticular ring at point of attachment of cirrus of *Opisthomum* to male genital canal; 8—granular secretion in sac of copulatory organ (original).

can force eversion of the tube to the exterior and outflow through the tube of spermatozoa accumulated in the sac (Fig. 184, *C*). A type of copulatory organ widespread among flatworms (Fig. 184, *D, E*) and known as a cirrus is thus developed. Strictly speaking, the name 'cirrus' applies to the eversible tube. Its duct is the sperm-ejecting duct, and its sac is the sac of the copulatory organ or, to be precise, the sac of the cirrus.

Another type of copulatory organ widespread among flatworms is the soft penis, a papilla traversed by a sperm-ejecting duct. This type may arise independently by overgrowing of the margins of a simple gonopore of the *Nemertoderma* type, or by reduction of the stylet in a copulatory organ of the *Stylochus* type.

Among Turbellaria, therefore, there are several types of male copulatory organs, dissimilar to and not homologous with each other, each of them having developed independently at least twice (in Polyclada and Acoela) or, rather, repeatedly and many times in these two orders and also to some extent, perhaps, in other orders of Turbellaria. With the appearance of internal fertilisation various adaptations for performing it have been poured out, as it were, from the cornucopia, and we see their independent appearance and rapid refinement in different groups of flatworms.

In the most primitive Acoela and Polyclada the male copulatory organ opens directly to the exterior. In most flatworms it opens into a deep invagination of the integument, of complex structure, known as the genital atrium, and sometimes into a male genital canal arising from the atrium. The atrium opens to the exterior by the gonopore, and at the moment of coitus the copulatory organ is caused to protrude from that orifice by protractor muscles.

A refinement of the male gonads in most flatworms higher than Acoela consists in formation of envelopes around the testes; at the same time permanent seminal ducts have developed, possessing their own walls and leading from the testes to the sac of the copulatory organ.

The female gonads have undergone extensive evolution in flatworms. *Xenoturbella* and *Oxyposthia* (Fig. 180, *C*) still have oogonia dispersed through the parenchyma, instead of compact ovaries. Acoela show all stages of transition from primitive, diffuse ovaries to a single pair of well-defined, compact ovaries (Fig. 183, *D*), and we find similar ovaries in Notandropora and Macrostomida. Polyclada, perhaps because of their larger size, have numerous ovarian follicles connected by fine ducts to two oviducts, instead of a single pair of compact ovaries. In spite of their different degrees of individualisation and concentration, the ovaries of all lower turbellarians correspond fully to those of all other animals: they produce ova containing a small amount of nutrient yolk and drops of shell secretion; after the ova are fertilised these drops emerge from the protoplasm to the exterior and, fusing together, form a thin shell around the ovum. The further evolution of the female gonads of flatworms has taken a unique course.

Already in *Hofstenia* (S. Bock, 1923) and several other Acoela each oocyte is surrounded by follicular cells, which provide it with nutrients until it matures, a phenomenon widespread outside Platyhelminthes. In *Gnosonesima* (Alloeocoela), although the oocyte forms yolk within itself it continues to be surrounded by follicular cells, also containing yolk, after fertilisation, and along with them it is covered by a shell: a compound ovum is formed, consisting of an oocyte and several yolk cells within a common shell (E. Reisinger, 1926). In *Prorhynchus* (Alloeocoela) the yolk cells are also formed from follicular cells, but they alone contain yolk and shell secretion, whereas the oocyte already lacks these (O. Steinböck, 1927). Reisinger and Steinböck believe that *Gnosonesima* and Prorhynchidae illustrate the original method of creation of vitelline glands in

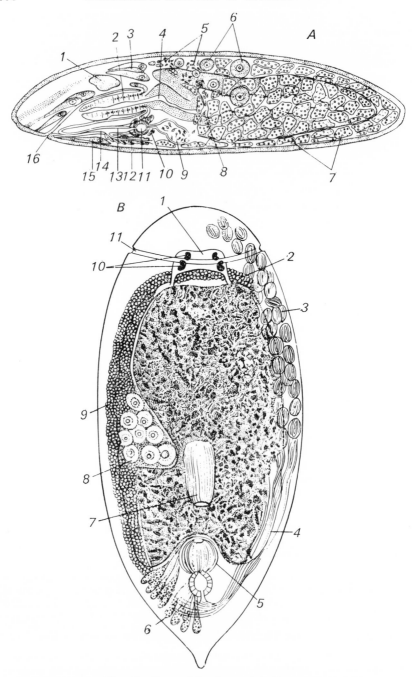

Fig. 185. *Types of female gonads in flatworms.*

A—Prolecithoplana lutheri (Alloeocoela), diagrammatic sagittal section (after Karling): 1—brain; 2—pharynx; 3—cephalic glands; 4—gullet; 5—testes; 6—oocytes; 7—yolk part of germovitellarium; 8—route of spermatozoids to

[*continued opposite*

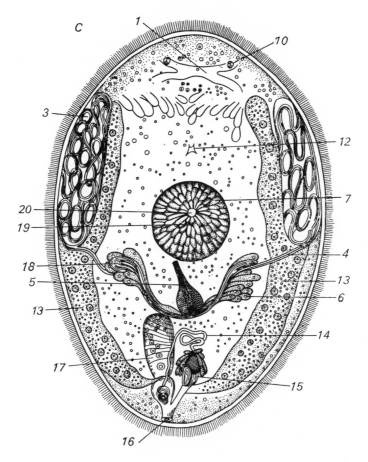

seminal vesicle; 9—seminal vesicle; 10—accessory male gland; 11—copulatory organ; 12—oviduct; 13—male genital atrium; 14—common genital atrium; 15—oral glands; 16—common opening of mouth and genital atrium. *B—Monoophorum triste* (Alloeocoela), germovitellaria shown on left and testes on right (after v. Graff). *C—Byrsophlebs graffi* (Rhabdocoela), unpaired germarium and, separate from it, paired yolk glands (after Jensen): 1—brain; 2—anterior union of the two germovitellaria; 3—testes; 4—seminal ducts; 5—copulatory organ; 6—accessory male glands; 7—pharynx; 8—germarium section; 9—yolk section of germovitellaria; 10—eyes; 11—annular ciliated groove; 12—mouth; 13—vitelline glands; 14—ductus spermaticus, connecting copulatory pouch with oviduct; 15—copulatory pouch; 16—female gonopore; 17—germarium; 18—seminal vesicle; 19—male gonopore; 20—pharyngeal orifice. (Female gonads of ovary type: see Fig. 183, *D*, etc.)

flatworms. At best, however, that is only one of the methods of development of vitelline glands, and most turbellarians have evidently taken other paths: the structure of the female gonads of *Prolecithoplana* (Alloeocoela Holocoela) points to one of these. *Prolecithoplana* (Fig. 185, *A*) possesses in the parenchyma, behind the brain, a centre of multiplication of undifferentiated genital cells; moving rearward, these become differentiated into oogonia, spermatogonia, and yolk cells. All three cell types lie freely in the parenchyma, where they develop into mature genital products. The spermatogonia lie dorsally, and the female genital cells of both kinds are closely apposed to the gut. Thus the gonads are just as diffuse as in the most primitive Acoela, and, as in these, there are no genital ducts; there are, however, a male copulatory organ and a female genital canal opening with their proximal ends directly into the parenchyma, and the genital products reach the efferent organs actively. The union of an oocyte with yolk cells and formation of an egg-capsule take place in the genital atrium (T. Karling, 1940). Thus *Prolecithoplana* is characterised by primitively-diffuse gonads, with fully-completed differentiation of the female genital cells into ova capable of development and yolk cells that are incapable of development and serve for the nutrition of the embryo.

Most Alloeocoela and all Rhabdocoela already have individualised gonads enclosed in envelopes. In some members of these orders the female gonads are differentiated into two sections: one, usually the anterior, producing yolk cells, and the other, usually the posterior, producing oocytes. When an oocyte matures it enters the female efferent duct; a certain number of yolk cells also enter there, surrounding the oocyte and becoming covered, with it, by a common shell. The female gonads, combining oocytes and yolk cells into a single gland, are called *germo-vitellaria* (Fig. 185, *B*). The great majority of Rhabdocoela, Alloeocoela, and Triclada possess separate *germaria* (ovaries) and *vitellaria* (yolk glands); the former produce oocytes capable of development but lacking reserve nutrients, and the latter produce only yolk cells, incapable of development and filled with yolk and shell secretion (Fig. 185, *C*). Yolk glands are usually several times as large as germaria. The yolk cells enclosed in a common shell with the fertilised ovum serve as food for the developing embryo. In this respect the yolk cells of higher flatworms differ from the so-called yolk (more precisely, feeding or alimentary) cells of other animals (Rotatoria, insects), which nourish the still-unfertilised oocyte.

If we look for analogues of the yolk cells of flatworms in other groups of Metazoa, we find them only in the food-ova of some marine animals: nemertines, Antarctic crinoids, many polychaetes (such as *Pygospio elegans* (Spiomorpha)) and Gastropoda Prosobranchia (such as *Nucella lapillus;* G. Thorson, 1946). In *Pygospio* a common egg-capsule contains several, sometimes very many, fully-viable ova, but some of the embryos outstrip the others in development and eat them. *Nucella*, like many other Prosobranchia, shows dimorphism of spermatozoids; ova fertilised by atypical spermatozoa are incapable of development and become food-ova (A. Portman, 1930). In all these cases, as in flatworms, some of the female genital products go to feed the embryos that develop from the others; but in flatworms specialised yolk cells, incapable of being fertilised, serve as food, whereas in polychaetes and gastropods fully-viable ova, and in polychaetes even embryos, are eaten.

Monogenea, Digenea, and lower cestodes (like Rhabdocoela, from which they are derived) possess separate vitellaria and germaria. Among higher cestodes the vitellaria are reduced in the order Cyclophyllidea (Fig. 187, *D*) and therefore the female gonad is represented only by the germaria, which might be taken for ovaries fully homologous with those of most Bilateria if comparison with other cestodes did not negate that possibility.

V. R. Veitsman (1939, 1953) asserts, however, that the germaria of Cyclophyllidea are vestigial organs that do not form ova capable of development; viable ova, according to him, are formed by the epithelium of the uterus. If that is confirmed, the situation will be analogous to the reproduction by vegetative larvae found in Scyphozoa, in particular in *Stygiomedusa fabulosa* (F. Russell and W. Rees, 1960).

Other relatives of Turbellaria (Gastrotricha and other Nematomorpha) always have true ovaries, more or less similar to those of lower Turbellaria such as Macrostomida.

The ovaries of Rotatoria contain, besides genital cells capable of development, nutrient cells that have lost that capacity, whose function is the supply of food material to the growing oogonia. That part of the ovary of Rotatoria is often erroneously called a vitellarium. The nutrient cells of Rotatoria and other Metazoa, however, differ from the vitellaria of flatworms in that they supply food material to the growing oogonia and not to the developing embryo. Among flatworms nutrient cells occur, as we have seen, in the ovaries of several Acoela.

The most primitive of Turbellaria, Xenoturbellida and some Acoela (*Haplodiscus, Childia, Nemertoderma, Hofstenia*, etc.), are entirely devoid of accessory parts of the female genital apparatus. Copulation probably takes place by attachment of spermatozoa to the skin of the partner (*Xenoturbella*) or by subcutaneous impregnation, when the spermatozoa are introduced by the male copulatory organ beneath the skin of another individual in any part of the body (Fig. 186, *A*). The spermatozoids inserted beneath the skin, or attached externally, make their way to mature ova through the parenchyma. This method of copulation is retained by many Polyclada (e.g. *Anonymus*) and Alloeocoela (e.g. *Baicalarctia gulo;* G. Fridman, 1933), which have no female copulatory organs. The latter are most simple in *Aphanostoma sanguineum* (Acoela; Fig. 186, *B*), which has a vagina in the form of a tubular fold of the integument, opening to the exterior at one end and merely into the parenchyma at the other (W. N. Beklemishev, 1915). The spermatozoids of another individual pass through the tube into the parenchyma, through which they move actively to mature ova.

In other species of *Aphanostoma* a closed vesicle is formed at the end of a similar vagina, the copulatory pouch (bursa copulatrix), which receives spermatozoa during copulation (Fig. 186, *C*). In this case ova are probably fertilised when being deposited, when already outside the mother's body, by some of the spermatozoa that are squeezed out of the pouch through the vagina. In very many Acoela the pouch has openings provided with hard nozzles; these openings lead into the parenchyma, and the

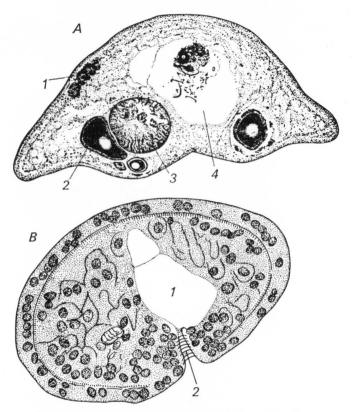

Fig. 186. *Types of female copulatory adaptations in Turbellaria.*
A—Hofstenia atroviridis (Acoela), cross-section, showing subcutaneous impregnation by spermatozoa: 1—mass of spermatozoa of another individual injected
beneath epithelium; 2—ovum surrounded by follicular epithelium; 3—mature
ovum; 4—digestive parenchyma (after Bock). *B—Aphanostoma sanguineum* (Acoela),
cross-section: 1—vacuoles in parenchyma; 2—vagina (after W. N. Beklemishev).
C—Aphanostoma pallidum (Acoela), cross-section: 1—ovary; 2—copulatory
pouch; 3—vagina (duct of pouch); 4—seminal ducts after) W. N. Beklemishev.
D—Polychoerus langerhansi (Acoela), diagram of copulatory apparatus in sagittal
section: 1—mature ovum; 2—mouth; 3—copulatory pouch; 4—its nozzles, dispensing spermatozoa as they pass to the ova; 5—male atrium; 6—male copulatory
organ (from Luther). *E—Gyratrix hermaphroditus* (Rhabdocoela), diagram of organisation: 1—testis; 2—vitellarium; 3—germarium; 4—spermatheca; 5—vagina
(dorsal orifice of spermatheca); 6—stylet of male copulatory organ; 7—receptacle
for granular secretion; 8—accessory (granular) male glands; 9—spermatheca
duct; 10—vitelline duct; 11—uterus; 12—seminal vesicle; 13—pharynx; 14—
proboscis; 15—brain (after Reisinger).

nozzles point towards the ovary. In copulation the spermatozoa enter the
pouch and thence gradually, in small batches, travel towards the ovary
(*Convoluta, Polychoerus;* Fig. 186, *D,* etc.).

Many Alloeocoela and Rhabdocoela have copulatory pouches homologous with those of Acoela, but opening into a common genital atrium
and not directly to the exterior. Sometimes these pouches end blindly, as in

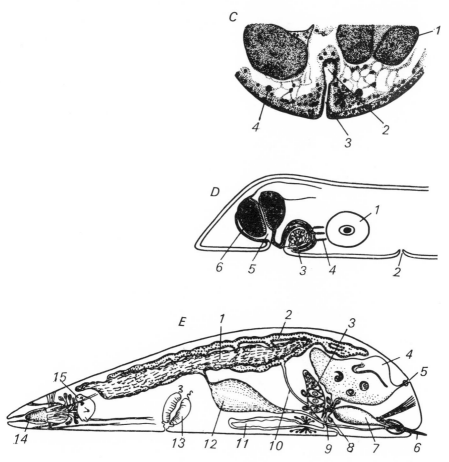

Fig. 186. (*continued*)

Aphanostoma (and in Dalyelliidae and Typhloplanidae among Rhabdocoela), and fertilisation takes place as the ovum passes through the efferent ducts of the ovary, by the return of some spermatozoa through the mouth of the pouch. In other cases the pouch has a nozzle, as in *Convoluta*, but on passing through the nozzle the spermatozoa enter, not the parenchyma, but a fine duct (ductus spermaticus) that connects the pouch with the oviduct or with a spermatheca, which is a widening of the oviduct (Anoplodiidae, Byrsophlebidae, and Proxenetidae among Rhabdocoela, Fig. 185, *C*; *Euxinia*, etc., among Alloeocoela). In all orders of Turbellaria there also occur copulatory appendages not homologous with those of Acoela; in Gyratricidae and Pterastericolidae (Rhabdocoela), for instance, the spermatheca opens by a small copulatory orifice directly to the exterior, and thus assumes the role of a copulatory pouch (Fig. 186, *E*).

Acoela, *Xenoturbella*, and Notandropora still have no efferent ducts for mature ova, which are expelled through the mouth or through a split in

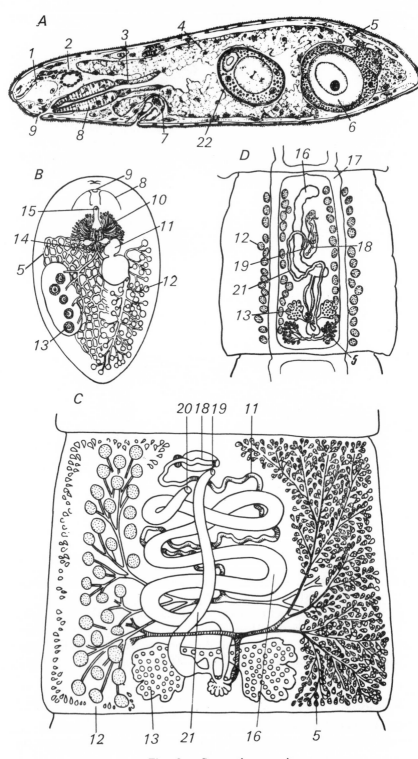

Fig. 187.—See caption opposite.

the body wall. Such ducts are also lacking in *Bresslauilla* among Rhabdocoela (E. Reisinger, 1929); there the ova enter the gut and are expelled through the mouth (Fig. 187, *A*). At the same time many Polyclada (e.g. *Anonymus;* A. Lang, 1884) and Alloeocoela (e.g. *Baicalarctia;* G. Fridman, 1933) possess female gonoducts used exclusively for expulsion of ova, while copulation takes place by subcutaneous impregnation. The female efferent ducts consist of paired oviducts, opening to the exterior or into a genital atrium by an unpaired ectodermal female genital canal; when vitellaria are present the oviducts are differentiated into oviducts and vitelline ducts. The female genital canal, like the genital atrium, the male ejaculatory duct, and, largely, the copulatory pouch, represent involutions of the external epithelium, as is evident from comparative-anatomical and (frequently) ontogenetic considerations (W. Beklemishev, 1916); the oviduct, yolk ducts, and spermatheca, on the other hand, are formed from the covering of the ovary, that is, they are mesodermal products, like the seminal ducts.

Usually some part of the female genital canal or the atrium serves as a *uterus*, i.e. a place of formation of egg-capsules. Sometimes the uterus is a separate sac (in Typhloplanidae, a pair of sacs) formed by invagination of the atrium wall. In many members of Polyclada and other orders the female genital canal and parturitional orifice are also used for admission of spermatozoa during copulation, undergoing various adaptive modifications. Thus the diversity of female copulatory organs increases still further. A number of forms in different orders of Turbellaria and Monogenea have a communication between the female genital ducts and the gut (communicatio genito-intestinalis). Taking *Bresslauilla* as an example, O. Steinböck and E. Reisinger believed that communication to be a relic of an original parturitional adaptation, whereby ova were expelled through the mouth. Actually it serves for ejection of surplus spermatozoa and yolk cells into the gut; and occasionally very different parts of the female ducts enter into communication with the gut, so that Steinböck and Reisinger's suggestion can be correct only in isolated cases.

Among the greatly-diversified genital apparatuses of higher turbellarians, the structure of the genital apparatus of the superfamily Dalyelliida (Rhabdocoela) is of particular interest, since in various families of that superfamily we find prototypes of the genital apparatuses of parasitic flatworms.

Fig. 187. *Parturitional adaptations of turbellarians and cestodes.*

A—Bresslauilla relicta (Rhabdocoela), sagittal section, ova deposited through mouth (after Reisinger). *B—Baicalarctia gulo* (Alloeocoela), diagram of organisation: female copulatory organs are lacking; copulation is by subcutaneous impregnation; female ducts are used solely for parturition (after G. M. Fridman). *C—Diphyllobothrium* (Cestodes, Pseudophyllidea), diagram of genital apparatus. *D—Mesocestoides lineatus* (Cestodes, Cyclophyllidea), mature proglottis with genital apparatus (after Fuhrmann). 1—frontal glands; 2—brain; 3—gullet; 4—gut; 5—vitellaria; 6—ova; 7—seminal vesicle; 8—pharynx; 9—mouth; 10—penis; 11—seminal duct; 12—testis; 13—germarium; 14—oviducts; 15—gonopore; 16—uterus; 17—nephridial canal; 18—cirrus; 19—vaginal orifice; 20—uterine orifice; 21—vagina; 22—mature egg-capsule in gut cavity.

Digenea, like Graffillidae (parasites of molluscs) and Pterastericolidae (parasites of starfishes), have no copulatory pouches. The short male genital canal and the very long female genital canal (here called the uterus and actually functioning as such) open into a small genital atrium. The oviduct, vitelline ducts, Laurer-Stieda canal, and spermatheca open into its proximal end (Fig. 188, *C*). The Laurer-Stieda canal opens directly to the exterior and is used mainly for expulsion of surplus yolk cells; its morphological significance is not quite clear. The rest of the structure of the genital ducts is very similar to that in Graffillidae and Pterastericolidae.

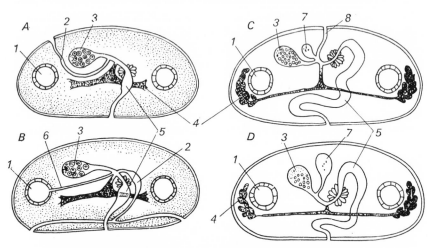

Fig. 188. *Parturitional and female copulatory adaptations of trematodes.*
Diagram of genital apparatus: *A—Dactylogyrus* (Monogenea); *B—Discocotyle* (Monogenea); *C—*Digenea, forms with Laurer-Stieda canal; *D—*Digenea, forms without Laurer's canal. 1—gut; 2—vagina; 3—germarium; 4—vitellarium; 5—uterus; 6—genito-intestinal canal; 7—spermatheca; 8—Laurer-Stieda canal (after Fuhrmann).

The structure found in Monogenea (Fig. 188, *A*) is similar in principle, but instead of a Laurer-Stieda canal they have a paired or unpaired vagina, also leading from the proximal end of the female genital canal to the exterior and used for admission of spermatozoa during copulation. Its external openings are always quite independent of the opening of the genital atrium. That adaptation is very similar to what we see in Pterastericolidae.

In cestodes the vagina always opens, together with the male genital apparatus, into a single common orifice, whereas the opening of the uterus is usually independent (Fig. 187, *C*). As in trematodes, the uterus is homologous with the female genital canal of Dalyelliida. The oviducts, vitelline ducts, and vagina open into its proximal end. We may surmise that the cestode vagina, like the ductus vaginalis of parasitic Dalyelliida of the family Anoplodiidae, is formed at its distal end from the copulatory pouch and at its proximal end from the spermatheca. This explains the permanent connection between its distal section and the male copulatory organ. A close topographical connection between the copulatory pouch and the male copulatory organ is characteristic of all Rhabdocoela. Both genital orifices (parturitional and copulatory) in cestodes are fully homologous with the two genital orifices of *Desmote* (Anoplodiidae). In Cyclophyllidea (Fig. 187, *D*) the uterine orifice is reduced; in other words, the female genital canal is cut off from the external environment and converted into a blind pouch, from which ova are freed only by the rupturing of proglottides.

Therefore there is no complete homology between the female ducts of Monogenea and Digenea and those of cestodes. Both represent extreme developments of two different modifications of the genital apparatus of Dalyelliida, and in this respect Monogenea are closer to Digenea than to cestodes (see Vol. 1, Chap. IV; cf. also L. Hyman, 1951). In spite of the great diversity of the genital apparatuses of parasitic flatworms, they are all derived from two types of genital apparatus found in the same superfamily of the order Rhabdocoela; therefore the differences in structure of the genital apparatus within the parasitic classes have a very much smaller morphological range than those within the class Turbellaria, and relate to characteristics of a different order. Only the modifications of the female gonads of cestodes are of great significance.

The genital apparatuses of Nemathelminthes are derived from a prototype close in structural plan to those of lower turbellarians such as Macrostomida, and Gastrotricha are closest to that prototype. All Nemathelminthes have simple ovaries (see page 403 regarding the ovaries of Rotatoria); the majority have well-developed gonads, with envelopes and ducts, and copulatory organs of relatively simple structure.

The possession of copulatory organs by Nemathelminthes has an ecological basis as well as a phylogenetic explanation (derivation from flatworms). Small animals have only small individual fecundity, small sexual productivity. Therefore external fertilisation, which requires a high concentration of genital products in the water, can be achieved by small forms only if there is a colossal number of individuals: Rotatoria, Gastrotricha, and small nematodes are among the most minute of Metazoa and usually produce not more than one ovum at a time. Internal fertilisation is therefore very appropriate for them. It is possible that Scolecida have been able to produce so many small and minute forms only because they acquired internal fertilisation at an early stage.

Regarding the connection between the genital and excretory apparatuses of Rotatoria, Acanthocephala, and Priapuloidea, see Chap. VII.

4. GENITAL APPARATUS OF NEMERTINES, ANNELIDS, AND MOLLUSCS

The primitive types of gonads that occur in lower Turbellaria and bring them close to sponges in level of development are not found outside the flatworm group. Copulatory organs have evidently arisen independently, however, in all the principal groups of the animal kingdom. Each of these groups contains primitive forms with external fertilisation and without copulatory organs, standing at the coelenterate level in that respect; at the same time very many groups have developed external-internal or internal fertilisation and the copulatory organs required for that purpose.

Adaptations for internal fertilisation are extremely diversified in different groups, and sometimes within a single group; many of them show considerable convergent similarity to the different types of copulatory organs and different methods of copulation observed in flatworms.

The genital apparatus of all nemertines is extremely simple (see Fig. 109). Their gonads, in the form of epithelial sacs filled with genital products, lie in two longitudinal rows in the body parenchyma. The gonoducts are formed at the time of maturation of genital products, and are short outgrowths of the gonad walls, opening independently to the exterior.

Fertilisation is usually external in nemertines; usually the genital products are expelled into a mucous mass jointly excreted by the male and the female (*Lineus, Amphiporus*); in *Micrura alaskensis* up to 50 individuals congregate into a ball and excrete common mucus, in which fertilisation and later development of the ova take place. In *Carcinonemertes* only spermatozoa are discharged into the mucus with which the male and the female surround themselves; from there they penetrate into the body of the female, and fertilisation is external-internal (W. Coe, 1943). Thus vivipary actually occurs in some nemertines (e.g. *Prosorhochmus;* V. Zalenskii, 1909) in spite of the absence of true copulation. Only a few pelagic nemertines possess male copulatory organs, in the form of small papillae on which the gonopores open (Fig. 66, *A*). Perhaps the presence of copulatory organs in these forms is not accidental but is due to the difficulty of pseudo-copulation within a mucous mass in planktonic living conditions. In the same way the pelagic Alciopidae (Phyllodocemorpha) are among the few polychaetes that possess copulatory organs.

We must point out that on the whole the structure of the genital apparatus represents one of the greatest organological differences between nemertines and flatworms.

In all Trochozoa the gonads develop on the coelom walls (Fig. 189, *A*), although, as we have seen, the genital elements are not originally connected with the coelomic epithelium, but settle there secondarily (Fig. 189, *B*). In discussing the coelom theory we came to the conclusion that the coelom is homologous with the peripheral sections of the gastrovascular apparatus of higher coelenterates and, like the radial chambers of Actiniaria or the meridional canals of Ctenophora, serves as the place of development of genital products. The ducts of the coelom, which are homologous with the intestinal pores of coelenterates, are used to expel genital products from its cavity, since expulsion through the mouth is no longer possible because of lack of communication between the coelom and the gut. The gonads proper occupy only a limited area on the coelom walls, and in this respect resemble the gonads of Actiniaria, which develop only on limited areas of the radial septa. Therefore as a result of the epithelisation of the peripheral phagocytoblast the structural plan of the genital apparatus of Trochozoa differs sharply from that of flatworms. Much information on the genital apparatuses of annelids and molluscs has already been given in the chapters on metamerism (Vol. 1, Chap. VI) and on the coelom (Vol. 2, Chap. V).

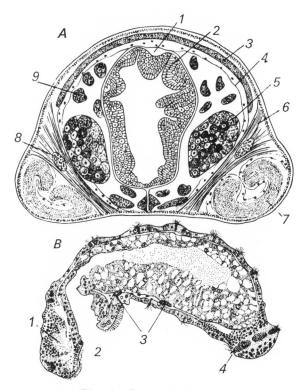

Fig. 189. *Gonads of polychaetes.*

A—Saccocirrus major, cross-section: 1—dorsal vessel; 2—gut; 3—dorsal longi-tudinal muscles; 4—coelomic epithelium; 5—ovary; 6—oblique muscles; 7—ventral longitudinal muscles; 8—bundle of spines; 9—mature ova (after Pieranton). *B*—sagittal section of larva (metatrochophore) of *Spio*, consisting only of larval segments: 1—brain; 2—mouth; 3—primary genital cells, which later creep into the post-larval segments, where they settle in the coelom wall and give rise to the gonads; 4—telotroch (after P. P. Ivanov).

External fertilisation occurs in the great majority of polychaetes. Genital products are expelled through the coelomoducts or, when these are reduced, through a split in the body wall (many Polygordiidae, Nereidae, Syllidae, etc.) directly into the water.

The ova are fertilised either in the water, or in a mucous mass excreted by the sexual partners, or in a dwelling-tube; in some cases the genital products are expelled with immediate contact between the male and the female (pseudocopulation). Evidently in certain forms the appearance of pseudocopulation provided the stimulus for development of true copulation by means of anatomically-distinct copulatory organs. In *Saccocirrus*, for instance, the nephromixia lying in the posterior, genital segments form seminal vesicles in the male, and their orifices lie on small conical papillae. The nephromixia of the female form spermathecae. On copulation the papillae of several segments of the male are inserted into the orifices of the nephromixia of the female. Something similar occurs in a few other families, e.g. in Hesionidae, Alciopidae, and Capitellidae. In some polychaetes, therefore, the parturitional orifices, which originally were coelomoduct orifices, are used also as copulatory orifices. The nature of the copulatory organs here repeats to some extent the type developed in nemertines, although the two groups are not closely related.

The sand-dwelling *Protodrilus rubropharyngeus* (G. Jägersten, 1952) employs external-internal fertilisation: pellets of spermatozoa are expelled through the coelomoducts of the male, which in this species open on the sides of the eleventh segment. On the way the spermatozoa are coated with a film of secretion, and they enter the sand as minute spermatophores, which disintegrate only some hours later. Each male ejects many spermatophores, and during the reproductive period the sand must contain vast numbers of them. If a female accidentally touches a spermatophore she turns over and applies to it one of the glandular-ciliated dorsal pits lying on the 17th and subsequent segments of her body; the spermatophores adhere by their mouths to the bottom of the pit, and the spermatozoa pass through its wall into the coelom and fertilise the ova there. This method of fertilisation, based on accidental encounter of the female with spermatophores scattered in the substrate, is possible only for small mobile animals in dense populations. We shall meet it again in soil-dwelling arthropods.

Unlike the great majority of polychaetes, all oligochaetes are hermaphrodites (Fig. 190). Their genital glands are restricted to a few definite segments; in Lumbricidae, for example, the male glands are in two segments and the female in one. The dissepiments of the segments containing the testes form evaginations, which fill with spermatozoa (seminal sacs). The genital products are expelled through typical coelomoducts, which develop only in the genital segments. When earthworms copulate two individuals surround their genital segments with a sleeve which is formed by the secretion of the clitellar glands, and the spermatozoa of each enter the spermatheca of the other, after which the partners separate. The spermathecae are small sacs, usually opening only to the exterior; rarely, in some Enchytraeidae and some other forms, the spermathecae also communicate with the gut cavity, displaying some analogy to the communicatio genito-intestinalis of flatworms. The ova are laid in a cocoon formed round the genital segments by the secretion of the same clitellar glands, and the spermatozoa received from the other individual during copulation are expressed into the cocoon from the spermathecae. Fertilisation in oligochaetes is still external, therefore, in spite of the existence of copulation, although it is uniquely modified on account of the lack of direct communication between the female copulatory organs and the female efferent ducts.

In this respect oligochaetes display a type of sexual behaviour similar to what we have seen in most species of *Aphanostoma* (Acoela) among turbellarians. In fact I regard the copulation of Lumbricidae, in which the two individuals are united by a secretion, and also the deposition and fertilisation of ova in a cocoon made of cutaneous-gland secretion, as a modification of the pseudo-copulation within a mass of joint secretion that we have seen in a number of polychaetes (*Scolecolepis*, etc.).

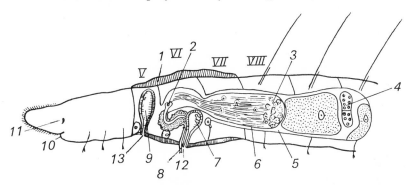

Fig, 190. *Genital apparatus of Oligochaeta (family Naididae), diagrammatic.*
1—clitellum; 2—funnel of coelomoduct, used as gonoduct; 3—testis; 4—ovary; 5—egg-sac, formed as a rearward evagination of the dissepiment between segments VI and VII of the coelom; 6—seminal sac, formed by the dissepiment between segments V and VI and invaginated into the egg-sac; 7—male atrium; 8—penial setae; 9—spermatheca; 10—mouth; 11—eye; 12—male external orifice; 13—female external orifice; at this stage the ovaries and testes are greatly reduced (after Piguet).

In leeches (Fig. 191), where the coelom is reduced to a canal system, the gonads have become separate sacs. They have their own ducts, formed partly of modified coelomoducts, opening to the exterior by two orifices, male and female. The 'testes' of *Acanthobdella* are outgrowths of the coelom of the tenth somite, projecting into the coelenchyma and gradually extending rearward along the sides of the gut as the animal grows older; they are homologous with the seminal sacs of Lumbricidae. On their walls the male gonads develop diffusely: the coelomoducts of the tenth somite serve as efferent ducts leading into an unpaired atrium or spermatophore chamber on the ventral side of the tenth segment. In other leeches elongated testes of *Acanthobdella* type have developed on separate, largely metamerically-arranged follicles; fine ducts joining them (vas deferens and vasa efferentia) are constricted parts of the same sac, while the terminal parts of the ducts, judging by their development, are homologous with the coelomoducts of the tenth somite. Therefore the multiplicity and metamerism of the testes of leeches are secondary. Similarly the ovaries of leeches are modified egg-sacs, i.e. individualised parts of the coeloms of the twelfth somite, enclosing the female gonads; the paired oviducts are usually united into an unpaired duct, which usually opens to the exterior on the twelfth body segment (N. Livanov, 1906).

Both the gonopores, male and female, are used by leeches for expulsion of genital products. Most Rhynchobdellea copulate by means of spermatophores formed from a glandular secretion entering the unpaired terminal

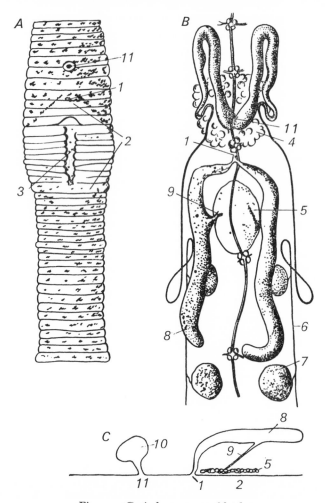

Fig. 191. *Genital apparatus of leeches.*

Piscicola geometra (Ichthyobdellidae): *A*—external aspect, *B*—internal structure (after Brumpt); *C*—diagrammatic longitudinal section (after Autrum). 1—parturitional orifice; 2—area copulatrix; 3—groove on area copulatrix; 4—glands of atrium, which form spermatophores; 5—mass of conducting tissue; 6—efferent seminal ducts; 7—testes; 8—ovaries; 9—cords of tissue through which spermatozoids pass from the mass of conducting tissue (5) that underlays the area copulatrix (2) to the ovaries (8); 10—male atrium; 11—male gonopore.

section of the male efferent ducts. The spermatophores attach themselves to the female's skin, and the spermatozoids penetrate into her parenchyma and reach the ovaries through it. In other words, most Rhynchobdellea employ subcutaneous impregnation, like many primitive turbellarians.

Some Ichthyobdellidae (e.g. *Piscicola*) have a 'copulatory zone' on the ventral surface of the clitellar region—an area of epithelium designed for attachment of spermatozoids (Fig. 191). Cords of 'conducting tissue', formed by the ovaries and serving as a route for spermatozoid penetration (E. Brumpt, 1900; V. Zelenskii, 1915) approach the skin at that point. In a few Rhynchobdellea (e.g. *Ozobranchus*) a female copulatory orifice and copulatory canals running from it to the ovaries are formed from the copulatory zone and the conducting tissue (A. Oka, 1904; V. Zelenskii, 1915). Thus Rhynchobdellea repeat independently, in a still clearer form, the same evolution from subcutaneous impregnation to preformed genital routes that we have seen in turbellarians. Most Gnathobdellea, unlike Rhynchobdellea, have a male copulatory organ, and fertilisation takes place through the female gonopore and the oviducts; like some Polyclada, therefore, Gnathobdellea use the parturitional passages for copulation.

Leeches deposit their ova in so-called cocoons, more precisely, egg-capsules; the latter are formed from sleeves of a protein secretion of the clitellar glands, slipped off over the head, as with earthworms. The ova are fertilised in the female gonoducts before entering the protein secretion. I believe that the oviposition method used by leeches is an extreme modification of that of oligochaetes, and represents the last link in the chain of evolution of oviposition methods that begins with polychaetes (Phyllodocidae, *Scolecolepis*, etc.).

In molluscs the gonads represent a more or less individualised section of the coelom. Very many molluscs are hermaphrodites. These include some Solenogastres, many Lamellibranchia, all Pulmonata and Opistho-branchia, and some Prosobranchia. Loricata, some Solenogastres (*Chaeto-derma*), all Scaphopoda and Cephalopoda, many Lamellibranchia, and most Prosobranchia are unisexual.

Pelseneer (1895) believes that hermaphroditism in molluscs is always phylogenetically secondary. M. Giese (1915) has confirmed that statement, particularly for Gastropoda, and at present that view is widely held. Pelseneer also believes that hermaphrodites are the result of modification of females.

In molluscs hermaphroditism sometimes consists in possession of separate male and female gonads, but more often in possession of herma-phroditic glands. Genital cells usually develop over all the gonad walls, but in Cephalopoda they are concentrated on a limited area of the wall of the genital section of the coelom. We have already discussed the morpho-logical significance of the gonoducts of molluscs in Vol. 1, Chap. VIII, and Vol. 2, Chap. V. In some Gastropoda, especially in hermaphroditic forms, their structure becomes very complex, as it is in Cephalopoda.

Copulatory organs are absent in Loricata, in *Neopilina*, in most Soleno-gastres, in all Lamellibranchia and Scaphopoda, and in most Prosobranchia Aspidobranchia (Fig. 192). It is known that most Loricata, Lamelli-branchia, and Aspidobranchia spawn directly into the water, although in Aspidobranchia sometimes the individuals of different sexes come close together for this purpose (G. Thorson, 1946). Copulatory organs have developed in other Gastropoda, in Cephalopoda, and in a few Solenogastres.

In view of the location of the gonopores of Cephalopoda and Prosobranchia in the mantle cavity, a copulatory organ formed there could scarcely function. Therefore in these classes the male copulatory organ is formed at a distance from the primary gonopore. A muscular penis has developed on the right side of the head of Prosobranchia; from the

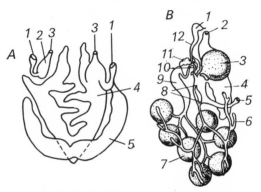

Fig. 192. *Genital apparatus of Gastropoda.*
A—Puncturella noachina (Prosobranchia Rhipidoglossa), an example of a primitive genital apparatus: the genital duct is a typical coelomoduct, and there are no copulatory organs (from A. V. Ivanov): 1—renopericardial orifices; 2—left kidney; 3—external renal orifices; 4—right kidney; 5—genital gland. *B—Limapontia depressa* (Opisthobranchia Nudibranchia Elysiomorpha), an example of the complex structure of the genital apparatus of higher Gastropoda (after Pelseneer): 1—penis; 2—oviduct orifice; 3—spermatheca; 4—mucus gland of oviduct; 5—female copulatory orifice; 6—albumen gland; 7—hermaphroditic genital gland; 8—albumen-gland duct; 9—hermaphroditic duct; 10—oviduct; 11—prostate gland; 12—seminal duct.

gonopore, which is located in the mantle cavity, a ciliated groove runs to the penis, extending along the latter to its tip and serving for transport of spermatozoa. In Stenoglossa and some Taenioglossa that groove closes into a canal, and then the male gonopore is at the end of the penis. It is curious that within the single subclass Prosobranchia fairly complex penes have apparently developed several times from different origins: the penes of most Pectinibranchia are innervated from the pedal ganglion and belong to the foot, but that of *Viviparus* is innervated from the cerebral ganglion and therefore belongs to the head, and that of *Ampullaria* is innervated from the pallial ganglion and therefore belongs to the mantle. The gonopore of the female remains in the mantle cavity. Hermaphroditic Prosobranchia consequently have two gonopores: female in the mantle cavity and male on the penis. Opisthobranchia (Fig. 192, *B*) and Pulmonata have independently followed the same evolutionary path. Among them we also find forms in which the penis is connected to the gonopore by a ciliated groove (*Scaphander* and other Bullomorpha, Aplysiomorpha) and forms with two gonopores (like hermaphroditic Prosobranchia), for instance *Actaeon* and, among Pulmonata, most Basommatophora; finally, in most Stylommatophora the male and female orifices are close together and a common genital atrium is formed. We must also point out that a separate female genital copulatory orifice has developed in some species in all orders.

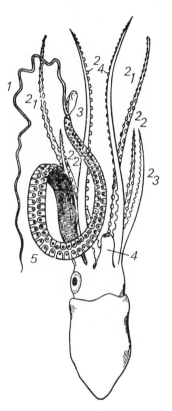

Fig. 193. *Male* Ocythoë catenulata (*Cephalopoda Octopoda*), *with hectocotylised arm on the right side.* 1—terminal filament of hectocotyle, derived from apical sac of hectocotyle; 2_1–2_4—arms; 3—apical sac of hectocotyle; 4—infundibulum; 5—hectocotyle (from Ray Lankester).

In Cephalopoda the arms have assumed the role of a copulatory organ. Usually one arm is adapted for that purpose (Fig. 193), and sometimes the two arms of one pair, the pair adapted differing from one genus to another. The most specialised are *Argonauta, Philonexis,* and *Tremoctopus* (Octopoda), whose copulatory arm bears a longitudinal canal filled with spermatophores, opening terminally to the exterior. That arm develops within a special sac that envelops it. On copulation it breaks off and remains within the female's mantle cavity, retaining its mobility and its capacity for reflex action for a long time; such arms were originally taken to be parasites and were called *Hectocotyle.* Now the name hectocotyle is kept as an anatomical term, denoting an arm of the above-named Octopoda modified for copulatory purposes. The less-perfect copulatory arms of other Cephalopoda, which do not break off during copulation, are called hectocotylised arms. In *Nautilus* one of the internal lateral lobes of the foot, left or right according to individual variation, is hectocotylised. Hectocotylised arms and actual hectocotyles represent a new type of male copulatory apparatus that we have not met before—a copulatory organ without direct communication with the male efferent ducts, formed far from the genital orifice, obtaining spermatozoa from the latter and transmitting them to the female. We shall find this type of copulatory organ again in arthropods.

On the whole, the genital apparatus attains great complexity in Cephalopoda and some Gastropoda. Without dwelling on details, we may point out that the diversity of methods of origin of the copulatory apparatus observed in these two classes proves that in them, as in annelids, that apparatus is not inherited from more primitive forms but has arisen independently in each class. The absence of a copulatory apparatus in some molluscs is just as primary as its absence in most polychaetes and nemertines. That is not surprising if one regards molluscs as an independent branch of Trochozoa, parallel to annelids and, like them, independently linked with a common prototype of all trochophore animals, which stood at the relatively low organisational level of coelenterates.

5. GENITAL APPARATUS OF ARTHROPODS AND ONYCHOPHORA

As we have seen in Chapter V, in Onychophora and all arthropods that possess a coelom during their development the gonad walls develop from the coelom walls, whereas the gonad cavities have different morphological significance in different cases. In some groups (Onychophora, Chilopoda) the gonad cavity is a vestige of the coelom; in others it corresponds to schizocoelic spaces compressed between coelomic pouches (*Lernaea* among Copepoda, Chelicerata); in others (insects) it is a new formation. The gonad cavity of insects (*Blatta* among Blattoidea, *Donacia* among Coleoptera) arises through separation of the cells of a ridge formed by the splanchnic wall of the dorsal section of the coelom, which gives rise to the gonad itself. Thus the gonad cavity of insects is apparently not homologous with either the coelomic or the schizocoelic cavities found in the gonads of some other arthropods.

In the crayfish (*Astacus*) the gonads appear comparatively late, when the abdominal coelomic pouches found in *Astacus* larvae no longer exist; the rudiments of the gonads are dense accumulations of mesodermal cells lying in pairs in the 14th, 15th, and 16th body segments; cavities form in them, and they fuse into one unpaired rudiment (H. Reichenbach, 1886). From their location these pouches may be homologous with the dorsal sections of the coelom, but that homology, like the homology of many other rudiments of gonads in arthropods, is still conjectural.

As a rule the gonoducts of arthropods are formed, in cases where the evidence is clear, partly from coelomoducts. Such gonoducts may be called morphologically-primary. At the same time secondary gonopores and secondary gonoducts of ectodermal origin often arise. Examples are the male gonopore of Gamasoidea (order Parasitiformes) (A. Zakhvatkin, 1952) and the gonoducts and gonopores of Pauropoda, Diplopoda, and Symphyla (O. Tiegs, 1947).

W. M. Schimkewitsch (1906) has analysed the problem of the origin of the gonads of Arachnoidea. He suggests that originally genital products developed in the coelom walls and entered the coelom cavity, and naturally passed to the exterior through the coelomoducts (Fig. 194, *A*). When the coelom disappeared the part of its walls where genital products were formed was retained: in Telyphones that was represented by layers of the ventral mesentery. They grew together, and the schizocoelic space enclosed in the

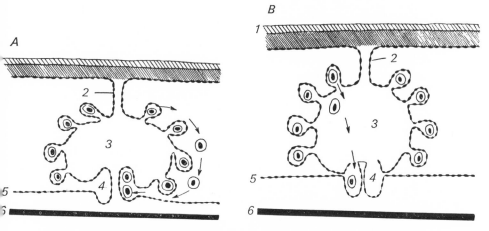

Fig. 194. *Diagram of origin of modern structure of genital apparatus of Telyphones* (*Chelicerata*).

A—annelid-like prototype with genital products entering coelom, from which they are expelled through coelomoducts. *B*—modern Telyphones with gonoducts of coelomic origin, opening into gonad cavity of retroperitoneal origin. 1—gut epithelium; 2—mesentery; 3—gonad cavity; 4—oviduct; 5—coelomic epithelium; 6—external covering (after Schimkewitsch).

mesentery constituted the cavity of the gonad so formed. The genital products began to be discharged into that cavity (i.e. in a retroperitoneal direction), and the coelomoducts that had opened into it by their infundibula also became attached to it (Fig. 194, *B*).

The above theory was worked out by W. M. Schimkewitsch for one particular case, but it is appropriate to ask the same questions regarding all arthropods: how have the gonads developed, and how have their relationships with the coelomoducts changed with the reduction of the coelom, on which the evolution of the higher Articulata has been based? We have seen that, in all the arthropods and Onychophora regarding which embryology enables us to form an opinion, the gonad walls represent part of the coelom walls, and the gonoducts (in their proximal parts at least) are formed from coelomoducts. Moreover, in all cases the coelomoducts have become attached to the gonads and have thus entered into a direct connection with them absent in the annelid prototype (Figs 195 and 196).

Otherwise great diversity is observed: the sections of the coelom (dorsal, ventral) from which the gonads are formed, and the serial numbers of the somites in which the gonads and their ducts are formed, vary; the gonad cavity also has different origins. This diversity suggests rather that in the evolution of arthropods the process of coelom reduction and the associated reconstruction of the genital apparatus took place independently, and took different courses, in different groups.

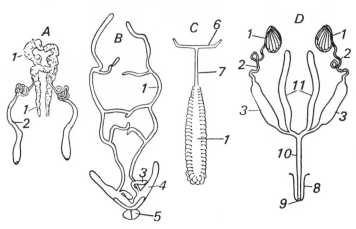

Fig. 195. *Male genital apparatus of arthropods.*

A—Palaemon (Crustacea Decapoda) (from Lang). *B—Scorpio maurus* (Scorpionoidea) (after Pavlovskii). *C—Glomeris marginata* (Diplopoda) (after Fabre). *D—* diagram of male genital apparatus of insects (from Snodgrass). 1—testes; 2—seminal duct; 3—seminal vesicles; 4—paraxial organ; 5—genital opercula; 6—paired efferent ducts; 7—unpaired seminal duct; 8—penis; 9—gonopore; 10—ejaculatory duct; 11—accessory glands of genital apparatus.

Both the gonads and the coelomoducts of polychaetes are paired and metameric, and in primitive forms occur in many segments of the body. Among modern arthropods only Pantopoda have metamerically-repeated paired gonopores: in primitive cases, in both males and females, they occur on all four pairs of legs (Fig. 197). The evolutionary trend was towards oligomerisation of the number of orifices, especially in males. The orifices disappear from front to rear. The process has been completed independently and in parallel in several branches of Pantopoda. All other arthropods have retained only one pair of gonoducts and one, usually unpaired, gonopore. As well as in Pantopoda, the primitive pairing of gonopores persists in most Crustacea; in Xiphosura, Diplopoda, and Pauropoda; and among insects in both sexes of Ephemeroptera, and in the males only of Protura and some Dermaptera (B. Shvanvich, 1949).

In the female genital apparatus of insects a primary type of structure appears in the existence of seven pairs of metameric rudiments of ovarioles in abdominal segments 2 to 8 (e.g. in *Lepisma;* A. Sharov, 1953); as the animal grows older the metameric arrangement is lost through shortening of the oviducts. In most insects the rudiments of the ovaries also lose their metameric arrangement, which later easily leads to polymerism of the ovarioles. The rudiments of the testes of *Lepisma* are also metameric, but are represented by only three pairs, in abdominal segments 3–4, 4–5, and 5–6.

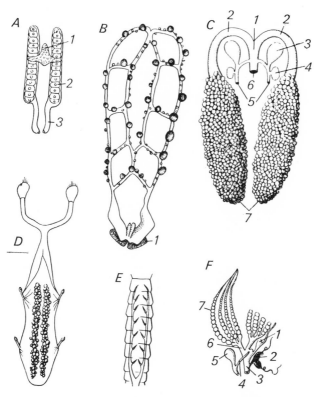

Fig. 196. *Female genital apparatus of arthropods.*

A—Mysis oculata relicta (Crustacea Mysidacea): 1—germinal zone of ovary; 2—its lateral sections; 3—oviducts. *B—Lychas variatus* (Scorpionoidea), tubular-network ovaries, paired oviducts: 1—genital opercula (after E. N. Pavlovskii). *C*—diagram of female genital apparatus of the spider *Aranea* (Araneina): 1—unpaired oviduct; 2—paired oviducts; 3—spermatheca; 4—its gland; 5—copulatory orifice; 6—parturitional orifice; 7—ovary (after J. Millot). *D—Glomeris marginata* (Diplopoda), paired oviducts running forward, egg-case opened, showing two ovaries (after Fabre). *E—Japyx* (Campodeoidea), with paired oviducts and meta-merically-arranged ovarioles (from Weber). *F*—diagram of female genital apparatus of butterfly (Lepidoptera): 1—spermatheca; 2—accessory glands; 3—unpaired oviduct; 4—ductus spermaticus; 5—copulatory pouch; 6—paired oviduct; 7—ovarioles (from Weber).

The form and structure of the gonads and the differentiation of the gonoducts are much diversified in arthropods, and are of great morphological interest in the separate classes, but discussion of them is outwith the scope of this book. We shall mention only some peculiarities in the evolution of the copulatory apparatus in arthropods. Arthropods are

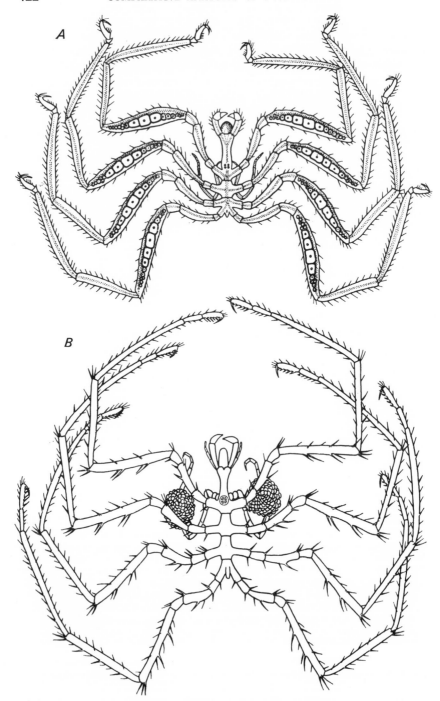

Fig. 197. *Genital apparatus of Pantopoda.*
A—Pallene brevirostris, female, showing location of mature ova in femurs, *B—Nymphon robustum,* male, showing clusters of ova attached to ovigers (after G. O. Sars).

derived from lower annelids, that is, from animals with external fertilisation and without copulatory organs. Therefore the appearance of the latter in all the groups we have discussed has been independent of their origin. The extreme diversity of copulatory adaptations in arthropods shows that they have arisen independently in the various branches of that subphylum.

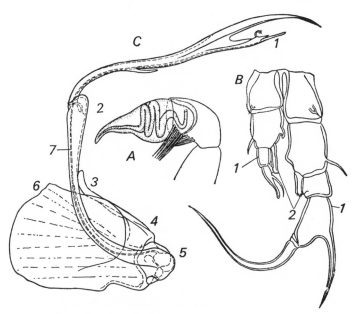

Fig. 198. *Male copulatory organs of some arthropods.*

A—Filistrata testacea (Araneina), last segment of pedipalp of male: spiral canal opening on sharp end of segment receives spermatozoa from the genital orifice proper, and transfers them to the female's spermatheca (from Lang). *B—Diaptomus gracilis* (Copepoda), fifth pair of legs of male; the right leg has a hook-like bend and is used to grasp the female; with the pincer-like left leg the male transfers spermatophores from the gonopore proper into the female's gonopore: 1—exopodite; 2—endopodite (after Rylov). *C*—leg of the seventh diplosegment of male *Bollmannia nodifrons* (Diplopoda), modified into a gonopod, lateral view: 1—orifice of canal containing spermatozoa; 2—distal end of femur; 3—outgrowth of coxa; 4—coxa of gonopod; 5—head of articulation of telopodite and coxa; 6—tracheal sac of gonopod segment; 7—canal containing spermatozoa (after Lohmander).

Among Chelicerata, Xiphosura still use external fertilisation: the female deposits her ova in a pit dug in the sand on the sea-shore, and the male pours semen over them (P. Iwanow, 1933). A similar type of external fertilisation occurs in Pantopoda: the male and the female appose their ventral sides (pseudocopulation); the female expels ova, and the male spermatozoa; the fertilised ova, adhering together, coil round the oviferous legs of the male and are carried there.

Some Crustacea have no male copulatory organs, and in copulation the two individuals simply appose their gonopores. Others have a true penis, e.g. Cirripedia, where it is formed from the rear end of the abdomen, and *Priapion* (Entoniscidae, order Isopoda), where it is a process of the seventh thoracic segment. In the crayfish (*Astacus*) the first pleopods are converted into copulatory organs, rod-shaped, with grooves, attached directly to the gonopore. The ova are fertilised outside the mother's gonoducts, because during copulation the spermatozoa adhere to the ventral surface of the female's eighth thoracic segment. In depositing the ova the female bends her abdomen, covering the ventral surface of the thorax, and a secretion from the oviducts dissolves the secretion that causes adherence of the spermatozoa; the latter are then freed and fertilise the ova as they emerge from the gonopore. This unique fertilisation mechanism, found in many Macrura, has evidently been developed within the order Decapoda and is a modification and complication of external fertilisation. In higher Decapoda (Brachyura) the spermatozoa are inserted directly into the female genital passages during copulation.

Among Copepoda, the fifth left thoracic leg of the male in Centropagidae is converted into pincers, the two halves of which are formed by the exopodite and the endopodite (Fig. 198, *B*). With these pincers the male takes spermatophores out of his own gonopore and transfers them to that of the female. The fifth right leg of the male is hook-shaped and is used to grasp the female. This type of male copulatory organ differs from what we have just described in *Astacus* and is similar in principle to those of Ixodidae and Solifugae, with the difference that in Centropagidae the limbs used for transferring spermatophores are specially modified for that function. Female Copepoda have spermathecae opening independently to the exterior, connected to the terminal section of the ovary by fine ducts. In this they resemble female spiders. We find similar female copulatory adaptations in many flatworms, leeches, and snails.

During the last 15 or 20 years many data on methods of fertilisation in terrestrial arthropods have been accumulated, enabling M. S. Gilyarov (1958) to build up a general picture of the evolution of that process. As we have seen in Chap. II, during the transition from aquatic to terrestrial life arthropods lived in soil, forest litter, and other protective habitats. It appears that many, if not all, primarily-soil-dwelling arthropods are characterised by a single common pattern of external-internal fertilisation, convergently resembling the fertilisation method of *Protodrilus rubropharyngeus* (Jägersten, see above). In one of the most primitive types the male affixes small claviform spermatophores to soil particles (Oribatei among Acariformes, *Campodea* among insects; F. Schaller, 1954), usually doing so regardless of whether a female is present. The female, on accidentally encountering a spermatophore, investigates it and then 'ingests' it with her gonopore. Or the male draws threads of secretion through the pores of the soil and deposits unprotected drops of semen on them; the female, encountering one of these threads and moving along it, arrives at the semen and takes it into her gonopore. We find this in Myriapoda: Geophilomorpha (J. Fabre, 1879); in *Polyxenus* (Diplopoda Pselaphognatha; K. Schömann and F. Schaller, 1954) (Fig. 199, *A*); in most Collembola (F. Schaller, 1952, 1953; D. Poggendorf, 1956; H. Mayer,

1956, 1957); and in Lepismatidae (G. Spenser, 1930; H. Sturm, 1956). Male Pselaphognatha possess an organ that is usually called a penis, but, since it is used not to introduce semen into the female's body but to place it on a filament laid down by the male, Gilyarov calls it a spermatopositor. Oribatei (Acariformes) also have a spermatopositor.

There are many divergences from that basic pattern of behaviour, in many respects parallel to one another. Male scorpions (H. Angermann, 1955; F. Schaller, 1955; Alexander, 1956) and Pseudoscorpiones attach spermatophores to the substrate: in some Pseudoscorpiones, regardless of whether a female is present; in most Pseudoscorpiones and all scorpions that have been studied, during courtship performance. In the latter case the male usually drags the female to the spermatophore, which she then takes up in her gonopore. The pectiniform organs (Vol. 1, Fig. 124, D) of scorpions play a large part in the attachment of spermatophores by the male and in the groping for them by the female. Similar types of fertilisation occur in *Scutigera* (H. Klingel, 1956) and Sminthuridae (Collembola) (H. Mayer, 1956).

The spermatopositor that has developed in Acariformes (e.g. Oribatei) is used as a penis by Acaridia, their relatives; this group has advanced to internal fertilisation. Internal fertilisation has also been observed in Arachnoidea in Opiliones of the suborder Palpatores, where the male has a long penis and the female a long ovipositor, both formed from the walls of the ectodermal genital atrium.

Another evolutionary trend is seen in *Machilis germanica:* the male, on meeting the female, affixes a filament to the substrate and deposits a drop of semen on it by means of his spermatopositor; the female stands parallel to the filament, and the male collects the semen with an antenna and places it in the region of the female's gonopore—a first step in the use for copulation of appendages not directly connected with the gonopore (in this case, the antenna). This principle has been little developed in insects, but in other arthropod groups it plays a substantial role. In Solifugae the male, in the presence of the female, deposits a mass of spermatozoa on the ground, and then picks it up with his chelicerae and places it in the female's gonopore. In Parasitiformes fertilisation takes place by means of spermatophores, which the male takes from his own gonopore with his chelicerae and transfers to the female's gonopore. The male's chelicerae are somewhat modified for that purpose. Araneina (spiders) discharge spermatozoa not upon the ground but into a specially-woven prenuptial web, from which they take them into the cavity of the much-modified last segment of the pedipalps (Fig. 198, A); later, on meeting a female, they insert them into her spermathecae. The spermathecae usually open to the exterior by ducts, and in addition are connected by fine ductus spermatici to the unpaired oviduct (see Fig. 196, C). In male Ricinulei the third pair of walking legs (converted into gonopodia) are adapted for copulation (Vol. 1, Fig. 125, H).

In the same way most Diplopoda transfer spermatozoa by means of appendages. Opisthandria transfer spermatozoa from the male to the female orifices (both paired) with their mandibles, and thus are at the same level as Parasitiformes in this respect. In Proterandria one pair (sometimes both) of the appendages of the seventh diplosegment are converted into gonopodia, often very complex in structure (Fig. 196, *D*). Sometimes the appendages of the sixth and eighth diplosegments also are modified. In the most complex cases the spermatozoa are taken up by one of the accessory pairs of gonopodia, which transfer them to another, and so on in strict order of succession until ultimately the front pair of appendages of the seventh displosegment transfer them to the female's gonopore.[1] Here we have an extreme application of the principle of use for copulation of appendages that are not directly connected with the gonopore but are merely able to reach it with their distal ends.

The external genital organs of both sexes of Thysanura are of rather complex structure. In the female they are represented by an ovipositor formed by the gonapophyses of the eighth and ninth abdominal segments (Vol. 1, Fig. 145). The male has an unpaired spermatopositor, and along its sides a pair of parameres: their homology is not fully established, but probably both are products of the gonapophyses of the ninth segment.

Higher insects (Pterygota) use internal fertilisation, but a number of transitional forms show that it originated from external-internal spermatophore fertilisation. Spermatophores persist tenaciously in insects, even with purely internal fertilisation, and disappear only in the highest orders, in Hymenoptera and in most Coleoptera and Diptera. In Diptera they are still retained, for instance, by *Culicoides* (family Heleidae; B. I. Pomerantsev, 1932). The genital appendages of Pterygota are homologous with those of Thysanura. The females of many orders have an ovipositor (converted into a sting in some Hymenoptera). The spermatopositor of Thysanura has become a penis, by change of function; grasping appendages are formed along its sides from the coxopodites and styles of the ninth segment and are used to hold the female during copulation. Ephemeroptera (Vol. 1, Fig. 145) and some Dermaptera have paired penes. In the various orders of insects both male and female copulatory organs display innumerable modifications, discussion of which would lead us far afield.

The types of copulation and copulatory organs in arthropods thus present a multitude of development series, some parallel, some divergent. These series begin with external fertilisation, still retained by Xiphosura and Pantopoda. In soil-dwelling Chelicerata and Atelocerata the dominant type is external-internal fertilisation, very similar to that of the sea-sand-dwelling *Protodrilus* (although performed by entirely different organs). It is not impossible that the ancestors of at least some terrestrial arthropods acquired that method of fertilisation while still marine-dwellers; that is suggested by the presence of pectiniform organs in the Silurian sea-scorpion *Palaeophonus* (Vol. 1, Fig. 124, *B*). All the fertilisation

---

[1] The adaptations of these appendages, mutually complementary and ensuring their joint action in a single function, represent a typical case of coaptation in the sense given to that term by A. Vandel (1948).

methods of other terrestrial arthropods are derived from that type, and different copulatory organs arose for carrying them out in different series. Male copulatory organs developed most often according to one of the following three types, which we find also in Crustacea: (*i*) from expansion and evagination of the margins of the gonopore, or of the walls of the

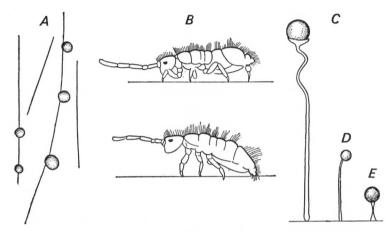

Fig. 199. *External-internal fertilisation in lower terrestrial arthropods.*
A—drops of semen on the cobweb-filaments of *Polyxenus lagurus* (Diplopoda, Pselaphognatha). *B*—fertilisation in *Orchesella villosa* (Collembola): above, the male depositing spermatophores on the substrate; below, the female picking up spermatophores from the substrate with her gonopore. *C, D, E*—spermatophores deposited on the substrate: *C—Belba* (Acariformes); *D—Orchesella*; *E—Campodea* (from M. S. Gilyarov).

genital atrium (e.g. in Opiliones); (*ii*) from appendages immediately adjoining the gonopore (e.g. in most Insecta Pterygota); and (*iii*) from appendages not adjoining the gonopore (e.g. in Solifugae, Araneina, most Diplopoda, etc.). In the first two cases, but not in the third, true copulatory organs apparently arose very often from spermatopositors through change of function (Oribatei-Acaridia, Thysanura-Pterygota).

The copulation methods and copulatory organs of Crustacea have still to be brought into the system.

Onychophora differ sharply from arthropods in their method of fertilisation: the male thrusts the spermatophore into the female's skin, and fertilisation consists in hypodermic impregnation, as in some leeches.

6. Genital Apparatus of Deuterostomia

The genital apparatus of Enteropneusta is very simple. It consists of a large number of paired gonads in the form of small sacs, opening to the exterior by short simple ducts. The gonads lie in the posterior half of the branchial section and behind it. Their orifices are on the dorsal side, forming two symmetrical and metameric rows. Their metamerism is due to the fact that the gonads alternate with the gill-pouches. New gonads are constantly being formed at the posterior end of each row. Fertilisation is external.

The genital apparatus of Pterobranchia differs only in the small number of their gonads (one pair in *Cephalodiscus* and *Atubaria*, one gonad in *Rhabdopleura*). Regarding the gonads of Pogonophora, see Vol. 1, Chap. IX.

At first glance the independence of the genital apparatus from the coelom in Hemichorda is surprising. Admitting the homology of the coelom of Deuterostomia with the peripheral section of the gastrovascular apparatus of coelenterates, we would naturally expect that in them the coelom would be the site for development of genital products. The history of the development of Enteropneusta fully supports that theory. Their gonads have in fact arisen from the trunk coelom (somatopleure), separating from the latter in the form of small vesicles.

The gonads of tunicates, because of the considerable reduction of their coelom, are also quite independent of the coelom, and are paired sacs, sometimes with fairly long ducts. Most tunicates are hermaphrodites. Ascidiae have either hermaphrodite glands, or a pair of ovaries and a pair of testes with simple ducts opening into the cloacal cavity. Appendicularia are protandrous hermaphrodites: their semen is expelled through ducts formed at the time of maturation of male genital products. The succeeding maturation of ova causes the animal's death, and the ova reach the exterior through disintegration of the body walls. Fertilisation is external in tunicates, except in Salpae and Pyrosomida, in which the single ovum is fertilised and develops inside the mother's body. In some Ascidiae (in the families Styelidae, Polyclinidae, and Didemnidae) the ova are fertilised in the mother's cloaca by spermatozoids of other individuals brought on water currents passing through the animals' siphons.

We have described the genital apparatus of echinoderms sufficiently in earlier chapters (Vol. 1, Chap. X and Vol. 2, Chap. V), and we have no need to return to it here. We shall merely remark that there also the gonads develop in close association with the coelom. Echinoderms have no copulatory organs, and fertilisation is usually external (see Fig. 179). Only in a few viviparous forms does fertilisation take place in the mother's body, but only by independent entrance of spermatozoids liberated into the water by males. This is the primitive type of fertilisation found in sponges and occurring sporadically in different groups of aquatic animals.

The gonads of *Amphioxus* are small paired metameric sacs pinched off from the coelom. They have no efferent ducts and are closely apposed to the walls of the peribranchial cavity. Spermatozoa are expelled into the peribranchial cavity through a short duct that develops in the males at the time of sexual maturation, as in Appendicularia; ova are expelled through a split in the wall of the gonad and in the wall of the peribanchial cavity adjacent to it. The genital products leave the peribranchial cavity along with the water current. Fertilisation is external.

The most primitive conditions among all Deuterostomia are found in lower vertebrates, in which the gonads develop on definite parts of the wall of the general coelomic cavity and the genital products are expelled through true coelomoducts, fertilisation being still external in most aquatic forms (Fig. 200). During the courtship performance of Urodela, which takes place in the water, the male deposits spermatophores on the bottom; the female 'ingests' them with the edges of the cloaca—a type of external-internal fertilisation closely resembling that seen in scorpions and *Scutigera*. In some specialised groups of the pisciforms and in Amniota, on the other hand, we find internal fertilisation, complex copulatory organs, and adaptations for vivipary.

On the basis of the material in this chapter we may assert that great complexity of the genital apparatus, the appearance of internal fertilisation, and the development of copulatory organs are characteristic of only the terminal branches of development and have arisen independently in each phylum of Bilateria and even in separate groups within each phylum.

Fig. 200. *The pelagic egg of the fish* Serranus cabrilla (*Teleostei*), *fertilised and developing in the water like the ova of many marine invertebrates* (after Lo Bianco).

We have seen the same independent appearance of organs and structures in different stems of the animal kingdom and during our survey of other functions and apparatuses: independent appearance of the liver, anus, nephridia, coelom, and circulatory vascular in different phyla; amazing regularity in the development of the neural apparatus, repeated in the most diverse stems; independent appearance and development of very similar sense organs in large and small groups of animals, etc. All these facts point to the tremendous role played, not only by inheritance from common ancestors, but also by parallel development and convergence in the creation of similarities between the higher members of different phyla, and they confirm the deductions in the architectonical section of this book: that the chief groups of Bilateria—Scolecida, Trochozoa, and Deuterostomia—all arose directly from Coelenterata, and that the same is probably true of the small isolated groups, Podaxonia, Brachiopoda, Chaetognatha, etc.

With regard to the course of phylogeny, the facts enable us to make two basic assumptions: first, the unity of origin of the animal kingdom, the close relationship of Metazoa to Flagellata, the undoubted relationship between coelenterates and sponges, the incontestably close relationship between Bilateria and coelenterates; and secondly, the very early divergence of all the main stems of the animal kingdom, and the extreme antiquity of these stems.

BIBLIOGRAPHY

RUSSIAN LIST

ABRIKOSOV, G. G. 1937. Class Kamptozoa. *Handbook of Zoology*. I. Moscow and Leningrad.

BECKER, E. G. 1926. Evolution of external skeleton and musculature of Atelocerata (Tracheata). 1. Tergal skeleton and dorsal longitudinal musculature of Chilopoda. *Zool. Zh.*, **6, 4.**

BECKER, E. G. 1949. Evolution of external skeleton and musculature of Tracheata (Atelocerata). 2. Pleural and sternal skeleton and musculature of Chilopoda Epimorpha. *Zool. Zh.*, **28,** No. 1.

BECKER, E. G. 1950. Locomotor organs and evolution of tracheal arthropods. *Vest. mosk. gos. Univ., Biol. i Pochvov.*, **10.**

BEKLEMISHEV, C. W. 1954. Discovery of siliceous formations in integument of lower crustaceans. *Dokl. Akad. Nauk SSSR, n. ser.*, **97,** No. 3.

BEKLEMISHEV, C. W. 1955. Predation among nemertines. *Priroda*, **44,** 9.

BEKLEMISHEV, W. N. 1915. Parasatic turbellarians of the Murmansk Sea. I. Acoela. *Trudy petrogr. Obshch. Estest.*, **43,** 4.

BEKLEMISHEV, W. N. 1916. Parasitic turbellarians of the Murmansk Sea. II. Rhabdocoela. *Trudy petrogr. Obshch. Estest.*, **45,** 4.

BEKLEMISHEV, W. N. 1925. Morphological problem of animal structures. *Izv. biol. nauchno-issled. Inst. perm. Univ.*, Appendix 3.

BEKLEMISHEV, W. N. 1931. Basic concepts of biocoenology as applied to the animal components of terrestrial associations. *Trudy Zashch. Rast.*, **1,** 2.

BEKLEMISHEV, W. N. 1937. Turbellaria. *Handbook of Zoology*. I. Moscow and Leningrad.

BEKLEMISHEV, W. N. 1944. *Principles of comparative anatomy of invertebrates.* 1st ed. Moscow. 'Sov. Nauka.'

BEKLEMISHEV, W. N. 1945. Principles of comparative parasitology as applied to bloodsucking arthropods. *Med. Parasitol.*, **14,** 1.

BEKLEMISHEV, W. N. 1950. Problem of individuality in biology. Colonies in bilaterally-symmetrical animals. *Usp. sovr. Biol.*, **29.**

BEKLEMISHEV, W. N. 1951. Constructing a system of animals. Deuterostomia, their origin and composition. *Usp. sovr. Biol.*, **32.**

BEKLEMISHEV, W. N. 1954. Turbellaria of the Caspian Sea. II. Triclada Maricola. *Byull. mosk. Obshch. Ispyt. Prir., ser. biol.*, **59,** 6.

BEKLEMISHEV, W. N., and MITROFANOVA, Yu. G. 1926. Ecology of the larva of *Anopheles maculipennis* Meig.: problem of distribution. *Izv. biol. nauchno-issled. Inst. perm. Univ.*, **4,** 7.

BELYAEV, G. M. 1950. Osmoregulatory capacity of lower crustaceans. *Trudy vses. gidrobiol. Obshch.*, **2.**

BELYAEV, G. M. 1951. Osmotic pressure of cavity fluid of aquatic invertebrates in lakes of differing salinity. *Trudy vses. gidrobiol. Obshch.*, **3.**

BOGOMOLOV, S. I. 1949. Problem of type of cleavage in Rhabdocoela. *Uchen. Zap. leningr. gos. Univ., ser. biol.*, **20.**

BOGOMOLOV, S. I. 1957. New data on cleavage of *Convoluta* in connection with the problem of the origin of the spiral type of development of animals. *Proc. 2nd Conf. Embryol. U.S.S.R.*

BYALINITSKII-BIRULYA, A. A. 1917. Scorpions. *Fauna of Russia and adjacent countries.* Arachnoidea. I. Petrograd.

BYKHOVSKII, B. E. 1937. Ontogeny and phylogenetic relationships of parasitic flatworms. *Izv. Akad. Nauk SSSR, ser. biol.,* 4.

CHERNYSHEV, B. I. 1934. Class Crustacea. In: Tsittel', K. *Principles of palaeontology* (palaeozoology), worked out by palaeontologists of the U.S.S.R., I. Leningrad.

DAVYDOV, K. N. 1905. Scientific results of travels in Java and other islands of the Malay Archipelago. III. Morphology of Archiannelida. Biological observations on an epitokous form of *Polygordius. Izv. s.-peterb. Akad. Nauk, ser.* 5, 22.

DETINOVA, T. S. 1945. Effect of internal-secretion glands on maturation of genital products and imaginal diapause in the common malaria mosquito. *Zool. Zh.,* 24.

DOGIEL, V. A. 1913. *Material on the history of the development of Pantopoda.* St. Petersburg.

DOGIEL, V. A. 1938, 1940. *Comparative anatomy of invertebrates.* I-II. Leningrad.

DOGIEL, V. A. 1940. Phylogeny of the phylum Mollusca. *Handbook of Zoology.* II. Moscow and Leningrad.

DOGIEL, V. A. 1947. The phenomena of polymerisation and oligomerisation of homologous organs in the animal kingdom and their evolutionary significance. *Izv. Akad. Nauk SSSR, ser. biol.,* 4.

DOGIEL, V. A. 1954. *Oligomerisation of homologous organs as one of the principal paths in animal evolution.* Leningrad.

DUBININ, V. B. 1951. Feather mites. I. Introduction to study of them. *Fauna of the U.S.S.R.,* Arachnoidea, vol. VI, 5. Moscow.

D'YAKONOV, A. M. 1914. *Ascothorax ophioctenis* n. g. et n. sp.—New parasite in the group Ascothoracida. *Trudy s.-peterb. Obshch. Estest.,* 45, 4.

FAUSEK, V. A. 1891. Studies in the history of the development and in the anatomy of Phalangidae (harvest spiders). *Trudy s.-peterb. Obshch. Estest.,* 22, 2.

FAUSEK, V. A. 1897. Investigation into the history of the development of Cephalopoda. *Trudy s.-peterb. Obshch. Estest.,* 28, 2.

FEDOTOV, D. M. 1915. *Protomyzostomum polynephris and its relationship to Gorgonocephalus eucnemis.* Petrograd.

FEDOTOV, D. M. 1923. Problem of the homology of the coeloms of Echinodermata, Enteropneusta, and Chordata. *Izv. biol. nauchno-issled. Inst. perm. Univ.,* 2, 1.

FEDOTOV, D. M. 1924. Relationships between Crustacea, Trilobita, Merostomata, and Arachnoidea. *Izv. russ. Akad. Nauk, ser.* 6, 18 (12–18).

FEDOTOV, D. M. 1928. Relationships among classes of Echinodermata according to zoological and palaeontological data. *Trudy osob. zool. Lab. sevast. St. Akad. Nauk SSSR, ser.* 2, 12.

FEDOTOV, D. M. 1935. Outline of the evolution of the animal world. *Trudy paleontol. Inst. Akad. Nauk SSSR,* 5.

FEDOTOV, D. M. 1939. Substitutions of secondary body cavities in Ophiuroidea. *Zool. Zh.,* 18, No. 6.

FEDOTOV, D. M. 1940. The phenomenon of organ reduction in ontogeny of *Operophthera brumata* L. *Coll. in memory of A. N. Severtsov.* Akad. Nauk SSSR. II. Moscow.

FEDOTOV, D. M. 1951. Phylum Echinodermata. *Handbook of zoology.* III. Leningrad and Moscow.

FILIP'EV, I. N. 1918–21. Free-living marine nematodes off Sevastopol. *Trudy osob. zool. Lab. sevast. St. russ. Akad. Nauk, ser.* 2, 4.

FRIDMAN, G. M. 1933. Anatomical structure of *Baicalarctia gulo* Fr. and its position in the system of Turbellaria. *Trudy baik. limnolog. St.,* 5.

GARYAEV, V. P. (1915) 1916. Structure of digestive tract of some Cephalopoda. *Trudy Obshch. Ispyt. Prir.*, **48**, 2.

GILEV, F. D. 1952. Sensory innervation of the gut of *Anodonta cellensis*. *Dokl. Akad. Nauk SSSR, n. ser.*, **87**, 1059–61.

GILYAROV, M. S. 1944. Functional significance of symmetry of organisms. *Zool. Zh.*, **23**.

GILYAROV, M. S. 1949. *Characteristics of soil as a living environment and its significance in insect evolution*. Moscow and Leningrad.

GILYAROV, M. S. 1958. Evolution of the character of fertilisation in terrestrial arthropods. *Zool. Zh.*, **37**, 5.

GOIKHER, G. B. 1949. Problem of passage of small organisms through the mantle cavity and the gut of filter-feeding bivalve molluscs. *Dokl. Akad. Nauk SSSR, n. ser.*, **65**, No. 5.

GUR'YANOVA, E. F. 1951. Amphipoda of the seas of the U.S.S.R. and adjacent waters (Amphipoda—Gammaroidea). *Guides to fauna of the U.S.S.R.*, publ. Zool. Inst. Akad. Nauk SSSR, Vol. 41. Moscow and Leningrad.

ISAEV, V. M. 1911. Anatomy of *Polyxenus (Lophoproctus) lucidus. Trudy s.-peterb. Obshch. Estest.*, **40**, 3.

IVANOV, A. V. 1946. Class Arachnoidea. *Practical course in the zoology of invertebrates*. II. Moscow.

IVANOV, A. V. 1946. Phylum Mollusca. *Practical course in the zoology of invertebrates*. II. Moscow.

IVANOV, A. V. 1951. The place of genus *Siboglinum* Caullery in the class Pogonophora. *Dokl. Akad. Nauk SSSR, n. ser.*, **76**, No. 5.

IVANOV, A. V. 1952a. Structure of *Udonella caligorum* Johnston, 1835, and place of Udonellidae in the system of flatworms. *Parasitol. coll., Zool. Inst. Akad. Nauk SSSR*, **14**.

IVANOV, A. V. 1952b. New Pogonophora in Far Eastern seas. *Zool. Zh.*, **31**, No. 3.

IVANOV, A. V. 1952c. Turbellaria Acoela of southern shore of Sakhalin. *Trudy Zool. Inst. Akad. Nauk SSSR*, **12**.

IVANOV, A. V. 1955a. External digestion in Pogonophora. *Dokl. Akad. Nauk SSSR*, **100**, No. 2.

IVANOV, A. V. 1955b. Principal features of organisation of Pogonophora. *Dokl. Akad. Nauk SSSR*, **100**, No. 1.

IVANOV, A. V. 1958a. Structure of genital system in Pogonophora. *Zool. Zh.*, **37**, 9.

IVANOV, A. V. 1958b. Neural system of Pogonophora. *Zool. Zh.*, **37**, 11.

IVANOV, A. V. 1960. Pogonophora. *Fauna of USSR*, new ser., No. 75. Moscow and Leningrad.

IVANOV, P. P. 1912. *Regenerative processes in Polychaeta and their relationship to the ontogeny and morphology of annelids*. St. Petersburg.

IVANOV, P. P. 1916. Regeneration and ontogeny in Polychaeta. *Zool. Vest.*, **1**.

IVANOV, P. P. 1937. *General and comparative embryology*. Moscow and Leningrad.

IVANOV, P. P. 1940. Embryonic development of Scolopendromorpha in relation to the embryology and morphology of Tracheata. *Izv. Akad. Nauk SSSR*, **6**.

IVANOV, P. P. 1945. *Handbook of general and comparative embryology*. Leningrad.

IVANOVA-KAZAS, O. M. 1959. Problem of origin and evolution of spiral cleavage. *Vest. leningr. gos. Univ.*, No. 9, ser. biol., 2.

IZOSIMOV, V. V. 1940. Class Oligochaeta. *Handbook of zoology*. Invertebrates, vol. II. Moscow and Leningrad.

KAMSHILOV, M. M. 1955. Feeding of *Beroë cucumis* Fab. *Dokl. Akad. Nauk SSSR*, **102**, 2.

KAUFMAN, Z. S. 1959. Structure of the stigmata of the myriapod *Geophilus proximus* C. L. Koch (Chilopoda). *Dokl. Akad. Nauk SSSR*, **129**, No. 3.

KHLOPIN, N. G. 1946. *General biological and experimental principles of histology.* Publ. Acad. Sci. U.S.S.R. Moscow.

KHOLODKOVSKII, N. A. 1891. Embryonic development of the cockroach (*Phyllodromia germanica*). *Trudy s.-peterb. Obshch. Estest.*, **22**, 1.

KHOLODKOVSKII, N. A. 1909. *Textbook of zoology and comparative anatomy for higher educational institutions.* St. Petersburg.

KHOLODKOVSKII, N. A. 1927–31. *Course in entomology.* I-III. 4th ed. Leningrad.

KNIPOVICH, N. M. 1892. Contribution to the knowledge of the group Ascothoracida. *Trudy s.-peterb. Obshch. Estest.*, **23**, 2.

KOLMOGOROVA, E. YA. 1956. Structure of the neural system of *Opisthorchis felineus. Referaty nauchn. Rab. perm. Otd. vses. Obshch. Anat., Gistol., Embriol.*, pp. 67–68.

KOLMOGOROVA, E. YA. 1959. Structure of central section of neural system of *Opisthorchis felineus. Zool. Zh.*, **38**, 11.

KOSHTOYANTS, KH. S. 1951. *Principles of comparative physiology.* I. Moscow and Leningrad.

KOVALEVSKII, A. O. 1867. Anatomy and history of development of *Phoronis.* Appendix No. 1 to Vol. 11 of *Zap. s.-peterb. Akad. Nauk.*

KOVALEVSKII, A. O. 1874. Development of Brachiopoda. *Zap. Obshch. Lyub. Estest. Etnogr.*, **14.**

KROTOV, A. I. 1956. Contribution to the knowledge of the physiology of motor reactions in Ascaridae. *Med. Parasitol. i parasitol. Bolezni*, **25**, 1.

KULAGIN, N. M. 1889. Natural history of *Pentastomum denticulatum* Lam. *Izv. mosk. sel'khoz. Inst.*, 15th year.

KUSHAKEVICH, S. 1906. Outline of theory of embryonic layers in its past and present forms. *Zap. novoross. Obshch. Estest.*, **29.**

LANGE, A. B. 1947. Morphology, history of development, and systematics of parasitic Acariformes of the family Laelaptidae. (Dissert.) Mosk. gos. Univ.

LANGE, A. B. 1954. Morphology of the tick *Zachvatkinella belbiformis* gen. n. et sp. n., a new member of the group Palaeacariformes (Acariformes). *Zool. Zh.*, **33**, No. 5.

LASTOCHKIN, D. A. 1914. Anatomy and physiology of Synaptidae. *Trudy s.-peterb. Obshch. Estest.*, **45**, 1.

LASTOCHKIN, D. A. 1915. Study of the ambulacral system of Synaptidae. *Trudy petrogr. Obshch. Estest.*, **46**, 1.

LASTOCHKIN, D. A. 1922. Regenerative phenomena in Naididae. *Izv. ivanovo-vosnesensk. politekh. Inst.*, **6.**

LERMONTOVA, E. V. 1934. Class Trilobita. In: Tsittel', K. *Principles of palaeontology* (palaeozoology), worked out by palaeontologists of the U.S.S.R., I. Leningrad.

LIGNAU, N. 1912. History of the embryonic development of *Polydesmus abchasius* Attems. Morphology of Diplopoda. *Zap. novoross. Obshch. Estest.*, **38.**

LIVANOV, N. A. 1905. *Acanthobdella peledina* Grube, 1851. Morphological investigation. *Uchen. Zap. kazan. Univ.*, **72.**

LIVANOV, N. A. 1910. Morphological studies of Hirudinea. Anatomy of circulatory system. *Biol. Zh.*, **1.**

LIVANOV, N. A. 1914. Boundary formations in Polychaeta. *Trudy Obshch. Estest. kazan. Univ.*, **46.**

LIVANOV, N. V. 1924. Studies of the anatomy of the neural system in a somite of Polychaeta. *Arkhiv Anat., Gistol., Embriol*, **3.**

LIVANOV, N. A. 1940. Class Polychaeta. Class Hirudinea. *Handbook of zoology.* II. Moscow and Leningrad.

LIVANOV, N. A. (1945), 1946. Paths of evolution in the animal world (analysis of organisation of phyla). *Uchen. Zap. kazan. gos. Univ.*, **105**, 3.

LIVANOV, N. A. 1955. Paths of evolution in the animal world. *Analysis of organisation of principal phyla of multicellular animals*. Moscow.

LIVANOV, N. A., and PORFIR'EVA, N. A. 1962. Organisation of Pogonophora. *Trudy Obshch. Estest. kazan. Univ.*, **65**, 102–20.

LYUBISHCHEV, A. A. 1924. Nephridial complexes of *Nephthys ciliata* and *Glycera capitata* (Polychaeta). *Zool. Zh.*, **4**, 1–2.

MARTYNOV, A. V. 1934. Class Insecta. In: Tsittel', K. *Principles of palaeontology* (palaeozoology), worked out by palaeontologists of the U.S.S.R. I. Leningrad.

MARTYNOV, A. V. 1937. Outlines of the geological history and phylogeny of insects (Pterygota). I. Palaeoptera and Neoptera-Polyneoptera. *Trudy paleontol. Inst. Akad. Nauk SSSR*, **7**.

MECHNIKOV, E. 1884. Investigations of intracellular digestion in invertebrates. *Russkaya Meditsina*, **9**.

MECHNIKOV, E., and MECHNIKOVA, L. 1870. Contribution to the knowledge of Siphonophora and Medusae. *Izv. mosk. Obshch. Lyub. Estest.*, **8**, 1.

MEDNIKOVA, M. V. 1952. The endocrine glands corpora allata and corpora cardiaca of mosquitoes (fam. Culicidae). *Zool. Zh.*, **31**, No. 5.

MEDVEDEVA, N. B. 1939. Evolution of humoral regulation of the functions of an organism. *Usp. sovr. Biol.*, **10**.

MEYER, E. A. 1898. Investigation of the development of annelids. *Trudy Obshch. Estest. kazan. Univ.*, **31**, 4.

MEN YAN-TSUN'. 1959. Fauna and ecology of nest and burrow gamasids in a focus of tick-borne encephalitis, and material on the biology of the acarid *Haemolaelaps casalis*. (Dissert.) Moscow.

MIRONOV, V. S. 1939. Behaviour of the taiga tick *Ixodes persulcatus* Schulze. *Med. Parasitol.*, **8**, 1.

MORDUKHAI-BOLTOVSKOI, D. D. 1936. Geometry of Radiolaria. *Uchen. Zap. rostov. Univ.*, **8**.

MYOLLER, V. 1878. Spirally-twisted Foraminifera in Carboniferous limestones in Russia. *Materialy geol. Rossii*, **8**.

NALIVKIN, D. V. 1925. Elements of symmetry of the organic world. *Izv. biol. nauchn.-issled. Inst. perm. Univ.*, **3**.

NASONOV, N. V. 1887. History of the development of the crustaceans *Balanus* and *Artemia*. *Izv. Obshch. Lyub. Estest., Antropol., Etnogr.*, **52**, 1.

NASONOV, N. V. 1901. *Course in Entomology*. Warsaw.

NAUMOV, D. V. 1950. Effect of water movement on the morphology of hydroids. *Dokl. Akad. Nauk SSSR, n. ser.*, **71**, No. 6.

NAUMOV, D. V. 1953. General problems of metagenesis in relation to the establishment of the primary generation in metagenetic hydroids. *Trudy Zool. Inst. Akad. Nauk SSSR*, **13**.

NAUMOV, D. V. 1960. Hydroids and hydromedusae of marine, brackish, and freshwater basins in the U.S.S.R. *Guides to fauna of the U.S.S.R.*, publ. Zool. Inst. Akad. Nauk SSSR, Vol. 70.

NEL'ZINA, E. N. 1951. The rat tick (*Bdellonyssus bacoti*). *Comparative-parasitological study*. Moscow.

NEVMYVAKA, G. A. 1947. Innervation of the gut in earthworms (*Allolobophora caliginosa*). *Dokl. Akad. Nauk SSSR*, **50**, No. 5.

NEVMYVAKA, G. A. 1956. Possible sources of origin of neural elements of the gut of invertebrates. *Referaty nauchn. Rab. perm. Otd. vses. Obshch. Anatom., Gistol., Embriol.*, pp. 97–98.

OKOROKOV, V. I. 1956. A new cestode species—*Tatria mathevossianae* (fam. Amabiliidae) from *Podiceps ruficollis* (Pallas). *Zool. Zh.*, **35**, 9.

ORLOV, J. A. 1924. Problem of histological structure of sympathetic nervous system in insects. *Izv. biol. nauchno-issled. Inst. perm. Univ.*, **3**.

PAVLOVSKII, E. N. 1917. *Material on the comparative anatomy and history of development of scorpions*. Petrograd.

PEDASHENKO, D. D. 1899. Embryonic development and metamorphosis of *Lernaea branchialis. Trudy s.-peterb. Obshch. Estest.*, **26.**

PETROCHENKO, V. I. 1952. Place of Acanthocephala in the zoological system (phylogenetic links of Acanthocephala with other groups of invertebrates). *Zool. Zh.*, **31,** No. 2.

PLOTNIKOVA, S. I. 1949. Comparative morphology of the vegetative nervous system. Location of ganglia and nerves in the digestive tract of the larva of the dragon-fly *Aeschna. Dokl. Akad. Nauk SSSR, n. ser.*, **68,** No. 5.

POLEZHAEV, N. N. 1890. Problem of the origin of mesoderm. *Vest. Estest.*, **1.**

POLOVODOVA, V. P. 1953. Innervation of the genital apparatus and hind-gut of the female malaria mosquito. *Zool. Zh.*, **32,** No. 4.

POMERANTSEV, B. I. 1932. Morphology of the genitals of *Culicoides* (Diptera, Nematocera). *Parasitol. sb. Akad. Nauk SSSR*, **3.**

REZVOI, P. D. 1937. Phylum Porifera (Spongia). *Handbook of zoology*. I. Moscow and Leningrad.

SHAROV, A. G. 1948. Triassic Thysanura from the Urals. *Dokl. Akad. Nauk SSSR, n. ser.*, **61,** No. 3.

SHAROV, A. G. 1953. Development of Thysanura (Apterygota) in connection with the problem of the phylogeny of insects. *Trudy Inst. morfologii zhivotnykh Akad. Nauk SSSR,* **8.**

SHAROV, A. G. 1957. Unique Palaeozoic wingless insects of the new order Monura (Insecta, Apterygota). *Dokl. Akad. Nauk SSSR*, **115,** 4.

SHAROV, A. G. 1958. Structure of appendages and methods of movement in Monura and Thysanura (Insecta, Apterygota). *Dokl. Akad. Nauk SSSR*, **122,** 4.

SHCHEGOLEV, G. G. 1951. Observations on movements of medicinal leeches in lakes. *Zool. Zh.*, **30,** No. 5.

SHCHEPOT'EV, A. 1907. *Pterobranchia*. St. Petersburg.

SHIMKEVICH, V. M. 1889. Observations on the fauna of the White Sea. 1. *Trudy s.-peterb. Obshch. Estest.*, **20,** 2.

SHIMKEVICH, V. M. 1900. Problem of the origin of Crustacea. *Trudy s.-peterb. Obshch. Estest.*, **30,** 4.

SHIMKEVICH, V. M. 1907. *Biological foundations of zoology*. St. Petersburg.

SHIMKEVICH, V. M. 1908. Metorisis as an embryological principle. *Izv. s.-peterb. Akad. Nauk, ser.* **6,** 2.

SHUBNIKOV, A. V. 1940. *Symmetry*. Moscow.

SHUBNIKOV, A. V. 1951. *Symmetry and antisymmetry in finite figures*. Moscow.

SHVANVICH, B. N. 1949. *Course of general entomology*. Moscow and Leningrad.

SKRYABIN, K. I. 1947. *Trematodes of animals and man*, Vol. 1. Moscow.

SKRYABIN, K. I., and MATEVOSYAN, E. M. 1945. *Tapeworms (Hymenolepididae) in domestic and game birds*. Moscow.

SMIRNOV, S. S. 1940. Phyllopoda. *Freshwater life*, Vol. 1. Moscow and Leningrad.

SOKOLOV, B. S. 1950. *Carboniferous Chaetetidae*. Leningrad and Moscow.

SOKOLOV, B. S. 1951. *Palaeozoic Tabulata of the European part of the U.S.S.R.* I-II. Leningrad and Moscow.

SOKOLOV, B. S. 1955. Palaeozoic Tabulata of the European part of the U.S.S.R. Introduction; general problems of the systematics and history of development of Tabulata (with description of morphologically-similar groups). *Trudy vses. neft. nauchno-issled. geologorazved. Inst., n. ser.*, **85.**

SOKOL'SKAYA, N. L. 1951. Motor apparatus of Oligochaeta. *Zool. Zh.*, **30,** No. 6.

SPASSKII, A. A. 1951. Metamerism of animals and the time factor. *Dokl. Akad. SSSR, n. ser.*, **79,** No. 3.

STUDITSKII, A. N. 1947. Histogeny and form-creation. *Zhurn. obshchei Biologii*, **8.**

SVETLOV, P. G. 1957. Primary heteronomy of body structure in vertebrates. *Arkhiv Anat., Gistol., Embriol.*, **34**, 2.

TIKHII, M. 1916. A planktonic hydroid of the Caspian Sea. *Trudy petrogr. Obshch. Estest.*, **47**, 1.

TIKHOMIROV, A. S. 1887. History of the development of hydroids. *Izv. Obshch. Lyub. Estest., Antropol., Etnogr.*, **50**, 2.

TIMOFEEV, S. 1923. Morphology of Polychaeta. III. Circulatory system. *Sb. trudov Prof. i Pred. irkutsk. Univ.*, **5**.

TRET'YAKOV, D. 1936. Neotenic character of ctenophores. *Zool. Zh.*, **15**, No. 2.

URBANOVICH, F. 1885. Development of *Cyclops*. *Varshavsk. Univ. Izv.*, **4**.

VAGIN, V. L. 1947. The secondary musculo-cutaneous sac and its phylogenetic significance. *Dokl. Akad. Nauk SSSR, n. ser.*, **58**, No. 3.

VAGNER, YU. 1894. History of the embryonic development of *Ixodes calcaratus* Bir. *Trudy s.-peterb. Obshch. Estest.*, **24**, 2.

VAGNER, YU. 1896. Some observations on the embryonic development of *Neomysis vulgaris* var. *baltica* Czern. *Trudy s.-peterb. Obshch. Estest.*, **26**, 2.

VEITSMAN, V. R. 1939. Development and reduction of the female genital organs of the pork tapeworm, *Taenia solium* (prelim. report). *Dokl. Akad. Nauk SSSR, n. ser.*, **22**, No. 9.

VEITSMAN, V. R. 1953. Development and reduction of the female genital system of the cat tapeworm, *Taenia crassicollis*. *Trudy Inst. Morf. Zhivotnykh Akad. Nauk SSSR*, **8**.

VEITSMAN, V. R. 1953. Development and reduction of the female genital system of *Taenia solium*. *Trudy Inst. Morf. Zhivot. Akad. Nauk SSSR*, **8**.

VERNADSKII, V. I. 1944. Some remarks on the noosphere. *Usp. sovr. Biol.*, **18**.

VETOKHIN, I. A. 1926. The work of the ciliated epithelium of the gastrovascular system of the medusa *Aurelia aurita* (L.) Lam. *Rab. murmansk. biol. St.*, **2**.

VINOGRADSKAYA, O. N. 1960. Morphology and physiology of the respiratory apparatus and water balance of winged bloodsucking mosquitoes in relation to their living conditions. (Dissert.) Moscow.

VLASTOV, B. V. 1923. Contractile elements. Contractile structure of the cercaria of *Bucephalus*. *Izv. biol. nauchn.-issled. Inst. perm. Univ.*, **2**.

VOLOGDIN, A. G. 1945. Colonial Archaeocyatha from the Middle Cambrian of Western Sayan. *Ezhegod. russk. paleontol. Obshch.*, **12**.

VOSKRESENSKII, K. A. 1948. Effect of organisms on water circulation. *Dokl. gos. okeanogr. Inst.*, **105**.

YAGUZHINSKAYA, L. V. 1954. New data on the physiology and anatomy of the heart of Diptera. *Byull. mosk. Obshch. Ispyt. Prir., otd. biol.*, **59**, 1.

YAKOVLEV, N. N. 1910. Origin of the characteristic features of Rugosa. *Trudy geol. Komiteta, n. ser.*, **6**.

YAKOVLEV, N. N. 1934. Order Tetracorallia Haeckel. In: Tsittel', K. *Principles of palaeontology* (palaeozoology), worked out by palaeontogists of the U.S.S.R. I. Leningrad.

ZAKHVATKIN, A. A. 1941. Tyroglyphoidea. *Fauna of the U.S.S.R.*, Arachnoidea, VI, I. Moscow.

ZAKHVATKIN, A. A. 1946. Nature of the blastular larvae of Metazoa. *Zool. Zh.*, **25**, No. 4.

ZAKHVATKIN, A. A. 1949. *Comparative embryology of lower invertebrates* (origins and methods of formation of individual development of Metazoa). Moscow.

ZAKHVATKIN, A. A. 1952. Division of Acarina into orders and their place in the system of Chelicerata. *Parasitol. sb. Zool. Inst. Akad. Nauk SSSR*, **14**.

ZAKHVATKIN, A. A. 1953a. *Investigation of Tyroglyphoidea by comparative morphology and history of development*. Coll. of scientific works. Izd. mosk. gos. Univ.

ZAKHVATKIN, A. A. 1953b. *Investigations of Tyroglyphoidea (Sarcoptiformes) by morphology and postembryonic development.* Coll. of scientific works. Izd. mosk. gos. Univ. Moscow.

ZAKHVATKIN, A. A., and LANGE, A. B. 1953. *Summary of Course on Acarology.* Coll. of scientific works. Izd. mosk. gos. Univ.

ZALENSKII, V. V. 1909. Embryonic development of *Prosorhochmus viviparus* Ulian (*Monopora vivipara*). *Zap. s.-peterb. Akad. Nauk, ser.* **6,** 3.

ZALESSKII, YU. M. 1949. Origin of wings and flight in insects in relation to their environmental living conditions. *Usp. sovr. Biol.,* **28,** 3 (6).

ZANINA, I. E., and POLENOVA, E. N. 1960. Ostracoda. In: Principles of palaeontology, arthropods—trilobites and crustaceans. Moscow. *Gos. nauchno-tekh. Izd. Lit. po geol. i okhrane nedr.*

ZAVARZIN, A. A. 1934. Evolutionary dynamics of tissues. *Arkhiv biol. Nauk,* **36,** 1.

ZAVARZIN, A. A. 1935. Comparative histology of blood and connective tissue. XI. Inflammatory formation of connective tissue in the earthworm (*Allolobophora caliginosa*). *Arkhiv biol. Nauk,* **37,** 3.

ZAVARZIN, A. A. 1941. *Outline of evolutionary histology of the nervous system.* Moscow and Leningrad.

ZAVARZIN, A. A. 1945. *Outline of evolutionary histology of the blood and connective tissue.* Moscow and Leningrad.

ZELENSKII, V. D. 1915. *Study of the morphology and systematics of Hirudinea.* I. Organisation of Ichthyobdellidae. Petrograd.

ZENKEVICH, L. A. 1937. History of the system of invertebrates. *Handbook of zoology.* I. Moscow and Leningrad.

ZENKEVICH, L. A. 1940. Class Priapuloidea (Priapulida). *Handbook of zoology.* II. Moscow and Leningrad.

ZENKEVICH, L. A. 1944. Outline of the evolution of the locomotor apparatus of animals. *Zh. obshch. Biol.,* **5,** No. 3.

ZHINKIN, L. N. 1949. Early stages of development of *Priapulus caudatus. Dokl. Akad. Nauk SSSR, n. ser.,* **65,** No. 3.

ZHINKIN, L. N. 1951. Characteristics of cleavage of ova in lower invertebrates. *Priroda,* No. 2.

ZHINKIN, L. N. 1955. Characteristics of development and systematic position of Priapuloidea. *Uchen. Zap. leningr. gos. ped. Inst. im. A. I. Gertsena,* **110.**

ZHINKIN, L. N., and KORSAKOVA, G. 1953. Early stages of development of *Halicryptus spinulosus. Dokl. Akad. Nauk SSSR, n. ser.,* **88,** No. 3.

ZHURAVLEVA, I. T. 1954. Instructions for collection and study of Archaeocyatha. *Instructions for collection and study of fossil organic remains.* V. Leningrad. Izd. paleontol. Inst. Akad. Nauk SSSR.

ZHURAVLEVA, I. T. 1959. Place of Archaeocyatha in the phylogenetic system. *Paleontol. Zh.,* No. 4.

NON-RUSSIAN LIST

AGASSIZ, A. 1876. On viviparous Echini from the Kerguelen Islands. *Proc. Amer. Acad. Arts Sci.,* **11** (N.S. 3).

ALEXANDER, A. J. 1956. Mating in scorpions. *Nature, Lond.,* **178,** 4538.

ALEXANDROWICZ, J. S. 1926. The innervation of the heart of the cockroach (*P. orientalis*). *J. Compar. Neurol.,* **41.**

ALEXANDROWICZ, J. S. 1932. The innervation of the heart of the Crustacea. I. Decapoda. *Q. J. microsc. Sci., N.S.,* **75.**

ALEXANDROWICZ, J. S. 1934. The innervation of the heart of the Crustacea. II. Stomatopoda. *Q. J. microsc. Sci., N.S.,* **75.**

ALEXANDROWICZ, J. S. 1952. Innervation of the heart of *Ligia oceanica*. *J. Mar. Biol. Assn. UK*, **31**.

ALEXANDROWICZ, J. S. 1953. Notes on the nervous system in the Stomatopoda. II. The system of dorsal trunks. III. Small nerve cells in motor nerves. *Pubbl. Staz. Zool., Napoli*, **24**.

ALEXANDROWICZ, J. S., and CARLISLE, D. B., 1953. Some experiments on the function of the pericardial organs in Crustacea. *J. Mar. Biol. Assn. UK*, **32**, 1.

ALTEN, H. VON. 1910. Zur Phylogenie des Hymenopterengehirnes. *Z. Naturw.*, **46**.

ALVERDES, F. 1923. Über Galvanotaxis und Flimmerbewegung. *Biol. Cbl.*, **43**.

ANDERSSON, K. A. 1907. Die Pterobranchier der Schwedischen Südpolarexpedition 1901 bis 1903. *Wiss. Ergebn. Schwed. Südpol. Exped.*, **5**.

ANDRÉ, M. 1927. Digestion extra-intestinale chez le rouget (*Leptus autumnalis* Shaw). *Bull. Muséum nat. histoire natur.*, **33**.

ANGERMANN, H. 1955. Indirekte Spermatophorenübertragung bei *Euscorpius italicus* (Hbst.). *Naturwissenschaften*, **42**.

ARVY, L. 1954. Sur l'existence de cellules neurosécrétrices chez quelques Annélides Polychètes sédentaires. *Compt. rend. Acad. Sci.*, **238**.

ARVY, L. 1954. Contribution à l'étude de la neurosécrétion chez les Annélides Polychètes sédentaires. *Bull. Labor. Dinard. f.*, **40**.

ARVY, L., and GABE, M. 1950. Données histophysiologiques sur les formations endocrines rétrocérébrales chez les Ecdyonuridae (Ephéméroptères). *Bull. Soc. zool. France*, **75**.

ARVY, L., and GABE, M. 1952. Données histophysiologiques sur la neuro-sécrétion chez quelques Ephéméroptères. *Cellule*, **55**.

ARVY, L., and GABE, M. 1952. Données histophysiologiques sur les formations endocrines rétrocérébrales de quelques Odonates. *Ann. Sci. natur. zool. biol. anim., sér.* **11**, 14.

ARVY, L., and GABE, M. 1953. Données histophysiologiques sur la neurosécrétion chez les Paléoptères (Ephéméroptères et Odonates). *Z. Zellforsch.*, **38**.

ARVY, L., and GABE, M. 1954. The intercerebralis-cardiacum-allatum system of some Plecoptera. *Biol. bull.*, **106**.

ATKINS, D. 1937, 1938. On the ciliary mechanisms and interrelationships of Lamellibranchs. Pt. I–VII. *Q. J. microsc. Sci.*, N.S., **79, 80**.

ATKINS, D. 1955. The cyphonautes larvae of Plymouth area and the metamorphosis of *Membranipora membranacea* (L.). *J. Mar. Biol. Assn UK*, **34**.

ATKINS, D. 1958. A new species of Kraussinidae (Brachiopoda) with a note on feeding. *Proc. Zool. Soc. London*, **131**.

ATTEMS, C. 1926. Myriapoda. In: Kükenthal, W., and Krumbach, Th. *Handbuch der Zoologie*, IV/1, 1–2. Berlin and Leipzig.

AX, P. 1956. Die Gnathostomulida, eine rätselhafte Wurmgruppe aus dem Meeressand. *Abhandl. Akad. Wiss. Mainz, Math.-Naturwiss. Kl.*, **8**.

AX, P. 1957. Ein chordoides Stützorgan des Entoderms bei Turbellarien. *Z. Morphol. Ökol. Tiere*, **46**.

AX, P. 1958. Vervielfachung des männlichen Kopulationsapparates bei Turbellarien. *Verhandl. Deutsch. Zool. Ges., Graz*, **1957**.

AX, P., and SCHULZ, E. 1959. Ungeschlechtliche Fortpflanzung durch Paratomie bei acoelen Turbellarien. *Biol. Zbl.*, **78**, 4.

AZEMA, M. 1937. Recherches sur le sang et l'excrétion chez les Ascidies. *Ann. Inst. Monaco*, **17**.

BACQ, Z. M., and FLORKIN, M. 1935. Mise en évidence dans le complèxe 'ganglion nerveux—glande neurale' d'une Ascidie (*Ciona intestinalis*) des principes pharmacologiquement analogues à ceux du lobe postérieur de l'hypophyse des Vertébrés. *Arch. internat. physiol.*, **40**.

BAGLIONI, S. 1910. Die Grundlagen der vergleichenden Physiologie des Nerven-systems und der Sinnesorgane. In: Winterstein, H. *Handbuch der verg-leichenden Physiologie*, IV, 1. Jena.

BAHL, K. N. 1938. On the significance of the enteronephric nephridial system found in Indian earthworms. I. Evidence from their habits and castings. *Q. J. microsc. Sci., N.S.,* **76.** London.

BAHL, K. N. 1946. Studies on the structure, development and physiology of the nephridia of Oligochaeta. Pt. VII. The enteronephric type of nephridial system in earthworms belonging to three species of *Megascolex* Templeton, and three species of *Travoscolides* Gates (*Megascolides* McCoy). *Q. J. microsc. Sci. N.S.,* **87.**

BAHL, K. N. 1947. Excretion in the Oligochaeta. *Biol. Revs Cambridge Philos. Soc.,* **22.**

BAHL, K. N., and LAL, M. B. 1938. On the occurrence of 'hepato-pancreatic' glands in the Indian earthworms of the genus *Eutyphoeus* Mic. *Q. J. microsc. Sci., N.S.,* **76.**

BAKER, E. W., and WHARTON, G. W. 1952. *An Introduction to Acarology.* New York.

BALFOUR, F. M. 1880–81. *A treatise of comparative embryology.* I–II. 1st ed. London.

BALFOUR, F. M. 1885. *A treatise of comparative embryology.* I–II. 2nd ed. London.

BALTZER, F. 1931. Priapulida. In: Kükenthal, W., and Krumbach, Th. *Handbuch der Zoologie*, II/2, 14. Berlin and Leipzig.

BARRINGTON, E. J. W., and FRANCHI, L. L. 1956. Organic binding iodine in the endostyle of *Ciona intestinalis. Nature, Lond.,* **177,** 4505.

BARROIS, J. 1924. Développement des Echinodermes, accompagné de quelques remarques sur l'origine des Procordés. *Ann. sci. natur. Zool. et biol. anim., sér.* **10, 7.**

BATESON, W. 1886a. The ancestry of the Chordata. *Q. J. microsc. Sci., N.S.,* **26.**

BATESON, W. 1886b. Lower stages in the development of *Balanoglossus kowalevskii*, and morphology of the Enteropneusta. *Q. J. microsc. Sci., N.S.,* **26.**

BATHAM, E. J., and PANTIN, C. F. A. 1950a. Muscular and hydrostatic action in the sea-anemone *Metridium senile* (L.). *J. Exptl Biol.,* **27.**

BATHAM, E. J., and PANTIN, C. F. A. 1950b. Inherent activity in the sea-anemone *Metridium senile* (L.). *J. Exptl Biol.,* **27.**

BATHAM, E. J., and PANTIN, C. F. A. 1951. The organisation of the muscular system of *Metridium senile. Q. J. microsc. Sci., N.* **5,** 92.

BATHER, F. A. 1915. *Studies in Edrioasteroidea.* I–IX. Wimbledon.

BATHER, F. A., GREGORY, J. W., and GOODRICH, E. S. 1900. The Echinoderma. In: Ray Lankester, E., *A treatise on Zoology*, III. London.

BAUNACKE, W. 1912. Statische Sinnesorgane bei den Nepiden. *Zool. Jahrb., Anat.,* **34.**

BEAUCHAMP, P. DE. 1907. Morphologie et variations de l'appareil rotateur. *Arch. Zool. Exper. géner.,* **36.**

BEAUCHAMP, P. DE. 1909. Recherches sur les Rotifères: les formations tégu-mentaires et l'appareil digestif. *Arch. zool. Exptl Gén. sér.* **4,** 10.

BEAUCHAMP, P. DE. 1910. Sur la présence d'un hémocoele chez *Dinophilus. Bull. Soc. zool. France,* **35.**

BEAUCHAMP, P. DE. 1929. Le développement des Gastrotriches. *Bull. Soc. zool. France,* **54.**

BECKER, G. 1937. Untersuchungen über den Darm und die Verdauung von Kamptozoen, Bryozoen und Phoronoiden. *Z. Morphol. Ökol. Tiere,* **33.**

BEDDARD, F. E. 1895. *A monograph of the order Oligochaeta.* Oxford.

BEIER, M. 1932. Pseudoscorpionida. In: Kükenthal, W., and Krumbach, Th. *Handbuch der Zoologie*, III/2, 5. Berlin and Leipzig.

BEKLEMISCHEV, W. N. 1927. Über die Turbellarienfauna des Aralsees. *Zool. Jahrb., Syst.*, **54.**

BEKLEMISCHEV, W. N. 1929a. Über den Bau der Drüsenstachel der Anaperiden. *Zool. Anz.*, **80.**

BEKLEMISCHEV, W. N. 1929b. Zur Kenntnis der Solenopharyngiden (Turbellaria, Rhabdocoela). *Pubbl. Staz. zool. Napoli*, **9.**

BENEDEN, P. J. VAN. 1849. Les Helminthes Cestoides, considérés sous le rapport de leurs métamorphoses, de leur composition anatomique et de leur classification, et mention de quelques espèces nouvelles de nos poissons Plagiostomes. *Bull. Acad. Belg.*, **16.**

BENHAM, W. 1896. Archiannelida, Polychaeta and Myzostomaria. In: Harmer, S. F., and Shipley, A. E. *Cambridge Natural History*, II. London.

BENHAM, W. 1901. The Platyhelmia. The Nemertini. In: Ray Lankester, E. *A treatise on zoology*, IV. London.

BERG, S. E. 1941. Die Entwicklung und Kolonienbildung bei *Funiculina quadrangularis* (Pallas). *Zool. Bidr., Uppsala*, **20.**

BERGERSEN, B., and BROCH, H. 1932. Rhabdopleuridae. In: Kükenthal, W., and Krumbach, Th. *Handbuch der Zoologie*, III/2, 2. Berlin and Leipzig.

BERGH, R. S. 1885. Die Exkretionsorgane der Würmer. *Kosmos*, **17.**

BERGH, R. S. 1899. Nochmals über die Entwicklung der Segmentalorgane. *Z. wiss. Zool.*, **66.**

BERLESE, A. 1909, 1912–25. *Gli Insetti*. I–II. Milano.

BERNARD, H. M. 1892. An endeavour to show that tracheae arose from setiparous sacs. *Zool. Jahrb., Anat.*, **5.**

BERRILL, N. J. 1955. *The origin of vertebrates*. Oxford.

BERTKAU, I. 1886. Die Augen der Spinnen. *Arch. Mikrobiol. Anat.*, **27.**

BETHE, A. 1927. Eigentümliche Formen und Mittel der Blutbewegung (*Phoronis, Tomopteris, Squilla*). *Z. vergl. Physiol.*, **5.**

BHATIA, M. L. 1956. Extraocular photoreceptors in the land leech, *Haemadipsa zeylanica agilis* (Moore) from Nainital, Almora (India). *Nature, Lond.*, **178,** 4530.

BICHAT, F. M. X. 1818. *Anatomie générale*, 2nd ed. Paris.

BIDDER, A. M. 1950. The digestive mechanism of the European squids *Loligo vulgaris, L. forbesi, Alloteuthis media* and *A. subulata*. *Q. J. microsc. Sci., N.S.*, **91.**

BINARD, A., and JEENER, R. 1928. Sur l'existence de cavités coelomiques dans le segment palpaire des Polychètes. *Rec. Inst. Zool. Torley-Rousseau.* **1.**

BLISS, D. E. 1953. Endocrine control of metabolism in the land crab, *Gecarcinus lateralis* (Frem.). I. Differences in the respiratory metabolism of sinusglandless and eyestalkless crabs. *Biol. Bull.*, **104.**

BLISS, D. E., and WELSH, J. H. 1952. The neurosecretory system of brachyuran Crustacea. *Biol. Bull.*, **103.**

BLOCHMANN, F. 1892. *Untersuchungen über den Bau der Brachiopoden.* Jena.

BLOWER, G. 1950. Aromatic tanning in the Myriapod cuticle. *Nature, Lond.*, **165,** 4197.

BLOWER, G. 1951. A comparative study of the Chilopod and Diplopod cuticle. *Q. J. microsc. Sci., N.S.*, **92.**

BLUMENTHAL, H. 1935. Untersuchungen über das Tarsalorgan der Spinnen. *Z. Morphol. Ökol. Tiere*, **29.**

BOBIN, G., and Durchon, M. 1952. Étude histologique du cerveau de *Perinereis cultrifera* Grube. Mise en évidence d'un complèxe cérébro-vasculaire. *Arch. anat. microsc. morphol. exptl*, **41,** 1.

BOCK, S. 1923. Eine neue marine Turbellariengattung aus Japan (*Hofstenia*). *Uppsala univ. årsskr.*, **1.**

BOCK, S. 1927. *Apidioplana*. Eine Polycladengattung mit muskulösen Drüsenorganen. *Göteborgs. Kgl. vet.-och vitterhets-samhäl. handl.*, *ser.* 4, 30.

BOEKE, J. 1935. The autonomic (enteric) nervous system of *Amphioxus lanceolatus*. *Q. J. microsc. Sci.*, *N.S.*, 77.

BOETTGER, C. R. 1955. Verwandschafftbeziehungen der primitiven Mollusken. *Verhandl Deutsch. Zool. Ges.* (Erlangen).

BÖHM, R. 1878. Helgolander Leptomedusen. *Z. Naturw.*, 12.

BOLZER, E. 1927. Untersuchungen über das Nervensystem der Coelenteraten. *Z. Zellforsch.*, 5.

BONE, Q. 1957. The problem of the 'Amphioxides' larva. *Nature, Lond.*, 180, 4600.

BORG, F. 1926. Studies on recent Cyclostomatous Bryozoa. *Zool. Bidr.*, *Uppsala*, 10.

BORG, F. 1947. Zur Kenntnis der Ökologie und des Lebenszyklus von *Electra crustulenta* (Bryozoa Chilostomata) nebst Bemerkungen über den sog. braunen Körper bei den Bryozoen. *Zool. Bidr.*, *Uppsala*, 25.

BÖRNER, C. 1901. Zur äußeren Morphologie von *Koenenia mirabilis* Grassi. *Zool. Anz.*, 24.

BÖRNER, C. 1909. Neue Homologien zwischen Crustaceen und Hexapoden. Die Beißmandibel der Insekten und ihre phylogenetische Bedeutung. Archi- und Metapterygota. *Zool. Anz.*, 34.

BORRADAILE, L. A. 1926. Notes on Crustacean limbs. *Ann. and Mag. Nat. Hist.*, *Ser.* 9, 17.

BOTAZZI, F. 1897. La pression osmotique du sang des animaux marins. *Arch. ital. biol.*, 28. Pisa.

BOUTAN, L. 1886. Recherches sur l'anatomie et le développement de la *Fissurella*. *Arch. zool. exptl gén.*, *ser.*, 2, 3.

BOUTAN, L. 1899. La cause principale de l'asymétrie des Mollusques Gastéropodes. *Arch. zool. exptl gén.*, *ser.*, 3, 7.

BOUVIER, E. 1893. Sur l'organisation des Actaeons. *Compt. rend. Soc. biol.*, 5.

BOVERI, TH. 1892. Die Nierencanälchen des *Amphioxus*. Ein Beitrag zur Phylogenie des Urogenitalsystems der Wirbeltiere. *Zool. Jahrb.*, *Anat.*, 5.

BOVERI, TH. 1893. Über die Entwicklung des Gegensatzes zwischen Geschlechtszellen und den somatischen Zellen bei *Ascaris megalocephala*. *S. B. Ges. Morphol. Physiol.*, 8.

BOVERI, TH. 1899. Die Entwicklung von *Ascaris megalocephala* mit besonderer Rücksicht auf die Kernverhältnisse. *Festschrift für Kupffer*. Jena.

BRACE, E. M. 1901. Notes on *Aeolosoma tenebrarum*. *J. Morphol.*, 17.

BRANDT, E. 1879a. Vergleichend-anatomische Skizze des Nervensystems der Insekten. *Horae Soc. Ent. Ross.*, *St. Pétersbourg*, 14, 15.

BRANDT, E. 1879b. Über die Metamorphose des Nervensystems der Insekten. *Horae Soc. Ent. Ross. St. Pétersbourg*, 14, 15.

BRANDT, E. 1879c. Vergleichend-anatomische Untersuchungen über das Nervensystem der Hymenopteren. *Horae Soc. Ent. Ross.*, *St. Pétersbourg*, 15.

BRANDT, E. 1879d. Vergleichend-anatomische Untersuchungen über das Nervensystem der Zweiflügler (Diptera). *Horae Soc. Ent. Ross.*, *St. Pétersbourg*, 14, 15.

BRESSLAU, E. 1909. Die Entwicklung der Acölen. *Verhandl. Deutsch. Zool. Ges.*, *Leipzig*, 19.

BRESSLAU, E. 1928–33. Turbellaria. In: Kükenthal, W., and Krumbach, Th. *Handbuch der Zoologie*, II/1, 1, 9, 16. Berlin and Leipzig.

BRESSLAU, E., and VOSS, H. VON. 1913. Das Nervensystem von *Mesostoma*. *Zool. Anz.*, 43.

BRIEN, P. 1948. Embranchement des Tuniciers, morphologie et reproduction. In: Grassé, P.-P. *Traité de zoologie*, XI. Paris.

BRIEN, P., and BREEDE, P. VAN DER. 1948. Le rôle des épicardes dans le bourgeonnement des Didemnidae. *Bull. Acad. Belg.*, *Cl. Sci.*, *sér.* 5, 34.

BROILI, F. 1928. Crustaceenfunde aus dem rheinischen Unterdevon. I. Über Extremitätenreste. *S. B. math.-naturw. Abt. Bayer. Akad. wiss. München, S.* **197–201.**

BROILI, F. 1931. Pantopoden aus dem rheinischen Unterdevon. *Forsch. Fortschr.*, **8.**

BROILI, F. 1932. *Palaeoisopus* ist ein Pantopod. *S. B. math.-naturw. Abt. Bayer. Akad. Wiss. München, S.* **45–60.**

BRONN, H. G. 1858. *Morphologische Studien über die Gestaltungs-Gesetze der Naturkörper überhaupt und der organischen insbesondere.* Leipzig and Heidelberg.

BROOKS, K. J. 1952. Primitive fossil Gastropods and their bearing on Gastropod classification. *Smiths. Misc. Coll.*, **117.**

BROOKS, W. K. 1886. Origin of medusae and significance of metagenesis. *Boston Soc. Natur. Hist., mem.* **3.**

BRUMPT, E. 1900. Reproduction des Hirudinés. *Mem. Soc. Zool. France*, **13.**

BRUNTZ, L. 1903. Sur la présence de reins labiaux et d'un organe phagocytaire chez les Diplopodes. *Compt. rend. Acad. Sci.*, **136.**

BRUNTZ, L. 1904. Contribution à l'étude de l'excretion chez les Arthropodes. *Arch. biol.*, **20.**

BRUNTZ, L. 1908. Sur la structure du réseau trachéen et des reins de *Machilis maritima. Compt. rend. Acad. Sci.*, **146.**

BÜCHERL, W. 1940. Sobre a musculature da *Scolopendra viridicornis* Newp. Uma contribução para o estudo comparativo da musculatura dos Quilopodos e Insetos. *Mem. Inst. Butantan*, **14.**

BÜCHNER, P. 1930. *Tier und Pflanze in Symbiose.* Berlin.

BÜCHNER, P. 1953. *Endosymbiose der Tiere mit pflanzlichen Mikroorganismen.* Basel and Stuttgart.

BUCK, J. B., and KEISTER, M. D. 1953. Cutaneous and tracheal respiration in the *Phormia* larva. *Biol. Bull.*, **105.**

BUDDENBROCK, W. VON. 1911. Untersuchungen über die Schwimmbewegungen und die Statocyste der Gattung *Pecten. S. B. Heidelberg. Akad. Wiss.*, **28.**

BULLOCK, T. H. 1940. The functional organisation of the nervous system of the Enteropneusta. *Biol. Bull.*, **79.**

BULLOCK, T. H. 1944. The giant nerve fibre system in *Balanoglossus. J. compar. Neurol.*, **80.**

BULLOCK, T. H. 1945. The anatomical organisation of the nervous system of Enteropneusta. *Q. J. microsc. Sci. N. S.*, **86.**

BULMAN, O. M. B. 1932, 1934, 1936. On the Graptolites prepared by Holm. *Arkiv zool.*, **24,** 8; **26,** 5; **28,** 17.

BURDON-JONES, C. 1951. Observations on the spawning behaviour of *Saccoglossus horsti* Brambell and Goodhart and of other Enteropneusta. *J. Mar. Biol. Assn UK*, **29.**

BÜRGER, O. 1897–1907. Nemertini. In: Bronn's *Klassen und Ordnungen des Tierreichs*, IV, 2, Suppl. Leipzig.

BURTON, M. 1949. Observations on littoral sponges, including the supposed swarming of larvae, movement and coalescence of immature individuals, longevity and death. *Proc. Zool. Soc.*, **118.**

BUTCHER, E. O. 1930. The pituitary in the ascidians (*Molgula manhattensis*). *J. exptl Zool.*, **57.**

BUTSCHINSKY, P. 1894. Zur Entwicklungsgeschichte von *Gebia litoralis. Zool. Anz.*, **17.**

BÜTSCHLI, O. 1877. Entwicklungsgeschichtliche Beiträge. I. Zur Entwicklungsgeschichte von *Paludina vivipara* Müll. *Z. wiss. Zool.*, **29.**

BÜTSCHLI, O. 1883. Über eine Hypothese bezüglich der phylogenetischen Herleitung des Blutgefäßapparates eines Teiles der Metazoen. *Morphol. Jahrb., Leipzig*, **8.**

BÜTSCHLI, O. 1887. Bemerkungen über die wahrscheinliche Herleitung der Asymmetrie der Gastropoden, spec. der Asymmetrie im Nervensystem der Prosobranchiaten. *Morphol. Jahrb., Leipzig*, **12.**

CALDWELL, W. H. 1883. Preliminary note on the structure, development and affinities of *Phoronis. Proc. Roy. Soc. London, B*, **34.**

CALDWELL, W. H. 1884. Blastopore, mesoderm, and metameric segmentation of *Phoronis. Q. J. microsc. Sci., N.S.*, **25.**

CALMAN, W. T. 1909. Crustacea. In: Ray Lankester, E. *A treatise on zoology*, VIII, 3, London.

CALMAN, W. T., and GORDON, I. 1933. A dodecapodous Pycnogonid. *Proc. Roy. Soc. London, B*, **113.**

CANNON, H. G. 1924. On the development of Estherid Crustaceans. *Phil. Trans. Roy. Soc. London, B*, **212.**

CANNON, H. G. 1925. On the segmental excretory organs of certain freshwater Ostracods. *Phil. Trans. Roy. Soc. London, B*, **214.**

CANNON, H. G. 1927. On the postembryonic development of the Fairy Shrimp (*Chirocephalus diaphanus*). *J. Linnean Soc., London*, **36.**

CANNON, H. G. 1928a. On the feeding mechanism of the fairy shrimp, *Chirocephalus diaphanus* Prevost. *Trans. Roy. Soc., Edinburgh*, **55.**

CANNON, H. G. 1928b. On the feeding mechanism of the Copepods *Calanus finmarchicus* and *Diaptomus gracilis. J. exptl Biol.*, **6.**

CANNON, H. G., and MANTON, S. M. 1927a. Notes on the segmental excretory organs of the Crustacea. *J. Linnean Soc. London*, **36.**

CANNON, H. G., and MANTON, S. M. 1927b. On the feeding mechanism of the Mysid Crustacean, *Hemimysis lamornae. Trans. Roy. Soc. Edinburgh*, **55.**

CANNON, H. G., and MANTON, S. M. 1929. On the feeding mechanism of the Syncarid Crustacea. *Trans. Roy. Soc. Edinburgh*, **56.**

CARLGREN, O. 1924. Die Larven des Ceriantharien, Zoantharien und Actiniarien. *Wiss. Ergebn. 'Valdivia'*, **19,** 8.

CARLISLE, D. B. 1953. Moulting hormones in *Leander* (Crustacea, Decapoda). *J. Mar. Biol. Assn UK*, **32.**

CARLISLE, D. B., and DOHRN, P. F. R. 1953. Studies on *Lysmata seticaudata* Risso (Crustacea, Decapoda). II. Experimental evidence for a growth- and moult-accelerating factor, obtainable from eyestalks. *Pubbl. Staz. zool. Napoli*, **24.**

CARLISLE, D. B., and KNOWLES, F. G. W. 1953. Neurohaemal organs in Crustaceans. *Nature, Lond.*, **172,** 4374.

CARLISLE, D. B., and PASSANO, L. M. 1953. The X-organ of Crustacea. *Nature, Lond.*, **171,** 1070–71.

CARPENTER, G. H. 1906. Notes on the segmentation and phylogeny of the Arthropoda, with an account of the maxillae in *Polyxenus lagurus. Q. J. microsc. Sci., N.S.*, **49.**

CAULLERY, M. 1914. Sur les Siboglinidae, type nouveau d'Invertébrés, recueilli par l'Expédition du 'Siboga'. *Compt. rend. Acad. sci.*, **158.**

CAULLERY, M. 1914. *Siboglinum Caullery* 1914, type nouveau d'invertébrés, d'affinités à preciser. *Siboga Expeditie, Monogr.* **25** bis.

CAZAL, P. 1948. Les glandes endocrines rétro-cérébrales des Insectes (étude morphologique). *Bull. biol. France Belgique*, **32,** suppl. Paris.

CHAPMAN, G. 1950. Of the movement of worms. *J. exptl Biol.*, **27.**

CHILD, C. M. 1900. The early development of *Arenicola* and *Sternapsis. Arch. Entwicklungsmech.*, **9.**

CHILD, C. M. 1911. The axial gradient in *Planaria dorotocephala* as a limiting factor in regulation. *J. exptl Zool.*, **10.**

CHRISTENSEN, B. 1959. Asexual reproduction in the Enchytraeidae (Olig.). *Nature, Lond.*, **184,** 4693.

CHUANG, S. H. 1959. The structure and function of the alimentary canal in *Lingula unguis* (L.) (Brachiopoda). *Proc. Zool. Soc. London*, **132**, 2.

CHUIN, T. T. 1928. Absence de strobilisation et persistance du bourgeonnement pendant l'hiver chez des scyphistomes alimentés artificiellement. *Compt. rend. Acad. sci.*, **186**.

CHUN, C. 1880. Die Ctenophoren des Golfes von Neapel und der angrenzenden Meeresabschnitte. In: *Fauna und Flora des Golfes von Neapel*. 1 Monogr. Leipzig.

CHUN, C. 1897. Die Siphonophoren. In: Hensen, V. *Ergebnisse der Plankton-Expedition der Humboldt-Stiftung*. II. Kiel and Leipzig.

CLARK, R. B. 1959. The neurosecretory system of the supra-oesophageal ganglion of *Nephthys* (Annelida, Polychaeta). *Zool. Jahrb., Physiol.*, **68**, 3.

CLARKE, A. H., and MENZIES, R. J. 1959. *Neopilina* (*Vema*) *ewingi*, a second living species of the Paleozoic class Monoplacophora. *Science*, **129**.

CLAUS, C. 1873. Zur Kenntnis des Baus und der Entwicklung von *Branchipus* und *Apus. Abhandl. Ges. Wiss. Göttingen*.

CLAUS, C. 1880–82. *Grundzüge der Zoologie*.

CLAUS, C. 1886a. Über die Entwicklung und den feinern Bau der Stilaugen von *Branchipus. Anz. Österr. Akad. Wiss. Math.-natur. Wiss.*, **23**.

CLAUS, C. 1886b. Untersuchungen über die Organisation und Entwicklung von *Branchipus* und *Artemia. Arb. Zool. Inst.*, **6**.

CLAUS, C. 1886c. Development and structure of pedunculated eyes of *Branchipus. J. Roy. microsc. Soc., ser.* **2**, 6.

CLAUS, C. 1888. Bemerkungen über marine Ostracoden aus der Familie der Cypriniden und Halocypriden. *Arb. Zool. Inst.*, **8**.

COE, W. R. 1943. Biology of the Nemerteans of the Atlantic Coast of North America. *Trans. Connect. Acad. Arts Sci.*, **35**.

COLOSI, G. 1927. Il popolamento delle terre emerse e i fattori delle grandi trasmigrazioni. *Universo*, **8**.

COLOSI, G. 1928. Über die Konstanz des respiratorischen Mediums (Lage des Problems und neue Beweise). *Zool. Anz.*, **77**.

COLWIN, A. L., and L. H. 1951. Relationships between the egg and larva of *Saccoglossus kowalevskii* (Enteropneusta): axes and planes; general prospective significance of the early blastomeres. *J. Exptl Zool.*, **117**.

CONKLIN, E. G. 1897. The embryology of *Crepidula. J. Morphol.*, **13**.

CONKLIN, E. G. 1902. The embryology of a Brachiopod, *Terebratulina septentrionalis* Conthony. *Proc. Amer. Philos. Soc.*, **41**.

CONKLIN, E. G. 1932. The embryology of *Amphioxus. J. Morphol.*, **54**.

COONFIELD, B. R. 1936. Apical dominance and polarity in *Mnemiopsis leidyi* Agassiz. *Biol. Bull.*, **70**.

CORI, C. I. 1929. Kamptozoa. In: Kükenthal, W., and Krumbach, Th. *Handbuch der Zoologie*, II/1, 6. Berlin and Leipzig.

CORI, C. I. 1932. Phoronoidea. In: Grimpe, G., and Wagler, E. *Die Tierwelt der Nord- und Ostsee*. VII, C 2.

CORI, C. I. 1936. Kamptozoa. In: Bronn's *Klassen und Ordnungen des Tierreichs*, IV/2, 4.

CORI, C. I. 1939. Phoronoidea. In: Bronn's *Klassen und Ordnungen des Tierreichs*, IV/1, 1. Leipzig.

CORRÊA, D. D. 1948. A embriologia de *Bugula flabellata* (J. V. Thomps). *Bol. Fac. Fil. Ciênc. Letr. S. Paulo. Zool.*, **13**.

CORRÊA, D. D. 1949. Sôbre o gênero *Zygantroplana. Zoologia*, **14**.

COX, L. R., and REES, W. J. 1960. A bivalve Gastropod. *Nature, Lond.*, **185**, 4715.

CUÉNOT, L. 1891. Etudes sur le sang et les glandes lymphatiques dans la série animale. *Arch. Zool. exptl gén., sér.* **2**, **9**.

Cuénot, L. 1902. Organes agglutinants et organes cilio-phagocytaires. *Arch. Zool. exptl gén., sér.* 3, **10**.

Cuénot, L. 1907. Fonction absorbante et excrétrice du foie des Céphalopodes. *Arch. Zool. exptl gén., sér.* 4, **7**.

Cuénot, L. 1948. Anatomie, éthologie et systématique des Echinodermes. In: Grassé, P.-P. *Traité de Zoologie*. XI. Paris.

Cuénot, L. 1949. Les Tardigrades. Les Pentastomides. In: Grassé, P.-P. *Traité de Zoologie*, IV. Paris.

Cuvier, G. 1817. *Le règne animal distribué d'après son organisation.* Paris.

Czwiklitzer, R. (1908). 1909. Die Anatomie der Larvae von *Pedicellina echinata. Arb. Zool. Inst. Univ. Wien,* **17**.

Dahl, E. 1952. Mystacocarida. Rept. Lund. Univ. Chile Exp. 1948–49. *Lund Univ. Årskr. Avd.* 2, **48**.

Dahl, E. 1956. Some crustacean relationships. In: Hanström, B. *Zool. Papers in honour of his 65th birthday.*

Dahl, E. 1957. Embryology of X-Organs in *Crangon allmanni. Nature, Lond.,* **179**, 4557.

Dahl, E. 1958. Influence of mode of life on brain topography in Crustacea. *Proc. XV Intern. Congr. Zool.* London.

Dawydoff, C. N. 1928a. Sur l'embryologie des Protonémertes. *Compt. rend. Acad. sci.,* **186**.

Dawydoff, C. N. 1928b. *Traité d'embryologie comparée des Invertébrés.* Paris.

Dawydoff, C. N. 1940. Les formes larvaires de Polyclades et de Némertes du plancton Indochinois. *Bull. biol. France Belgique,* **74**, 4.

Dawydoff, C. N. 1946. Contribution à la connaissance des Cténophores pélagiques des eaux de l'Indochine. *Bull. biol. France Belgique,* **80**, 2.

Dawydoff, C. N. 1948. Embranchement des Stomocordés. In: Grassé, P.-P. *Traité de Zoologie,* XI. Paris.

Dawydoff, C. N. 1949. Développement embryonnaire des Arachnides. In: Grassé, P.-P. *Traité de Zoologie,* VI. Paris.

Dawydoff, C. N. 1953. Contribution à nos connaissances de *l'Hydroctena. Compt. rend. Acad. sci.,* **237**.

De Beer, R. 1949. La pédomorphose, mode d'évolution progressive. *XIII Congr. Internat. zool.,* Compt. rend., Paris.

De Beer, G. R. 1955. The continuity between the cavities of the premandibular somites and of Rathke's pocket in *Torpedo. Q. J. microsc. Sci., N.S.,* **96**.

Dechaseaux, C. 1953. Classe des Crustacés (Crustacea Pennant 1777). In: Pivetau, J. *Traité de Paléontologie,* III. Paris.

Dehl, E. 1934. Morphologie von *Lindia. Z. wiss. Zool.,* **145**.

Deineka, D. 1908. Das Nervensystem von *Ascaris. Z. wiss. Zool.,* **89**.

Delage, Y. 1898. Sur la place des Spongiaires dans la classification. *Compt. rend. Acad. sci.,* **126**.

Delage, Y., and Hérouard, E. 1897–99, 1901, 1913. *Traité de Zoologie concrète.* II, III, V, VIII. Paris.

Delaunay, H. 1934. Le métabolisme de l'ammoniaque d'après les recherches relatives aux Invertébrés. *Ann. Physiol. physicochim. biol.,* **10**.

Demoll, R. 1909. Augen von *Alciopa cantrainii. Zool. Jahrb. Anat.,* **27**.

Demoll, R. 1914. Die Augen von *Limulus. Zool. Jahrb. Anat.,* **38**.

Dennel, R. 1947. The occurrence and significance of phenolic hardening in the newly formed cuticle of Crustacea Decapoda. *Proc. Roy. Soc. London, B,* **134**.

Denton, E. J., and Gilpin-Brown, J. B. 1959. Buoyancy of the cuttlefish. *Nature, Lond.,* **184**, 4695.

Dethier, V. G. 1955. The physiology and histology of the contact chemoreceptors of the blowfly. *Q. Rev. Biol.,* **30**.

DETHIER, V. G., EVANS, D. R., and RHONDES, M. V. 1956. Some factors controlling the ingestion of carbohydrates by the blowfly. *Biol. Bull.*, **111**, 2.

DOBELL, C. 1911. The principles of protistology. *Arch. Protistenkunde*, **23**.

DOGIEL, A. S. 1903. Das periphere Nervensystem des *Amphioxus. Anatom. Hefte*, **1**, 21.

DOGIEL, V. A. 1911a. Ein interessanter Fall von atavistischer Mißbildung bei einer Pantopodenlarve. *Zool. Anz.*, **38**.

DOGIEL, V. A. 1911b. Untersuchungen zur Entwicklungsgeschichte der Pantopoden. Das Nervensystem und die Drüsen der Larve der Pantopoden. *Z. wiss. Zool.*, **99**.

DOGIEL, V. A. 1913. Embryologische Studien an Pantopoden. *Z. wiss. Zool.*, **107**.

DOGIEL, V. A. 1929. Polymerisation als ein Princip der progressiven Entwicklung bei Protozoen. *Biol. Zbl.*, **49**.

DOHRN, A. 1875. *Der Ursprung der Wirbelthiere und das Prinzip des Functionswechsels*. Leipzig.

DOHRN, A. 1904. Studien zur Urgeschichte des Wirbelthierkörpers. 24, Die Prämandibularhöhle. *Mitt. Zool. Stat. Neapel*, **17**. Berlin.

DOLLO, L. 1893. Les lois de l'évolution. *Bull. Soc. belge géol.*, **7**.

DRACH, P. 1948. Embranchement des Céphalocordés. In: Grassé, P.-P. *Traité de Zoologie*, XI. Paris.

DRESEL, E. I. B., and MOYLE, V. 1950. Nitrogenous excretion of Amphipods and Isopods. *J. exptl Biol.*, **27**.

DRZEWINA, A., and BOHN, G. 1916. Sur un changement du type de symétrie (symétrie métabolique) chez un Hydraire, *Stauridium productum. Compt. rend. Soc. biol.*, **79**.

DUBOSCQ, O., and GRASSÉ, P. 1933. L'appareil parabasal des Flagellés, avec des remarques sur la trophosponge, l'appareil Golgi, les mitochondries et le vacuome. *Arch. zool. exptl. gén.*, **73**.

DUBOSCQ, O., and TUZET, O. 1933. Quelques structures des amphiblastules d'éponges calcaires. *Compt. rend. Acad. sci.*, **197**. Paris.

DUBOSCQ, O. 1935. Un nouveau stade du développement des éponges calcaires. *Compt. rend. Acad. sci.*, **200**.

DUBOSCQ, O. 1937. L'ovogénèse, la fécondation et les premiers stades du développement des éponges calcaires. *Arch. zool. exptl gén.*, **79**.

DURCHON, M. 1948. Epitoquie expérimentale chez deux Polychètes: *Perinereis cultrifera* et *Nereis irrorata. Compt. rend. Acad. sci.*, **227**.

DURCHON, M. 1949. Inhibition de l'épitoquie par le prostomium chez les Néréidiens. *Compt. rend. Acad. sci.*, **229**.

DURCHON, M. 1951. L'ablation du prostomium provoque, chez les Néréidiens, la maturation précoce des produits génitaux mâles. *Compt. rend. Acad. sci.*, **232**.

EGGERS, F. 1920. Das thoracale bitympanale Organ einer Gruppe der Lepidoptera Heterocera. *Zool. Jahrb., Anat.*, **41**.

EGGERS, F. 1937. Zur hypothetischen Homologie verschiedensegmentiger Tympanalorgane. *Zool. Anz.*, **118**.

EGGERS, F. 1943. Die Gehörorgane und das Gehör der Schmetterlinge. *Entomol. Z.*, **56**.

EHRENBERG, CH. G. 1838. *Die Infusionsthierchen als vollkommene Organismen*. Leipzig.

EISIG, H. 1887. Monographie der Capitelliden des Golfes von Neapel. In: *Fauna und Flora des Golfes von Neapel*, 16 Monogr. Leipzig.

EISIG, H. 1899. Zur Entwicklungsgeschichte der Capitelliden. *Mitt. Zool. Stat. Neapel*, **13**.

ELTRINGHAM, H. 1933. *The senses of insects*. London.

EVANS, R. 1902. On the Malayan species of Onychophora. The development of *Eoperipatus weldoni*. *Q. J. microsc. Sci., N.S.*, **45**.

FABRE, J. 1879. *Souvenirs entomologiques*. Paris.

FAGE, L. 1906. Recherches sur les organes segmentaires des Annélides Polychètes. *Ann. sci. natur. biol. anim.*, ser. **9, 3**.

FAGE, L., and BARROS MACHADO, A. DE. 1951. Quelques particularités remarquables de l'anatomie des Ochyrocératides (Araneae). *Arch. zool. exptl gén.*, **87**.

FAHLANDER, K. 1938. Beiträge zur Anatomie und systematische Einteilung der Chilopoden. *Zool. Bidr., Uppsala*, **17**.

FAHLANDER, K. 1938–40. Die Segmentalorgane der Diplopoda, Symphyla und Insecta Apterygota. *Zool. Bidr., Uppsala*, **18**.

FAUSSEK, W. 1900. Was ist das Coelom? Untersuchungen über die Entwicklung der Cephalopoden. *Mitt. Zool. Stat. Neapel*, **14**.

FAUSSEK, W. 1911. Vergleichend-embryologische Studien. *Z. wiss. Zool.*, **98**.

FEDELE, M. 1927. Ancora sulla organizzazione e le caratteristiche funzionale dell' attivita nervosa dei Tunicati. III. Il systemo nervoso viscerale. *Atti Acad. naz. Lincei. Mem. sci. fis., mat. natur.*, ser. **6, 6**.

FEDOROW, B. 1926. Zur Anatomie des Nervensystems von *Peripatus*. I. Das Neurosomit von *Peripatus tholloni*. *Zool. Jahrb., Anat.*, **48**.

FEDOROW, B. 1929. Zur Anatomie des Nervensystems von *Peripatus*. II. Das Nervensystem des vorderen Körperendes und seine Metamerie. *Zool. Jahrb., Anat.*, **50**.

FEDOTOV, D. M. 1914. Die Anatomie von *Protomyzostomum polynephris*. *Z. wiss. Zool.*, **109**.

FEDOTOV, D. M. 1924. Zur Morphologie des axialen Organkomplexes der Echinodermen. *Z. wiss. Zool.*, **123**.

FEDOTOV, D. M. 1926. The plan of structure and systematic status of the Ophiocystia (Echinoderma). *Proc. Zool. Soc. London*, part 4.

FEDOTOV, D. M. 1926. Die Morphologie der Euryalae. *Z. wiss. Zool.*, **127**.

FEDOTOV, D. M. 1927. Morphologische Studien an Euryalen. *Z. Morphol. Ökol. Tiere*, **9**.

FEDOTOV, D. M. 1929. Beiträge zur Kenntnis der Morphologie der Myzostomiden. *Z. Morphol. Ökol. Tiere*, **15**.

FELIX, J. 1906. Die Entwicklung der Harn- und Geschlechtsorgane. In: Hertwig, O. *Handbuch der vergleichenden und experimentellen Entwicklungslehre der Wirbeltiere*. III, 1. Jena.

FERNANDEZ, M. 1904. Zur mikroskopischen Anatomie des Blutgefäßsystems der Tunicaten. Nebst Bemerkungen zur Phylogenese des Blutgefäßsystems im allgemeinen. *Z. Naturw.*, **39**.

FEWKES, J. W. 1883–85. On the development of certain worm larvae. *Bull. Museum Compar. Zool.*, **5**.

FILIPJEV, I. N. 1912. Zur Kenntnis des Nervensystems bei den freilebenden Nematoden. *Compt. rend. Soc. Imp. Nat. St. Pétersbourg*, **43**.

FISCHER, W. K. 1904. The anatomy of *Lottia gigantea* Gray. *Zool. Jahrb., Anat.*, **20**.

FISH, G. R. 1955. Digestion and the production of sulfuric acid by Mollusca. *Nature, Lond.*, **175**, 4460.

FLOREY, E. VON. 1951. Reizphysiologische Untersuchungen an der Ascidie *Ciona intestinalis* L. *Biol. Zbl.*, **70**.

FOUQUE, G. 1953. Contribution à l'étude de la glande pylorique des Ascidiacés. *Ann. Inst. océanogr.*, **28**, 5.

FOWLER, G. H. 1887. The anatomy of Madreporaria. II. *Madrepora*. *Q. J. microsc. Sci.*, **17**.

FRETTER, V., and GRAHAM, A. 1954. Observations on the Opisthobranch Mollusc *Actaeon tornatilis* (L.). *J. Mar. Biol. Assn UK*, **33**.

FREUDENTHAL, H. D. 1962. *Symbiodinium* gen. nov. and *S. adriaticum* sp. nov., a zooxanthella: taxonomy, life cycle and morphology. *J. Protistol.*, **9.**

FRINGS, H., and FRINGS, M. 1949. The loci of contact chemoreceptors in insects. *Amer. Midland Naturalist*, **41.**

FRISCH, K. VON. 1914. Der Farbensinn und Formensinn der Biene. *Zool. Jahrb.*, *Allg.*, **35.**

FUHRMANN, O. 1914. Turbellariés d'eau douce de Colombie. *Mém. Soc. sci. natur.*, **5.**

FUHRMANN, O. 1928–33. Trematoda. In: Kükenthal, W., and Krumbach, Th. *Handbuch der Zoologie*, II/1, 3, 7. Berlin and Leipzig.

FUHRMANN, O. 1930–31. Cestoidea. *Handbuch der Zoologie*, 7. 11.

GABE, M. 1949. Sur la présence de cellules neuro-sécrétrices chez *Dentalium entale* Deshayes. *Compt. rend. Acad. sci.*, **229.**

GABE, M. 1951. Données histologiques sur la neurosécrétion chez les Pterotracheidae (Heteropodes). *Rev. canad. biol.*, **10.**

GABE, M. 1952. Sur l'emplacement et les connexions des cellules neuro-sécrétrices dans les ganglions cérébroides de quelques Chilopodes. *Compt. rend. Acad. sci.*, **235.**

GABE, M. 1953a. Données histologiques sur les glandes endocrines céphaliques de quelques Thysanoures. *Bull. Soc. zool. France*, **78.**

GABE, M. 1953b. Particularités histologiques sur les glandes endocrines céphaliques de *Scutigera coleoptrata*. *Bull. Soc. zool. France*, **78.**

GABE, M. 1953c. Données histologiques sur la neuro-sécrétion chez les Arachnides. *Arch. anat. microsc. morphol. exptl.*, **44.**

GABE, M. 1953d. Données histologiques sur la neuro-sécrétion chez quelques Sipunculiens. *Bull. Labor. Dinard, Fasc.* **38.**

GABE, M. 1953e. Particularités morphologiques des cellules neuro-sécrétrices chez quelques Prosobranches monotocardes. *Compt. rend. Acad. Sci.*, **236.**

GABE, M. 1953f. Particularités histologiques des cellules neuro-sécrétrices chez quelques Gastropodes opisthobranches. *Compt. rend. Acad. Sci.*, **236.**

GABE, M. 1953g. Sur l'existence chez quelques Crustacés Malacostracés d'un organe comparable à la glande de la mue des Insectes. *Compt. rend. Acad. Sci.*, **237.**

GABE, M. 1954a. Emplacement et connexions des cellules neurosécrétrices chez quelques Aranéides. *Compt. rend. Acad. Sci.*, **238.**

GABE, M. 1954b. Emplacement et connexions des cellules neurosécrétrices chez quelques Diplopodes. *Compt. rend. Acad. Sci.*, **239.**

GABE, M. 1954c. La neuro-sécrétion chez les Invertébrés. *Année biol.*, **30.**

GABE, M. 1955. Données histologiques sur la neuro-sécrétion chez les Arachnides. *Arch. anat. microsc. morphol. exptl.*, **44.**

GARSTRANG, W. 1896. The morphology of the Mollusca. *Sci. Progr.*, **5.**

GASKELL, J. F. 1919. Adrenalin in Annelids. *J. Gen. Physiol.*, **5.**

GASKELL, W. H. 1896. Address to the physiological section on: The origin of vertebrates. *Rept. Brit. Ass. Adv. Sci., Meet. Liverpool*, **66**; *Proc. Cambridge Philos. Soc.*, **9.**

GASKELL, W. H. 1908. *The origin of vertebrates*. London.

GATENBY, J. B., DALTON, A. J., and FELIX, M. D. 1955. The contractile vacuole of Parazoa and Protozoa and the Golgi apparatus. *Nature, Lond.*, **176,** 4476.

GEBHARDT, H. 1951. Lokalisatorischer Nachweis von Thermorezeptoren bei *Dorcus parallelepipedus* L. und *Pyrrhocoris apterus* L. *Experientia*, **7.**

GEGENBAUR, C. 1854. Zur Lehre von Generationswechsel und der Fortpflanzung bei Medusen und Polypen. *Verhandl. Phys. Med. Ges. Würzburg*, **4.**

GELEI, J. VON. 1930. Echte freie Nervenendigungen. *Z. Morphol. Ökol. Tiere*, **18.**

GELEI, J. VON. 1937. Der schraubige Körperbau in der Ciliatenwelt im Vergleich zu den Symmetrieverhältnissen der vielzelligen Tiere. *Arch. Protistenkunde*, **88.**

GEMMILL, J. F. 1914. The development and certain points in the structure of the adult starfish of *Asterias rubens* L. *Phil. Trans. Roy. Soc., London, B,* **205.**

GEMMILL, J. F. 1915. On the ciliation of *Asterias*, and on the question of ciliary nutrition in certain species. *Proc. Zool. Soc., London,* **74.**

GEOFFROY SAINT-HILAIRE, E. 1822. *Philosophie anatomique.* Paris.

GERHARDT, U. 1935. Merostoma. In: Kükenthal, W., and Krumbach, Th. *Handbuch der Zoologie,* III/2, 8. Berlin and Leipzig.

GERHARDT, U., and KÄSTNER, A. 1937-38. Araneae. In: Kükenthal, W., and Krumbach, Th. *Handbuch der Zoologie,* 11-12.

GEROULD, J. H. 1906. The development of *Phascolosoma. Zool. Jahrb., Anat.,* **23.**

GEROULD, J. H. 1929. History of the discovery of periodic reversal of heart-beat in Insects. *Biol. Bull.,* **56.**

GERSCH, M. 1959. Neurohormone bei wirbellosen Tieren. *Verhandl. Deutsch. Zool. Ges. Frankfurt a. M.,* 1958.

GERSCH, M., and SCHEFFEL, H. 1958. Sekretorisch tätige Zellen im Nervensystem von *Ascaris. Naturwissenschaften,* **45,** 14.

GERWERZHAGEN, A. 1913. Beiträge zur Kenntnis der Bryozoen. 1. Das Nervensystem von *Cristatella mucedo* Cuv. *Z. wiss. Zool.,* **107.**

GIARD, A. 1888. Sur les *Nephromyces*, genre nouveau de champignons parasites du rein des Molgulidées. *Compt. rend. Acad. sci.,* **106.**

GIESBRECHT, W. 1913. Crustacea. In: Lang, A. *Handbuch der Morphologie der wirbellosen Tiere.* IV. Jena.

GIESE, M. 1913. Gonopericardialgang und Umbildung der Geschlechtwege im Zusammenhang mit Protandrie bei *Calyptraea sinensis. Zool. Anz.,* **42.**

GIESE, M. 1915. Der Genitalapparat von *Calyptraea sinensis* Lin., *Crepidula unguiformis* Lam. und *Capulus hungaricus* Lam. *Z. wiss. Zool.,* **114.**

GILCHRIST, F. G. 1937. Budding and locomotion in the scyphistomas of *Aurelia. Biol. Bull.,* **72.**

GISLÉN, T. 1924. Echinoderm studies. *Zool. Bidr., Uppsala,* **9.**

GISLÉN, T. 1930. Affinities between the Echinodermata, Enteropneusta and Chordonia. *Zool. Bidr., Uppsala,* **12.**

GISLÉN, T. 1947. On the Haplozoa and the interpretation of *Peridionites. Zool. Bidr., Uppsala* **25.**

GLAESSNER, M. F. 1957. Evolutionary trends in Crustacea (Malacostraca). *Evolution,* **11,** 2.

GODEAUX, J. 1954. Observations sur la glande pylorique des Thaliacés. *Ann. Soc. roy. zool. Belgique,* **85.**

GODEAUX, J. 1957-58. Contribution à la connaissance des Thaliacés (*Pyrosoma* et *Doliolum*). *Ann. Soc. roy. zool. Belgique,* **88.**

GOETTE, A. 1884. *Abhandlungen zur Entwicklungsgeschichte der Thiere.* 2. Hamburg.

GOETTE, A. 1907. Vergleichende Entwicklungsgeschichte der Geschlechtsindividuen der Hydropolypen. *Z. wiss. Zool.,* **87.**

GOLDSCHMIDT, R. 1908, 1909. Das Nervensystem von *Ascaris lumbricoides* und *megalocephala.* Ein Versuch in dem Aufbau eines einfachen Nervensystems einzudringen. I, II. *Z. wiss. Zool.,* **90, 92.**

GOLDSCHMIDT, R. 1910. Das Nervensystem von *Ascaris lumbricoides* und *megalocephala.* Ein Versuch in den Aufbau eines einfachen Nervensystems einzudringen. III. *Festschr. für R. Hertwig.* II. Jena.

GONTCHAROFF, M. 1948. Note sur l'alimentation de quelques Némertes. *Ann. sci. natur. zool. biol. anim.,* 11 *ser.,* **10.**

GOODRICH, E. S. 1897, 1898, 1900. On the nephridia of the Polychaeta. Parts I, II and III. *Q. J. microsc. Sci., N.S.,* **40, 41, 43.**

GOODRICH, E. S. 1902. On the structures of the excretory organs of *Amphioxus. Q. J. microsc. Sci., N.S.,* **45.**

GOODRICH, E. S. 1903. On the body-cavities and nephridia of the Actinotrocha larva. *Q. J. microsc. Sci., N.S.*, **47.**

GOODRICH, E. S. 1917. 'Proboscis pores' in Craniate Vertebrates, a suggestion concerning the premandibular 'somites' and hypophysis. *Q. J. microsc. Sci., N.S.* **62.**

GOODRICH, E. S. 1934. The early development of the nephridia in *Amphioxus:* Introductory and Pt. I. Hatschek's nephridium. Pt. II. The paired nephridia. *Q. J. microsc. Sci., N.S.*, **76.**

GOODRICH, E. S. 1945. The study of nephridia and genital ducts since 1895. Pt. I. *Q. J. microsc. Sci., N.S.*, **86.**

GOODRICH. E. S. 1946. The study of nephridia and genital ducts since 1895. Pt. II. *Q. J. microsc. Sci., N.S.*, **87.**

GOOSSEN, H. 1949. Untersuchungen an Gehirnen verschieden grosser, jeweils verwandter Coleopteren- und Hymenopterenarten. *Zool. Jahrb. Abt. Anat.*, **62, 1.**

GORVETT, H. 1946. The tegumental glands in the land Isopoda. A. The rosette glands. *Q. J. miscrosc. Sci., N.S.*, **87.**

GOSSE, P. H. 1859. *Evenings at the microscope;* or researches among the minuter organs and forms of animal life. London (also in: Russell, F. S. 1953. *The medusae of the British Isles.* Cambridge).

GRÄBER, H. 1933. Über die Gehirne der Amphipoden und Isopoden. *Z. Morphol. Ökol. Tiere*, **26.**

GRAFF, L. VON. 1882, 1889. *Monographie der Turbellarien.* I. Rhabdocoelida. II. Triclada Terricola. Leipzig.

GRAFF, L. VON. 1889. *Enantia spinifera. Mitt. Naturwiss. Vereines Steiermark*, **26.**

GRAFF, L. VON. 1891. *Die Organisation der Turbellaria Acoela.* Leipzig.

GRAFF, L. VON. 1904–08. Acoela und Rhabdocoelida. In: Bronn's *Klassen und Ordnungen des Tierreichs.* IV, 1. Leipzig.

GRAFF, L. VON. 1912–17. Triclada. *Klassen und Ordnungen des Tierreichs.* IV. 1. Leipzig.

GRAHAM, A. 1931. On the morphology, feeding mechanism and digestion of *Ensis siliqua* (Schumacher). *Trans. Roy. Soc. Edinburgh*, **56.**

GRAHAM, A. 1932. On the structure and function of the alimentary canal of the limpet. *Trans. Roy. Soc. Edinburgh*, **57.**

GRAHAM, A. 1938. On a ciliary process of food collecting in the gastropod *Turritella communis* Risso. *Proc. Zool. Soc. London*, **108.**

GRAHAM, A. 1939. On the structure of the alimentary canal of style bearing Prosobranchs. *Proc. Zool. Soc. London, B*, **109.**

GRAHAM, A. 1953. Form and function in the Molluscs. *Proc. Linnean Soc., London*, **164.**

GRANDJEAN, F. 1936, 1937. Le genre *Pachygnathus* Dugès (*Alycus* Koch), Acariens. Partie 1–5. *Bull. Museum nat. histoire natur., ser. 2*, **8, 9.**

GRANDJEAN, F. 1937. Remarques sur la terminologie des divisions du corps chez les Acariens. *Bull. Museum nat. histoire natur., ser. 2*, **9.**

GRANDJEAN, F. 1939. Quelques genres d'Acariens appartenant au groupe des Endeostigmata. *Ann. sci. natur. zool. biol. anim., ser. 9*, **2.**

GRASSÉ, P.-P. 1948. *Traité de Zoologie, Echinodermes, Stomocordés, Procordés.* XI. Paris.

GRASSÉ, P.-P. 1949. *Traité de Zoologie, Onychophores*—Tardigrades, Arthropodes— Trilobitomorphes Chélicérates. VI. Paris.

GRAVIER, C. 1909. Contribution à l'étude de la régéneration de la partie antérieure du corps chez les Polychètes. *Ann. sci. natur. zool. biol. anim., ser. 9*, **9.**

GREGORY, J. W. 1900. See Bather, F. A., Gregory, J. W., and Goodrich, E. S.

GRELCK, M. 1958. Über einen abnormen weiblichen Geschlechtsapparat von *Loligo forbesi* (Steenstr.). *Zool. Anz.*, **160,** 3–4.

GRIESBACH, H. 1889. Beiträge zur Histologie des Blutes. *Arch. Mikrobiol.*, *Anatomie*, 37.

GROBBEN, C. 1905. *Lehrbuch der Zoologie, begründet von C. Claus.* Marburg.

GROBBEN, C. 1908. Die systemische Eintheilung des Tierreichs. *Verhandl. zool.-bot. Ges.*, *Wien*, 58.

GRÖNBERG, G. 1898. Beiträge zur Kenntnis der Gattung *Tubularia*. *Zool. Jahrb.*, *Anat.*, 11.

GUIART, J. 1901. Contribution à l'étude des Gastropodes Opisthobranches et en particulier des Céphalaspidés. *Mém. Soc. Zool. France*, 14.

GUSTAFSON, G. 1930. Anatomische Studien über die Polychäten-Familien Amphinomidae und Euphrosynidae. *Zool. Bidr.*, *Uppsala*, 12.

HAAS, F. 1929–55. Bivalvia. In: Bronn's *Klassen und Ordnungen des Tierreichs*, III, 3. Leipzig.

HÄCKER, V. 1895. Die spätere Entwicklung der *Polynoë*-Larven. *Zool. Jahrb.*, *Anat.*, 8.

HADENFELDT, D. 1929. Das Nervensystem von *Stylochoplana maculata* und *Notoplana atomata*. *Z. wiss. Zool. A*, 133.

HADŽI, J. 1944. Turbelarijska teorija Knidarijev. *Razpr. Mat.-Prir. Akad.*, *Ljubljana*, 3. 1.

HAECKEL, E. 1866. *Generelle Morphologie der Organismen.* Berlin.

HAECKEL, E. 1874. Die Gastraea-Theorie, die phylogenetische Classification des Thierreichs und die Homologie der Keimblätter. *Z. Naturw.*, 8.

HAECKEL, E. 1862, 1887, 1888. *Die Radiolarien* (Rhizopoda Radiaria). Eine Monographie. I–IV. Berlin.

HAECKEL, W. 1960. Zellknorpel am Zentralnervensystem einer Gastropode [*Strophocheilus rosaceus* (King and Broderip)] (Pulmonata). *Zool. Anz.*, 164, 1–2.

HAFFNER, K. VON. 1959. Über den Bau und den Zusammenhang der wichtigsten Organe des Kopfendes von *Hyalinoecia tubicola* Malmgren (Polychaeta, Eunicidae, Onuphidinae), mit Berücksichtigung der Gattung *Eunice*. *Zool. Jahrb.*, *Anat.*, 77, 2.

HAMAKER, J. 1898. The nervous system of *Nereis virens*. *Bull. Museum Compar. Zool.*, 32.

HAMANN, O., and LUDWIG, H. 1889–1907. See Ludwig, H., and Hamann, O. 1889–1907.

HAMMARSTEN, O. 1913. Beiträge zur Entwicklung von *Halicryptus spinulosus*. *Zool. Anz.*, 41.

HAMMARSTEN, O. 1915. Zur Entwicklungsgeschichte von *Halicryptus spinulosus* Sier. *Z. wiss. Zool.*, 112.

HAND, O., and HENDRICKSON, J. R. 1950. A two-tentacled, commensal Hydroid from California (Limnomedusae, Proboscidactyla). *Biol. Bull.*, 99.

HANDLIRSCH, A. 1907. Hexapoda. In: Steinmann, G. *Einführung in die Paläontologie.* Leipzig.

HANDLIRSCH, A. 1925. Paläontologie; Phylogenie oder Stammesgeschichte; Systematische Übersicht. In: Schröder, Ch. *Handbuch der Entomologie*, III. Jena.

HANSEN, H. J. 1893. Zur Morphologie der Gliedmaßen und Mundtheile bei Crustaceen und Insekten. *Zool. Anz.*, 16.

HANSTRÖM, B. 1926. Eine genetische Studie über die Augen und Sehzentren von Turbellarien, Anneliden und Arthropoden (Crustacea, etc.). *Kgl. Svenska vetenskaps akad. handl.*, ser. 3, 4, 1.

HANSTRÖM, B. 1927. Das zentrale und periphere Nervensystem des Kopflappens einiger Polychaeten. *Z. Morphol. Ökol. Tiere*, 7.

HANSTRÖM, B. 1928a. *Vergleichende Anatomie des Nervensystems der wirbellosen Tiere unter Berücksichtigung seiner Funktion*, Berlin.

HANSTRÖM, B. 1928b. Die Beziehungen zwischen dem Gehirn der Polychaeten und dem der Arthropoden Z. Morphol. Ökol., 11.

HANSTRÖM, B. 1939. Hormones in Invertebrates. Oxford.

HANSTRÖM, B. 1940. Inkretorische Organe, Sinnesorgane und Nervensystem des Kopfes einiger niederer Insektenordnungen. Kgl. svenska vetenskaps akad. handl., ser. 3, 18, 8.

HANSTRÖM, B. 1941. Einige Parallelen im Bau und in der Herkunft der inkretorischen Organe der Arthropoden und der Vertebraten. Förh. fysiogr. Sällsk. Lund., N.F., 42.

HANSTRÖM, B. 1947. The brain, the sense organs, and the incretory organs of the head in the Crustacea Malacostraca. Acta Univ. Lund., N.F., Avd. 2, 43.

HANSTRÖM, B. 1953. Neurosecretory pathways in the head of Crustaceans, Insects and Vertebrates. Nature, Lond., 171, 4341.

HARADA, I. 1931. Das Nervensystem von Bolbosoma turbinella (Diesing). Japan. J. Zool., 3.

HARLEY, M. B. 1950. Occurrence of a filter-feeding mechanism in the Polychaete Nereis diversicolor. Nature, Lond., 165, 4201.

HARM, K. 1902. Die Entwicklungsgeschichte von Clava squamata. Z. wiss. Zool., 73.

HARMER, S. F. 1917. On Phoronis ovalis Strethill. Q. J. microsc. Sci., N.S., 62.

HARMER, S. F. 1923. On Cellularinae and other Polyzoa. J. Linnean Soc., London (Zool.), 35.

HARMER, S. F. 1926. The Polyzoa of the Siboga Expedition. II. Cheilostomata Anasca. Siboga Exped., 28.

HARMS, J. W. 1934. Wandlungen des Artgefüges. Leipzig.

HASKELL, P. T., and BELTON, P. 1956. Electrical responses of certain Lepidopterous tympanal organs. Nature, Lond., 177, 4499.

HATSCHEK, B. 1878. Studien über die Entwicklungsgeschichte der Anneliden. Arbeiten Zool. Inst., Wien, 1.

HATSCHEK, B. 1882. Studien über Entwicklung des Amphioxus. Arbeiten Zool. Inst., Wien, 4.

HATSCHEK, B. 1888–91. Lehrbuch der Zoologie. I–III. Jena.

HATSCHEK, B. 1911. Das neue zoologische System. Leipzig.

HAUSENSCHILD, C. 1956a. Hormonale Hemmung der Geschlechreife und Mondphase bei den Polychaeten Platynereis dumerilii. Z. Naturforsch., 15.

HAUSENSCHILD, C. 1956b. Weitere Versuche zur Frage des Juvenilhormons bei Platynereis. Z. Naturforsch., 15.

HAYWOOD, C. A., and MOON, H. P. 1953. Reversal of the heart beat in Tunicates. Nature, Lond., 172, 4366.

HEATH, H. 1898. The development of Ischnochiton. Zool. Jahrb., Anat., 12.

HEATHCOTE, F. G. 1886. The early development of Julus terrestris. Q. J. microsc. Sci., N.S., 26.

HEDGPETH, J. W. 1950. Pycnogonida of the United States Navy Antarctic Expedition, 1947–48. Proc. U.S. Nation. Museum, 100.

HEEGAARD, P. 1945. Remarks on the phylogeny of the Arthropods. Arkiv. Zool., 37, 3.

HEEGAARD, P. 1948. Discussion of the mouth appendages of the Copepods. Arkiv. Zool., 40, 1.

HEIDENHAIN, M. 1907. Plasma und Zelle, Jena.

HEIDER, C. 1879. Die Gattung Lernanthropus. Arbeiten Zool. Inst. Wien, 2, 3.

HEIDER, K. 1927. Vom Nervensystem der Ctenophoren. Z. Morphol. Ökol. Tiere, 9.

HELMCKE, J. G. 1939. Brachiopoda. In: Kükenthal, W., and Krumbach, Th. Handbuch der Zoologie, III/2, 13. Berlin and Leipzig.

HEMPELMANN, F. 1911. Zur Naturgeschichte von Nereis dumerilii Aud. et Edw. Zoologica, 62.

HEMPELMANN, F. 1931. Archiannelida, Polychaeta. In: Kükenthal, W., and Krumbach, Th. *Handbuch der Zoologie*, II/2, 13. Berlin and Leipzig.

HENKE, K., and RÖNSCH, G. 1951. Über Bildungsgleichheiten in der Entwicklung epidermaler Organe und die Entstehung des Nervensystems im Flügel der Insekten. *Naturwissenschaften*, **38.**

HENRIKSEN, K. L. 1926. The segmentation of the Trilobite's head. *Medd. Dansk. Geol. Foren., København*, **7.**

HENRIKSEN, K. L. 1928. Critical notes upon some Cambrian Arthropods described by Charles D. Walcott. *Vidensk. Medd. Dansk. Naturhist. For.*, **86.**

HENTSCHEL, E. 1923–24. Porifera. In: Kükenthal, W., and Krumbach, Th. *Handbuch der Zoologie*, I/2–3. Berlin and Leipzig.

HERLANT-MEEWIS, H. 1950a. Les lois de la scissiparité chez *Aeolosoma hemprichi* (Ehrenberg). *Contr. Inst. Zool. Univ. Montreal*, **24.**

HERLANT-MEEWIS, H. 1950b. Cyst formation in *Aeolosoma hemprichi* (Ehrb.). *Biol. Bull.*, **99.**

HERLANT-MEEWIS, H. 1956. Croissance et neurosécrétion chez *Eisenia foetida* (Sav.). *Ann. sci. natur. zool. biol. anim., ser.* 11, **28.**

HERTWIG, O., and R. 1878. *Das Nervensystem und Sinnesorgane der Medusen*. Leipzig.

HERTWIG, O., and R. 1879. Die Actinien, anatomisch und histologisch mit besonderer Berücksichtigung des Nervenmuskelsystems untersucht. *Z. Naturw.*, **13.**

HERTWIG, O., and R. 1881. Die Cölomtheorie. *Z. Naturw.*, **15.**

HERTWIG, R. 1880. *Über den Bau der Ctenophoren*. Jena.

HERTWIG, R. 1907. *Lehrbuch der Zoologie*, 8. Jena.

HESCHELER, K. 1900. Mollusken. In: Lang, A. *Lehrbuch der vergleichenden Anatomie der wirbellosen Thiere*. III. Jena.

HESSE, R. 1896–1902. Untersuchungen über die Organe der Lichtempfindung bei niederen Tieren. I–VIII. *Z. wiss. Zool.*, **61, 62, 63, 65, 68, 70, 72.**

HESSLE, Ch. 1917. Zur Kenntnis der Terebellomorphen Polychaeten. *Zool. Bidr., Uppsala*, **5.**

HEYMONS, R. 1895. *Die Embryonalentwicklung von Dermapteren und Orthopteren unter besonderer Berücksichtigung der Keimblätterbildung, monographisch bearbeitet.* Jena.

HEYMONS, R. 1899. Beiträge zur Morphologie und Entwicklungsgeschichte der Rhynchoten. *Acta Acad. German.*, **74.**

HEYMONS, R. 1899. Der morphologische Bau des Insektenabdomens. Eine kritische Zusammenstellung der wesentlichsten Forschungsergebnisse auf anatomischem und embryologischem Gebiete. *Zool. Cbl.*, **6.**

HEYMONS, R. 1901. Entwicklungsgeschichte der Scolopender. *Bibl. Zool.*, **33.**

HEYMONS, R. 1935. Pentastomida. In: Bronn's *Klassen und Ordnungen des Tierreichs*, V/4, 1.

HICKMAN, V. V. 1936. The embryology of the Syncarid crustacean, *Anaspides tasmaniae*. *Pap. and Proc. Roy. Soc., Tasmania.*

HICKSON, S. J. 1891. The Medusae of *Millepora murrayi* and the gonophores of *Allopora* and *Distichopora*. *Q. J. microsc. Sci., N.S.*, **32.**

HIRST, S. 1923. On some Arachnid remains from the Old Red Sandstone (Rhynie Chert Bed, Aberdeenshire). *Ann. Mag. Natur. History, ser.* 9, **12.**

HOEK, P. P. C. 1881. Nouvelles études sur les Pycnogonides. *Arch. zool. exptl. gén., sér. 1*, **9.**

HOFFMANN, H. 1939–40. Opisthobranchia. In: Bronn's *Klassen und Ordnungen des Tierreichs*. III/2, 3. Leipzig.

HOLM, Å. 1940–41. Studien über die Entwicklung und Entwicklungsbiologie der Spinnen. *Zool. Bidr., Uppsala*, **19.**

HOLMES, W. 1942. The giant myelenated nerve fibres of the prawns. *Phil. Trans. Roy. Soc., London, ser. B*, **231.**

HOLMES, W. 1953. The atrial nervous system of *Amphioxus* (*Branchiostoma*). *Q. J. microsc. Sci. N.S.*, **94.**

HOLMGREN, N. 1916. Zur vergleichenden Anatomie des Gehirns von Polychaeten, Onychophoren, Xiphosuren, Arachniden, Crustaceen, Myriapoden und Insekten. *Kgl. svenska vetenskaps akad. handl.*, *N.F.*, **56**, 1.

HOLSTE, G. 1910. Das Nervensystem von *Dytiscus marginalis*. Ein Beitrag zur Morphologie des Insektenkörpers. *Z. wiss. Zool.*, **96.**

HOPKINS, A. E. 1936. Pulsation of blood vessels in Oysters, *Ostrea lurida and O. gigas*. *Biol. Bull.*, **70.**

HORRIDGE, G. A. 1956a. A thorough-conducting system co-ordinating the protective retraction of *Alcyonium* (Coelenterata). *Nature, Lond.*, **178**, 4548.

HORRIDGE, G. A. 1956b. Responses of *Heteroxenia*. *J. exptl Biol.*, **33.**

HORRIDGE, G. A. 1957. Coordination of protective reaction of coral polyps. *Phil. Trans. Roy. Soc. London, ser. B*, **240.**

HORST, C. 1939. Hemichordata. In: Bronn's *Klassen und Ordnungen des Tierreichs*. IV/4, 2. Leipzig.

HORST, C. 1954. Enteropneusta. In: Kükenthal, W., and Krumbach, Th. *Handbuch der Zoologie*, III/2, 18. Berlin.

HOSKINS, W. M., and CRAIG, R. 1935. Recent progress in Insect physiology. *Physiol. Revs.*, **15.**

HOTZ, H. 1938. *Protoclepsis tesselata* (O. F. Müller). Ein Beitrag zur Kenntnis von Bau und Lebensweise der Hirudineen. *Rev. suisse zool.*, **45**, Suppl.

HOVASSE, R. 1934. Sur l'existence d'un appareil parabasal dans les cellules flagellées des larves nageantes chez l'oursin, *Paracentrotus lividus*. *Compt. rend. Acad. sci.*, **199.**

HOVASSE, R. 1935a. L'apparition des parabasaux chez les jeunes blastulas de l'oursin, *Paracentrotus lividus*. *Compt. rend. Soc. biol.*, **118.**

HOVASSE, R. 1935b. Deux Péridiniens parasites convergents: *Oodinium poucheti* (Lemm.), *Protoodinium chattoni* g. n., sp. n. *Bull. biol. France Belgique*, **69.**

HOVASSE, R. 1937. Les zooxanthelles sont des dinoflagellés. *C.R. Acad. Sci.*, **205.**

HOVASSE, R. 1958. Appeared in R. Rasmont, *Proc. XV Intern. Congr. Zool.* London.

HUBENDICK, B. 1947. Phylogenie und Tiergeographie der Siphonariidae. Zur Kenntnis der Phylogenie in der Ordnung Basommatophora und des Ursprungs der Pulmonatengruppe. *Zool. Bidr.*, *Uppsala*, **24.**

HUBL, H. 1953. Die inkretorischen Zellelemente im Gehirn der Lumbriciden. *Arch. Entw. mech.*, **146.**

HUBRECHT, A. A. W. 1883. On the ancestral forms of the Chordata. *Q. J. miscrosc. Sci.*, *N.S.*, **23.**

HUBRECHT, A. A. W. 1887. The relation of the Nemertea to the Vertebrata. *Q. J. microsc. Sci.*, *N.S.*, **26.**

HUGHES, G. M. 1953. 'Giant' fibres in dragonfly nymphs. *Nature, Lond.*, **172**, 4341.

HUPÉ, P. 1953. Classe des Trilobites. In: Pivetau, J. *Traité de Paléontologie*, III. Paris.

HUUS, J. 1937. Ascidiaceae. In: Kükenthal, W., and Krumbach, Th. *Handbuch der Zoologie*, V/2, 6. Berlin and Leipzig.

HUXLEY, Th. H. 1878. *Grundzüge der Anatomie der wirbellosen Thiere*. Leipzig.

HYMAN, L. H. 1940–60. *The invertebrates*, v. I–V. McGraw-Hill, New York, Toronto, London.

IHERING, H. VON. 1877. Vergleichende Anatomie des Nervensystems und Phylogenie der Mollusken. *Z. Naturw.*, **11**, *N.F.* 4.

IHERING, H. VON. 1922. Phylogenie und System der Mollusken. *Abhandl. Arch. Molluskenk.*, *Frankfurt a. M.*, **1.**

IHLE, J. E. W. 1913. Die Appendicularien. Ergebn. Fortschr. Zool., 3. Jena.

IHLE, J. E. W. 1935. Desmomyaria. In: Kükenthal, W., and Krumbach, Th. *Handbuch der Zoologie*, V/2, 5. Berlin and Leipzig.

ILLIES, J. 1960. Penturoperlidae, eine neue Plecopteren Familie. *Zool. Anz.*, **164**, 1–2.

IMMS, A. 1937. *A general textbook of entomology.* 2nd ed. London.

IVANOV, A. V. 1957. Neue Pogonophora aus dem nordwestlichen Teil des Stillen Ozeans. *Zool. Jahrb. Syst.*, **85**, 4/5.

IVANOVA-KASAS, O. M. 1959. Die embryonale Entwicklung der Blattwespe *Pontania capreae* L. (Hymenoptera, Tenthredinidae). *Zool. Jarb., Anat.*, **77.**

IWANOW, P. P. 1903. Die Regeneration von Rumpf- und Kopfsegmenten bei *Lumbriculus variegatus. Z. wiss. Zool.*, **75.**

IWANOW, P. P. 1928. Die Entwicklung der Larvalsegmente bei den Anneliden. *Z. Morphol. Ökol. Tiere*, **10.**

IWANOW, P. P. 1933. Die embryonale Entwicklung von *Limulus moluccanus. Zool. Jahrb., Anat.*, **56.**

JÄGERSTEN, G. 1939. Zur Kenntnis der Larvenentwicklung bei *Myzostomum. Arkiv zool.*, **31**, 11.

JÄGERSTEN, G. 1940. Zur Kenntnis der äußeren Morphologie, Entwicklung und Ökologie von *Protodrilus rubropharyngeus* n. sp. *Arkiv. zool.*, **32a**, 16.

JÄGERSTEN, G. 1942–44. Über den Bau des Kopulationsapparates und den Kopulationsmechanismus bei *Dinophilus. Zool. Bidr., Uppsala*, **22.**

JÄGERSTEN, G. 1944. Zur Kenntnis der Morphologie, Enzystierung und Taxonomie von *Dinophilus. Kgl. Svenska vetenskaps akad. handl.*, ser. 3, **21**, 2.

JÄGERSTEN, G. 1947. On the structure of the pharynx of the Archiannelida with special reference to there-occurring muscle cells of aberrant type. *Zool. Bidr., Uppsala*, **25.**

JÄGERSTEN, G. 1952. Studies on the morphology, larval development and biology of *Protodrilus. Zool. Bidr., Uppsala*, **29.**

JÄGERSTEN, G. 1955. On the early phylogeny of the Metazoa (The Bilaterogastraea theory). *Zool. Bidr., Uppsala*, **30.**

JÄGERSTEN, G. 1956. Investigations on *Siboglinum ekmani* n. sp. encountered in Skagerrak. *Zool. Bidr., Uppsala*, **31.**

JÄGERSTEN, 1957. On the larva of *Siboglinum* with some remarks on the nutrition problem of the Pogonophora. *Zool. Bidr., Uppsala*, **32.**

JÄGERSTEN, 1959. Further remarks on the early phylogeny of the Metazoa. *Zool. Bidr., Uppsala*, **33.**

JAKOVLEV, N. N. 1922. *Bothriocidaris* und die Abstammung der Seeigel. *Z. Dtsch. geol. Ges.*, **74.**

JAKOVLEV, N. N. 1923. The relationship of the Rugosa to the Hexacorallia. *Geol. Mag.*, **60.**

JANETSCHEK, H. 1957. Über die mögliche phyletische Reversion eines Merkmals bei Felsenspringern mit einigem Bemerkungen über die Natur der Styli der Thysanuren (Ins.). *Broteria (Lisbôa), Sér. Ciênc. Natur.*, **26** (53), 1.

JANICKI, C. 1921. Grundlinien einer 'Cercomer' Theorie zur Morphologie der Trematoden und Cestoden. *Festschrift Zschokke*, **30.**

JANICKI, C. 1928. Die Lebensgeschichte von *Amphilina foliacea* G. Wag., Parasiten des Wolga-Sterlets nach Beobachtungen und Experimenten. *Arbeiten Biol. Wolgastat., Saratow*, **10.**

JATSU, N. 1902. On the development of *Lingula anatina. J. Fac. Sci. Univ. Tokyo*, **17.**

JATZENKO, A. T. 1928. Die Bedeutung der Mantelhöhlenflüssigkeit in der Biologie der Süßwasserlamellibranchier. *Biol. Zbl.*, **48.**

JAWOROWSKI, A. 1891. Über die Extremitäten bei den Embryonen der Arachniden und Insekten. *Zool. Anz.*, **14.**

JENNER, CH. E. 1956. The occurrence of a crystalline style in the marine snail, *Nassarius obsoletus. Biol. Bull.*, **111**, 2.

JEPPS, M. W. 1947. Contribution to the study of the Sponges. *Proc. Roy. Soc. London, B*, **134.**

JOHANSON, L. 1896. Über den Blutumlauf bei *Piscicola* und *Callobdella. Festsk. Lilljeborg, Uppsala.*

JOHANSSON, G. 1933. Beiträge zur Kenntnis der Morphologie und Entwicklung des Gehirns von *Limulus polyphemus. Acta zool.*, **14.**

JOHANSSON, K. E. 1927. Beiträge zur Kenntnis der Polychaeten-Familien Hermellidae, Sabellidae und Serpulidae. *Zool. Bidr.*, *Uppsala*, **11.**

JOHANSSON, K. E. 1937. Über *Lamellisabella zachsi* und ihre systematische Stellung. *Zool. Anz.*, **117.**

JOHANSSON, K. E. 1939. *Lamellisabella zachsi* Uschakow, ein Vertreter einer neuen Tierklasse Pogonophora. *Zool. Bidr.*, *Uppsala*, **18.**

JORGENSEN, C. B., and GOLDBERG, E. D. 1953. Particle filtration in some Ascidians and Lamellibranchs. *Biol. Bull.*, **105.**

JÖSTING, E. A. 1942. Die Innervierung des Skelettmuskelsystems des Mehlwurms (*Tenebrio molitor* L. Larve). *Zool. Jahrb., Anat.*, **67.**

JULIN, C. 1881. Recherches sur l'organisme des Ascidies simples, sur l'hypophyse, etc. *Arch. Biol.*, **2.**

KAHN, M., CELESTIN, W., and OFFENHAUSER, W. 1945. Recording of sounds produced by certain disease-carrying mosquitoes. *Science*, **101.**

KAISER, J. 1893. Die Acanthocephalen und ihre Entwicklung. *Bibl. Zool.*, *Stuttgart*, **7.**

KARLING, T. G. 1940. Zur Morphologie und Systematik der Alloeocoela Cumulata und Rhabdocoela Lecithophora (Turbellaria). *Acta Zool. fennica*, **26.**

KASSIANOW, N. 1908. Untersuchungen über das Nervensystem der Alcyonaria. *Z. wiss. Zool.*, **90.**

KASSIANOW, N. 1914. Die Frage über den Ursprung der Arachnoideenlungen aus den Merostomenkiemen (*Limulus*-Theorie). *Biol. Cbl.*, **34.**

KÄSTNER, A. 1932, 1933, 1935, 1937, 1940. Pedipalpi, Palpigradi, Solifugae, Ricinulei, Opiliones. In: Kükenthal, W., and Krumbach, Th. *Handbuch der Zoologie*, III/2, 3, 4, 5, 14. Berlin und Leipzig.

KÄSTNER, A. 1954–55, 1956, 1959. *Lehrbuch der Speziellen Zoologie.* Jena.

KAWAGUTI, S. 1944. On the physiology of reef corals. Zooxanthellae of reef corals in *Gymnodinium* sp., Dinoflagellata; its culture in vitro. *Palao trop. biol. stud.*, **2.**

KENNEDY, C. H. 1927. The exoskeleton as a factor in limiting and directing the evolution of Insects. *J. Morphol.*, **44.**

KENNEL, J. 1884, 1888. Entwicklungsgeschichte von *Peripatus edwardsii* Blanch, und *Peripatus torquatus* n. sp. *Arbeiten Zool. Inst. Würzburg*, **7, 8.**

KENT, S. 1880, 1881. *Manual of infusoria*, I–II. London.

KIRTISINGHE, P. 1940. The myenteric nerve-plexus in some lower Chordates. *Q. J. microsc. Sci.*, *N.S.*, **81**, London.

KIRTISINGHE, P. 1952. Giant nerve cells in *Lingula. Nature, Lond.*, **170**, 4318.

KISHINOUYE, K. 1890. On the development of Araneina. *J. Coll. Sci. Japan*, **4.**

KLAWE, W. L., and DICKIE, L. M. 1957. Biology of the bloodworm, *Glycera dibranchiata* Ehlers and its relation to the bloodworm fishery of the Maritime provinces. *Bull. Fish. Res. Board Canada*, **115.**

KLEINENBERG, N. 1872. *Hydra. Eine anatomisch-entwicklungsgeschichtliche Untersuchung.* Leipzig.

KLEINENBERG, N. 1886. Die Entstehung des Annelids aus der Larve von *Lopadorhynchus. Z. wiss. Zool.*, **44.**

KLINGEL, H. 1956. Indirekte Spermatophorenübertragung bei Chilopoden, beobachtet bei der Spinnenassel, *Scutigera coleoptrata* Latz. *Naturwissenschaften*, **43,** 13.

KOLBE, H. 1893. *Einführung in die Kenntnis der Insekten.* Berlin.

KOLISKO, A. 1939. Über *Conochilus unicornis* und seine Kolonienbildung. *Internat. Rev. Hydrobiol.*, **39.**

KOMAI, T. 1920. Notes on *Coeloplana bocki* n. sp. and its development. *Annot. zool. Japan*, **9.**

KOMAI, T. 1922. *Studies on two aberrant Ctenophores*, Coeloplana *and* Gastrodes. Kyoto, Japan.

KOMAI, T. 1936. The nervous system in some coelenterate types. I. *Coeloplana. Mem. Coll. Sci. Univ. Kyoto, B*, **11.**

KORN, H. 1959a. Zum Nervensystem des Ctenophore *Pleurobrachia pileus* O. Müller. *Zool. Anz.*, **163.**

KORN, H. 1959b. Vergleichend-embryologische Untersuchungen an *Harmothoë* Kinberg. Organogenese und Neurosekretion. *Z. wiss. Zool.*, **161,** 3–4.

KOROTNEFF, A. 1886. *Ctenoplana kowalevskii. Z. wiss. Zool.*, **43.**

KORSCHELT, E. 1882. Über Bau und Entwicklung des *Dinophilus apatris. Z. wiss. Zool.* **37.**

KORSCHELT, E., and HEIDER, K. 1890, 1892, 1893. *Lehrbuch der vergleichenden Entwicklungsgeschichte der wirbellosen Thiere, Specieller Theil.* I–III. Jena.

KORSCHELT, E., and HEIDER, K. 1902, 1903, 1909. *Lehrbuch der vergleichenden Entwicklungsgeschichte der wirbellosen Thiere.* Allgemeiner Theil. I–IV. Jena.

KOWALEVSKY, A. 1866. Entwicklungsgeschichte der einfachen Ascidien. *Mém. Acad. Sci.*, *St. Pétersbourg, ser.* 7, **10,** 15.

KOWALEVSKY, A. 1867. Entwicklungsgeschichte des *Amphioxus lanceolatus. Mém. Acad. Sci.*, *St. Pétersbourg, ser.* 7, **11,** 4.

KOWALEVSKY, A. 1871. Weitere Studien über die Entwicklung der einfachen Ascidien. *Arch. Mikrobiol. Anat., Bonn.* **7.**

KOWALEVSKY, A. 1877. Weitere Studien über die Entwicklungsgeschichte des *Amphioxus lanceolatus* nebst einem Beitrag zur Homologie des Nervensystems der Würmer und Wirbelthiere. *Arch. Mikrobiol. Anat., Bonn.* **13.**

KOWALEVSKY, A. 1880. Über *Coeloplana metschnikowi. Zool. Anz.*, **3.**

KOWALEVSKY, A. 1889. Ein Beitrag zur Kenntnis der Exkretionsorgane. *Biol. Zbl.*, **9.**

KOWALEVSKY, A. 1892. Einige Beiträge zur Bildung des Mantels der Ascidien. *Mém. Acad. Sci. St. Pétersbourg, ser.* 7, **38,** 10.

KOZŁOWSKI, R. 1947. Les affinités des Graptolithes. *Biol. Revs. Cambridge Phil. Soc.*, **22.**

KOZŁOWSKI, R. 1942. Les Graptolithes et quelques nouveaux groupes d'animaux du Tremadoc de la Pologne. *Palaeontol. polon.*, **3.**

KOZŁOWSKI, R. 1949. Découverte du Ptérobranche *Rhabdopleura* à l'état fossile dans le Crétacé supérieur en Pologne. *Compt. rend. Acad. sci.*, **228.**

KOZŁOWSKI, R. 1961. Découverte d'un Rhabdopleuridé ordovicien. *Acta Palaeont. Polonica*, **6,** 1.

KRAMP, P. L. 1943. On development through alternating generations, especially in Coelenterata. *Vidensk Medd. naturh. Foren.*, **107.**

KRISHNAN, G. 1955. Nature of the cuticle in the Pycnogonida. *Nature, Lond.*, **175,** 4464.

KROGH, A. 1939. *Osmotic regulation in aquatic animals.* Cambridge.

KRULL, H. 1933. Die Aufhebung der Chiastoneurie bei den Pulmonaten. *Zool. Anz.*, **105.**

KRULL, H. 1935. Anatomische Untersuchungen an einheimischen Prosobranchiern und Beiträge zur Phylogenie der Gastropoden. *Zool. Jahrb. Anat.*, **60.**

KRUMBACH TH. 1925. Ctenophora. In: Kükenthal, W., and Krumbach, Th. *Handbuch der Zoologie*, I, 7. Berlin and Leipzig.

KÜHN, A. 1913. Entwicklungsgeschichte und Verwandschaftsbeziehungen der Hydrozoen. I. Die Hydroiden. *Ergebn. Fortsch. Zool.*, **4.**

KÜHNE, H. 1959. Die neurosecretorische Zellen und der retrocerebrale neuro-endokrine Komplex von Spinnen (Araneae, Labidognatha) unter Berück-sichtigung einiger histologischerkennbarer Veränderungen während des postembryonalen Lebenslaufes. *Zool. Jahrb., Anat.*, **77**, 4.

KÜKENTHAL, W. 1923–25. Coelenterata. In: Kükenthal, W., and Krumbach, Th. *Handbuch der Zoologie*, I, 3. Berlin and Leipzig.

LACAZE-DUTHIERS, H. 1859. Mémoire sur le système nerveux de l'Haliotide (*H. tuberculata* et *H. lamellosa* Lam.) *Ann. sci. natur. Zool. et biol. anim.*, ser. *4*, **12**.

LACAZE-DUTHIERS, H. 1874. Les Ascidies simples des côtes de France. *Arch. zool. exptl. gén.*, **3**.

LAMEERE, A. 1926. Les organes segmentaires des Polychètes. *Ann. Soc. roy. Zool. Belgique*, **56**.

LANG, A. 1881. Der Bau von *Gunda segmentata* und die Verwandschaft der Plathel-minthen mit Coelenteraten und Hirudineen. *Mitt. Zool. Stat. Neapel*, **3**.

LANG, A. 1884. Die Polycladen des Golfes von Neapel und der angrenzenden Meeresabschnitte. In: *Fauna und Flora des Golfes von Neapel*, 11 Monogr. Leipzig.

LANG, A. 1888, 1889, 1892, 1894. *Lehrbuch der vergleichenden Anatomie der wir-bellosen Thiere*. I–IV. Jena.

LANG, 1891. Versuch einer Erklärung der Asymmetrie der Gastropoden. *Viertel-jahrschr. Naturforsch. Ges., Zürich*, **36**.

LANG, 1903. Beiträge zu einer Trophocoeltheorie. *Naturwissenschaften*, **38**.

LANG, 1912. Allgemeine Lehre vom zelligen Aufbau des Metazoenkörpers. In: *Handbuch der Morphologie der wirbellosen Tiere*. II. Jena.

LANG, K. 1950. A contribution to the morphology of *Stratiodrilus platensis* Cordeio (Histriobdellidae). *Arkiv. zool.*, **42**, 4.

LANG, K. 1953. Die Entwicklung des Eies von *Priapulus caudatus* Lam. und die systematische Stellung der Priapuliden. *Arkiv zool.*, **5**.

LEBOUR, M. 1937. The eggs and larvae of the British Prosobranchs with special reference to those living in the plankton. *J. Marine Biol. Assn UK*, **22**.

LEES, A. D. 1947. Transpiration and the structure of the epicuticle in ticks. *J. exptl Biol.*, **23**.

LEGENDRE, R. 1953a. Recherches sur les glandes prosomatiques des Araignées du genre *Tegenaria*. *Ann. Univ. Saraviensis*, **2**.

LEGENDRE, R. 1953b. Le système sympathique stomatogastrique (organe de Schneider) des Araignées du genre *Tegenaria*. *Compt. rend. Acad. Sci.*, **223**.

LEGENDRE, R. 1954a. Sur la présence de cellules neuro-sécrétrices dans le système nerveux central des Aranéides. *Compt. rend. Acad. Sci.*, **238**.

LEGENDRE, R. 1954b. Données anatomiques sur le complexe neuro-endocrine rétrocérébral des Aranéides. *Ann. sci. natur. Zool. biol. anim.*, ser. 11, **16**.

LEGENDRE, R. 1956a. Les éléments neurosécréteurs de la masse nerveuse et leur cycle d'activité chez les Aranéides. *Compt. rend. Acad. sci.*, **242**.

LEGENDRE, R. 1956b. Sur l'origine embryologique et la répartition métamérique des cellules neurosécrétrices chez les Araignées. *Compt. rend. Acad. sci.*, **242**.

LEGENDRE, R. 1956 (1957). Sur le genèse du pont stomodéal chez les araignées. *Bull. soc. zool. France*, **81**, 5–6.

LEGROS, R. 1896. Sur la morphologie des glandes sexuelles de *l'Amphioxus lanceo-latus*. *3 Congrès Internat. Zool., Leyde*, 1895. Compt. rend. du Congrès.

LEGROS, R. 1897, 1898. Développement de la cavité buccale de *l'Amphioxus lanceolatus*. Contribution à l'étude de la morphologie de la tête. *Arch. anat. microsc. morphol. exptl*, **1, 2**.

LEMCHE, H. 1940. The origin of winged insects. *Vidensk. medd. Dansk naturhistor. foren*, **104**.

LEMCHE, H. 1957. A new living deep-sea Mollusc of the Cambro-Devonian Class Monoplacophora. *Nature, Lond.*, **179**, 4556.

LEMCHE, H. 1960. A possible central place for *Stenothecoides* Resser 1939 and *Cambridium* Horny 1957 (Mollusca Monoplacophora) in invertebrate phylogeny. *Intern. Geol. Congr., XXI sess., Norden*, pt. XXII. Copenhagen.

LEMCHE, H., and WINGSTRAND, K. G. 1958. The comparative anatomy of *Neopilina galatheae* Lemche 1957 (Mollusca, Monoplacophora). *XV Intern. Congr. Zool.*, Sect. 4, pap. 24.

LEMCHE, H., and WINGSTRAND, K. G. 1959. The anatomy of *Neopilina galatheae* Lemche 1957 (Mollusca Tryblidiacea). *Galathea report*, **3**.

LEUCKART, R. 1853. *Zoologische Untersuchungen*. 1. Siphonophoren. Gießen.

LEUCKART, R. 1856. *Die Blasenbandwürmer und ihre Entwicklung*. Gießen.

LEUCKART, R. 1860. *Bau und Entwicklungsgeschichte der Pentastomen*. Leipzig.

LEWIS, G. T. 1954. Contact chemoreceptors of blowfly tarsi. *Nature, Lond.*, **173**, 4394.

LIEBMAN, E. 1946. On trophocytes and trophocytosis; a study on the role of leucocytes in nutrition and growth. *Growth*, **10**.

LINDER, F. 1941. Contributions to the morphology and taxonomy of the Branchiopoda Anostraca. *Zool. Bidr., Uppsala*, **20**.

LINDER, F. 1945. Affinities within the Branchiopoda, with notes on some dubious fossils. *Arkiv zool.*, **37**, 4.

LINDER, F. 1947. Abnormal body-rings in Branchiopoda Notostraca. *Zool. Bidr., Uppsala*, **25**.

LINDROTH, A. 1938. Studien über die respiratorischen Mechanismen von *Nereis virens* Sars. *Zool. Bidr., Uppsala*, **17**.

LIPIN, A. 1910. Die Morphologie und Biologie von *Polypodium hydriforme* Uss. *Zool. Jahrb., Anat.*, **31**.

LIVANOW, N. 1903. Untersuchungen zur Morphologie der Hirudineen. I. Das Neuro- und Myosomit der Hirudineen. *Zool. Jahrb., Anat.*, **19**.

LIVANOW, N. 1904. Untersuchungen zur Morphologie der Hirudineen. II. Das Nervensystem des vorderen Körperendes und seine Metamerie. *Zool. Jahrb., Anat.*, **20**.

LIVANOW, N. 1906. *Acanthobdella peledina* Grube, 1851. *Zool. Jahrb., Anat.*, **22**.

LOCHHEAD, J. H., and RESNER, R. 1958. Function of the eyes and neurosecretion in Crustacea Anostraca. *Proc. XV Intern. Congr. Zool.* London.

LÖHNER, L. 1910. Untersuchungen über *Polychoerus caudatus* Mark. *Z. wiss. Zool.*, **95**.

LOVÉN, S. 1874. Etude sur les échinoidées. *Kgl. Svenska vetenskaps akad. handl.*, *n. ser.*, **11**, 7.

LOWE, E. 1936. On the anatomy of a marine Copepod, *Calanus finmarchicus* (Gunnerus). *Trans. Roy. Soc., Edinburgh*, **58**.

LOWY, J., and BOWDEN, J. 1955. The lamellibranch muscle (Contractile mechanism. Innervation). *Nature, Lond.*, **176**, 4477.

LUBISCHEV, A. 1912. Beiträge zur Histologie der Polychäten. *Mitt. Zool. Stat. Neapel*, **20**.

LUDWIG, H., and HAMANN, O. 1889–1907. Echinodermen. In: Bronn's *Klassen und Ordnungen des Tierreichs*, II, 3. Leipzig.

LÜLING, K. H. 1940. Über die Entwicklung des Urogenitalsystems der Priapuliden (Ein Beitrag zur Anatomie und Histologie dieser Tiere.) *Z. wiss. Zool.*, **153**.

LUTAUD, G. 1955. Sur la ciliature du tentacule chez les bryozoaires Chilostomes. *Arch. zool. exptl gen.*, **92** (notes et revues), 1.

LUTHER, A. 1904. Die Eumesostominen. *Z. wiss. Zool.*, **77**.

LUTHER, A. 1912. Studien über acöle Turbellarien aus dem Finnischen Meerbusen. *Acta Soc. fauna flora fennica*, **36**, 5.

LUTHER, A. 1923. Über das Vorkommen von *Protohydra leuckarti* Greeff, bei Tvärminne. *Acta Soc. fauna flora fennica*, **52**, 3.

LUTHER, A. 1955. Die Dalyelliiden. Eine Monographie. *Acta zool. fennica*, **87**, XI.

MAAS, C. 1909. Die Scyphomedusen. *Ergebn. Fortsch. Zool.*, **1**.

MACBRIDE, E. W. 1910. The formation of layers in *Amphioxus* and its bearing on the interpretation of the early ontogenetic process in other Vertebrates. *Q. J. microsc. Sci., N.S.*, **54.**

MACBRIDE, E. W. 1914. *Text book of Embryology.* I. Invertebrata. London.

MACBRIDE, E. W. 1933. Habit and structure in starfishes. *Nature, Lond.*, **132.**

MACBRIDE, E. W., and SPENCER, W. K. 1938. Two new Echinoidea, *Aulechinus* and *Ectechinus* and an adult plated Holothurian, *Eothuria*, from the upper Ordovician of Girvan, Scotland. *Phil. Trans. Roy. Soc., London, B,* **229.**

MACGINITIE, G. E. 1937. The use of mucus by marine plankton feeders. *Science,* **86.**

MACGINITIE, G. E. 1939a. The method of feeding of *Chaetopterus. Biol. Bull.,* **77.**

MACGINITIE, G. E. 1939b. The method of feeding of Tunicates. *Biol. Bull.,* **77.**

MACGINITIE, G. E., and MACGINITIE, N. 1949. *Natural History of Marine Animals.* New York.

MACKIE, G. O. 1960. Studies on *Physalia physalis* (L.). Pt. 2. Behaviour and histology. *Discovery Repts.,* **30**, 371–408.

MACKINTOSH, N. A. 1925. The crystalline style in Gastropods. *Q. J. microsc. Sci., N.S.,* **69.**

MACLEOD, J. 1884. Recherches sur la structure et la signification de l'appareil respiratoire chez les Arachnides. *Arch. biol.,* **5.**

MAIN, R. J. 1928. Observations of the feeding mechanism of a Ctenophore, *Mnemiopsis leidyi. Biol. Bull.,* **55.**

MALPIGHI, M. 1705. *Opera omnia.* 2 ed. Amsterdam.

MANTON, S. M. 1928. On the embryology of the Mysid Crustacean, *Hemimysis lamornae. Phil. Trans. Roy. Soc., London, B,* **216.**

MANTON, S. M. 1929. On some points in the anatomy and habits of the Lophogastrid Crustacea. *Trans. Roy. Soc., Edinburgh,* **56.**

MANTON, S. M. 1934. On the embryology of the Crustacean, *Nebalia bipes. Phil. Trans. Roy. Soc., London, B,* **223.**

MANTON, S. M. (Mrs. Harding, J. R.). 1949. Studies on the Onychophora. VII. The early embryonic stages of *Peripatopsis*, and some general considerations concerning the morphology and phylogeny of the Arthropoda. *Phil. Trans. Roy. Soc., London,* **233.**

MANTON, S. M. 1950–61. The evolution of arthropodan locomotory mechanisms. Pts 1–7. *J. Linnean Soc. (Zool.),* **41, 42, 43, 44.**

MANTON, S. M. 1958. Habits of life and evolution of body design in Arthropods. *J. Linnean Soc. (Zool.),* **44.**

MARCUS, E. 1926. 1. Beobachtungen und Versuche an lebenden Meeresbryozoen. 2. Beobachtungen und Versuche lebenden Süßwasserbryozoen. *Zool. Jahrb. Syst.,* **52.**

MARCUS, E. 1934. Über *Lophopus crystallinus* (Pall.). *Zool. Jahrb., Anat.,* **58.**

MARCUS, E. 1937, 1938, 1939. Bryozoarios marinhos brasileiros. I, II, III. *Bol. Fac. Fil. Ciênc. Letr. S. Paulo., Zool.,* **1, 2, 3.**

MARCUS, E. 1946. Sobre Turbellaria Brasileiros. *Zoologia,* **11.**

MARCUS, E. DU BOIS-REYMOND. 1949. *Phoronis ovalis* from Brazil. *Zoologica (Bol. Fac. Fil. Ciênc. Letr.),* **14.**

MARCUS, E., and MACNAE, W. 1954. Architomy in a species of *Convoluta. Nature, Lond.,* **173,** 4394.

MARTINI, E. 1909. Studien über die Konstanz histologischer Elemente. I. *Oikopleura longicauda. Z. wiss. Zool.,* **92.**

MARTINI, E. 1912. Studien über die Konstanz histologischer Elemente. III. *Hydatina senta. Z. wiss. Zool.*, **102.**

MARTINI, E. 1913. Über die Stellung der Nematoden im System. *Verhandl. Deutsch. Zool. Ges. Leipzig,* **23.**

MARTINI, E. 1916. Die Anatomie der *Oxyuris curvula.* 1, 2. *Z. wiss. Zool.*, **116.**

MARTINI, E. 1923. Über Beeinflussung der Kiemenlänge von Aedeslarven durch das Wasser. *Verhandl. Intern. Ver. Limnol., Kiel,* **1** (1922).

MASTERMAN, A. T. 1894. On the nutritive and excretory processes in Porifera. *Ann. Mag. Natur. Hist., ser.* 6, **13.**

MASTERMAN, A. T. 1897. On the Diplochorda. 2. The structure of *Cephalodiscus. Q. J. microsc. Sci., N.S.*, **40.**

MASTERMAN, A. T. 1900. On the Diplochorda. 3. The early development and anatomy of *Phoronis buskii. Q. J. microsc. Sci., N.S.* **43.**

MATTHAI, G. 1926. Colony formation in Astraeid corals. *Phil. Trans. Roy. Soc., London, B,* **214.**

MATTOX, N. T., and CROWELL, S. 1951. A new commensal Hydroid of the mantle cavity of an oyster. *Biol. Bull.*, **101.**

MAYER, A. G. 1910. Medusae of the world. *Carnegie Inst. Pub.*, **109.**

MAYER, A. M. 1874. Experiments on the supposed auditory apparatus of the mosquito. *Amer. Naturalist,* **8.**

MAYER, H. 1956. Vergleichende Untersuchungen zur Fortpflanzungsbiologie symphytopleoner Collembolen. *Naturwissenschaften,* **43.**

MAYER, H. 1957. Zur Biologie und Ethologie einheimischer Collembolen. *Zool. Jahrb., Syst.,* **85,** 6.

MAYRAT, A. 1955. Mise en évidence de tendons chez les Crustacés dans le muscle attracteur du sinciput de *Praunus flexuosus* O. F. Müller. *Bull. Soc. zool. France,* **80,** 2–3.

McINDOO, N. 1923. Segmental blood vessels of the American cockroach, *Periplaneta americana. J. Morphol,* **65.**

McINDOO, N. 1945. Innervation of Insect hearts. *J. compar. Neurol.*, **83.**

McINTOSH, W. C. 1894. A contribution to our knowledge of the Annelida. On some points in the structure of *Euphrosyne*, on certain young stages in *Magelona* and on Claparède's unknown larval *Spio. Q. J. microsc. Sci., N.S.*, **36.**

McMURRICH, J. P. 1891, 1892. Contributions on the morphology of the Actinozoa. *J. Morphol.*, **4, 5.**

MEEWIS, H. 1939. Etude comparative des larves d'éponges siliceuses. *Ass. Franc. Avanc. sci.,* Liège.

MEISENHEIMER, I. 1902. Beiträge zur Entwicklungsgeschichte der Pantopoden. I. *Z. wiss. Zool.,* **72.**

MEIXNER, J. 1928. Der Genitalapparat der Tricladen und seine Beziehungen zu ihrer allgemeinen Morphologie, Phylogenie, Ökologie und Verbreitung. *Z. Morphol. Ökol. Tiere,* **11.**

MELLANBY, H. N. 1935. Early embryological development of *Rhodnius prolixus. Q. J. microsc. Sci., N.S.*, **78.** London.

MELLANBY, H. N. 1936. The later embryology of *Rhodnius prolixus. Q. J. microsc. Sci., N.S.,* **79.** London.

MERESCHKOWSKY, C. 1877. On a new genus of Hydroids from the White Sea with short description of other new Hydroids. *Ann. Mag. Natur. History, ser.* 4, **13.**

MERGNER, H. 1957. Die Ei- und Embryonalentwicklung von *Eudendrium racemosum* Cavolini. *Zool. Jahrb., Anat.,* **76,** 1.

METALNIKOFF, S. 1900. *Sipunculus nudus. Z. wiss. Zool.*, **68.**

METSCHNIKOFF, E. 1869. Studien über die Entwicklung der Echinodermen und Nemertinen. *Mém. Acad. Sci. St. Pétersbourg,* **14,** 8.

METSCHNIKOFF, E. 1871a. Über die Metamorphose einiger Seethiere. II. Mitraria. *Z. wiss. Zool.*, **21**. Leipzig.

METSCHNIKOFF, E. 1871b. Über die Entwicklung einiger Coelenterata. *Mél. biol. Acad. Sci. St. Pétersbourg*, **7**, 3.

METSCHNIKOFF, E. 1874. Studien über die Entwicklung der Medusen und Siphono phoren. *Mél. biol. Acad. Sci. St. Pétersbourg*, **24**.

METSCHNIKOFF, E. 1880. Über die intracelluläre Verdauung bei Coelenteraten. *Zool. Anz.*, **3**.

METSCHNIKOFF, E. 1882. Vergleichend-embryologische Studien. 3. Über die Gastrula einiger Metazoen. *Z. wiss. Zool.*, **37**.

METSCHNIKOFF, E. 1883. Untersuchungen über die intracelluläre Verdauung bei wirbellosen Thieren. *Arbeiten Zool. Inst.*, **5**.

METSCHNIKOFF, E. 1885. Vergleichend-embryologische Studien. 4. Über die Gastrulation und Mesodermbildung der Ctenophoren. *Z. wiss. Zool.*, **42**.

METSCHNIKOFF, E. 1886. *Embryologische Studien an Medusen*. Wien.

MEYER, ANNA. 1913. Das Renogenitalsystem von *Puncturella noachina* L. *Biol. Zbl.*, **33**.

MEYER, A. 1926. Die Segmentalorgane von *Tomopteris catharina* (Gosse) nebst Bemerkungen über das Nervensystem die rosettenförmigen Organe und die Cölombewimperung. Ein Beitrag zur Theorie der Segmentalorgane. *Z. wiss. Zool. A*, **127**.

MEYER, A. 1927. Über Cölombewimperung und cölomätische Kreislaufsysteme bei Wirbellosen. Ein Beitrag zur Histophysiologie der secundären Leibeshöhle und ökologischen Bedeutung der Flimmerbewegung. *Z. wiss. Zool. A*, **129**.

MEYER, A. 1929. Über Cölombewimperung und cölomätische Kreislaufsysteme bei Wirbellosen. II. Teil (Sipunculoidea, Polychaeta Errantia). Ein Beitrag zur Histophysiologie und Phylogenese des Cölomsystems. *Z. wiss. Zool. A*, **135**.

MEYER, A. 1932–33. Acanthocephala. In: Bronn's *Klassen und Ordnungen des Tierreichs*, IV/2, 2. Leipzig.

MEYER, E. 1886–87. Studien über den Körperbau der Anneliden. I–III. Mitt. *Zool. Stat. Neapel*, **7**.

MEYER, E. 1888. Studien über den Körperbau der Anneliden. IV. *Zool. Stat. Neapel*, **8**.

MEYER, E. 1890. Die Abstammung der Anneliden (der Ursprung der Metamerie und die Bedeutung des Mesoderms). *Biol. Zbl.*, **10**.

MEYER, E. 1901. Studien über den Körperbau der Anneliden. V. *Mitt. Zool. Stat. Neapel*, **14**.

MICHAILOFF, S. 1921. Système nerveux cellulaire périphérique des Céphalopodes. *Bull. Inst. océanorg.*, **402**.

MILLAR, R. H. 1956. Structure of the ascidian *Octacnemus* Moseley. *Nature, Lond.*, **178**, 4535.

MILLOT, J. 1949. Ordre des Aranéides. In: *Traité de Zoologie*, Ed. P. Grassé vol. 6.

MILLOT, J., and VACHON, M. 1949. Ordre des Solifuges. In: *Traité de Zoologie*, Ed. P. Grassé, vol. 6.

MINCHIN, E. A. 1896. Note on the larval and the postlarval development of *Leucosolenia variabilis. Proc. Roy. Soc. London*, **60**.

MINCHIN, E. A. 1900. Sponges—Phylum Porifera, In: Ray Lankester, E. *A treatise on zoology*, II. London.

MINNICH, D. E. 1921. An experimental study of the tarsal chemoreceptors of two Nymphalid butterflies. *J. exptl Zool.*, **33**.

MINNICH, D. E. 1930. The chemical sensitivity of the legs in the blow-fly, *Calliphora vomitoria*, to various sugars. *Z. Vergl. Physiol.*, **2**.

MONTGOMERY, TH. H. 1904. The development and structure of the larva of *Paragordius*. *Proc. Acad. Natur. Sci.*, **56.**

MORRISON, P. K. 1946. Physiological observations on water loss and oxygen consumption in *Peripatus. Biol. Bull.*, **91.**

MORSE, E. S. 1878. On Japanese *Lingula* and shell mounds. *Amer. J. Sci. Arts*, ser. 3, **15.**

MORTENSEN, TH. 1912. Über eine sessile Ctenophore, *Tjalfiella tristoma* Mrtsn. *Verhandl. Dtsch. Zool. Ges., Leipzig*, **22.**

MORTENSEN, TH. 1912. Ctenophora. *Danish Ingolf-Expedition*, **5.**

MORTENSEN, TH. 1920. Studies on the development of Crinoids. *Pap. Dep. Marine Biol. Carnegie Inst.*, **16.**

MORTENSEN, TH. 1921. *Studies of the development and larval forms of Echinoderms.* Copenhagen.

MORTENSEN, TH. 1928, 1935, 1940, 1943, 1948, 1950, 1951. *A monograph of the Echinoidea.* I–V. Copenhagen and London.

MORTON, J. E. 1951. The ecology and digestive system of the Struthiolariidae (Gastropoda). *Q. J. microsc. Sci., N.S.*, **92.**

MORTON, J. E. 1953. The functions of the Gastropod stomach. *Proc. Linnean Soc., London*, **164.**

MORTON, J. E. 1955. The evolution of Vermetid Gastropods. *Pacific Sci.*, **9.**

MORTON, J. E., and HOLME, N. A. 1955. The occurrence at Plymouth of the Opisthobranch *Akera bullata*, with notes on its habits and relationships. *J. Marine biol. Assn UK*, **34.**

MOSER, F. 1924. Siphonophora. In: Kükenthal, W., and Krumbach, Th. *Handbuch der Zoologie*, I, 3.

MÜLLER, F. 1864. *Für Darwin.* Leipzig.

MÜLLER, F. 1875. Beiträge zur Kenntnis der Termiten. *Z. Naturw.*, **7.**

MÜLLER, H. G. 1936. Untersuchungen über spezifische Organe niederer Sinne bei Rhabdocoelen Turbellarien. *Z. vergl. Physiol.*, **23.**

MÜLLER, J. 1846. Bericht über einige neue Thierformen der Nordsee. *Müller's Arch. Anat.*, Berlin.

NABERT, A. 1913. Die Corpora allata der Insekten. *Z. wiss. Zool.*, **104.**

NACHTWEY, R. 1925. Keimbahn, Organogenese und Anatomie von *Asplanchna*. *Z. wiss. Zool.*, **126.**

NAEF, A. 1911–13. Studien zur generellen Morphologie der Mollusken. I. Über Torsion und Asymmetrie der Gastropoden. *Ergebn. Fortschr. Zool.*, **3.**

NAEF, A. 1921–23, 1928. Die Cephalopoden. In: *Fauna und Flora des Golfes von Neapel*, 35 Monogr., Teil I, I–II. Napoli.

NAIR, K. B. 1941. On the embryology of *Squilla. Proc. Indian Acad. Sci.*, B, **14.**

NAIR, K. B. 1949. The embryology of *Caridina laevis Heller. Proc. Indian Acad. Sci.*, B, **29.**

NARASIMHAMURTI, N. 1933. The development of *Ophiocoma nigra. Q. J. microsc. Sci., N.S.*, **76.**

NAUMANN, E. 1921. Spezielle Untersuchungen über Ernährungsbiologie des tierischen Limnoplanktons. I. Über die Technik des Nahrungserwerbs bei den Cladoceren und ihre Bedeutung für die Biologie der Gewässertypen. *Lunds univ. årsskr., N.F., Avd. 2*, **17.**

NEEDHAM, J. 1935. Problems of nitrogen catabolism in Invertebrates. II. Correlation between uricotelic metabolism and habitat in the phylum Mollusca. *Biochem. J.*, **29.**

NELSON, I. A. 1904. Early development of *Dinophilus. Proc. Acad. Natur. Sci.*, **56.**

NEUMANN, G. 1913. Die Pyrosomen. In: Bronn's *Klassen und Ordnungen des Tierreichs*. III. Suppl. 2, 1. Leipzig.

NEUMANN, G. 1935. Cyclomyaria. In: Kükenthal, W., and Krumbach, Th. *Handbuch der Zoologie*, V/2, 4. Berlin and Leipzig.

NEWELL, B. S. 1953. Cellulolytic activity in the Lamellibranch crystalline style. *J. Mar. Biol. Assn UK*, **32.**

NEWELL, G. E. 1950. The role of the coelomic fluid in the movements of earthworms. *J. exptl. Biol.*, **27.**

NICOL, E. A. T. 1931. The feeding mechanism, formation of the tube and physiology of digestion in *Sabella pavonina*. *Trans. Roy. Soc., Edinburgh*, **56.**

NICOL, J. A. C. 1951. Giant axons and synergic contractions in *Branchiomma vesiculosum*. *J. exptl Biol.*, **28.**

NICOL, J. A. C. 1952. Studies on *Chaetopterus variopedatus* (Renier). I–III. *J. Mar. Biol. Assn UK*, **30, 31.**

NICOL, J. A. C. 1953. Luminiscence in Polynoid worms. *J. Mar. Biol. Assn UK*, **32.**

NIERSTRASS, H. F. 1909. Die Amphineura. I. Solenogastres. *Ergebn. Fortschr. Zool.*, **1.**

NILSSON, D. 1912. Beiträge zur Kenntnis des Nervensystems der Polychäten. *Zool. Bidr., Uppsala*, **1.**

NORMAN, A. M. 1913. *Synagoga mira*, a crustacean of the order Ascothoracica. *Trans. Linnean Soc., Zool., London*, **11.**

NOVAK, V. J. A. 1960. Juvenile hormone and morphogenesis. 11. *Internat. Congress für Entomologie*, **1.**

NÜESCH, H. 1957. Die Morphologie des Thorax von *Telea polyphemus* Cr. (Lepid.). II. Nervensystem. *Zool. Jahrb., Anat. Ont.*, **75, 4.**

NUTTING, W. L. 1951. A comparative anatomical study of the heart and accessory structures of the Orthopteroid insects. *J. Morphol.*, **89.**

NYHOLM, K. G. 1942–44. Zur Entwicklung und Entwicklungsbiologie der Ceriantharien und Aktinien. *Zool. Bidr., Uppsala*, **22.**

NYHOLM, K. G. 1947. Studies in the Echinoderida. *Arkiv zool.*, **39, 14.**

ODHNER, N. 1912. Morphologische und phylogenetische Untersuchungen über die Nephridien der· Lamellibranchien. *Z. wiss. Zool.*, **100.**

OKA, A. 1894. Beiträge zur Anatomie der *Clepsine. Z. wiss. Zool.*, **58.**

OKA, A. 1902. Über das Blutgefäßsystem der Hirudineen. *Annot. zool. Japan*, **4.**

OKA, A. 1904. Über den Bau von *Ozobranchus. Annot. zool. Japan*, **5.**

OLMSTED, J. M. D. 1922. The role of the nervous system in the locomotion of certain marine Polyclads. *J. exptl Zool.*, **36.**

OMER-COOPER, J. 1957. *Protohydra* and Kinorhyncha in Africa. *Nature, Lond.*, **179,** 4557.

ÖPIK, A. A. 1959. Genal caeca of Agnostids. *Nature, Lond.*, **183,** 4677.

ORLOV, J. 1925. Die Innervation des Darmes des Flußkrebses. *Z. Mikroskop. Anat. Forsch.*, **4.**

ORLOV, J. 1927. Das Magenganglion des Flußkrebses. (Ein Beitrag zur· vergleichenden Histologie des sympatischen Nervensystems). *Z. Mikroskop. Anat. Forsch.*, **8.**

ORLOV, J. 1930. Über die Innervation des Schlundes bei Arthropoda. *Z. Mikroskop. Anat. Forsch.*, **20.**

ORTON, J. H. 1912. The mode of feeding in *Crepidula* (with an account of the current-producing mechanism in the mantle cavity, and some remarks on the mode of feeding in Gastropods and Lamellibranchs). *J. Marine Biol. Assn UK, N.S.*, **9.**

ORTON, J. H. 1913. The ciliary mechanisms on the gill and the mode of feeding in *Amphioxus*, Ascidians and *Solenomya togata. J. Marine Biol. Assn UK, N.S.*, **10.**

ORTON, J. H. 1914. On ciliary mechanisms in Brachiopods and some Polychaetes, with a comparison of the ciliary mechanisms on the gills of Mollusca, Protochordata, Brachiopods, and Cryptocephalous Polychaetes, and an account of the endostyle of *Crepidula* and its allies. *J. Marine Biol. Assn UK, N.S.*, **10.**

ORTON, J. H. 1922. Feeding of *Aurelia*. *Nature, Lond.*, **110**.

OSCHE, G. 1959. Arthropodencharaktere bei einem Pentastomiden Embryo (*Reighardia sternae*). *Zool. Anz.*, **163**, 5–6.

OWEN, G. 1956. Observation on the stomach and digestive diverticula of the Lamellibranchia. II. Nuculidae. *Q. J. Microsc. Sci.*, **97**, 541–68.

OWEN, G., TRUEMAN, E. R., and YONGE, C. M. 1953. The ligament in the Lamellibranchia. *Nature, Lond.*, **171**, 4341.

PALM, N. B. 1952. Storage and excretion of vital dyes in Insects. With special regard to trypan blue. *Arkiv Zool. a. s.*, **3**, 2.

PALM, N. B. 1954. The elimination of injected vital dyes from the blood in Myriapods. *Arkiv Zool. a. s.*, **6**, 3–4.

PALM, N. B. 1956. Neurosecretory cells and associated structures in *Lithobius forficatus* L. *Arkiv Zool. a. s.*, **9**, 4.

PALOMBI, A. 1936. *Eugymnanthea inquilina*, nuova leptomedusa derivante di un atecato idroide ospite interno di *Tapes decussatus* L. *Pubbl. Staz. Zool., Napoli*, **15**.

PANOUSE, J. B. 1943. Influence de l'ablation du pédoncule oculaire sur la croissance de l'ovaire chez la crevette *Leander serratus*. *Compt. rend. Acad. sci.*, **217**.

PANOUSE, J. B. 1944. L'action de la glande du sinus sur l'ovaire de la crevette *Leander*. *Compt. rend. Acad. sci.*, **218**.

PANTIN, C. F. A. 1940. The primitive nervous system. *Boll. Soc. ital. biol. sperim.*, **15**.

PANTIN, C. F. A. 1950. Locomotion in British terrestrial Nemertines and Planarians: with a discussion on the identity of *Rhynchodemus bilineatus* (Mecznikow) in Britain, and on the name *Fasciola terrestris* O. F. Müller. *Proc. Linnean Soc., London*, **162**.

PAPI, F., and SWEDMARK, B. 1959. Un Turbellario con lo scheletro: *Acanthomacrostomum spiculiferum* n. sp. *Monit. zool. ital.*, **66**.

PAPPENHEIM, R. 1903. Entwicklungsgeschichte von *Dolomedes fimbriatus* (Gehirn und Augen). *Z. wiss. Zool.*, **74**.

PARKER, G. H. 1916. The effector systems of Actinians. *J. Exptl Zool.*, **21**.

PARKER, G. H., and ALSTYNE, M. A. VAN. 1932. The control and discharge of nematocysts, especially in *Metridium* and *Physalia*. *J. Exptl Zool.* **63**.

PATTEN, W. 1891. On the origin of Vertebrates from Arachnids. *Q. J. microsc. Sci., N.S.*, **31**.

PAVLOVSKY, E. N., and ZARIN, E. J. 1926. On the structure and ferments of the digestive organs of Scorpions. *Q. J. microsc. Sci., N.S.*, **70**.

PELSENEER, P. 1892. A propos de l'asymétrie des Mollusques univalves. *J. Conchyliol.*, **32**.

PELSENEER, P. 1895. Hermaphroditism in Mollusca. *Q. J. microsc. Sci., N.S.*, **37**.

PELSENEER, P. 1898–99. Recherches morphologiques et phylogénétiques sur les Mollusques archaiques. *Mém. Cour. Acad. roy. Belg.*, **57**.

PELSENEER, P. 1906. Mollusca. In: Ray Lankester, E. *A treatise on zoology*, V. London.

PELSENEER, P. 1911. Lamellibranches de l'Expédition du Siboga. Partie anatomique. *Siboga Exped.*, **53**.

PELSENEER, P. 1931. Quelques particularités d'organisation chez les Pectinacea. *Ann. Soc. roy. zool. Belg.*, **61**.

PENNAK, R. W., and ZINN, D. J. 1943. Mystacocarida, a new order of Crustacea from intertidal beaches in Massachusetts and Connecticut. *Smithson. Misc. Coll.*, **103**.

PÉRÈS, M., and PICARD, J. 1955. Observations biologiques effectuées au large de Toulon avec le Bathyscaphe 'F. N. R. S. III' de la Marine Nationale. *Bull. Inst. océanogr. (Monaco)*, **1061**.

PEREYASLAWTZEWA, S. 1901. Développement embryonnaire des Phrynes. *Ann. sci. natur. Zool. et biol. Anim.*, sér. 8, **13**.

PETRUNKEVICH, A. 1922. The circulatory system and segmentation in Arachnida. *J. Morphol.*, **36.**

PETRUNKEVICH, A. 1949. Palaeozoic Arachnida. *Trans. Connect. Acad. Arts Sci.*, **37.**

PFITZNER, I. 1958. Die Bedingungen der Fortbewegung bei den deutschen Landplanarien. *Zool. Beiträge, N.F.*, **3,** 2.

PFLUGFELDER, O. 1934. Bau und Entwicklung der Spinndrüse der Blattwespen. *Z. wiss. Zool.*, **145.**

PFLUGFELDER, O. 1948. Entwicklung von *Paraperipatus amboinensis* n. sp. *Zool. Jahrb., Abt. Anat., Ontog. der Tiere*, **69.**

PFLUGFELDER, O. 1950. Die Funktion der Pericardialdrüsen der Insekten. *Verhandl. Dtsch. Zool. Ges.* (1949), **43.**

PFLUGFELDER, O. 1952. *Entwicklungsphysiologie der Insekten.* Leipzig.

PFLUGFELDER, P. 1929. Histogenetische und organogenetische Prozesse bei der Regeneration polychäter Anneliden. *Z. wiss. Zool.*, **133.**

PHILIPTSCHENKO, J. 1906. Anatomische Studien über Collembola. *Z. wiss. Zool.*, **85.**

PHILIPTSCHENKO, J. 1908. Beiträge zur Kenntnis der Apterygoten. II. Über die Kopfdrüsen der Thysanuren. *Z. wiss. Zool.*, **91.**

PHILIPTSCHENKO, J. 1912. Beiträge zur Kenntnis der Apterygoten. III. Die Embryonalentwicklung von *Isotoma cinerea* Nic. *Z. wiss. Zool.* **103.**

PICARD, J. 1955. Les Hydroides Pteronematidae, origine des 'Siphonophores' Chondrophoridae. *Bull. Inst. océanogr.* (*Fondation Albert I, prince de Monaco*), **1059.**

PIELOU, D. P. 1940. The humidity behaviour of the mealworm beetle, *Tenebrio molitor* L. II. The humidity receptors. *J. exptl Biol.*, **17.**

PIERANTONI, O. 1908. *Protodrilus. Fauna und Flora des Golfes von Neapel*, **31.**

PIETSCHKER, H. 1911. Das Gehirn der Ameisen. *Z. Naturw.*, **47.**

PLATE, L. 1895. Bemerkungen über die Phylogenie und die Entstehung der Asymmetrie der Mollusken. *Zool. Jahrb., Anat.*, **9.**

POCOCK, R. 1901. The Scottish Silurian scorpion, V. *Q. J. microsc. Sci., N.S.*, **44.**

POGGENDORF, D. 1956. Über rhythmische sexuelle Aktivität und die Beziehung zur Häutung und Haarbildung bei Arthropleoner Collembolen. *Naturwissenschaften*, **43.**

POKROWSKY, S. 1899. Noch ein Paar Kopfhöcker bei den Spinnenembryonen. *Zool. Anz.*, **22.**

POLÉJAEFF, N. 1893. Sur la signification systématique du feuillet moyen et de la cavité du corps. *Congr. Intern. Zool., Moscou*, **2,** 2.

PORTMANN, A. 1927. Die Nähreierbildung durch atypische Spermien bei *Buccinum undatum* L. *Z. Zellforsch.*, **5.**

PORTMANN, A. 1930. Die Entstehung der Nähreier bei *Purpura lapillus* durch atypische Befruchtung. *Z. Zellforsch.*, **12.**

POTTS, F. A. 1923. The structure and function of the liver of *Teredo*, the shipworm. *Proc. Cambridge Phil. Soc.*, **1.**

POURBAIX, N. 1933. Recherches sur la nutrition des Spongiaires. *Inst. Espan. Oceanogr. Notas Résum.*, **3,** 69.

PRATT, E. M. 1906. The digestive organs of the Alcyonaria and their relation to the mesogloeal cell-plexus. *Q. J. microsc. Sci., N.S.*, **49.**

PRENANT, M. 1924. Recherches histologiques sur les peroxydases animales. *Arch. morphol. exptl. gén., fasc.* **21.**

PRENANT, M. 1927. Topographie du plexus nerveux cutané d'*Ephesia gracilis* Rathke. *Bull. Soc. zool. France*, **52.**

PREUSS, G. 1951. Die Verwandschaft der Anostraca und Phyllopoda. *Zool. Anz.*, **147.**

PROSSER, C. L. 1950. Nervous system. In: *Comparative animal physiology.* Philadelphia and London.

PROSSER, C. L., BROWN, F. A., and others. 1950. *Comparative animal physiology*. Philadelphia and London.

PRUVOT, G. 1885. Recherches anatomiques et morphologiques sur le système nerveux des Annélides Polychètes. *Arch. zool. exptl. gén., sér.* 2, 3.

PRYOR, M. G. M. 1940. On the hardening of the ootheca of *Blatta orientalis*. *Proc. Roy. Soc. London, ser. B*, 128.

PRZIBRAM, H. 1905. Die Heterochelie bei Decapoden Crustaceen. *Arch. Entwickelungsmech.*, 19.

PRZIBRAM, H. 1907. Die 'Scherenumkehr' bei Decapoden Crustaceen (zugleich experimentelle Studien über Regeneration. Vierte Mitteilung). *Arch. Entwickelungsmech.*, 25.

PURCELL, W. F. 1909. Development and origin of the respiratory organs in Araneae. *Q. J. microsc. Sci., N.S.*, 54.

PURCHON, R. D. 1955. The structure and function of the British Pholadidae (rockboring Lamellibranchia). *Proc. Zool. Soc.*, 124.

PYEFINCH, A. 1949. The larval stages of *Balanus crenatus* Brugière. *Proc. Zool. Soc.*, 118.

QUIEL, G., 1915. Anatomische Untersuchungen an Collembolen. *Z. wiss. Zool.*, 113.

RACOVITZA, E. G. 1896. Le lobe céphalique et l'encéphale des Annélides Polychètes. *Arch. zool. exptl. gén., sér.* 3, 4.

RAMSAY, J. A. 1952. *A Physiological approach to the lower animals*. Cambridge.

RASMONT, R. 1958. Ultra-structure des choanocytes d'Eponges. *Proc. XV. Intern. Congr. Zool.* London.

RATHKE, H. 1842. *Beiträge zur vergleichenden Anatomie und Physiologie*. Danzig.

RATTENBURY, J. C. 1954. The embryology of *Phoronopsis viridis*. *J. Morphol.*, 95.

RAY LANKESTER, E. 1875. On some new points in the structure of *Amphioxus* and their bearing on the morphology of Vertebrata. *Q. J. microsc. Sci., N.S.*, 15.

RAY LANKESTER, E. 1877. Notes on the embryology and classification of the Animal Kingdom. *Q. J. microsc. Sci., N.S.*, 17.

RAY LANKESTER, E. 1881. *Limulus* an Arachnid. *Q. J. microsc. Sci., N.S.*, 21.

RAY LANKESTER, E. 1884. On the skeleto-trophic tissues and coxal glands of *Limulus*, *Scorpio* and *Mygale*. *Q. J. microsc. Sci., N.S.* 24.

RAY LANKESTER, E. 1885. A new hypothesis as to the relationship of the lung-book of *Scorpio* to the gillbook of *Limulus*. *Q. J. microsc. Sci., N.S.*, 25.

RAY LANKESTER, E. 1889. Contribution to the knowledge of *Amphioxus lanceolatus* Yarrell. *Q. J. microsc. Sci., N.S.*, 29.

RAY LANKESTER, E. 1900, 1901, 1903, 1906, 1909. *A treatise on Zoology*. I–V, VII, IX. London.

RAY LANKESTER, E. 1904. The structure and classification of the Arthropoda. *Q. J. microsc. Sci., N.S.*, 47.

RAY LANKESTER, E., BENHAM, W., and BECKE, E. 1885. On the muscular and endoskeletal systems of *Limulus* and *Scorpio*. *Trans. Zool. Soc.*, 11.

RÉAUMUR, R. DE. 1734–42. *Mémoires pour servir à l'histoire des Insectes*. Paris.

REDFIELD, A. C., and FLORKIN, M. 1931. The respiratory function of blood of *Urechis caupo*. *Biol. Bull.*, 61.

REES, W. J. 1936. On a new species of Hydroid, *Staurocoryne filiformis*, with a revision of the genus *Staurocoryne* Rotch, 1872. *J. Mar. Biol. Assn UK, N.S.*, 21.

REES, W. J. 1938. Observations on British and Norwegian Hydroids and their medusae. *J. Mar. Biol. Assn UK, N.S.*, 23.

REES, W. J. 1957. Evolutionary trends in the classification of Capitate Hydroids and Medusae. *Bull. Brit. Museum (Nat. Hist.), Zool.*, 4, 9.

REGEN, J. 1901. Das tympanale Sinnesorgan von *Thamnotrizon apterus* ♂ als Gehörapparat experimentell nachgewiesen. *S. B. Akad. Wiss., Wein*, 117.

REGNÉLL, G. 1948. Swedish Hybocrinida (Crinoidea Inadunata Disparata: Ordovician—Lower Silurian). *Arkiv Zool.*, **40**, 9.

REHM, E. 1939. Die Innervation der inneren Organe von *Apis mellifica*, zugleich ein Beitrag zur Frage des sogenannaten sympathischen Nervensystems der Insekten. *Z. Morphol. Ökol. Tiere*, **36**.

REICHENBACH, H. 1886. Studien zur Entwicklungsgeschichte des Flußkrebses. *Abhandl. Senkenberg. Naturforsch. Ges.*, **14**.

REISINGER, E. 1922. Untersuchungen über Bau und Funktion des Excretionsapparates bei rhabdocölen Turbellarien. *Zool. Anz.*, **54**, 200–209.

REISINGER, E. 1923. *Protomonotresis centrophora* n. g., n. sp., eine Süßwasseralloeocoele aus Steiermark. *Zool. Anz.*, **58**.

REISINGER, E. 1925. Untersuchungen am Nervensystem der *Bothrioplana semperi* Braun. *Z. Morphol. Ökol. Tiere*, **5**.

REISINGER, E. 1926. Zur Turbellarienfauna der Antarktis. *Deutsche Südpolar Exped., Ser. Zool.*, **18**.

REISINGER, E. 1929. Zum Ductus Genito-Intestinalis-Problem. I. Über primäre Geschlechtstrakt-Darmverbindungen bei rhabdocoelen Turbellarien. *Z. Morphol. Ökol. Tiere*, **16**.

REISINGER, E. 1931. Vermes Polymera. In: Kükenthal, W., and Krumbach, Th. *Handbuch der Zoologie*, II/2, 12. Berlin and Leipzig.

REMANE, A. 1926. Morphologie und Verwandtschaftsbeziehungen der aberranten Gastrotrichen. I. *Z. Morphol. Ökol. Tiere*, **5**.

REMANE, A. 1929. Gastrotricha. In: Kükenthal, W., and Krumbach, Th. *Handbuch der Zoologie*, II/1, 6. Berlin and Leipzig.

REMANE, A. 1929, 1930. Kinorhyncha. *Handbuch der Zoologie*, **8**.

REMANE, A. 1936. *Monobryozoon ambulans* n. g., n. sp., ein eigenartiges Bryozoon des Meeressandes. *Zool. Anz.*, **113**.

REMANE, A. 1950. Die Entstehung der Metamerie der Wirbellosen. *Verhandl. Deut. Zool. Ges.*, Mainz.

REMANE, A. 1951. Die Bursa-Darmverbindung und das Problem des Enddarmes bei Turbellarien. *Zool. Anz.*, **146**.

RICHARDS, A. G. 1951. *The integument of Arthropods*. Minneapolis, London.

RICHARDS, A. G. 1952. Studies on Arthropod cuticle. VIII. The antennal cuticle of honeybees, with particular reference to the sense plates. *Biol. Bull.*, **103**.

RICHTERS, F. 1902. Marine Tardigraden. *Verhandl. Deut. Zool. Ges.*, **19**.

RICHTERS, F. 1926. Tardigrada. In: Kükenthal, W., and Krumbach, Th. *Handbuch der Zoologie*, III/1, 1. Berlin and Leipzig.

RIEDL, R. 1960. Über einige nordatlantische und mediterrane *Nemertoderma*-Funde. *Zool. Anz.*, **165**, 5–6.

RILLING, G. 1960. Zur Anatomie des braunen Steinlaufers *Lithobius forficatus* L. (Chilopoda). Skelettmuskelsystem, periferes Nervensystem und Sinnesorgane des Rumpfes. *Zool. Jahrb., Anat.*, **78**, 1.

RIPPER, W. 1931. Versuch einer Kritik der Homologiefrage der Arthropodentracheen. *Z. wiss. Zool.*, **138**.

RISBEC, J. 1955. Considération sur l'anatomie comparée et la classification des Gastéropodes prosobranches. *J. Conchyliol.*, **95**, 2.

RISLER, H. 1953. Das Gehörorgan des Männchen von *Anopheles stephensi* Liston (Culicidae). *Zool. Jahrb., Anat.*, **73**.

RITTER, W. E. 1900. *Harrimania maculosa*, a new genus and species of Enteropneusta from Alaska, with special regard to the character of its notochord. Papers from the Harriman Alaska Expedition. II. *Proc. Washington Acad. Sci.*, **2**.

ROBERTSON, J. D. 1957. Osmotic and ionic regulation in aquatic Invertebrates. *Recent Adv. Invert. physiol.* Oregon.

ROBSON, E. A. 1957. A sea-anemone from brackish water. *Nature, Lond.*, **179**, 4563.

ROHDENDORF, B. B. 1961. The description of the first winged insect from the Devonian beds of the Timan. *Rev. Entomol. USSR*, **40**, 1, 3.

ROMIEU, M. 1922. Méthode de coloration élective du système nerveux chez quelques Invertébrés. *Comp. rend. Acad. sci.*, **175**.

ROONWALL, M. L. 1937. Studies on the embryology of *Locusta migratorioides*. II. Organogeny. *Phil. Trans. Roy. Soc. London, B*, **227**.

ROSA, D. 1906. Sui nefridi con sbocco intestinale commune dell' *Allolobophora antipae. Arch. zool.*, **3**.

ROTH, M. 1948. A study of mosquito behaviour. An experimental laboratory study of the sexual behaviour of *Aëdes aegypti* (L.). *Amer. Midland Naturalist*, **40**.

ROTH, M., and WILLIS, E. R. 1951. Hygroreceptors in adults of *Tribolium* (Coleoptera, Tenebrionidae). *J. exptl Zool.*, **116**.

RULLIER, F. 1950. Rôle de l'organe nucal des Annélides Polychètes. *Bull. Soc. zool. France*, **75**.

RUSSELL, F. S., and REES, W. J. 1960. The viviparous scyphomedusa *Stygiomedusa fabulosa* Russell. *J. Mar. Biol. Assn UK*, **39**.

RUSZKOWSKI, J. S. 1932. Sur les larves de *Gyrocotyle. Bull. Internat. Acad. Polon.*, *sér. B*, **2**.

SACHS, J. 1874. *Lehrbuch der Botanik.* 4th ed. Leipzig.

SACHWATKIN, A. A. 1956. *Vergleichende Embryologie der niederen Wirbellosen* (Ursprung und Gestaltungswege der individuellen Entwicklung der Vielzeller). Berlin.

SACKS, M. 1955. Observations on the embryology of a Gastrotrich, *Lepidoderella squamata. J. Morphol.*, **96**, 3.

SALENSKY, W. W. 1882. Etudes sur le développement des Annélides. II. *Nereis. Arch. biol.*, **3**.

SALENSKY, W. W. 1894. Beiträge zur Entwicklungsgeschichte der Synascidien. I. Über die Entwicklung von *Diplosoma listeri. Mitt. Zool. Stat. Neapel*, **11**.

SALENSKY, W. W. 1907. Morphogenetische Studien an Würmern. II. Über die Anatomie der Archianneliden nebst Bemerkungen über den Bau einiger Organe des *Saccocirrus papillocercus.* III. Über die Metamorphose des *Polygordius ponticus* n. sp. mihi. IV. Schlußbetrachtungen. *Mém. Acad. St. Pétersbourg, sér.* 8, **19**, 11.

SALENSKY, W. W. 1908. Radiata und Bilateralia. *Biol. Zbl.*, **28**.

SALENSKY, W. W. 1912. Morphogenetische Studien an Würmern. I. Entwicklungsgeschichte der Nemertine im Inneren des Pilidiums. *Mem. Acad. Sci. St. Pétersbourg, sér.* 8, **30**, 10.

SANDERS, H. L. 1955. The Cephalocarida, a new subclass of Crustacea from Long Island Sound. *Proc. Nat. Acad. Sci., U.S.A.*, **4** (1).

SANDERS, H. L. 1957. The Cephalocarida and Crustacean phylogeny. *Systematic Zoology*, **6**, 3.

SATO, T. 1936. Vorläufige Mitteilung über *Atubaria heterolopha*, gen. nov., sp. nov., eines in freiem Zustand aufgefundenen Pterobranchier aus dem Stillen Ozean. *Zool. Anz.*, **115**.

SAYLES, L. P. 1936. Regeneration in the Polychaete *Clymenella torquata.* III. Effect of level of cut on type of new structures in anterior regeneration. *Biol. Bull.*, **70**.

SCHALLER, F. 1952. Das Fortpflanzungsverhalten Apterygoter Insekten (Collembolen und Machiliden). *Verhandl. Deut. Zool. Ges. Freiburg*, **19**.

SCHALLER, F. 1953. Untersuchungen zur Fortpflanzungsbiologie Arthropleoner Collembolen. *Z. Morphol. Ökol. Tiere*, **41**.

SCHALLER, F. 1954. Indirekte Spermatophorenübertragung bei *Campodea* (Apterygota, Diplura). *Naturwissenschaften*, **41**, 17.

SCHALLER, F. 1955. Zwei weitere Fälle indirekter Spermatophoren Übertragung: Skorpione und Silber-Fischchen. *Forsch. und Fortwchritte*, **29**, 9.

SCHARRER, B. 1936. Über 'Drüsen-Nervenzellen' im Gehirn von *Nereis virens* Sars. *Zool. Anz.*, **113**.

SCHARRER, B. 1937. Über sekretorisch tätige Nervenzellen bei wirbellosen Tieren. *Naturwissenschaften*, **25**.

SCHARRER, B. 1952. Further studies of the intercerebralis-cardiacum-allatum system of Insects. *Biol. Bull.*, **103**.

SCHARRER, B., 1953. Comparative physiology of invertebrate endocrines. *Ann. Rev. Physiol.*, **15**.

SCHARRER, B., and SCHARRER, E. 1944. Neurosecretion. VI. A comparison between the intercerebralis-cardiacum-allatum system of the insects and the hypothalamo-hypophyseal system of the Vertebrates. *Biol. Bull.*, **87**.

SCHARRER, E., and SCHARRER, B. 1954. Neurosecretion. In: Möllendorff, W. *Handbuch der mikroskopischen Anatomie des Menschen*. VI, 5. Berlin.

SCHEPOTIEFF, A. 1905. Über Stellung der Graptolithen im zoologischen System. *Neues Jahrb. Mineral.*, **2**.

SCHEPOTIEFF, A. 1906. Die Pterobranchier. Anatomische und histologische Untersuchungen über *Rhabdopleura normanii* Allman und *Cephalodiscus dodecalophus* M'Int. 1. *Rhabdopleura normanii* Allman. Abschn. 1. Die Anatomie von *Rhabdopleura*. *Zool. Jahrb.*, *Anat.*, **23**.

SCHEPOTIEFF, A. 1909. Die Pterobranchier des Indischen Ozeans. *Zool. Jahrb.*, *Syst.*, **28**.

SCHEUERING, L. 1913. Die Augen der Arachnoideen. I. Scorpiones. *Zool. Jahrb.*, *Anat.*, **33**.

SCHEUERING, L. 1914. Die Augen der Arachnoideen. II. Phalangiden und Araneiden. *Zool. Jahrb.*, *Anat.*, **37**.

SCHEWIAKOFF, W. T. 1889. Beiträge zur Kenntnis des Acalephenauges. *Morphol. Jahrb.*, **15**.

SCHIMKEWITSCH, W. M. 1887. Etudes sur le développement des Araignées. *Arch. Biol.*, **6**.

SCHIMKEWITSCH, W. M. 1893. Sur la structure et sur la signification de l'endosternite des Arachnides. *Zool. Anz.*, **16**.

SCHIMKEWITSCH, W. M. 1894. Über Bau und Entwicklung des Endosternits der Arachniden. *Zool. Jahrb.*, *Anat.*, **8**.

SCHIMKEWITSCH, W. M. 1895. Zur Kenntnis des Baues und der Entwicklung des *Dinophilus* vom Weißen Meere. *Z. wiss. Zool.*, **59**.

SCHIMKEWITSCH, W. M. 1896. Studien über parasitische Copepoden. *Z. wiss. Zool.*, **61**.

SCHIMKEWITSCH, W. M. 1906. Über die Entwicklung von *Thelyphonus caudatus* (L.), vergleichen mit derjenigen einiger anderer Arachniden. *Z. wiss. Zool.*, **81**.

SCHIMKEWITSCH, W. M. 1908. Über die Beziehungen zwischen den Bilateria und Radiata. *Biol. Zbl.*, **28**.

SCHINDEWOLF, O. H. 1939. Stammesgeschichliche Ergebnisse an Korallen. *Paläontol. Z.*, **21**.

SCHLEIDEN, M. 1842. *Grunzüge der wissenschaftlichen Botanik*. Leipzig.

SCHLOTTKE, E. 1933. Histologische Beobachtungen über die intrazelluläre Verdauung bei *Dendrocoelum lacteum* (Müll.) und *Euscorpius carpathicus* (L.). *S.B. Naturforsch. Ges., Rostock*, **3**, 3.

SCHLOTTKE, E. 1933. Darm und Verdauung bei Pantopoden. *Z. Mikroskop. Anat. Forsch.*, **32**.

SCHLOTTKE, E. 1935. Biologische, physiologische und histologische Untersuchungen über die Verdauung von *Limulus*. *Z. vergl. Physiol.*, **22**.

SCHMIDT, A. 1938. Geschmacksphysiologische Untersuchungen an Ameisen. *Z. vergl. Physiol.*, **25**.

SCHMIDT, W. J. 1920. *Sphaerobactrum wurduae*, ein kettenbildender Ciliat. *Arch. Protistenkunde*, **40.**

SCHNEIDER, A. 1892. Systéme stomatogastrique des Aranéides. *Tabl. Zool.*, **2.**

SCHNEIDER, K. C. 1902. *Lehrbuch der vergleichenden Histologie der Tiere.* Jena.

SCHNEIDER, K. C. 1908. *Histologisches Praktikum der Tiere.* Jena.

SCHNEIDERMAN, H. A. 1956. Spiracular control of discontinuous respiration in Insects. *Nature, Lond.,* **177, 4521.**

SCHNEIDERMAN, H. A., and WILLIAMS, C. M. 1954a. Physiology of Insect diapause. VIII. Qualitative changes in the metabolism of the *Cecropia* silkworm during diapause and development. *Biol. Bull.,* **106.**

SCHNEIDERMAN, H. A., and WILLIAMS, C. M. 1954b. Physiology of insect diapause. IX. The cytochrome oxydase system in relation to the diapause and development of the *Cecropia* silkworm. *Biol. Bull.,* **106.**

SCHÖMANN, K., and SCHALLER, F. 1954. Das Paarungsverhalten von *Polyxenus lagurus* L. *Verhandl. Deutsch. Zool. Ges. Tübingen,* **33.**

SCHRÖDER, CH. 1925, 1928, 1929. *Handbuch der Entomologie.* I–III. Jena.

SCHROFF, FR. 1957. Zur Ernährungsphysiologie der Ammocoeteslarven der Cyclostomen. *Zool. Anz.,* **159,** 3–4.

SCHULZ, E. 1895. Über den Process der Exkretion bei Holothurien. *Biol. Zbl.,* **15.**

SCHULTZE, P. 1932. Über die Körpergliederung der Zecken, die Zusammensetzung des Gnathosoma und die Beziehung der Ixodoidea zu den fossilen Anthracomarti. *S. B. Naturforsch. Ges., Rostock,* **3,** 3.

SCHULTZE, P. 1937. Trilobita, Xiphosura, Acarina. Eine morphologische Untersuchung über Plangleichheit zwischen Trilobiten und Spinnentieren. *Z. Morphol. Ökol. Tiere,* **32.**

SCHULTZE, P. 1941. Das Geruchsorgen der Zecken. *Z. Morphol. Ökol. Tiere,* **37.**

SCHWANN, TH. 1839. *Mikroskopische Untersuchungen über die Übereinstimmung in der Struktur und dem Wachstum der Thiere und Pflanzen.* Berlin.

SCOURFIELD, D. J. 1926. On a new type of Crustacean from the Old Red Sandstone (Rhynie Chert Bed, Aberdeenshire)—*Lepidocaris rhyniensis*, gen. et sp. nov. *Phil. Trans. Roy. Soc., London, B,* **214.**

SCOURFIELD, D. J. 1939–40. The oldest known fossil insect (*Rhyniella praecursor* Hirst and Maulik). Further details from additional specimens. *Proc. Linnean Soc., London,* **152.**

SCOURFIELD, D. J. 1940. Two new and nearly complete specimens of young stages of the Devonian fossil Crustacean *Lepidocaris rhyniensis.* *Proc. Linnean Soc., London,* **152.**

SCRIBAN, I., and AUTRUM, H. 1932, 1934. Hirudinea. In: Kükenthal, W., and Krumbach, Th. *Handbuch der Zoologie,* II/2, 15, 17. Berlin and Leipzig.

SEDGWICK, A. 1884. On the origin of metameric segmentation and some other morphological questions. *Q. J. microsc. Sci., N.S.,* **24.**

SEILERN-ASPANG, F. 1957. Die Entwicklung von *Macrostomum appendiculatum* (Fabr.). *Zool. Jahrb., Anat.,* **47.**

SELYS LONGCHAMPS, M. DE. 1907. *Phoronis.* In: *Fauna und Flora des Golfes von Neapel,* 30 Monogr. Berlin.

SEMBRAT, K. 1958. Hormones and evolution. *Proc. XV Intern. Congr. Zool.* London.

SEMPER, O. 1872. Zoologische Aphorismen (*Trochosphaera*). *Z. wiss. Zool.,* **22.**

SERÈNE, R. 1951. Sur la circulation d'eau à la surface du corps des Stomatopodes. *Bull. Soc. Zool. France,* **76.**

SHAPEERO, W. L. 1961. Phylogeny of Priapulida. *Science (Washington, D.C.),* **133.**

SHIPLEY, A. E. 1909. Pentastomida s. Linguatulida. In: Harmer, S. F., and Shipley, A. E. *The Cambridge Natural History.* London.

SIEBOLD, C. TH. 1850. Über den Generationswechsel der Cestoden nebst einer Revision der Gattung *Tetrarhynchus. Z. wiss. Zool.,* **2.**

SIEWING, R. 1956. Untersuchungen zur Morphologie der Malacostraca (Crustacea). *Zool. Jahrb., Anat.*, **75**.

SIEWING, R. 1958. Anatomie und Histologie von *Thermosbaena mirabilis*. Ein Beitrag zur Phylogenie der Reihe Pancarida (Thermosbaenacea). *Akad. Wiss. Liter. Mainz, Abhandl.* Math.-Nat. wiss. Kl. Jg. 1957, No. 7.

SIEWING, R. 1959. Syncarida. In: Bronn's *Klassen und Ordnungen des Tierreichs*, V/1, 4.

SIEWING, R. 1963. Zum Problem der Arthropodenkopfsegmentierung. *Zool. Anz.*, **170**, 11/12.

SILÉN, L. 1935. Zur Kenntnis des Polymorphismus der Bryozoen. Die Avicularien der Cheilostomata Anasca. *Zool. Bidr., Uppsala*, **17**.

SILÉN, L. 1942. Carnosa and Stolonifera (Bryozoa), collected by Prof. Dr. Sixten Bock's Expedition to Japan and the Bonin Islands, 1914. *Arkiv Zool.*, **34**, 8.

SILÉN, L. 1942–44. Origin and development of the Cheilo–Ctenostomatous stem of Bryozoa. *Zool. Bidr., Uppsala*, **22**.

SILÉN, L. 1944. On the division and movements of the alimentary canal of the Bryozoa. *Arkiv Zool.*, **35**, 12.

SILÉN, L. 1945. The main features of the development of the ovum, embryo and oecium in the oeciferous Bryozoa Gymnolaemata. *Arkiv Zool.*, **35**, 17.

SILÉN, L. 1947. Conescharellinidae (Bryozoa Gymnolaemata) collected by Prof. Dr. Sixten Bock's Expedition to Japan and the Bonin Islands, 1914. *Arkiv Zool.*, **39**, 9.

SILÉN, L. 1950a. On the nervous system of *Glossobalanus marginatus* Meek (Enteropneusta). *Acta zool.*, **31**.

SILÉN, L. 1950b. On the mobility of entire zooids in Bryozoa. *Acta zool.*, **31**.

SILÉN, L. 1954. On the nervous system of *Phoronis*. *Ark. Zool.*, **6**.

ŠIMKEVIČ, L., and W. 1911. Ein Beitrag zur Entwicklungsgeschichte der Tetrapneumones. I–III. *Bull. Acad. Sci., St. Pétersbourg, ser.* 6, **5**.

SIMROTH, H. 1898. Über die mögliche oder wahrscheinliche Herleitung der Asymmetrie der Gastropoden. *Biol. Zbl.*, **18**.

SIMROTH, H. 1896–1907. Gastropoda Prosobranchia. In: Bronn's *Klassen und Ordnungen des Thierreichs*, III/2 1. Leipzig.

SIMROTH, H. 1908–28. Pulmonata. In: Bronn's *Klassen und Ordnungen des Thierreichs*, III/2, 2. Leipzig.

SKRAMLIK, E. VON 1938. Über den Kreislauf bei den niedersten Chordaten. *Ergebn. Biol.*, **15**.

SMITH, J. E. 1937. On the nervous system of the starfish *Marthasterias glacialis* (L.). *Phil. Trans. Roy. Soc., London, B*, **227**.

SMITH, J. E. 1940. The reproductive system and associated organs of *Ophiothrix fragilis*. *Q. J. microsc. Sci.*, **82**.

SNODGRASS, R. E. 1928. Morphology and evolution of the insect head and its appendages. *Smithson. Misc. Coll.*, **81**.

SNODGRASS, R. E. 1935. *Principles of insect morphology.* New York.

SNODGRASS, R. E. 1938. Evolution of the Annelida, Onychophora and Arthropoda. *Smithson. Misc. Coll.*, **97**.

SNODGRASS, R. E. 1948. Feeding organs of Arachnida. *Smithson. Misc. Coll.*, **110**.

SNODGRASS, R. E. 1952. *A Textbook of Arthropod Anatomy.* Ithaca, New York.

SÖDERSTRÖM, A. 1930. Über segmental wiederholte 'Nuchalograne' bei Polychäten. *Zool. Bidr., Uppsala*, **12**.

SOLLAUD, E. 1923. Recherches sur l'embryogénie des Crustacés Décapodes de la sousfamille des Palaemoninae. *Bull. biol. France et Belgique*, **5**.

SOLLAUD, E. 1933. Le blastopore et la question du 'prostomium' chez les Crustacés. *Compt. rend. Assoc. Franc. Sci.*, **57**.

SOUTHWARD, A. J. 1949. Ciliary mechanisms in *Aurelia aurita*. *Nature, Lond.*, **163**, 4144.

SPENCER, G. J. 1930. The firebrat *Thermobia domestica* in Canada. *Canad. Entomologist*, **62,** 1–2.

SPENCER, W. K. 1951. Early Palaeozoic starfish. *Phil. Trans. Roy. Soc., London, B,* **235.**

SPENGEL, J. W. 1893. Die Enteropneusten des Golfes von Neapel und des angrenzenden Meeres-Abschnittes. In: *Fauna und Flora des Golfes von Neapel,* 18 Monogr., Leipzig.

STANNIUS, H. 1831. Über den innern Bau der *Amphinome rostrata. Isis.* Jena and Leipzig.

STAUBER, A. 1950. The fate of India ink injected intracardially into the oyster, *Ostrea virginica* Gmelin. *Biol. Bull.,* **98.**

STEINBÖCK, O. 1924. Untersuchungen über die Geschlechtstrakt-Darmverbindung bei Turbellarien, nebst einem Beitrag zur Morphologie des Trikladendarmes. *Z. Morphol. Ökol. Tiere,* **2.**

STEINBÖCK, O. 1925. Zur Systematik der Turbellaria metamerata, zugleich ein Beitrag zur Morphologie des Tricladen-Nervensystems. *Zool. Anz.,* **64.**

STEINBÖCK, O. 1927. Monographie der Prorhynchidae. *Z. Morphol. Ökol. Tiere,* **8.**

STEINBÖCK, O. 1931. *Nemertoderma bathycola* nov. gen. nov. sp. *Vidensk. Medd. Naturh. Foren., København,* **90.**

STEINBÖCK, O. 1954. Sobre la misión del 'plasmodio digestivo' en la regeneracion de *Amphiscolops* (Turbellaria Acoela). *Publ. Inst. Biol. Apl., Barcelona,* **17.**

STEINBÖCK, O. 1955. Regeneration azöler Turbellarien. *Verhandl. Deut. Zool. Ges.,* **1954** (Zool. Anz., Suppl. 18).

STEINBÖCK, O. 1958. Zur Phylogenie der Gastrotrichen. *Verhandl. Deut. Zool. Ges.,* (Graz, 1957).

STEINMANN, G. 1907. *Einführung in die Paläontologie.* 2nd ed. Leipzig.

STEPHENS, G. J. 1952. Mechanisms regulating the reproductive cycle in the crayfish, *Cambarus.* I. The female cycle. *Physiol. Zool.,* **25.**

STEPHENSON, T. A. 1929. A contribution to actinian morphology: the genera *Phellia* and *Sagartia. Trans. Roy. Soc., Edinburgh,* **56.**

STEPHENSON, T. A. 1935. *British sea anemones.* 2. Ray Society.

STORCH, O. 1912. Zur vergleichenden Anatomie der Polychaeten. *Verhandl. zool. bot. Ges., Wein,* **42.**

STORCH, O. 1924. Morphologie und Physiologie des Fangapparates der Daphniden. *Ergebn. Fortschr. Zool.,* **6.**

STORCH, O. 1924–25. Der Phyllopoden-Fangapparat. *Internat. Rev. Hydrobiol.,* **12–13.**

STØRMER, L. 1934. Merostomata from the Downtonian sandstone of Ringerike, Norway. *Skr. Norske vidensk.-akad. Oslo,* **1,** 10.

STØRMER, L. 1939. Studies on Trilobite morphology. I. The thoracic appendages and their phylogenetic significance. *Norsk geol. tidsskr.,* **19.**

STØRMER, L. 1944. On the relationships and phylogeny of fossil and recent Arachnomorpha. *Skr. Norske vidensk. akad. Oslo,* **1,** 5.

STØRMER, L. 1949. Classe des Trilobites. Classe des Merostomoidea, Marellomorpha et Pseudocrustacea. In: Grassé, P.-P. *Traité de Zoologie,* VI. Paris.

STØRMER, L. 1956. A Lower Cambrian merostome from Sweden. *Arkiv zool.,* **9,** 25.

STRAELEN, V. VAN. 1928. Sur les Crustacés Décapodes triassiques et sur l'origine d'un phylum de Brachyoures. *Bull. cl. sci. Acad. roy. Belgique,* **14.**

STRUM, H. 1956. Die Paarung des Silberfischens (*Lepisma saccharina* L.). *Verhandl. Deut. Zool. Ges., Erlangen. Zool. Anz., Suppl.* **19.**

STUNKARD, H. W. 1954. The life history and systematic relations of the Mesozoa. *Q. Rev. Biol.,* **29.**

SUKATSCHOFF, B. 1900. Beiträge zur Entwicklungsgeschichte der Hirudineen. I. Zur Kenntnis der Urnieren von *Nephelis vulgaris* Moqu.-Tand. und *Aulostomum gulo* Moqu.-Tand. *Z. wiss. Zool.,* **67.**

Šulc, K. 1927. Das Tracheensystem von *Lepisma* (Thysanura) und Phylogenie der Pterygogenea. *Acta Soc. Sci. Nat. Morav.*, **4.**

Sutton, M. F. 1958. Salp reproduction and the evolution of the Chordates. *Proc. XV Intern. Congr. Zool.* London.

Svetlov, P. G. 1924. On the appendices of the ambulacral ring in Echinoidea (the Tiedemann's bodies of Echinoidea). *Rev. zool. Russe, Moskva*, **4,** 1–2.

Svetlov, P. G. 1935. Regulationserscheinungen an *Cristatella*-Kolonien. Zum Individualitätsproblem. *Z. wiss. Zool.*, **147.**

Tannreuther, G. W. 1915. The embryology of *Bdellodrilus philadelphicus*. *J. Morphol.*, **26.**

Teissier, G. 1930. Polarité morphologique et polarité physiologique de l'embryon des Hydraires. *Compt. rend. Soc. biol.*, **105.**

Teissier, G. 1931. Etude expérimentale du développement de quelques Hydraires. *Ann. sci. natur., sér.* 10, **14.**

Thiele, J. 1891. Die Stammesverwandtschaft der Mollusken. *Z. Naturw.*, **25.**

Thiele, J. 1902. Die systematische Stellung der Solenogastren und die Phylogenie der Mollusken. *Z. wiss. Zool.*, **72.**

Thiele, J. 1929, 1931, 1935. *Handbuch der systematischen Weichtierkunde.* I–III. Jena.

Thompson, D'Arcy W. 1942. *On growth and form.* Cambridge.

Thorson, G. 1946. Reproduction and larval development of Danish marine bottom Invertebrates, with special reference to the planktonic larvae in the Sound (Øresund). *Medd. Komm. Danmarks Fisk. Havundersøgelser, Ser. Plankton*, **4,** 1.

Thorson, G. 1950. Reproductive and larval ecology of marine bottom Invertebrates. *Biol. Revs Cambridge Phil. Soc.*, **25.**

Tiegs, O. W. 1940. The embryology and affinities of the Symphyla, based on a study of *Hanseniella agilis*. *Q. J. miscrosc. Sci., N.S.*, **82.**

Tiegs, O. W. 1947. The development and affinities of the Pauropoda, based on a study of *Pauropus silvaticus*. Part I–II. *Q. J. microsc. Sci., N.S.*, **88.**

Tischner, H. 1953. Über den Gehörsinn von Stechmücken. *Acustica*, **3.**

Totton, A. K. 1954a. Siphonophora of the Indian Ocean. *Discovery Rep.*, **27.**

Totton, A. K. 1954b. Egg-laying in Ctenophora. *Nature, Lond.*, **174,** 4425.

Totton, A. K. 1960. Studies on *Physalia physalis* (L.). Part 1. Natural history and morphology. *Discovery Rep.*, **30.**

Turner, R. S. 1946. Observations on the central nervous system of *Leptoplana acticola*. *J. Compar. Neurol.*, **85.**

Tuzet, O. 1948. Les premiers stades du développement de *Leucosolenia bothryoides* Ellis et Solander et de *Clathrina* (*Leucosolenia*) *coriacea* Mont. *Ann. Sci. natur. Zool. biol. anim., sér.* 11, **10.**

Tuzet, O. 1950. Le spermatozoïde dans la série animale. *Rev. suisse zool.*, **57.**

Uljanin, B. 1884. *Doliolum. Fauna und Flora des Golfes von Neapel*, Monogr. X. Berlin.

Ulrich, W. 1949. Über die systematische Stellung einer neuen Tierklasse (Pogonophora K. E. Johansson). *Sitzungsber. Deut. Akad. Wiss. Berlin, Math.-Naturw. Kl.*, **2.**

Urbanowicz, F. 1885. Przyczynek do embryologji raków widłonogich. *Kosmos* (*Polska*), **10.**

Uschakow, P. 1933. Eine neue Form aus der Familie Sabellidae (Polychaeta). *Zool. Anz.*, **104.**

Vachon, M. 1949. Ordre des Pseudoscorpions. In: Grassé, P.-P. *Traité de Zoologie*, VI. Paris.

Vagin, V. L. 1937. Die Stellung der Ascothoracida ord. nov. (Cirripedia Ascothoracica Gruvel 1905) im System der Entomostraca. *Compt. rend. Acad. Sci. USSR. Moscou n. sér.*, **15.**

VALKANOFF, A. 1936. Beitrag zur Anatomie und Morphologie der Rotatoriengattung *Trochosphaera* Semper. *Trav. Soc. Bulg. Sci. Nat.*, **17**.

VANDEL, A. 1948. Les Isopodes valvationnels exoantennés et la génèse de leurs coaptations. *Bull. biol. France et Belgique*, **82**.

VANDEL, A. 1949. Embranchement des Arthropodes: généralités, composition de l'embranchement. In: Grassé, P.-P. *Traité de Zoologie*, VI. Paris.

VERHOEFF, K. W. 1925. Chilopoda. In: Bronn's *Klassen und Ordnungen des Tierreichs*, V/2, 1. Leipzig.

VERHOEFF, K. W. 1933–34. Symphyla und Pauropoda. *Klassen und Ordnungen des Tierreichs*, 3.

VERNADSKY, V. I. 1945. The Biosphere and the Noosphere. *Amer. Scientist*, **33**.

VICQ D'AZYR, F. 1805. *Oeuvres*. Paris.

VIRCHOW, R. 1858. *Die Cellularpathologie*. Berlin.

VITZTHUM, H. G. 1931. Acari. In: Kükenthal, W., and Krumbach, Th. *Handbuch der Zoologie*, III/2, 1. Berlin and Leipzig.

WALCOTT, CH. D. 1911a. Cambrian geology and paleontology. II. No. 2. Middle Cambrian Merostomata. *Smithson. Misc. Coll.*, **57**.

WALCOTT, CH. D. 1911b. Cambrian geology and paleontology. II. No. 3. Middle Cambrian Holothurians and Medusae. *Smithson. Misc. Coll.*, **57**.

WALCOTT, CH. D. 1912. Cambrian geology and paleontology. II. No. 6. Middle Cambrian Branchiopoda, Malacostraca, Trilobita and Merostomata. *Smithson. Misc. Coll.*, **57**.

WALLSTABE, P. 1908. Entwicklungsgeschichte der Araneiden. *Zool. Jahrb., Anat.*, **26**.

WALSHE, B. M. 1950. The function of haemoglobin in *Chironomus* under natural conditions. *J. exptl Biol.*, **27**.

WATERLOT, G. 1953. Généralités sur les Arthropodes. In: Piveteau, J. *Traité de Paléontologie*, III. Paris.

WEBER, H. 1933. *Lehrbuch der Entomologie*. Jena.

WEBER, H. 1938. Beiträge zur Kenntnis der Überordnung Psocoidea. I. Die Labialdrüse der Copeognatha. *Zool. Jahrb., Anat.*, **64**.

WEEL, P. B. VAN. 1937. Die Ernährungsbiologie von *Amphioxus lanceolatus*. *Pubbl. Staz. Zool., Napoli*, **16**.

WEILL, R. 1936. Existence des larves polypoides dans le cycle de la Trachyméduse *Olindias phosphorica* Della Chiaje. *Compt. rend. Acad. sci.*, **203**.

WEISSERMEL, W. 1897. Die Gattung *Columnaria* und Beiträge zur Stammesgeschichte der Cyathophilliden und Zaphrenitiden. *Z. Dtsch. geol. Ges.*, **49**.

WEISSMANN, A. 1864. Die nachembryonale Entwicklung der Musciden. *Z. wiss. Zoll.*, **14**.

WELLS, G. P. 1949. Respiratory movements of *Arenicola marina* L., intermittent irrigation of the tube, and intermittent aerial respiration. *J. Mar. Biol. Assn UK*, **28**.

WELLS, G. P. 1950. The anatomy of the body wall and appendages in *Arenicola marina* L., *Arenicola claparedii* Levinsen and *Arenicola ecaudata* Johnston. *J. Mar. Biol. Assn UK*, **29**.

WELLS, G. P. 1951. On the behaviour of *Sabella*. *Proc. Roy. Soc., London, B*, **138**.

WELLS, G. P. 1952. The proboscis apparatus of *Arenicola*. *J. Mar. Biol. Assn UK*, **31**.

WELLS, G. P., and DALES, R. PH. 1951. Spontaneous activity patterns in animal behaviour: the irrigation of the burrows in the Polychaetes *Chaetopterus variopedatus* Renier and *Nereis diversicolor* O. F. Müller. *J. Mar. Biol. Assn UK*, **29**.

WELSH, J. H. 1957. Neurohormones or transmitter agents. *Rec. Adv. Invert. physiol.* Oregon.

WENZ, W. 1938. Gastropoda. Teil I: Allgemeiner Teil und Prosobranchia. In: Schindewolf, O. H. *Handbuch der Paläozoologie*. VI. Berlin.

WENZ, W. 1940. Ursprung und Stammesgeschichte der Gastropoden. *Arch. Molluskenkunde*, **72,** 1.

WERMEL, E. 1926. Cytologische Studien an *Hydra. Z. Zellforsch. Mikrosk. Anat.*, **4.**

WESENBERG-LUND, C. 1899. Danmarks Rotifera. I. *Vid. medd. Dansk naturhistor. foren. København*, **5,** 10.

WESTBLAD, E. 1923. Zur Physiologie der Turbellarien. *Lunds univ. årsskr.*, N.F., Avd. 2, **18.**

WESTBLAD, E. 1937. Die Turbellariengattung *Nemertoderma* Steinböck. *Acta Soc. fauna et flora fennica*, **60.**

WESTBLAD, E. 1940, 1942, 1945, 1946, 1948. Studien über skandinavische Turbellaria Acoela. I–V. *Arkiv zool.*, **32,** 20; **33,** 14; **36,** 5; **38,** 1; **41,** 7.

WESTBLAD, E. 1947. Notes on hydroids. I. The organization, reproduction and systematic position of *Boreohydra simplex* mihi. II. Some observations and systematic remarks on *Acaulis primarius* Stimpson. *Arkiv zool.*, **39,** 5.

WESTBLAD, E. 1949. *Xenoturbella bocki* n. g., n. sp., a peculiar, primitive Turbellarian type. *Arkiv zool., a. s.*, **1.**

WEYGOLD, P. 1958. Die Embryonalentwicklung des Amphipoden *Gammarus pulex pulex* (L.). *Zool. Jahrb., Anat.*, **77.**

WEYGOLD, P. 1960. Embryologische Untersuchungen an Ostrakoden. Die Entwicklung von *Cyprideis litoralis* (Brady) (Ostracoda, Podocopa, Cythereidae). *Zool. Jahrb., Anat.*, **78,** 3.

WHITEAR, M. 1960. Chordotonal organs in Crustacea. *Nature, Lond.*, **187,** 4736.

WHITEHOUSE, F. W. 1941. The Cambrian faunas of North-Eastern Australia, early Cambrian Echinodermes, similar to the larval stage of recent forms. *Mem. Queensl. Mus.*, **12.**

WIDMANN, E. 1908. Über den feineren Bau der Augen einiger Spinnen. *Z. wiss. Zool.*, **90.**

WIESER, W. 1954. On the morphology of the head in the family Leptosomatidae (marine free living Nematodes). *Arkiv zool., a. s.*, **6,** 3.

WIETRZYKOWSKI, W. 1912. Recherches sur le développement des Lucernaires. *Arch. zool. exptl. gén., ser.* 5, **10.**

WIGGLESWORTH, V. B. 1936. The function of the corpus allatum in the growth and reproduction of *Rhodnius prolixus* (Hemiptera). *Q. J. microsc. Sci.*, N.S., **79.**

WIGGLESWORTH, V. B. 1947. The epicuticle in an insect, *Rhodnius prolixus* (Hemiptera). *Proc. Roy. Soc., B*, **134.**

WIGGLESWORTH, V. B. 1950. *The principles of insect physiology.* 4th ed. London and New York.

WIGGLESWORTH, V. B. 1954. Neurosecretion and the corpus cardiacum of insects. *Pubbl. Staz. zool., Napoli*, **24,** suppl.

WIJHE, J. W. VAN. 1901. Beiträge zur Anatomie der Kopfregion des *Amphioxus lanceolatus. Petrus Camper, Haarlem*, **1.**

WIJHE, J. W. VAN. 1914. Studien über *Amphioxus.* I. Mund und Darmkanal während der Metamorphose. *Verhandel. Kon. nederl. akad. wet. Afd. natuurkunde. II reeks.* 18. Amsterdam.

WILHELMI, J. 1909. Tricladen. In: *Fauna und Flora des Golfes von Neapel*, 32 Monorg. Leipzig.

WILKE, U. 1954. Mediterrane Gastrotrichen. *Zool. Jahrb., Syst.*, **82.**

WILLIAMS, C. M. 1946. Physiology of insect diapause. The role of the brain in the production and termination of pupal dormancy in the giant silkworm, *Platysamia cecropia. Biol. Bull.*, **90.**

WILLIAMS, C. M. 1947. Physiology of insect diapause. II. Interaction between the pupal brain and prothoracic glands in the metamorphosis of the giant silkworm, *Platysamia cecropia. Biol. Bull.*, **93.**

WILLIAMS, C. M. 1948. Physiology of insect diapause. III. The prothoracic glands in the *Cecropia* silkworm, with special reference to their significance in embryonic and postembryonic development. *Biol. Bull.*, **94**.

WILLIAMS, C. M. 1949. The prothoracic glands of insects in retrospect and in prospect. *Biol. Bull.*, **97**.

WILLIAMS, C. M. 1952. Physiology of insect diapause. IV. The brain and pro-thoracic glands as an endocrine system in the *Cecropia* silkworm. *Biol. Bull.*, **103**.

WILLIAMS, C. M. 1956. Physiology of insect diapause. X. An endocrine mechanism for the influence of temperature on the diapausing pupa of the *Cecropia* silkworm. *Biol. Bull.*, **110**.

WILSON, D. P. 1932. On the mitraria larva of *Owenia fusiformis* D. Ch. *Phil. Trans. Roy. Soc., London, B*, **221**.

WILSON, E. B. 1881. The origin and significance of the metamorphosis of actino-trocha. *Q. J. microsc. Sci., N.S.*, **21**.

WILSON, E. B. 1892. The cell-lineage of *Nereis. J. Morphol.*, **6**.

WILSON, E. B. 1904. Experimental studies on germinal localisation. I. The germ regions in *Dentalium*. II. Experiments on the cleavage mosaic in *Patella* and *Dentalium. J. exptl Zool.*, **1**.

WINTERSTEIN, H. 1910–12. *Handbuch der vergleichenden Physiologie*. I–IV. Jena.

WOLTERECK, R. 1902. Trochophora-Studien. I. Über die Histologie der Larve und Entstehung des Annelids bei den *Polygordius*-Arten des Nordsee. *Zoologica*, **34**.

WOLTERECK, R. 1904a. Beiträge zur praktischen Analyse der *Polygordius*-Entwicklung nach dem 'Nordsee' und dem 'Mittelmeertypus'. *Arch. Entwicklungsmech.*, **18**.

WOLTERECK, R. 1904b. Wurmkopf, Wurmrumpf und Trochophora. *Zool. Anz.*, **28**.

WOLTERECK, R. 1905. Zur Kopffrage der Anneliden. *Vernandl. Dtsch. Zool. Ges.*, **15**.

YALVAC, S. 1939. Histologische Untersuchungen über die Entwicklung des Zeckenadultus in der Nymphe. *Z. Morphol. Ökol. Tiere*, **35**.

YONGE, C. M. 1923. Studies on the comparative physiology of digestion. I. The mechanism of feeding, digestion and assimilation in the Lamellibranch *Mya. J. exptl Biol.*, **1**.

YONGE, C. M. 1926. Structure and physiology of the organs of feeding and digestion in *Ostrea edulis. J. Mar. Biol. Assn UK, N.S.*, **14**.

YONGE, C. M. 1928a. Structure and function of the organs of feeding and digestion in the Septibranchs *Cuspidaria* and *Poromya. Phil. Trans. Roy. Soc., London, B*, **216**.

YONGE, C. M. 1928b. Feeding mechanisms in the invertebrates. *Biol. Revs. Cambridge Phil. Soc.*, **3**.

YONGE, C. M. 1930, 1931. Studies on the physiology of Corals. I, III. *Sci. Reps. Great Barrier Reef Exped.*, 1928–1929. **1, 2, 4**.

YONGE, C. M. 1931. Digestive processes in marine invertebrates and fishes. *J. Conseil perman. internat. explorat. mer*, **6**.

YONGE, C. M. 1932. Notes on feeding and digestion in *Pteroceras* and *Vermetus* with a discussion on the occurrence of the crystalline style in the Gastropoda. *Sci. Reps. Great Barrier Reef Exped.*, 1928–1929. **1, 10**.

YONGE, C. M. 1935. On some aspects of digestion in ciliary feeding animals. *J. Mar. Biol. Assn UK, N.S.*, **20**.

YONGE, C. M. 1936. The evolution of the swimming habit in the Lamellibranchia. *Mém. Mus. Roy. Hist. Nat. Belg., ser. 2*, **3**.

YONGE, C. M. 1937. The biology of *Aporrhais pes-pelicani* (L.) and *A. serresiana* (Mich.). *J. Mar. Biol. Assn UK, N.S.*, **21**.

YONGE, C. M. 1938. Evolution of ciliary feeding in the Prosobranchia, with an account of feeding in *Capulus ungaricus*. *J. Mar. Biol. Assn UK, N.S.*, **22.**

YONGE, C. M. 1941. The Protobranchiate Mollusca; a functional interpretation of their structure and evolution. *Philos. Trans. Roy. Soc., London, B*, **230.**

YONGE, C. M. 1946. On the habits of *Turritella communis* Rosso. *J. Mar. Biol. Assn UK*, **26.**

YONGE, C. M. 1949. On the structure and adaptations of the Tellinacea, deposit-feeding Eulamellibranchia. *Phil. Trans. Roy. Soc., London, B*, **234.**

YONGE, C. M. 1954. The Monomyarian condition in the Lamellibranchia. *Trans. Roy. Soc. Edinburgh*, **62.**

YONGE, C. M. 1956. Ecology and physiology of reef-building corals. In: Buzzati-Traverso, A.-A. *Perspectives in marine biology.*

YONGE, C. M. 1958. Form and function in the Mollusca. *Nature, Lond.*, **182,** 4641.

YONGE, C. M., and NICHOLS, A. G. 1930, 1931, 1932. Studies on the physiology of Corals. II, IV, V, VI. *Sci. Reps. Great Barrier Reef Exped.*, 1928–1929. **1,** 3, 6, 7, 8.

YOUNG, J. Z. 1936. The giant nerve fibre and epistellar body of Cephalopods. *Q. J. microsc. Sci., N.S.*, **78.**

YOUNG, J. Z. 1944. Les fibres nerveuses géantes. *Endeavour*, **3.**

ZACHER, F. 1933. Onychophora. In: Kükenthal, W., and Krumbach, Th. *Handbuch der Zoologie*, III/2, 6. Berlin and Leipzig.

ZADDACH, E. G. 1841. De *Apodis cancriformis* Schaeff, anatome et historia evolutionis. *Dissertatio inauguralis zootomica*. Bonn.

ZAWARZIN, A. A. 1912. Histologische Studien über Insekten. II. Das sensible Nervensystem der Aeschnalarven. *Z. wiss. Zool.*, **100.**

ZAWARZIN, A. A. 1924. Histologische Studien über Insekten. VI. Zur Morphologie der Nervenzentren. Das Bauchmark der Insekten. Ein Beitrag zur vergleichenden Histologie. *Z. wiss. Zool.*, **122.**

ZAWARZIN, A. A. 1925. Der Parallelismus der Strukturen als ein Grundprinzip der Morphologie. *Z. wiss. Zool.*, **124.**

ZIEGLER, H. 1898. Über den derzeitigen Stand der Coelomfrage. *Verhandl. Dtsch. Zool. Ges.*, **8.**

INDEX OF TERMS

Roman numerals indicate Volume No., and Arabic numerals page numbers. References to pages where a term is defined or explained are italicised.

INDEX OF SCIENTIFIC NAMES

Volume No. (I or II) in Roman numerals; page numbers given in small Roman numerals (i to xxx in Volume I) or Arabic numerals.